Artificial Intelligence in Music Production:

Innovations, Practices, and Industry Implications

Randy Joy Magno Ventayen
https://orcid.org/0000-0002-0952-7795
Pangasinan State University, Phillippines

IGI Global
Scientific Publishing
Publishing Tomorrow's Research Today

Published in the United States of America by
IGI Global Scientific Publishing
701 East Chocolate Avenue
Hershey, PA, 17033, USA
Tel: 717-533-8845 | Fax: 717-533-7115
Website: https://www.igi-global.com E-mail: cust@igi-global.com

Library of Congress Cataloging-in-Publication Data

Names: Ventayen, Randy Joy Magno, 1984- editor
Title: Artificial intelligence in music production : innovations, practices, and industry implications / edited by Randy Joy Magno Ventayen.
Description: Hershey, PA : IGI Global Scientific Publishing, 2026. | Includes bibliographical references and index. | Summary: "This book explores how Artificial Intelligence is reshaping music production, from composition to distribution. It highlights emerging tools, creative possibilities, and industry challenges, offering critical insights into innovation, ethics, intellectual property, and the evolving role of human artistry in an AI-driven music landscape"-- Provided by publisher.

Identifiers: LCCN 2025039399 (print) | LCCN 2025039400 (ebook) | ISBN 9798337362793 hardcover | ISBN 9798337362809 paperback | ISBN 9798337362816 ebook
Subjects: LCSH: Artificial intelligence--Musical applications | Music--Production and direction
Classification: LCC ML74.3 .A78 2026 (print) | LCC ML74.3 (ebook) | DDC 780.285/63--dc23/eng/20250827
LC record available at https://lccn.loc.gov/2025039399
LC ebook record available at https://lccn.loc.gov/2025039400

British Cataloguing in Publication Data
A Cataloguing in Publication record for this book is available from the British Library.

Table of Contents

Detailed Table of Contents

in these systems and to advocate for practices that respect both the complexity of musical traditions and the transformative potential of technological innovation. The convergence of AI and musicology thus presents both new opportunities and new obligations, positioning the field at a pivotal intersection of technology, creativity, and culture. Thus, the purpose of this study is to explore the discipline of musicology from its near history to its new form that's formed with the invention of AI.

This chapter explores the role of artificial intelligence (AI), in redesigning music, as a tool for experimentation and innovation, which connects traditional composition with technology. The interconnection of AI with human creativity gives rise to various musical opportunities. By using AI tools, artists are pushing beyond established boundaries and providing new experiences to listeners. The main point is that AI is not a replacement for humans rather it's an addition to them. The goal of this chapter is to examine the evolving method of AI and its influence on different areas of music such as songwriting, composition and performance etc. By analyzing previous and existing studies this chapter provides a summary of creative co-partnership between artificial intelligence and human intelligence, while also addressing its cultural and ethical implications. The chapter visualize a future in which artificial intelligence (AI) preserves the human touch while enabling real-time improvisational performance, emotional composition, humanized instruction and innovative teaching practices.

Artificial intelligence (AI) is transforming music creation, distribution, and discovery, offering new opportunities but also reinforcing bias. When trained on datasets dominated by specific genres or demographics, AI systems often amplify mainstream trends while marginalizing niche artists and underrepresented communities. This chapter explores how such bias emerges through skewed data and algorithmic design, and how it narrows musical diversity by favoring "safe" and commercially viable

The chapter focuses on the changes in the role of artificial intelligence in sound design, synthesis and music creation by tracing the development of technology starting with the earliest forms of algorithmic systems and ending with more modern deep learning. It describes the growth of creative opportunities enabled by neural networks, generative methods, and advanced synthesis tools, alongside helping in mixing, mastering, and audio restoration. The chapter also emphasizes that AI can help to create new timbres, textures, and compositional forms to allow professional and novice creators to experiment with the larger soundscapes. Moral issues such as authorship, copyright, and training data, and artistic authenticity are also critically evaluated to explain the role of developers and creators. The chapter concentrates on the idea that AI can be a collaborative partner and a tool that improves creativity, democratizes production, and affects the way people interact with machines in the sonic arts in the future with the help of practical tools and case studies.

This chapter delves into the innovative use of Artificial Intelligence (AI) in the field of audio composition, focusing on its growing influence in film scoring, game soundtracks, and commercial jingles. Once the exclusive domain of composers and audio engineers, music and sound design are now increasingly being shaped by algorithms and neural networks. Through curated real-world case studies, this chapter highlights how AI is not merely automating audio creation but expanding the boundaries of human creativity in media production.

on computational creativity, digital transformation, and the evolving role of AI as a collaborative infrastructure in contemporary music production ecosystems.

Foreword

The junction of AI as well as production of music signifies one of the intense cultural as well as technological shifts of the 21st century. Until now, for many centuries, it was believed that creation of music as exclusive attempt of humans only, which is based on foundation of lived experiences of humans, along with cultural memories, intuition in addition to emotions. However, with technological developments in the areas of computational creativity, NNs & MLs, the world has reached the stage where the process of music production is being reshaped including distribution, consumption, production, as well as composition of music. In the said situation, Artificial Intelligence in Music Production: Innovations, Practices, and Industry Implications come as an influential in addition to well-timed contribution which catches not only the AI promises but also the complications of the transformation of music production by AI technologies. The said volume comes at a crucial time when AI is not limited to labs but implanted in conventional creative practice, allowing AI systems to master recordings, design sonic textures, produce melodies, compositions harmonization & listener behavior analysis with outstanding elegance. Apart from technological capability, AI is changing professional workflows, music's broader cultural meaning, IP frameworks, & artistic agency. This book comprises of chapters which cooperatively explain such components with not only scholarly rigor but practical insights, to give an inclusive understanding of musical landscape redefined by AI.

The major strength of this book is considering multidisciplinary viewpoint from diverse areas like sound engineering, musicology, CS, psychology, industry practice & education, which shows the innate hybrid nature of music production aided by AI, needing artistic sensibility along with technical novelty. This book incorporated applied research with theoretical analysis along with real-world case studies to showcase new accounts of collaboration amid creativity of humans & intelligence of machines, while covering opportunities along with challenges. Though AI gives exceptional tools for performance, composition, & distribution, but comes with the cost of authenticity, cultural representation, authorship, bias, and future of creative labor. The contributors carefully considered such issues, acknowledging technical progress should come with responsible governance & ethical reflection, because AI produced music gradually becomes identical to the human-created works & the worldwide music industry contends with novel legal & economic realisms. The contributors engage thoughtfully with these issues, recognizing that technological advancement must be accompanied by ethical reflection and responsible governance. Such discussions are particularly important as AI-generated music becomes increasingly indistinguishable from human-created works and as the global music industry grapples with new legal and economic realities. The upcoming policies associated with algorithmic accountability, IP rights, in addition to the governance of data together with initiatives corresponding to international standards, are progressively explaining the limits within with the operation of music production led by Artificial Intelligence could take place.

The book shows an intense perception of the developing ecosystem in the operation of music production, where data-driven strategies, streaming services, digital platforms, and referred algorithms allowing the transformation of music & AI making it highly personal for users. Then, it is also essential to accept the present constraints of AI systems in the production of music. This includes problems such as absence of contextual understanding, innate biases which are algorithmic ones, dependency on the dataset, as well as limitations in reproducibility persist in affecting not only the diversity but also the quality of output produced by Artificial Intelligence. Discussing such constraints remains a crucial area which must be covered in future development as well as research.

Another major feature of this book is its global orientation, showing the developing tools can be implemented to local contexts & assist in global innovation with cultural diversity in music. For educators and students, the book gives a modern resource to aid in pedagogical practice & curriculum development in technology programs, media and music, in accordance with execution of computational methods & digital production tools in university courses, where this book allows such technologies within ethical, social & artistic frameworks. For researchers, it finds developing areas at the crossroads of creative arts & AI, whereas for industry professionals, it gives insights into strategic accounts, practical applications, & possible innovation trajectories. It emphasizes that future of music production is by the advancing bond amid human imagination & technology, where artists retain agency & purpose, whereas AI aids in increasing the creative possibilities.

In assembling this volume, the editor(s) and contributors have produced a work of significant scholarly and practical value. It documents a transformative era in which the boundaries between composer and programmer, performer and algorithm, studio and software are becoming increasingly fluid. Readers will find in these pages both a record of current developments and a roadmap for future exploration. We have no doubt that Artificial Intelligence in Music Production: Innovations, Practices, and Industry Implications will become an essential reference for academics, practitioners, policymakers, and anyone interested in the evolving relationship between technology and the arts. It stands as a testament to the power of interdisciplinary collaboration and to the enduring capacity of music to adapt, innovate, and inspire even in the face of rapid technological change, whereas also helping in laying the grounds to shape comprehensive, progressive, and responsible routes in the time of AI.

Aditya Singh
Amrita Vishwa Vidyapeetham, India

Preface

INTRODUCTION

As Editor, it is both a privilege and a responsibility to present *Artificial Intelligence in Music Production: Innovations, Practices, and Industry Implications*, a volume conceived at a moment of profound transformation in the global music ecosystem. The aftermath of the pandemic has accelerated technological adoption across creative industries, and nowhere is this more evident than in music production. What was once considered the exclusive domain of human intuition, artistry, and technical expertise is now increasingly shaped by the integration of Artificial Intelligence (AI). From composition and sound design to mixing, mastering, and distribution, intelligent systems are not merely assisting but actively redefining creative workflows.

This rapid evolution calls for careful reflection. While new AI tools and research initiatives continue to emerge at an unprecedented pace, the discourse surrounding their implications often remains fragmented across disciplines. The need for a consolidated, interdisciplinary resource—one that bridges theory, practice, and critical inquiry—has never been more urgent. This volume is an attempt to meet that need. It brings together scholars, practitioners, and industry voices to examine not only the technological innovations driving AI in music production, but also their broader consequences for creativity, labor, ownership, and cultural value.

The objectives of this book are threefold. First, it seeks to provide a foundational understanding of the core AI technologies shaping contemporary music production, including generative models, machine learning techniques for audio processing, and deep learning architectures. Second, it highlights current innovations and practical applications across the production pipeline, offering insights into how AI is being used in composition, synthesis, mixing, and mastering. Third—and perhaps most importantly—it engages critically with the wider implications of these developments. Questions surrounding intellectual property, authorship, economic models, and ethical responsibility are no longer peripheral; they are central to the future of music as both an art form and an industry.

This collection has been carefully curated with a diverse audience in mind. It is intended for researchers and academics across fields such as music technology, computer science, media studies, and digital humanities, as well as for students seeking to understand the evolving landscape of AI in creative practice. At the same time, it speaks directly to practitioners—producers, audio engineers, composers, and sound designers—who are navigating the integration of AI into their workflows. Industry professionals, policymakers, and legal experts will also find in these pages a critical resource for understanding the shifting terrain of rights, regulation, and market structures in the age of intelligent systems.

Ultimately, this book does not claim to offer definitive answers. Rather, it aims to open a space for informed dialogue, critical reflection, and collaborative exploration. The intersection of AI and music production is not merely a technical development; it is a cultural and philosophical turning point. As Editor, I hope this volume will serve as both a guide and a catalyst—supporting readers in engaging thoughtfully with the opportunities and challenges that lie ahead, and contributing to a more nuanced understanding of what it means to create music in an age increasingly shaped by machines.

ORGANIZATION OF THE BOOK

As Editor, I am pleased to introduce a collection of chapters that together map the evolving terrain of artificial intelligence in music production, each contributing a distinct perspective to this rapidly developing field.

Chapter 1 opens the volume by addressing one of the most pressing and enduring questions: how to balance human creativity with technological innovation. The contributors offer a nuanced exploration of the relationship between artists and AI systems, drawing on empirical listener studies and artist-centered insights. Their analysis reveals a landscape in which AI is increasingly embraced as a creative partner, yet persistent concerns regarding authenticity, ownership, and sustainability continue to shape both artistic practice and audience reception.

Building on this foundation, Chapter 2 situates AI within the broader discipline of musicology. It traces the evolution of musicological inquiry from its historical roots to its contemporary intersection with intelligent systems. In doing so, the chapter highlights the critical role of musicologists in interrogating the cultural, ethical, and epistemological dimensions of AI, positioning the field as an essential mediator between tradition and technological transformation.

Chapter 3 further develops the theme of collaboration by framing AI as a tool for innovation and experimentation. Rather than displacing human creators, AI is presented as an enabler of new musical possibilities, fostering hybrid practices that extend the boundaries of composition, performance, and pedagogy. The chapter underscores the importance of preserving the human element while embracing the creative affordances of machine intelligence.

In Chapter 4, attention turns to the issue of bias within AI systems and its implications for musical diversity. Through compelling case studies and critical analysis, the authors demonstrate how algorithmic design and dataset limitations can reinforce dominant cultural trends while marginalizing underrepresented voices. The chapter calls for greater accountability and interdisciplinary collaboration to ensure that AI serves as a force for inclusivity rather than homogenization.

Chapter 5 introduces a more technical perspective by examining the use of Markov chain models in AI-assisted songwriting. The study provides valuable insights into the capabilities and limitations of probabilistic approaches, showing that while such models can generate musically viable motifs, they struggle with long-term structure. This contribution highlights the continued importance of human intervention in shaping coherent musical narratives.

Advancing the discussion of generative systems, Chapter 6 presents an integrated framework combining recurrent neural networks and genetic algorithms to produce personalized pop songs. By incorporating user feedback and vocal synthesis, the chapter illustrates how AI can move beyond fragment generation toward complete, user-tailored compositions, signaling a significant step toward accessible, end-to-end music production systems.

Chapter 7 explores the analytical potential of AI through the development of a convolutional neural network model for chord recognition. Centered on the Chord-Craft dataset, this work demonstrates how deep learning techniques can uncover complex harmonic patterns, offering new tools for music analysis while also supporting creative and educational applications.

In Chapter 8, the focus shifts to sound design and synthesis, tracing the evolution of AI-driven tools from early algorithmic methods to advanced deep learning systems. The chapter emphasizes the expansion of creative possibilities alongside the ethical challenges that accompany them, presenting AI as both a powerful collaborator and a subject of critical scrutiny.

Chapter 9 extends the application of AI into media production contexts, examining its role in film scoring, game audio, and commercial composition. Through real-world case studies, the authors illustrate how AI is not merely automating processes but actively reshaping creative practices across industries, expanding the scope of what is musically and commercially possible.

Chapter 10 introduces a novel interdisciplinary perspective by examining AI-driven music within the context of femtech and mental wellbeing. The study provides empirical evidence of the impact of personalized, biometrically informed music interventions, highlighting the growing relevance of AI in health-related applications and user-centered design.

In Chapter 11, the discussion turns to the critical issue of royalty distribution, with a focus on the Indonesian context. The chapter explores how AI can address longstanding challenges in transparency and accuracy, offering comparative insights from international systems while emphasizing the need for ethical governance and institutional alignment.

Chapter 12 broadens the scope to consider the structural transformation of the music industry in the AI era. By comparing traditional and popular music systems, the authors illuminate shifts in production, distribution, and cultural value, while also addressing concerns related to authenticity, cultural preservation, and economic equity.

Chapter 13 focuses on education, examining how AI tools are being integrated into music production curricula. The chapter highlights pedagogical strategies that balance technical proficiency with creative development, preparing students to navigate a future in which human and machine collaboration is central to artistic practice.

In Chapter 14, attention is given to cloud-based AI music tools and their implications for IT and multimedia professionals. Through empirical analysis, the chapter explores how scalable, cloud-native systems are transforming workflows, skill requirements, and ethical considerations, underscoring the growing convergence of music production and digital infrastructure.

Finally, Chapter 15 concludes the volume with a practical perspective on home music production. Focusing on independent artists, it examines how AI tools are reshaping decentralized creative environments. The findings highlight the importance of perceived augmentation and creative empowerment, reinforcing the concept of AI as a collaborative infrastructure that enhances, rather than replaces, human creativity.

Taken together, these chapters offer a comprehensive and multifaceted exploration of artificial intelligence in music production. As Editor, I believe this collection not only documents current advancements but also provides a critical framework for understanding the opportunities and challenges that will define the future of music in the age of intelligent systems.

CONCLUSION

As this volume draws to a close, it becomes evident that the integration of artificial intelligence into music production is not a transient trend, but a defining transformation of our time. The chapters presented in this collection collectively illuminate a landscape in which creativity, technology, and industry are increasingly intertwined, reshaping not only how music is produced, but also how it is conceptualized, valued, and experienced.

From foundational explorations of human–AI collaboration to advanced technical frameworks and critical industry analyses, this book has sought to capture both the promise and the complexity of this evolving domain. A recurring theme throughout these contributions is the notion that artificial intelligence, while powerful, is not an autonomous creative force operating in isolation. Rather, it is deeply embedded within human systems—trained on human data, guided by human intent, and ultimately evaluated through human perception. The future of music production, therefore, lies not in the replacement of human creativity, but in its augmentation.

At the same time, the discussions presented here underscore that innovation must be accompanied by responsibility. Issues of bias, authorship, intellectual property, cultural representation, and economic equity demand careful and sustained attention. As AI systems become more sophisticated and more deeply integrated into creative workflows, the need for transparent, ethical, and inclusive practices becomes increasingly urgent. The challenge before us is not only to harness the capabilities of AI, but to shape its development in ways that reflect and respect the diversity and richness of global musical cultures.

Equally is the recognition that this transformation extends beyond the studio. It is reshaping education, redefining professional roles, and altering the structural dynamics of the music industry. As highlighted throughout this volume, new competencies, interdisciplinary collaboration, and adaptive frameworks will be essential for navigating this changing environment. Whether one is a researcher, practitioner, educator, policymaker, or artist, engagement with AI is no longer optional—it is integral to the future of the field.

As Editor, I hope that this book serves not only as a record of current advancements, but also as a catalyst for ongoing inquiry and dialogue. The intersection of artificial intelligence and music production remains a space of exploration—one that invites experimentation, critical reflection, and collective responsibility. It is my sincere expectation that the insights gathered here will inspire readers to engage thoughtfully with the opportunities and challenges ahead, and to contribute meaningfully to the shaping of a more innovative, equitable, and expressive musical future.

Randy Joy Magno Ventayen
Pangasinan State University, Phillippines

Acknowledgement

The editor gratefully acknowledges the institutions and individuals whose support, collaboration, and encouragement made the completion of this book, Artificial Intelligence in Music Production: Innovations, Practices, and Industry Implications, possible.

Sincere appreciation is extended to Pangasinan State University for providing a strong academic foundation and an environment that promotes research, innovation, and interdisciplinary engagement. The editor likewise expresses heartfelt thanks to the Philippine Technical Institute for its institutional support throughout the development of this work.

The editor is deeply grateful to INTI International University, Malaysia, and UNITAR International University, Malaysia, where the editor serves as a Research Fellow. These institutions have provided valuable academic engagement, international perspective, and research collaboration that enriched the intellectual depth of this book. Warm appreciation is also extended to Bina Bangsa University, Banten, Indonesia, and Universitas Muhammadiyah Prof. Dr. HAMKA (UHAMKA), Indonesia, where the editor has served as a Visiting Lecturer. The academic exchanges, collaborative opportunities, and professional interactions with faculty and students greatly contributed to the global outlook reflected in this work. The editor further acknowledges partner institutions and friend universities across the region for their continued cooperation, knowledge sharing, and support of scholarly initiatives that advance research and innovation in technology and the arts.

Deep gratitude is also given to the anonymous reviewers whose constructive feedback, professional guidance, and rigorous evaluation significantly improved the clarity, quality, and scholarly value of this manuscript. The editor would like to recognize the music artists, producers, and creative practitioners whose works, experiences, and insights were included in this book.

Special thanks are extended to colleagues, collaborators, and research assistants who provided technical assistance, shared expertise, and offered encouragement throughout the writing process. Appreciation is likewise given to the broader academic and professional communities in music technology and artificial intelligence whose prior research laid the foundation for this work.

Lastly, the editor expresses sincere gratitude to family and close friends for their patience, understanding, and unwavering moral support during the long journey of research and writing.

Chapter 1
Balancing Human Creativity and Technological Innovation in AI Music Production

Randy Joy Magno Ventayen
http://orcid.org/0000-0002-0952-7795
UNITAR International University, Philippines

Elmar B. Noche
http://orcid.org/0009-0008-4363-5600
Pangasinan State University, Philippines

Maricel Aqui
Philippine Technical Institute, Philippines

ABSTRACT

Artificial intelligence (AI) is reshaping the landscape of music production by introducing automated composition systems, digital performers, and hybrid creative workflows. This chapter examines the evolving relationship between AI and human creativity by exploring artist perceptions, authenticity preservation strategies, and listener acceptance of AI-generated music. Through interdisciplinary analysis combining empirical listener studies, artist-centered reflections, and industry discourse, this chapter presents a examination of AI's impact on music production. The findings reveal that while artists increasingly adopt AI as a creative support tool, concerns remain regarding artistic ownership, emotional authenticity, and economic sustain-

DOI: 10.4018/979-8-3373-6279-3.ch001

ability. Listener studies demonstrate growing acceptance of AI-generated music, particularly when technical quality is achieved; however, emotional authenticity remains a key determinant of audience trust.

1. INTRODUCTION

The Emergence of Artificial Intelligence in Music Production

Artificial intelligence (AI) has moved from being a niche research topic to a mainstream production force across creative industries, and music is one of the most visibly disrupted domains. Modern generative systems can now support (and sometimes automate) core production tasks—melodic idea generation, arrangement drafting, timbre design, and even vocal simulation—reshaping how music is created, distributed, and evaluated. From a research perspective, AI music generation has evolved from earlier algorithmic or rule-based approaches toward data-driven models capable of producing stylistically coherent outputs at scale (Sturm & Maruri-Aguilar, 2021).

This shift is not simply technical; it changes the meaning of "who creates music." AI systems can synthesize patterns learned from large corpora and generate outputs that approximate human musical conventions. That capability accelerates production and lowers barriers for non-specialists, but it also intensifies debates about authorship, originality, and cultural value—especially once AI outputs enter real commercial distribution environments. In artist-centered discussions, AI's mainstream presence is increasingly framed as a "paradigm shift" that challenges long-standing assumptions about creative labor and identity.

In parallel, the music ecosystem has become platform-driven: attention, recommendation systems, and release volume strongly influence success. When AI-generated tracks scale quickly, concerns arise that the market may become saturated with high-volume outputs optimized for platform performance rather than meaning. This phenomenon has been discussed in industry and research discourse as "AI music slop," where volume and polish may outpace artistic intention or emotional grounding. Whether one agrees with that label or not, the underlying issue is serious: AI changes competitive dynamics by making production speed and stylistic imitation cheap.

AI as Creative Collaboration, Not Just Automation

A more realistic view is that AI is not a single replacement tool but a spectrum of production supports—ranging from idea generation to audio manipulation—often used within human-led workflows. Many creators treat AI as an assistive system: it can propose motifs, suggest arrangements, or generate drafts that are then curated, edited, and emotionally shaped by a human producer. This aligns with the broader "human-in-the-loop" view of creative AI, where the machine's role is generative support rather than artistic agency.

However, even when AI is used as a collaborator, the creative relationship raises philosophical questions. Human artistry is typically associated with intention, lived experience, and expressive meaning; AI systems generate outputs without consciousness or personal narrative. This difference matters because music is not judged solely on technical structure. It is also evaluated through perceived intent, authenticity, and the listener's belief about the creator. Empirical work shows that people can evaluate identical or near-identical outputs differently depending on whether they believe a piece was created by AI or a human (Shank et al., 2022). In other words, origin information can shape aesthetic judgment.

This matters even more in production contexts where AI mimics known artists' styles or vocal characteristics. The music domain has increasingly seen controversy around voice cloning, style imitation, and the training of models using copyrighted catalogues. Scholarly and legal discussions point to unresolved governance gaps around consent, compensation, and the "commons versus ownership" tension in music AI (Drott, 2021). Technical systems can already support sophisticated production pipelines, but the policy and ethics landscape is still catching up.

Authenticity and Artistic Identity Under AI Pressure

Authenticity is a central concept in music culture because listeners often treat music as evidence of a human story: a lived experience, a point of view, or a vulnerable emotional act. That belief doesn't vanish just because AI can generate convincing audio. Instead, authenticity becomes a sharper filter—especially in a digital environment where content is abundant. In your artist-focused synthesis, authenticity is positioned as a defining feature of human artistry that AI can imitate sonically but not embody as lived meaning.

Research and commentary increasingly describe a tension: AI can be technically impressive, but "perfect" outputs can also be perceived as emotionally sterile or lacking personal intention. This connects with broader creative cognition arguments: music is not only a structure of notes; it's also an interaction between performer intent, cultural context, and audience interpretation. Studies in music and emotion emphasize that emotional communication in music is tied to human perception and context, not just acoustic patterns (Juslin & Sloboda, 2001). Meanwhile, cross-cultural research suggests certain emotional cues in music can be widely recognized (Fritz et al., 2009), which raises a key question for AI: can a non-human creator reliably produce signals that feel emotionally authentic across contexts—or does it merely simulate the surface?

From the artist point of view, authenticity often becomes a strategy, not just a value. Musicians differentiate themselves by leaning into what AI lacks: embodied performance, imperfect nuance, personal narrative, and community connection . This is not nostalgia—it's market logic. When anything can be generated, what becomes scarce is "realness" that audiences can trust.

Listener Acceptance of AI Music

Listener acceptance is the pressure test for AI music's long-term industry impact. Even if AI can generate audio that meets technical standards, the audience ultimately decides whether it belongs in everyday listening—and under what conditions. Recent experimental and applied research suggests two things can be true at once:

1. People may struggle to distinguish AI from human music in certain settings, especially when the genre and production style align with algorithmic strengths.
2. People may still evaluate AI music differently once they know it is AI, because authenticity is partly psychological and cultural, not purely auditory.

This matches findings in the broader literature on AI-authorship bias: listeners tend to like music less when they believe it was composed by AI (Shank et al., 2022). Expectancy and framing also matter: when audiences' expectations are violated (e.g., learning a track is AI-made), their perceptions can shift (Hong et al., 2021).

Your listener study on AI music covers captures this exact tension: participants often praised technical quality while raising concerns about emotional depth and authenticity—especially for emotional genres—yet many still expressed openness to AI becoming mainstream as a tool for innovation rather than a replacement for human artists. That combination (impressed + cautious) is essentially the audience mood of the current era.

Genre also appears to shape acceptance. Where music is expected to be polished, synthetic, or pattern-driven (some pop/electronic production contexts), AI may fit more naturally. Where audiences expect raw vulnerability (ballads, acoustic-centered genres), AI's perceived lack of lived experience may become more visible. This aligns with discussions that emphasize AI's strengths in technical imitation and weaknesses in embodied emotional meaning.

Purpose and Structure of the Chapter

This chapter investigates AI in music production through a merged ecosystem lens: artists, authenticity, and audience acceptance. Rather than treating these as separate debates, it frames them as interdependent forces shaping the future of AI music.

The chapter proceeds as follows:

- Section 2 examines artist perspectives on AI music and AI artists, focusing on creative opportunity versus economic, legal, and identity risks.
- Section 3 analyzes authenticity strategies used by real artists to remain culturally and economically relevant in AI-saturated environments.
- Section 4 presents listener acceptance evidence, highlighting how technical quality, emotional resonance, and disclosure influence audience preferences.
- Section 5 introduces a Human–AI Music Ecosystem Framework linking stakeholder dynamics across creation, distribution, and consumption.
- Section 6 discusses ethical, economic, and policy implications (copyright, consent, platform governance, compensation).
- Section 7 outlines future directions for responsible human–AI collaboration in music production.

This structure supports both scholarly contribution and practical relevance, positioning AI not as a single tool but as a system-level transformation shaping production practice, artistic identity, and listener trust.

2. ARTISTS' PERSPECTIVES ON AI MUSIC AND DIGITAL ARTISTS

The Expanding Role of Artificial Intelligence in Creative Music Workflows

The integration of artificial intelligence into music production has altered traditional artistic workflows, prompting musicians to reassess their roles as composers, performers, and producers. AI technologies now support a wide range of production processes, including automated composition, vocal synthesis, beat generation, arrangement assistance, and real-time sound design. These tools offer efficiency and creative flexibility, but they also challenge the long-standing perception of music as a fundamentally human-driven art form.

From a technological standpoint, modern AI music systems rely on deep learning architectures trained on large-scale musical datasets to generate stylistically coherent outputs. These systems can replicate melodic structures, harmonic progressions, and performance characteristics with a level of technical sophistication previously associated with trained musicians (Kwiecień et al., 2024). As AI systems become more accessible through commercial platforms, musicians increasingly encounter AI not only as a production tool but also as a potential competitor.

Artist responses to AI adoption are complex and multidimensional. Many musicians acknowledge AI's ability to expand creative experimentation and accelerate production timelines. Others express concern that widespread AI adoption may alter the economic structure of music industries by reducing demand for certain forms of creative labor. These conflicting viewpoints highlight a broader transition within creative industries, where technological innovation simultaneously creates new opportunities and introduces professional uncertainty.

Recent qualitative analyses of artist discourse reveal that musicians frequently position AI as a form of creative augmentation rather than a complete replacement for human creativity. Musicians often describe AI as an advanced instrument capable of generating musical ideas while requiring human interpretation and emotional refinement. This perspective aligns with co-creative frameworks in human–computer interaction research, which emphasize collaborative relationships between AI systems and human creators (Newman et al., 2022).

However, even artists who embrace AI collaboration often emphasize that creative intention remains inherently human. Professional musicians argue that AI systems can generate musical patterns but cannot develop artistic identity or personal perspective. This distinction reinforces the idea that music is not solely defined by technical structure but by expressive meaning shaped through lived experience.

AI as Creative Opportunity: Expanding Musical Innovation

Many artists view AI technologies as tools that enhance creative exploration by allowing rapid experimentation with new musical ideas. AI-assisted composition systems can generate melodic variations, suggest harmonic progressions, and simulate instrumental arrangements, enabling musicians to explore sonic possibilities that might otherwise require extensive time and resources. In production environments, AI can function as a brainstorming assistant, offering alternative creative directions that inspire new compositions.

Musicians working within electronic and digital production genres often demonstrate higher acceptance of AI-assisted workflows. In these contexts, music production already relies heavily on digital manipulation, synthesis, and automated processing, making AI integration a natural extension of existing technological practices. AI-generated sound design and automated mixing tools allow producers to focus on higher-level artistic decisions rather than repetitive technical tasks.

Research examining human–AI co-creative music production demonstrates that musicians frequently treat AI outputs as raw creative material rather than finished compositions. Artists often edit, rearrange, or reinterpret AI-generated content to align with personal artistic style. In this sense, AI operates as a catalyst that accelerates the exploratory phase of music production while preserving human authorship (Newman et al., 2022).

Some musicians also highlight AI's potential to democratize music production by reducing technical barriers for new creators. Individuals without formal musical training can use AI tools to produce compositions that meet professional production standards. This accessibility expands participation in music creation, potentially increasing cultural diversity and artistic experimentation. However, democratization also introduces challenges related to content saturation and artistic valuation within digital distribution platforms.

2.3 AI as Creative Threat: Economic and Labor Concerns

While AI offers creative advantages, many artists express concern regarding its economic and labor implications. One of the most frequently cited fears involves the potential displacement of professional musicians whose roles may be partially replaced by generative AI systems. Session instrumentalists, background vocalists, and music arrangers are particularly vulnerable, as AI systems can replicate stylistic performance characteristics without requiring human participation.

Interview-based research examining musician perceptions reveals widespread anxiety about AI's influence on job security and income stability (Herington et al., 2025). Many artists worry that AI-generated music may reduce demand for human-produced content by providing cheaper and faster alternatives for commercial production. In streaming-driven music economies where revenue is often distributed based on algorithmic exposure, high-volume AI-generated releases could potentially overshadow human artists' visibility.

Economic concerns extend beyond employment displacement to include issues related to compensation and intellectual property ownership. AI models frequently rely on training datasets containing copyrighted musical works, raising ethical questions about whether artists should receive compensation when their stylistic signatures contribute to AI-generated outputs. Scholars argue that existing copyright frameworks struggle to address these emerging complexities, particularly when AI-generated music closely resembles identifiable artistic styles (Drott, 2021).

The use of voice cloning technologies has intensified these debates. Several high-profile incidents involving AI-generated songs imitating well-known artists have sparked industry-wide discussions about consent and artistic ownership. Musicians increasingly advocate for regulatory frameworks requiring explicit permission for the use of vocal likeness and stylistic imitation in AI-generated content. These concerns highlight the need for transparent governance models supporting ethical AI integration within music industries.

2.4 AI Artists and the Redefinition of Musical Authorship

The emergence of fully AI-generated performers represents one of the most controversial developments in modern music production. AI-generated artists—digital personas capable of producing and performing music—challenge traditional assumptions regarding artistic identity and creative authorship. Unlike human musicians, AI performers do not possess personal narratives, emotional experiences, or cultural backgrounds. Nevertheless, AI artists have gained significant popularity within digital streaming platforms and virtual performance environments.

The rise of AI artists has generated philosophical debates regarding whether algorithmic systems can be considered legitimate creators. Some scholars argue that artistic value requires intentionality and emotional communication, attributes currently absent in AI-generated music. Others suggest that artistic authorship may evolve toward hybrid creative models in which AI-generated performers coexist alongside human artists.

Audience perception research indicates that listeners often evaluate AI-generated music differently depending on their awareness of its origin. When listeners believe music is created by AI, they may assign lower authenticity and emotional value even if sonic quality remains comparable to human-produced compositions (Shank et al., 2022). These findings suggest that artistic authorship influences audience perception independently of technical performance quality.

Musicians themselves frequently emphasize that artistic identity extends beyond technical skill to include personal narrative, social context, and cultural participation. Artists argue that AI-generated performers lack these elements, making them fundamentally different from human creators. Nevertheless, some musicians acknowledge that AI artists may contribute to entertainment diversity and experimental music forms, provided that their role remains transparent and ethically regulated.

2.5 Creative Agency and Human Control in AI-Assisted Music Production

Despite concerns regarding AI's disruptive potential, many musicians emphasize the importance of maintaining human creative agency within AI-assisted workflows. Artists increasingly advocate for "human-in-the-loop" production models that position AI as a support tool rather than an autonomous creator. In these models,

AI generates musical ideas or technical outputs, but final artistic decisions remain under human control.

Maintaining creative agency is essential for preserving artistic authenticity and professional identity. Musicians argue that creative expression involves intentional storytelling, emotional vulnerability, and cultural communication—qualities that cannot be fully automated through algorithmic generation. By integrating AI as a collaborative partner rather than a replacement, artists can leverage technological innovation while preserving the human elements that define musical artistry.

Industry recommendations frequently emphasize transparency regarding AI involvement in music production. Clear labeling of AI-assisted or AI-generated music helps maintain audience trust and ensures ethical recognition of human contributions. Transparent disclosure also supports emerging legal frameworks addressing intellectual property ownership in AI-generated creative works.

Furthermore, artists increasingly emphasize the importance of technological literacy in navigating AI-driven music production environments. Understanding how AI systems generate music enables musicians to maintain creative control and negotiate ethical usage of their artistic materials. Educational programs integrating AI music technology training may play a critical role in preparing future musicians for hybrid creative environments.

2.6 Artistic Adaptation in the Era of Algorithmic Creativity

Rather than resisting AI adoption entirely, many musicians are adapting their creative strategies to coexist with algorithmic music production. Artists increasingly emphasize emotional authenticity, live performance, and personal storytelling as differentiating factors that distinguish human creativity from AI-generated content. Live concerts, interactive fan engagement, and behind-the-scenes creative transparency allow artists to build emotional relationships with audiences that AI performers cannot replicate.

Artists also explore hybrid production approaches combining AI-generated content with human performance elements. These collaborative methods enable musicians to experiment with AI-generated sounds while maintaining personal artistic identity. Research suggests that hybrid creative models may represent the most sustainable pathway for integrating AI technologies into music production without diminishing human artistic value.

The evolution of AI-assisted music production ultimately reflects a broader transformation within creative industries. AI technologies challenge traditional definitions of creativity while simultaneously expanding artistic possibilities. Musicians' responses to AI adoption demonstrate that technological disruption does not necessarily eliminate human creativity; instead, it reshapes artistic roles and encourages new forms of creative expression.

3. AUTHENTICITY AND ARTISTIC IDENTITY IN AI-DRIVEN MUSIC CREATION

3.1 Authenticity as a Defining Element of Musical Expression

Authenticity has long served as a fundamental principle in music creation and evaluation. Beyond technical proficiency, audiences often interpret music as a reflection of human emotion, cultural identity, and personal storytelling. Within traditional artistic frameworks, authenticity is associated with sincerity, originality, and emotional expression grounded in lived experience. The rapid integration of artificial intelligence into music production challenges these assumptions by introducing musical works generated through computational pattern recognition rather than human experience.

Scholars emphasize that authenticity in music extends beyond acoustic quality and encompasses psychological and cultural perception. Listeners frequently evaluate musical works based on their beliefs about the creator's emotional intention and personal narrative. Emotional authenticity contributes to audience engagement, trust, and long-term artist–listener relationships. Research in music psychology suggests that emotional communication in music is closely linked to human expressive behavior and contextual interpretation rather than purely structural musical patterns (Juslin & Sloboda, 2001).

Artificial intelligence introduces a fundamental shift in how authenticity is interpreted. AI systems can replicate stylistic elements of musical genres and mimic human performance characteristics with high technical accuracy. However, AI lacks consciousness, emotional memory, and cultural upbringing, raising questions regarding whether algorithmically generated music can be considered authentic artistic expression. Artist-centered research suggests that authenticity remains strongly associated with human intention, emotional vulnerability, and cultural storytelling, characteristics that AI-generated music may struggle to replicate fully.

The tension between technical realism and perceived authenticity forms one of the central debates surrounding AI music production. While AI-generated compositions may demonstrate sonic sophistication, audiences often seek emotional resonance and narrative meaning that extend beyond musical structure. This distinction highlights authenticity as both an artistic value and a competitive advantage for human musicians in AI-driven creative environments.

3.2 Emotional Embodiment and Human Artistic Expression

One of the most significant differences between human-created and AI-generated music lies in emotional embodiment. Human musicians draw upon personal experiences, social interactions, and emotional memories to shape musical expression. These lived experiences influence compositional decisions, performance dynamics, and interpretative nuance. Emotional embodiment allows artists to communicate complex emotional states such as vulnerability, nostalgia, or cultural identity through musical performance.

AI music systems generate compositions by identifying statistical relationships between musical elements within training datasets. Although these systems can produce outputs that evoke emotional responses, they do not experience emotions themselves. The absence of emotional consciousness limits AI's ability to develop intentional artistic expression. Studies examining listener perception indicate that audiences often recognize subtle differences between emotionally performed music and algorithmically generated compositions, particularly in genres emphasizing expressive performance such as acoustic ballads or vocal-centered music.

Cross-cultural research on musical emotion demonstrates that humans can recognize emotional cues such as happiness, sadness, and fear through musical structures across diverse cultural contexts (Fritz et al., 2009). However, while AI systems can replicate these structural cues, they do not possess the subjective experience that informs emotional interpretation. Artists frequently argue that emotional expression emerges from human vulnerability and personal storytelling, qualities that are difficult to reproduce through computational generation alone.

Artist narratives emphasize that emotional authenticity involves imperfection, spontaneity, and expressive nuance. Slight variations in vocal delivery, timing fluctuations, and improvisational performance elements contribute to the perception of human authenticity. In contrast, AI-generated music often prioritizes structural precision and stylistic consistency. While technical accuracy may enhance production quality, excessive perfection can sometimes reduce emotional relatability. Some musicians describe AI-generated music as technically impressive yet emotionally detached, suggesting that emotional embodiment remains a defining characteristic of human artistry.

3.3 Cultural Identity and Social Context in Music Authenticity

Music is deeply embedded within cultural and social contexts. Musical genres frequently reflect cultural traditions, regional identities, and social experiences. Human musicians develop artistic identity through participation in cultural communities, collaborative networks, and social storytelling. Cultural authenticity contributes to the meaning and symbolic significance of musical works, shaping audience interpretation and emotional engagement.

AI-generated music presents challenges in replicating cultural authenticity. Although AI systems can analyze musical patterns associated with specific cultural genres, they lack cultural lived experience and social participation. Cultural expression often involves shared historical narratives, social values, and collective emotional memory, elements that cannot be fully captured through dataset-driven pattern recognition. Scholars in ethnomusicology emphasize that music functions as a cultural communication medium linking artists and audiences through shared identity and community participation.

Artist-centered analyses highlight that musicians maintain cultural authenticity by engaging with local performance communities, collaborating with culturally diverse artists, and integrating social narratives into musical production. Community engagement allows artists to develop personal identity and artistic influence through social interaction, reinforcing authenticity through relational experience. AI systems, lacking social agency and community participation, face limitations in replicating these sociocultural dimensions of music creation.

Furthermore, cultural authenticity influences audience trust. Listeners often connect with artists whose musical works reflect shared social experiences or cultural heritage. AI-generated music may successfully imitate genre characteristics but may struggle to convey the deeper cultural narratives that shape authentic artistic identity. This distinction reinforces the importance of cultural storytelling as a strategy for maintaining human artistic relevance within AI-driven music production environments.

3.4 Personal Branding and Narrative Identity in the Digital Music Economy

The digital transformation of music distribution has expanded the role of personal branding in artistic success. Modern audiences frequently engage with artists through social media platforms, streaming services, and digital fan communities. Artistic identity is increasingly shaped by narrative storytelling, personal values, and interactive audience engagement. Musicians cultivate brand identity by sharing creative processes, personal experiences, and social perspectives that strengthen emotional connections with listeners.

AI-generated artists challenge traditional models of narrative identity by presenting digital personas without personal biography or lived experience. Although AI performers can be designed with fictional narratives, these constructed identities differ from authentic human storytelling. Audience perception research indicates that listeners often value transparency and emotional sincerity when forming connections with artists. Music that reflects genuine personal experiences tends to generate stronger audience loyalty and long-term engagement.

Personal branding also contributes to artistic differentiation in competitive digital environments. As AI-generated music increases content availability, artists increasingly emphasize individuality and personal storytelling as distinguishing factors. Transparency regarding creative processes, including disclosure of AI-assisted production, can strengthen audience trust and reinforce authenticity. Industry discussions suggest that audiences are more receptive to AI integration when artists maintain clear authorship and artistic intention.

3.5 Imperfection as a Marker of Human Authenticity

Within contemporary music production, imperfection has emerged as an important indicator of authenticity. Human performance frequently includes expressive deviations from technical perfection, such as vocal cracks, timing variations, or spontaneous improvisation. These imperfections contribute to emotional realism and artistic individuality. Music cognition research suggests that listeners often perceive expressive irregularities as indicators of emotional sincerity and artistic presence.

AI-generated music typically prioritizes structural accuracy and stylistic consistency, producing outputs that align closely with genre expectations. While precision can enhance production quality, excessive uniformity may reduce emotional relatability. Musicians often intentionally preserve expressive imperfections during recording and performance to maintain authenticity. Live performances, in particular, highlight the unpredictability and emotional spontaneity associated with human artistry.

Industry commentary suggests that embracing imperfection represents a strategic response to AI-generated music. Artists emphasize live performance, improvisation, and acoustic recording techniques as methods of reinforcing human authenticity. By highlighting expressive individuality, musicians create experiences that differentiate human performance from algorithmic generation.

3.6 Authenticity as a Sustainability Strategy for Human Musicians

The increasing integration of AI into music production has prompted artists to redefine authenticity as a sustainability strategy rather than solely an artistic value. As AI-generated music expands content volume and technical accessibility, human musicians emphasize emotional storytelling, cultural identity, and audience engagement as competitive advantages. Authenticity becomes a form of artistic differentiation that reinforces long-term audience loyalty.

Research examining listener responses to AI-generated music suggests that emotional authenticity remains a significant determinant of audience acceptance. While listeners may appreciate AI-generated compositions for technical quality, emotional sincerity and human storytelling often influence long-term music preference. These findings indicate that authenticity functions as both a cultural value and an economic strategy within music production industries.

Hybrid creative models combining AI-assisted production with human performance elements represent an emerging approach to balancing technological innovation with artistic authenticity. Musicians increasingly incorporate AI tools for sound design, composition assistance, and production efficiency while maintaining human-centered storytelling and performance expression. These hybrid workflows demonstrate that AI integration does not necessarily diminish artistic authenticity but may reshape how authenticity is expressed in digital music environments.

3.7 The Future of Authenticity in Human–AI Music Collaboration

The concept of authenticity is likely to evolve alongside advancements in AI music production. Rather than representing a fixed artistic standard, authenticity may become a dynamic interaction between human creativity and technological innovation. Future music production environments may emphasize transparency, collaborative creativity, and ethical AI integration as defining components of authentic artistic expression.

Researchers increasingly suggest that the future of music authenticity may involve hybrid authorship models in which AI contributes to technical production while human artists provide emotional meaning and narrative interpretation. Such collaborative models require ethical frameworks that recognize human creative contribution and ensure transparent disclosure of AI involvement. As audiences become more familiar with AI-assisted music, perceptions of authenticity may shift toward evaluating artistic intention rather than production method alone.

Despite technological advancements, human emotional storytelling, cultural identity, and community participation remain central to musical authenticity. AI systems may continue to expand creative possibilities, but the cultural significance of music is likely to remain deeply connected to human expression. Understanding how authenticity evolves within human–AI collaboration will play a crucial role in shaping the future of music production and audience engagement.

4. LISTENER ACCEPTANCE AND EMOTIONAL PERCEPTION OF AI MUSIC

4.1 The Importance of Audience Acceptance in AI Music Production

The integration of artificial intelligence into music production cannot be fully evaluated without examining listener acceptance. Music functions not only as an artistic output but also as a social and emotional communication medium shaped by audience perception. The commercial success, cultural relevance, and long-term sustainability of AI-generated music depend heavily on listener trust and engagement. As AI technologies become increasingly capable of generating high-quality musical content, understanding audience reactions becomes essential for predicting the future role of AI in music industries.

Listener acceptance is influenced by multiple factors, including perceived audio quality, emotional authenticity, artistic originality, and beliefs about authorship. Research in human–computer interaction suggests that individuals evaluate AI-generated creative content differently depending on their awareness of the creator's identity. When audiences are informed that music is generated by artificial intelligence, their perception of emotional authenticity and artistic value may shift, even if the audio quality remains unchanged (Shank et al., 2022).

This phenomenon reflects broader psychological patterns in creative evaluation, where the perceived origin of artistic work influences emotional engagement. In music consumption, listeners often interpret songs as expressions of human experience. When music is perceived as lacking personal narrative or emotional intention, listeners may experience reduced emotional connection. Consequently, audience perception plays a critical role in determining whether AI-generated music is accepted as legitimate artistic expression or treated as technological novelty.

4.2 Technical Quality Versus Emotional Authenticity in Listener Evaluation

Empirical studies examining listener perception frequently highlight the distinction between technical production quality and emotional authenticity. AI-generated music systems excel at replicating structural musical features, including harmonic progression, rhythmic stability, and genre-specific arrangement patterns. These technical strengths often result in positive initial listener reactions, particularly when listeners evaluate music based on sonic clarity and production precision.

Research demonstrates that listeners often express surprise regarding the technical sophistication of AI-generated compositions. In controlled listening experiments, participants frequently report that AI-generated music sounds polished and professionally produced. Many listeners initially struggle to distinguish AI-generated music from human-created compositions, particularly in production-heavy genres such as electronic dance music or contemporary pop music.

However, listener evaluations frequently reveal nuanced concerns regarding emotional authenticity. Emotional authenticity refers to the listener's perception that music reflects genuine human emotional experience. While AI-generated music may evoke emotional responses, audiences sometimes describe these responses as lacking depth or vulnerability. Psychological research suggests that emotional engagement in music is strongly influenced by perceived artistic intention and expressive performance dynamics (Juslin & Sloboda, 2001).

Expectancy violation theory further explains audience reactions to AI-generated music. When listeners encounter music that violates their expectations regarding artistic origin, their evaluation of the music may change significantly. Experimental research indicates that listeners may rate identical musical pieces differently depending on whether they believe the composition was created by AI or by human musicians (Hong et al., 2021). These findings demonstrate that emotional authenticity is not solely determined by audio characteristics but also by contextual beliefs about the creator.

4.3 Empirical Evidence of Listener Responses to AI Music Covers

Experimental investigations into listener perception provide valuable insights into audience acceptance of AI-generated music. Controlled listening studies involving AI-generated music covers demonstrate that listeners frequently evaluate AI-generated compositions positively in terms of technical accuracy, arrangement quality, and genre consistency. Participants often describe AI-generated covers as sounding realistic and professionally produced, indicating that AI technologies have reached a level of technical proficiency comparable to human production standards.

However, empirical findings reveal that listener acceptance is influenced by genre expectations and emotional context. In studies examining AI-generated music covers, participants frequently reported that AI-generated music performed well in genres emphasizing production precision and stylistic consistency. In contrast, genres requiring expressive vocal performance or emotionally driven storytelling often revealed stronger listener preference for human-created music.

Listener responses also demonstrate that emotional connection remains a critical factor influencing music preference. Participants frequently acknowledge that AI-generated covers can replicate musical structure and genre style but may lack the expressive vulnerability associated with human performance. Some listeners describe AI-generated music as emotionally consistent but lacking spontaneity or emotional rawness. These perceptions suggest that audiences evaluate authenticity not only through musical structure but through perceived artistic intention and performance expressiveness.

Interestingly, empirical studies indicate that many listeners remain open to AI-generated music as a complementary creative tool rather than a replacement for human artists. Audience acceptance tends to increase when AI-generated music is framed as collaborative innovation rather than artistic substitution. This suggests that transparency regarding AI involvement in music production may influence listener trust and acceptance.

4.4 Genre-Specific Differences in Listener Acceptance

Listener acceptance of AI-generated music varies significantly across musical genres. Production-focused genres such as electronic dance music, synth-pop, and ambient music often demonstrate higher audience acceptance of AI-generated compositions. These genres typically emphasize sonic texture, rhythmic consistency, and digital sound design, areas in which AI systems demonstrate strong performance capabilities.

In contrast, genres emphasizing lyrical storytelling, acoustic instrumentation, and vocal expressiveness often reveal stronger listener preference for human musicians. Acoustic ballads, folk music, and traditional cultural genres rely heavily on emotional storytelling and performance nuance, characteristics that audiences associate with human artistic expression. When AI-generated music attempts to replicate these expressive elements, listeners may perceive the performance as technically accurate but emotionally limited.

Cross-cultural research supports the importance of emotional cues in music perception. Studies demonstrate that listeners across different cultural backgrounds can recognize emotional expression in music through variations in tempo, pitch, and timbre (Fritz et al., 2009). While AI systems can reproduce these acoustic indicators, they may struggle to replicate the contextual storytelling and emotional depth associated with culturally embedded musical traditions.

Genre-based differences in listener acceptance highlight the importance of understanding audience expectations within specific musical contexts. AI-generated music may achieve broader acceptance in production-driven genres while remaining more controversial in performance-driven genres. These distinctions suggest that AI adoption in music production will likely vary across artistic communities rather than following a uniform industry transformation.

4.5 The Role of Transparency and Disclosure in Audience Trust

Transparency regarding AI involvement in music production has emerged as a critical factor influencing listener acceptance. Audience trust is often shaped by perceptions of honesty and artistic integrity. When listeners discover that music they believed to be human-created was generated by AI, they may experience decreased trust in the artist or production process. Conversely, transparent disclosure of AI involvement may strengthen audience acceptance by framing AI as a creative support tool rather than a deceptive substitute.

Industry discussions increasingly emphasize the importance of labeling AI-generated or AI-assisted music. Transparent labeling allows audiences to make informed listening choices and helps maintain ethical recognition of human artistic contribution. Research examining consumer attitudes toward AI-generated content suggests that transparency reduces negative perceptions associated with algorithmic creativity, particularly when audiences understand how AI contributes to artistic production.

Musicians themselves often advocate for transparency as a strategy for maintaining authenticity and audience trust. By openly discussing AI-assisted workflows, artists can demonstrate creative agency while embracing technological innovation. Transparent disclosure may also help differentiate collaborative human–AI creativity from fully automated content production, reinforcing the cultural value of human artistic expression.

4.6 Audience Familiarity and Technological Acceptance

Audience acceptance of AI-generated music is influenced by broader technological familiarity and cultural attitudes toward artificial intelligence. As AI technologies become integrated into everyday digital experiences, listeners may develop increased comfort with AI-assisted creative outputs. Younger audiences, in particular, often demonstrate higher acceptance of AI-generated content due to greater exposure to digital media technologies and algorithm-driven entertainment platforms.

Technology acceptance models suggest that perceived usefulness and ease of interaction significantly influence audience adoption of AI-generated creative products. When listeners perceive AI-generated music as innovative, entertaining, or creatively valuable, they are more likely to accept its integration into mainstream music culture. Conversely, concerns regarding artistic authenticity, job displacement, or ethical misuse of AI technologies may reduce acceptance among more traditional audiences.

Media representation of AI-generated music also influences public perception. Positive portrayals of AI as a creative collaboration tool may encourage audience curiosity and acceptance. Negative portrayals emphasizing job displacement or artistic imitation may reinforce skepticism toward AI-generated music. Public discourse surrounding AI creativity therefore plays a significant role in shaping audience attitudes and adoption patterns.

4.7 Listener Acceptance as a Driver of Industry Transformation

Listener acceptance directly influences how music industries integrate AI technologies into production, distribution, and marketing strategies. Streaming platforms, record labels, and digital distribution services monitor audience engagement metrics to determine which types of content receive commercial investment. If audiences demonstrate sustained acceptance of AI-generated music, industry stakeholders may expand AI-driven production models and digital artist development strategies.

However, listener preference for human authenticity suggests that AI adoption will likely follow hybrid production models rather than complete automation. Industry strategies increasingly emphasize human–AI collaboration, combining algorithmic efficiency with human emotional storytelling. These hybrid models allow music industries to leverage technological innovation while preserving audience trust and cultural relevance.

Audience acceptance may also influence future regulatory frameworks governing AI music production. Consumer demand for transparency and ethical content production may encourage industry standards requiring disclosure of AI involvement and protection of artistic intellectual property. Listener attitudes therefore play a crucial role in shaping ethical and commercial guidelines for AI integration within music production ecosystems.

4.8 The Future of Listener Perception in AI-Driven Music Environments

As AI technologies continue to evolve, listener perception of AI-generated music is likely to become more nuanced. Rather than viewing AI-generated music as inherently inferior or superior to human-created music, audiences may evaluate music based on creative intention, emotional authenticity, and entertainment value. Hybrid creative models blending human performance and AI-assisted production may gradually redefine how authenticity is interpreted in digital music environments.

Future research suggests that audience perception of AI-generated music may shift as listeners become more familiar with AI-assisted creative processes. Educational initiatives, transparent production practices, and collaborative artistic experimentation may contribute to broader audience acceptance. Nevertheless, emotional authenticity and human storytelling are likely to remain central factors influencing listener engagement with music, regardless of technological advancement.

Understanding listener acceptance provides essential insight into the cultural sustainability of AI music production. By examining how audiences evaluate AI-generated music, researchers and industry stakeholders can develop responsible integration strategies that balance technological innovation with preservation of human artistic expression.

5. THE HUMAN–AI MUSIC ECOSYSTEM FRAMEWORK

5.1 The Need for an Integrated Conceptual Model

The rapid evolution of artificial intelligence in music production has created complex interactions between technological innovation, artistic identity, and audience perception. While existing studies often examine AI music from isolated perspectives—focusing separately on technical capabilities, artist adaptation, or listener acceptance—there is a growing need for a comprehensive framework that explains how these dimensions interact within a unified creative ecosystem.

Music production functions as a multidimensional system involving creators, audiences, distribution platforms, and technological infrastructures. Artificial intelligence introduces new stakeholders and alters traditional relationships between these components. Understanding AI's influence on music production therefore requires a holistic analytical model that captures the dynamic interaction between human creativity and algorithmic generation.

Conceptual frameworks play a critical role in interdisciplinary research by providing structured explanations of complex social and technological phenomena. In creative AI research, models help explain how technological tools influence artistic practice, cultural perception, and industry economics. Building upon findings from artist perception studies, authenticity research, and listener acceptance analyses, this chapter proposes the *Human–AI Music Ecosystem Framework* as an integrative model describing AI's role within contemporary music production environments.

5.2 Core Components of the Human–AI Music Ecosystem Framework

The Human–AI Music Ecosystem Framework consists of four primary stakeholder domains: *Human Artists*, *Artificial Intelligence Systems*, *Audience Communities*, and *Music Industry Platforms*. These domains interact continuously, shaping the production, distribution, and consumption of music within AI-driven creative environments.

5.2.1 Human Artists

Human artists remain central to the music ecosystem as primary sources of emotional storytelling, cultural expression, and artistic identity. Despite technological advancements, human creativity continues to define music's cultural and emotional significance. Artists contribute interpretative nuance, lived experience, and social narrative, which form essential elements of musical authenticity.

Within the ecosystem framework, human artists perform multiple roles, including composition, performance, production supervision, and audience engagement. Artists also function as cultural mediators, translating social experiences into musical expression. Research suggests that audiences often associate authenticity and emotional sincerity with human artistic presence, reinforcing the continued importance of human musicians in AI-assisted production environments (Juslin & Sloboda, 2001).

Human artists additionally act as curators of AI-generated content. In collaborative production workflows, musicians select, modify, and interpret AI-generated musical ideas. This curatorial role preserves artistic intention while leveraging algorithmic creativity as a supportive production resource.

5.2.2 Artificial Intelligence Systems

Artificial intelligence systems function as technological collaborators within the music ecosystem. AI tools contribute to composition assistance, vocal synthesis, sound design, mixing automation, and music recommendation algorithms. Machine learning models analyze extensive musical datasets to generate stylistically consistent compositions and replicate genre-specific performance characteristics.

AI systems enhance creative productivity by accelerating production processes and expanding experimental sonic possibilities. Research demonstrates that AI-assisted production tools enable musicians to explore musical variations more efficiently, supporting creative brainstorming and compositional experimentation (Kwiecień et al., 2024).

However, AI systems also introduce ethical and cultural challenges related to authorship, intellectual property ownership, and artistic originality. The framework recognizes AI not as an autonomous artistic agent but as a technologically mediated creative partner. Human oversight remains essential in ensuring ethical usage and artistic accountability within AI-assisted music production.

5.2.3 Audience Communities

Audience communities represent the evaluative and interpretative dimension of the music ecosystem. Listener perception determines commercial success, cultural relevance, and long-term sustainability of AI-generated music. Audience acceptance is influenced by emotional authenticity, perceived artistic intention, and transparency regarding AI involvement in music production.

Research indicates that listener perception of AI-generated music is shaped by contextual beliefs about artistic authorship. When audiences perceive music as human-created, they often attribute higher emotional authenticity and artistic value. Conversely, disclosure of AI involvement may alter listener evaluation even when audio quality remains identical (Shank et al., 2022).

Audience communities also influence artistic adaptation. Musicians frequently adjust creative strategies in response to listener expectations, emphasizing emotional storytelling, live performance, and cultural authenticity. This reciprocal interaction between artists and audiences highlights the importance of social perception in shaping AI music adoption.

5.2.4 Music Industry Platforms

Music industry platforms function as distribution and economic mediators within the ecosystem. Streaming services, digital distribution networks, social media platforms, and music production software companies influence how AI-generated music is produced, marketed, and monetized. Algorithmic recommendation systems play a significant role in determining content visibility and audience exposure.

Industry platforms increasingly incorporate AI technologies into music discovery and production tools. Recommendation algorithms shape listening behavior by promoting content based on user preferences and engagement metrics. AI-generated music may benefit from these algorithmic distribution systems, particularly when production efficiency allows high-volume content generation.

However, industry platforms also influence ethical governance and copyright regulation. Legal frameworks governing AI-generated music remain under development, particularly regarding ownership of algorithmically generated compositions and compensation for artists whose works contribute to AI training datasets (Drott, 2021).

5.3 Interaction Dynamics Within the Ecosystem

The Human–AI Music Ecosystem Framework emphasizes dynamic interactions between stakeholder domains rather than isolated relationships. These interactions form feedback loops that continuously shape music production practices and cultural perception.

5.3.1 Artist–AI Collaboration

Human artists and AI systems interact through co-creative production workflows. Artists utilize AI-generated outputs as compositional inspiration, while AI systems learn from human-created musical datasets. This reciprocal relationship enables technological innovation while preserving human creative control. Research in human–computer interaction suggests that co-creative systems enhance creative productivity by combining algorithmic generation with human interpretative decision-making (Newman et al., 2022).

5.3.2 Artist–Audience Interaction

Artists maintain emotional and cultural connections with audiences through performance, storytelling, and community engagement. Audience feedback influences artistic adaptation and production strategy. Authenticity and emotional sincerity play significant roles in strengthening artist–listener relationships. These interactions reinforce the cultural value of human musicians within AI-assisted production environments.

5.3.3 AI–Audience Interaction

AI-generated music interacts with audiences through algorithmic distribution platforms and digital streaming environments. Listener acceptance of AI-generated music influences technological adoption and industry investment. Audience perception research suggests that transparency regarding AI involvement significantly affects listener trust and emotional engagement.

5.3.4 Industry–Technology Interaction

Music industry platforms mediate the integration of AI technologies into production and distribution systems. Industry investment in AI development influences technological innovation and regulatory policy formation. Platform governance decisions regarding AI-generated content labeling, copyright enforcement, and algorithmic transparency shape the ethical landscape of AI music production.

5.4 Feedback Mechanisms and Ecosystem Adaptation

The Human–AI Music Ecosystem Framework emphasizes adaptive feedback mechanisms connecting stakeholder domains. Audience acceptance influences industry investment in AI technologies. Industry platform algorithms shape listener exposure to AI-generated music. Artist adaptation strategies respond to audience expectations and technological capabilities. These feedback loops create a continuously evolving creative ecosystem.

Technological diffusion theory suggests that innovation adoption occurs through gradual social acceptance and cultural integration. AI-generated music adoption will likely depend on sustained audience engagement and ethical governance frameworks. Hybrid human–AI production models may emerge as dominant industry practice, balancing technological efficiency with preservation of human artistic authenticity.

5.5 Conceptual Framework Diagram Description

The Human–AI Music Ecosystem Framework can be visually represented as a four-node interaction model:

Figure 1. Human–AI music ecosystem framework

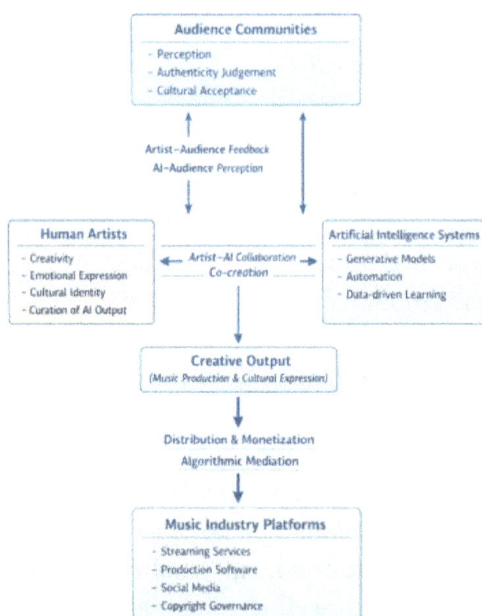

Bidirectional arrows connect each domain, representing collaborative interaction, feedback influence, and cultural adaptation. The center of the model represents *Creative Output*, symbolizing music production as a shared outcome shaped by stakeholder interaction.

This conceptual model highlights AI as a transformative but interconnected component of music production rather than an isolated technological replacement for human creativity.

5.6 Theoretical and Practical Implications of the Framework

The Human–AI Music Ecosystem Framework contributes to interdisciplinary research by integrating technological innovation, artistic identity, and audience perception into a unified analytical model. The framework provides theoretical guidance for understanding how AI reshapes creative industries while preserving human artistic agency.

Practically, the framework offers strategic insights for musicians, industry stakeholders, and policymakers. Artists can utilize the framework to design sustainable creative strategies emphasizing authenticity and technological collaboration. Industry platforms can develop ethical AI integration policies that balance innovation with intellectual property protection. Policymakers can use the framework to design regulatory systems addressing AI-generated creative ownership and artistic labor sustainability.

5.7 Future Research Directions

Future research may expand the Human–AI Music Ecosystem Framework by examining cross-cultural differences in AI music acceptance, longitudinal changes in listener perception, and emerging hybrid performance models combining virtual and physical artistic presence. Investigating educational strategies for AI-assisted music training may also provide valuable insights into preparing future musicians for collaborative creative environments.

6. ETHICAL, ECONOMIC, AND POLICY IMPLICATIONS OF AI IN MUSIC PRODUCTION

6.1 The Ethical Landscape of AI-Generated Music

The rapid integration of artificial intelligence into music production has generated significant ethical concerns regarding authorship, consent, compensation, and creative ownership. Unlike traditional music production, where authorship is clearly attributed to identifiable composers and performers, AI-generated music complicates the definition of creative origin. When a machine learning model generates a composition based on patterns extracted from vast musical datasets, questions arise regarding whether the output constitutes original work or derivative transformation.

One of the most debated ethical issues involves the training data used by AI systems. Many generative AI models are trained on large corpora of copyrighted music without explicit consent from original creators. Scholars argue that this practice challenges existing copyright norms, particularly when AI-generated outputs replicate identifiable stylistic elements of specific artists (Drott, 2021). If AI systems benefit from patterns derived from human-created music, ethical considerations demand clearer guidelines regarding attribution and compensation.

Another ethical issue concerns voice cloning and style imitation. AI technologies can now generate vocal performances that closely resemble identifiable artists. Several high-profile cases involving AI-generated songs mimicking well-known musicians have sparked public debate about artistic consent and digital identity protection. Voice represents not only a musical instrument but also a personal biometric identifier. Unauthorized replication of vocal likeness may constitute infringement on artistic rights and personal identity.

Transparency further emerges as a key ethical requirement in AI music production. Listeners increasingly expect clear disclosure regarding whether music is fully human-created, AI-assisted, or fully AI-generated. Transparency influences audience trust and ethical consumption behavior. Research suggests that disclosure of AI involvement can alter listener perception, underscoring the importance of honest labeling practices (Shank et al., 2022).

Ethical governance in AI music production therefore requires clear standards addressing consent, attribution, and disclosure. Without such standards, technological advancement risks undermining trust between artists, audiences, and industry stakeholders.

6.2 Intellectual Property and Copyright Challenges

Copyright law historically protects original works created by human authors. AI-generated music complicates this framework because copyright statutes in many jurisdictions assume human authorship as a prerequisite for protection. When a machine autonomously generates a composition, determining copyright ownership becomes legally ambiguous.

Several jurisdictions have begun addressing the question of AI authorship. In many cases, copyright protection is granted only if substantial human creative input can be demonstrated. This human-in-the-loop requirement reinforces the importance of maintaining human agency within AI-assisted production workflows. However, purely autonomous AI-generated music may fall into legal gray areas where ownership is unclear or contested.

The issue of derivative similarity further complicates copyright evaluation. AI systems trained on existing music may produce outputs that resemble identifiable compositions or stylistic features of specific artists. Determining whether such outputs constitute infringement requires complex analysis of musical similarity and originality. Scholars emphasize that copyright frameworks must evolve to address the unique characteristics of generative AI systems (Drott, 2021).

Additionally, licensing mechanisms for AI training datasets remain underdeveloped. Policymakers may consider introducing mandatory licensing systems that compensate artists whose works are included in AI training corpora. Such systems could function similarly to mechanical royalties in traditional music distribution, ensuring that artists benefit financially from the use of their creative contributions in AI development.

6.3 Economic Disruption and Labor Implications

Artificial intelligence introduces significant economic implications for musicians and creative professionals. As AI systems become capable of generating high-quality music at low cost, concerns arise regarding labor displacement and income instability within the music industry. Session musicians, background vocalists, and entry-level producers may face reduced demand if AI tools can replicate their functions efficiently.

Economic research on automation across creative industries suggests that technological disruption does not necessarily eliminate employment but may shift the nature of creative labor. Musicians may transition from purely performance-based roles to supervisory or curatorial roles overseeing AI-assisted production workflows. However, this transition requires digital literacy and access to technological resources, potentially creating inequality between technologically equipped artists and those without access to advanced AI tools.

Streaming-based revenue models further complicate economic sustainability. High-volume AI-generated content may saturate streaming platforms, reducing visibility for human artists. Algorithm-driven recommendation systems prioritize engagement metrics, which may favor frequent content release rather than artistic depth. If AI-generated music increases overall content supply, individual artists may experience reduced revenue per stream.

Nevertheless, AI technologies may also create new economic opportunities. Artists can utilize AI tools to reduce production costs, accelerate content creation, and experiment with new sonic identities. Independent musicians may gain competitive advantages by leveraging AI for marketing, sound design, and distribution analytics. Hybrid production models combining human creativity with AI efficiency may enable more sustainable economic strategies.

Economic sustainability in AI-driven music production will depend on regulatory frameworks, platform governance policies, and audience preferences. Listener demand for authenticity may limit complete automation and preserve economic roles for human artists.

6.4 Platform Governance and Algorithmic Responsibility

Music industry platforms play a central role in mediating AI integration within production and distribution systems. Streaming services, social media platforms, and digital marketplaces determine which music reaches audiences through algorithmic recommendation systems. These algorithms significantly influence listener exposure and commercial success.

As AI-generated music increases in volume, platform governance becomes critical in ensuring fair competition between human and AI-generated content. Platforms may need to implement content labeling systems that clearly identify AI-generated music. Such transparency measures can protect consumer trust and enable informed listening decisions.

Algorithmic recommendation systems must also address potential biases that may favor high-volume AI-generated releases over human-created music. If recommendation algorithms prioritize engagement metrics without considering content origin, AI-generated music may disproportionately dominate playlists and discovery channels. Responsible algorithm design requires balancing efficiency with ethical considerations regarding artistic diversity and labor sustainability.

Governments and regulatory bodies may consider introducing standards for AI-generated content disclosure and training data transparency. These standards could mirror existing regulations in advertising and media disclosure, ensuring that consumers are aware of algorithmic involvement in creative outputs.

6.5 Policy Recommendations for Responsible AI Integration

Developing responsible AI governance frameworks requires collaboration between policymakers, industry stakeholders, and creative communities. Several policy recommendations emerge from the ethical and economic analysis of AI music production:

1. Mandatory Transparency Standards

Platforms should require disclosure of AI involvement in music production. Clear labeling distinguishes between fully AI-generated, AI-assisted, and human-created content.

2. Consent-Based Training Frameworks

AI developers should obtain explicit consent for using copyrighted music in training datasets. Compensation mechanisms may be established to reward artists whose works contribute to AI training models.

3. Human Authorship Clarification

Copyright policies should define minimum thresholds of human creative contribution required for ownership protection.

4. Algorithmic Fairness Audits

Streaming platforms should conduct regular audits to ensure recommendation systems do not unfairly disadvantage human artists.

5. Educational Integration

Music education programs should integrate AI literacy training, preparing musicians to engage ethically and effectively with AI technologies.

These recommendations support balanced integration of AI innovation while preserving artistic agency and economic sustainability.

6.6 Balancing Innovation and Human Creativity

The ethical and economic implications of AI in music production underscore the need for balanced technological integration. AI offers unprecedented creative efficiency and experimental potential, yet unregulated deployment risks undermining artistic labor and cultural authenticity. Sustainable AI adoption requires frameworks that protect human creativity while embracing technological advancement.

The Human–AI Music Ecosystem Framework introduced in Part 5 provides a foundation for understanding how ethical governance and economic policy influence stakeholder interactions. Audience expectations regarding authenticity, artist adaptation strategies, and platform governance decisions collectively shape AI's role in music production.

Ultimately, the future of AI in music will depend not solely on technological capability but on collective decisions regarding ethical responsibility, cultural value, and artistic sustainability.

7. CONCLUSION

Artificial intelligence is not merely a technical enhancement in music production; it represents a structural transformation of the creative ecosystem. This chapter has examined AI's influence across three interdependent dimensions: artist perspectives, authenticity preservation, and listener acceptance. By synthesizing these domains, the chapter demonstrated that AI's integration into music production cannot be understood solely through technological capability. Instead, AI reshapes relationships between creators, audiences, and industry platforms.

Artist perspectives reveal a dual narrative. On one hand, AI provides unprecedented creative flexibility, enabling rapid experimentation, production efficiency, and new sonic possibilities. On the other hand, musicians express concerns regarding authorship ambiguity, labor displacement, and erosion of emotional authenticity. These tensions illustrate that AI's impact extends beyond technical production to the social and economic foundations of artistic identity.

Authenticity emerged as a central theme across creative and evaluative dimensions. Human musicians differentiate themselves through emotional embodiment, cultural storytelling, performance nuance, and relational engagement with audiences. While AI systems can replicate stylistic patterns with increasing sophistication, they do not possess lived experience or emotional intentionality. As demonstrated in Parts 3 and 4, audience acceptance is often shaped not only by audio quality but by perceived artistic origin. Emotional authenticity continues to function as a critical determinant of long-term listener engagement.

Listener studies indicate that AI-generated music is frequently perceived as technically impressive but emotionally moderated by disclosure and contextual framing (Shank et al., 2022). Acceptance increases when AI is framed as collaborative innovation rather than artistic substitution. This finding reinforces the importance of transparency and ethical governance within AI-driven music ecosystems.

The Human–AI Music Ecosystem Framework presented in Part 5 integrates these insights into a stakeholder model linking human artists, AI systems, audience communities, and music industry platforms. The framework emphasizes that AI does not operate in isolation but functions within feedback loops shaped by perception, governance, and economic mediation. Responsible AI integration requires alignment between technological innovation and preservation of human creative agency.

AI's long-term role in music production will ultimately depend on collective decisions made by artists, listeners, industry stakeholders, and policymakers. If guided by ethical frameworks and transparency, AI may enhance creative diversity and expand musical participation. If implemented without safeguards, AI risks destabilizing artistic labor and eroding audience trust.

The future of music production is unlikely to be purely human or purely algorithmic. Rather, it will be hybrid—shaped by collaborative interaction between emotional storytelling and computational generation.

7.1 Future Directions

Several research and industry pathways emerge from this analysis:

7.1.1 Longitudinal Listener Perception Studies

Future research should examine how listener acceptance evolves over time as audiences become more familiar with AI-generated music. Longitudinal studies may reveal shifts in authenticity perception and emotional evaluation.

7.1.2 Cross-Cultural Comparative Research

Cultural context significantly influences musical authenticity and genre interpretation. Comparative research across cultural communities can assess whether AI acceptance varies by regional music traditions.

7.1.3 Hybrid Performance Models

Emerging hybrid concerts combining live human performance with AI-generated soundscapes present opportunities to explore co-creative authenticity models.

7.1.4 Regulatory Development Studies

Legal scholarship should continue examining copyright adaptation, dataset licensing frameworks, and digital identity protection in AI-generated music contexts (Drott, 2021).

7.1.5 AI Literacy in Music Education

Music education curricula should integrate AI-assisted production training to prepare future musicians for hybrid creative environments.

ACKNOWLEDGMENT

The researcher would like to thank Philippine Technical Institute and Pangasinan State University administrators for the conduct of several studies related to this topic. This research was supported by the Pangasinan State University.

REFERENCES

Drott, E. (2021). Copyright, compensation, and commons in the music AI ecosystem. *Creative Industries Journal, 27*(6), 1–16. DOI: 10.1080/17510694.2020.1839702

Fritz, T., Jentschke, S., Gosselin, N., Sammler, D., Peretz, I., Turner, R., Friederici, A. D., & Koelsch, S. (2009). Universal recognition of three basic emotions in music. *Current Biology, 19*(7), 573–576. DOI: 10.1016/j.cub.2009.02.058 PMID: 19303300

Herington, M. J.. (2025). Artificial intelligence and labor displacement in creative industries. *AI & Society*. Advance online publication. DOI: 10.1007/s00146-025-02601-6

Hong, J., Kim, J., & Kim, Y. (2021). When artificial intelligence creates art: The role of expectancy violation. *New Media & Society, 23*(12), 1–19. DOI: 10.1177/1461444820925798

Juslin, P. N., & Sloboda, J. A. (2001). *Music and emotion: Theory and research.* Oxford University Press., DOI: 10.1093/oso/9780192631886.001.0001

Kwiecień, J., Skrzyński, P., Chmiel, W., Dąbrowski, A., Szadkowski, B., & Pluta, M. (2024). Technical, musical, and legal aspects of an AI-Aided algorithmic music production system. *Applied Sciences (Basel, Switzerland), 14*(9), 3541. DOI: 10.3390/app14093541

Newman, L. J., Stewart, S. E., Freeman, N. C., & Thompson, G. (2022). A systematic review of music interventions to support parent–child attachment. *Journal of Music Therapy, 59*(4), 430–459. DOI: 10.1093/jmt/thac012 PMID: 36995776

Shank, D. B., DeSanti, A., & Maninger, T. (2022). When artificial intelligence authors art: The effects of AI authorship on aesthetic judgment. *Journal of Experimental Psychology. Applied, 28*(3), 1–15. DOI: 10.1037/xap0000447 PMID: 36006713

Sturm, B. L., & Maruri-Aguilar, H. (2021). Artificial intelligence and music: Open questions of copyright law and engineering praxis. *Journal of Cultural Management and Cultural Policy, 7*(1), 1–24. DOI: 10.5920/jcms.950

KEY TERMS AND DEFINITIONS

AI-Assisted Music: Music created through collaborative workflows in which human artists use AI tools for compositional or production support while retaining creative control.

AI-Generated Music: Musical compositions or performances produced primarily through algorithmic systems without direct human performance input.

Algorithmic Authorship: The attribution of creative output to computational systems capable of generating original compositions.

Artificial Intelligence (AI) in Music Production: The application of machine learning and generative algorithms to compose, arrange, synthesize, or assist in the creation of musical works.

Authenticity in Music: The perceived sincerity, emotional truthfulness, and cultural grounding of a musical work as interpreted by listeners.

Creative Agency: The capacity of a human artist to intentionally shape and control artistic output.

Digital Artist: A virtual or AI-generated persona that produces and distributes music through digital platforms.

Emotional Embodiment: The expression of lived emotional experience through musical performance or composition.

Human–AI Collaboration: A co-creative process in which human musicians and artificial intelligence systems interact to produce musical output.

Chapter 2
Musicology in Artificial Intelligence

Evren Idil Yazan
http://orcid.org/0009-0001-2259-4407
Akdeniz (Mediterranean) University, Turkey

ABSTRACT

Musicology is the scholarly, scientific, and humanistic study of music. Historical musicology studies the development of musical works, styles, and practices over time, using archival research and contextual analysis. Archival research involves examining original documents, manuscripts, letters, and musical works, while contextual analysis considers the social, cultural, and historical circumstances of the music (Levy & Emmery, 2021). As the boundaries between human and machine creativity continue to blur, musicologists are uniquely positioned to interrogate the values embedded in these systems and to advocate for practices that respect both the complexity of musical traditions and the transformative potential of technological innovation. The convergence of AI and musicology thus presents both new opportunities and new obligations, positioning the field at a pivotal intersection of technology, creativity, and culture. Thus, the purpose of this study is to explore the discipline of musicology from its near history to its new form that's formed with the invention of AI.

THE REDEFINITION OF MUSICOLOGY

Musicology is the scholarly, scientific, and humanistic study of music. It explores music's history, structure, theory, social roles, cultural meanings, and the technologies, acoustics, and cognitive processes that shape musical experience (American Musicological Society, n.d.). The term of musicology was originally defined by

DOI: 10.4018/979-8-3373-6279-3.ch002

a Belgian music theorist named François-Joseph Fétis a century ago. He simply called it *Musikwissenschaft, which meant the science of music. The revolution of this scholar's approach was his attempt to define the discipline of musicology free from dogmatic historical perspectives and narratives. He claimed musicology as an independent field of study with its own unique methods (Hibberd, 1959). Later, Guido Adler wrote a seminal essay, and he made a categorial distinction between historical and* systematic musicology. Adler divided musicology into historical and systematic branches, creating a framework that still guides the field. Historical musicology studies the development of musical works, styles, and practices over time, using archival research and contextual analysis. Archival research involves examining original documents, manuscripts, letters, and musical works, while contextual analysis considers the social, cultural, and historical circumstances of the music (Levy & Emmery, 2021). For example, a scholar might study a composer's manuscripts and letters to understand how the music reflected the cultural and social environment of its time.

Systematic musicology examines general principles of music, including acoustics, cognition, performance, cultural function, and the structural elements that make music intelligible, such as melody, harmony, rhythm, form, and texture. Together, these branches provide a complete approach: one focuses on the history of music, the other on its underlying structures, patterns, and effects, offering both a temporal and analytical understanding of the art. Within the context of systematic musicology, Adler indeed used the term comparative musicology to describe the study of music across different cultures (Adler, 1885). He wanted to compare musical traditions from around the world to find similarities and differences. In this way, his purpose was to understand the principles of music regarding how rhythm, melody, harmony, and instruments work across cultures. The inference is that Adler did this because he believed that studying music comparatively would reveal universal patterns and rules in music, helping scholars move beyond individual composers or styles and see music as a shared human phenomenon. Therefore, it can be noted that historical musicology investigates music's development through time, using close study of manuscripts, archival sources, notation systems, and compositional practices (Parncutt, 2007; Popović, 2020). Adler's model was later refined, and the system he described evolved: what he termed "comparative musicology" eventually branched into ethnomusicology, while systematic musicology itself diversified into subfields such as music psychology, acoustics, and cognitive studies.

Following these pioneering scholars, the term of 'musicology' has revolved into a more inclusive and interdisciplinary field that embraces diverse musical traditions and methodologies. From this perspective, contemporary musicology integrates technological advances, critical theory, and cross-cultural perspectives, moving beyond traditional Western music to include popular, electronic, and global musics.

This shift reflects the discipline's responsiveness to social change, digital media, and the complex ways music interacts with identity, politics, and globalization (Born & Devine, 2015; Clayton, 2020). Today, musicology is then best understood as a deeply interdisciplinary field.

The continuous evolution of the field is characterized by a "digital turn," which has profoundly reshaped how researchers access, analyze, and disseminate musical knowledge. Digital humanities have equipped musicologists with tools for large-scale data analysis of scores, performance data, and archival records, moving away from purely close-reading methods toward quantitative approaches (Hentschel et al., 2023). This shift is particularly evident in the rise of Computational Musicology, a field that uses machine learning and statistical modeling to explore patterns in vast musical corpora, revealing structural relationships and stylistic markers that would be invisible through traditional manual analysis alone (Eck, 2023; Morgado Vega et al., 2025). The exploration of these massive datasets has spurred new conversations concerning what constitutes a "musical work," how genre boundaries are defined, and how consumption patterns influence creation.

Furthermore, the expansion of musicology has been fueled by its increasing engagement with critical theory and socio-political inquiry. Musicology now routinely intersects with race studies, gender and postcolonial studies, addressing how power structures are reflected in or challenged by musical practices. For instance, scholars might analyze the mechanisms by which certain musical histories—particularly those originating outside the Western canon—have been marginalized or how gender roles are enforced or subverted in performance and consumption (Champion, 2022; Crowdus, 2023; Hsieh & Olley, 2025). This focus ensures that the discipline remains responsive to contemporary social issues and acknowledges music as a powerful site for negotiating identity and politics.

Thus, at present, musicology is best defined as a comprehensive discipline that uses a variety of methods to study music as a historical, cognitive, cultural, and aesthetic phenomenon. It remains true to Adler's dual vision yet goes beyond it, embracing multiple approaches to capture the rich complexity of musical life in its diverse forms. The recent integration of artificial intelligence highlights this diversity, offering new tools to analyze, interpret, and create music across cultural and historical boundaries. As musicology continues to expand its methods and perspectives, it opens exciting opportunities for research and creativity in the digital era because of machine learning and AI (Deruty et al., 2022).

THE ROLE OF AI IN CONTEMPORARY MUSICOLOGY

Artificial Intelligence (AI) has changed the course of almost all settled disciplines. In musicology, it has moved the study of music from relying solely on human analysis and rule-based systems to using computers. music-AI programs used simple step by step rules of instructions written by humans. To illustrate, a rule used to say: "After this note, play this note next," or "Use these three chords in this specific order." While following these steps as instructed, the programs did not have the ability to understand the song as a whole (Rowe, 1993). The computers started to learn patterns and structures in music on their own, which was absolutely revolutionary. First, this allowed musical experts to explore melodies, harmonies, rhythms, and textures in ways that were even impossible before. This shift has not only transformed analytical approaches but has also expanded the kinds of questions musicologists can ask to explore the discipline further. With AI, researchers can now study patterns across vast collections of music that would be impossible to analyze manually. In this way, they analyze trends in melody, harmony, rhythm, and instrumentation across centuries or even cultures. AI also allows scholars to investigate deeper relationships within music that were previously difficult to detect. For instance, it can uncover how changes in rhythm correlate with shifts in harmony or how recurring melodic motifs appear across different periods and regions. Moreover, AI can model hypothetical scenarios, such as exploring how a composer might have adapted their style when exposed to unfamiliar musical traditions. These capabilities expand the possibilities of musicological research, bridging traditional scholarship with computational methods and opening pathways for new discoveries in understanding music as a cognitive phenomenon (Razvan et al., 2023; Yang & Lee, 2024).

In addition to perceiving musicology as a cognitive phenomenon, AI allows for the analysis of compositional processes for the purpose of investigating how different musical elements interact and evolve over time. Furthermore, it has created unique opportunities like no other for collaboration between musicologists and provided an integrated understanding of music as a cultural artifact. Blending traditional research with computational methods now allows human understanding and machine insights to work together effectively (Lupker & Turkel, 2021; Yang & Lee, 2024). In recent years, this development has also encouraged musicologists to investigate questions that involve areas that were previously difficult to address with traditional methods alone. AI now makes it possible to examine how musical features shift across thousands of recordings, how stylistic boundaries blur in digital environments, or how listener behavior influences the circulation of musical ideas. These emerging research directions reflect a broader change in the field, where music is increasingly understood not only as an artistic product but also as a dynamic system shaped by patterns of production, distribution, and interaction.

Moreover, the use of deep learning because of AI has changed the course of music and how it is perceived in the world today. Deep learning is a type of artificial intelligence that uses networks of connected "virtual neurons" to automatically learn complex patterns from large datasets (Goodfellow et al., 2016). Virtual neurons are like tiny computer "thinking units" that work together, passing information to each other to help the system recognize patterns and understand data, similar to how brain cells (neurons) in a human brain communicate (Nielsen, 2015). Because of AI, musicologists now have new kinds of evidence—large-scale, pattern-based insights that cannot be discovered by listening or reading scores alone, expanding the ways they can interpret and explain music. These advanced AI systems are not just following rules: they can produce creative musical ideas in original ways. In this way, modern generative models can compose polyphonic music with surprising harmony and texture, showing novelty, value, and even conceptual "leaps" that resemble human creativity (Cadiz et al., 2021). These creative capabilities suggest that machines are not merely tools, but also creative partners, enabling explorations into musical imagination that were once limited to human composers.

CREATIVITY IN MUSIC: COMPARING HUMANS AND AI

Have you ever considered how a composer adapts their unique style to existing forms of traditions when creating a piece that is entirely novel? Hypothetical scenarios modeled by AI changed the way music is practiced and perceived. These intelligent capabilities expand the possibilities of musicological research, bridging traditional scholarship with computational methods and opening pathways for new discoveries in understanding music as a cognitive phenomenon (Eraslan, 2025b). Moreover, understanding music as a cognitive phenomenon emphasizes not just its aesthetic qualities but its deeply embedded role in mental processes. Computational and cognitive theories increasingly consider music to be a dynamic system that engages prediction, memory, and affect. For instance, the 4E (embodied, embedded, enacted, extended) cognitive framework has been applied to music, conceptualizing musical experience as a tool for feeling that is shaped by both brain dynamics and sociocultural contexts (Merlini, 2024). Meanwhile, novel AI-driven studies show that music practitioners' creative perception is significantly influenced by their interaction with artificial intelligence tools, suggesting that AI not only augments creativity but reshapes the cognitive processes underlying musical generation (Ma et al.,2025).

Furthermore, the ability of AI systems to explore hypothetical variations in compositional style offers unprecedented opportunities for both historical and ethnomusicological inquiry. By manipulating parameters within generative models, researchers can explore alternative evolutionary courses of musical genres and styles, thereby testing hypotheses about cultural transmission and influence with greater precision than previously possible (Wang & Zhang, 2025). This method facilitates a dynamic engagement with music not as a static artifact but as an evolving system shaped by complex interactions between individual creativity and sociocultural powers (Bryant & Smaldino, 2025). This new intelligent paradigm then encourages musicologists to reconsider the boundaries of human creativity. While human composers rely on intuition, emotion, and personal experience to generate musical ideas, AI systems approach composition by identifying patterns across vast musical datasets. These systems can synthesize styles from different eras or genres, creating pieces that might combine Baroque counterpoint with contemporary electronic textures or merge global folk traditions with modern jazz harmonies (Vear et al., 2023). Unlike humans, AI can analyze vast amounts of works simultaneously by detecting correlations and structures that would be impossible to grasp or analyze manually.

Despite these strengths, the nature of machine creativity remains fundamentally different from human creativity. Machines lack subjective experience, emotional intentionality, and cultural memory. Their "innovations" emerge from recombination and statistical inference rather than lived experience. On the other hand, this difference does not diminish the value of AI-generated music; instead, it highlights a complementary relationship. Human musicians and musicologists can guide AI, selecting or refining outputs, while AI can propose possibilities beyond the limits of human imagination (Zhang, 2025). This ability allows researchers to ask questions such as:

- How have harmonic progressions evolved across centuries?
- What rhythmic motifs appear most frequently across cultures?

First of all, recent computational and cross-cultural research offers emerging insights into both how harmonic progression has evolved over time in terms of human creativity. Historically, harmonic progressions in Western music show both continuity and periods of experimentation: Baroque music often adhered to functional harmonic rules emphasizing cadences like V–I, while Classical composers expanded modulation and chromaticism, and Romantic composers introduced chromatic mediants and ambiguous tonal centers to increase expressive range (Kopp, 2002; Temperley, 2010). Here, the term V–I refers to a chord progression where the "V" chord (dominant) creates tension that naturally resolves to the "I" chord (tonic), the home chord of a key. To explain simply, the I chord feels like "home base," while the V chord feels like a note that wants to return home. For example, in C major, G major (V chord: G–B–D) naturally resolves to C major (I chord: C–E–G), creating a satisfying sense of completion. This progression is one of the most common ways composers make music feel resolved and complete.

In addition to the exploration of this process, which rhythmic motifs are common across cultures is noteworthy to investigate. The results are quite complicated since the data over four years showed that harmonic changes and transition became novel yet uncertain in certain historical periods. Computational studies suggest that such periods of harmonic novelty often corresponded to cultural or stylistic transitions, where composers deliberately experimented with unexpected chord sequences, thus increasing expressive variety in musical composition (González Espinoza & Plotkin, 2023). This suggests that composers did not always follow predictable chord patterns: in some eras, they experimented more creatively by reflecting harmonic complexity

Research into rhythm perception has also demonstrated that culture strongly influences rhythm reproduction: people from different cultural backgrounds tap and reproduce beats in ways that reflect their familiar musical rhythms (Cameron et al., 2015). On the rhythmic side, cross-cultural studies confirm a shared cognitive preference for simple integer-ratio rhythms (like 1:1:1), even though different societies emphasize different rhythmic structures depending on their musical traditions (Jacoby et al., 2024; Dolan, 2024). Large-scale AI-assisted analyses of ethnomusicological databases indicate that while simple integer-ratio rhythms are widespread, complex rhythmic patterns such as 7/8 or 9/8 cycles frequently appear in region-specific traditions, highlighting the balance between universal perceptual tendencies and cultural specificity (Agawu, 2003; Holzapfel & Bozkurt, 2012). To illustrate, a large-scale cross-cultural study led by researchers from MIT and the Max Planck Institute found that while there is a general preference for simple rhythmic patterns (such as integer ratio rhythms), the specific rhythmic "priors" differ significantly between societies.

TECHNOLOGIES, AI MODELS AND THE MUSICOLOGICAL TURN

The landscape of contemporary music production and analysis has been profoundly reshaped by advancements in artificial intelligence (AI), presenting both technical innovation and new theoretical challenges for musicology. But, how are these technologies shaped and used so far? Central to this transformation are generative AI models, including transformers, recurrent neural networks (RNNs), and generative adversarial networks (GANs), which allow for the autonomous creation of musical elements such as melody, harmony, rhythm, and texture. These models are trained on vast datasets of musical works, learning statistical relationships between notes, chords, rhythmic patterns, and structural motifs. RNNs, for instance, are particularly adept at modeling sequential data, capturing temporal dependencies in musical phrases, while transformers excel in learning long-range correlations, enabling them to generate harmonically coherent compositions that extend across entire pieces (Huang et al., 2018; Vaswani et al., 2017). GANs, on the other hand, employ a dual-network system—a generator and a discriminator—that iteratively refines outputs to produce music indistinguishable from human-composed material, effectively simulating compositional decision-making through adversarial training (Brunner et al., 2018).

AI-driven lyric generation has further expanded the scope of creative possibilities. Language models, such as GPT-based architectures, can produce textual content that aligns with thematic, emotional, or structural requirements of a musical piece. These systems are capable of generating verses and choruses that maintain semantic and stylistic coherence while adapting to melodic constraints or rhyming patterns, offering composers a tool to explore new lyrical directions that might have been impractical through manual composition alone (Hopkins & Kiela, 2017). When combined with melody and harmony generators, these systems can produce integrated musical compositions, raising questions about the intersection between textual meaning, musical form, and cognitive reception.

Sound design, long a domain of experimental practice, has similarly been transformed by deep learning. Neural networks can synthesize entirely new timbres, manipulate spectral envelopes, or interpolate between existing instruments to create novel sonic textures. By learning from massive datasets, these models capture subtle acoustic features that human sound designers may overlook, enabling the exploration of musical possibilities that extend beyond traditional instrument categories (Engel et al., 2017). From a musicological perspective, this raises fascinating questions about the ontology of sound: if timbre can be algorithmically generated, to what extent does it retain cultural or historical significance, and how should it be analyzed in terms of its perceptual and emotional impact on listeners?

The implications of these technological innovations intersect directly with musicological concerns regarding form, harmony, texture, timbre, and genre. A generative model capable of producing "plausible harmonic progressions" without human intervention invites reconsideration of traditional theories of harmonic function and voice leading. Similarly, AI tools that autonomously adjust dynamics or spectral content compel scholars to rethink analytical frameworks that presume human intention as a central determinant of musical structure. These developments suggest a necessary expansion of musicology to include not only historical and systematic methodologies but also computational and algorithmic perspectives, fostering a more comprehensive understanding of music as both a cultural and technological phenomenon (Ji et al., 2023; Zhao et al., 2025).

On the other hand, the deployment of AI in music is not neutral. The underlying datasets, algorithms, and design decisions embed cultural biases and value judgments that can significantly shape output. Many generative systems are trained primarily on Western classical, popular, or electronic music, which overrepresents certain harmonic conventions, rhythmic structures, and melodic idioms while marginalizing non-Western or vernacular traditions (Mehta et al., 2025). As a result, AI-generated compositions may reflect a narrow cultural perspective, inadvertently reinforcing dominant musical norms and aesthetic hierarchies. Musicologists must therefore consider how these biases influence both creative practice and scholarly interpretation, interrogating questions of representation, accessibility, and cultural preservation.

The musicological turn towards AI also necessitates methodological innovation. Traditional analysis—score study, performance observation, or archival research—must be complemented by computational techniques capable of handling the scale and complexity of AI-generated music. Network analysis, statistical modeling, and pattern recognition offer tools for understanding emergent structures in algorithmic compositions, while ethnographic approaches can elucidate how musicians interact with AI tools in creative practice. In this way, musicology becomes both reflective and proactive, engaging critically with the technologies that shape contemporary music production (Herrington et al., 2025).

A particularly salient example is AI-generated cross-genre synthesis. Systems trained on diverse musical corpora can produce compositions that blend Baroque counterpoint with jazz improvisation, or integrate folk motifs from multiple global traditions into a single piece. These hybrid outputs demonstrate both the potential for creative expansion and the risk of homogenization, as algorithms may prioritize statistical likelihood over cultural nuance. Musicologists must navigate these tensions, evaluating how algorithmic outputs reflect or distort historical, stylistic, and cultural knowledge (Vear et al., 2023).

In addition to creativity, AI introduces new dimensions of interactivity and pedagogy. Intelligent systems can provide real-time feedback to performers, suggest compositional alternatives, or simulate ensemble contexts, enabling a deeper understanding of musical form, style, and technique. These applications not only enhance practical musicianship but also provide novel opportunities for research into cognition, perception, and learning. By bridging computational, cognitive, and cultural perspectives, AI facilitates a richer and more multidimensional approach to musicology (Corbelli, 2024).

Ultimately, the integration of AI into music production and scholarship exemplifies a broader musicological turn toward interdisciplinarity. It demands engagement with computer science, data ethics, cognitive psychology, and cultural studies while retaining a commitment to core musicological principles: analysis, contextualization, and interpretive understanding. Generative models, intelligent mixing, deep learning sound design, and lyric generation are not merely technical novelties—they reshape what it means to study, create, and experience music. Musicology, in this sense, becomes an evolving practice, attentive to technological innovation, social responsibility, and the dynamic interplay between human and machine creativity.

In conclusion, the convergence of AI technologies and musicology represents both an opportunity and a challenge. The technical capabilities of generative models, deep learning systems, and intelligent tools redefine the possibilities for music creation, analysis, and pedagogy, while simultaneously raising questions of authorship, cultural bias, and ethical responsibility. Musicologists must therefore adopt a dual perspective: one that embraces computational innovation while critically examining the social, cultural, and aesthetic implications of these tools. By doing so, the discipline can maintain its relevance in the digital era, offering insights that honor both human creativity and technological potential. As AI continues to develop, the field is poised to explore increasingly complex musical systems, hybridized forms, and collaborative processes, ultimately expanding the scope and depth of musicological inquiry in the twenty-first century.

Ethical concerns extend beyond cultural representation to include intellectual property and authorship. Generative AI systems often rely on massive corpora that include copyrighted material, raising complex legal questions about the ownership of AI-generated compositions. If an algorithm produces a melody or harmonic sequence learned from copyrighted works, to what extent does the human programmer, the AI model, or the original composer hold creative or legal credit? This intersection of technology, law, and aesthetics demands careful scrutiny within musicological inquiry, highlighting the discipline's expanding remit into ethical and social analysis (de Berardinis et al., 2025).

In parallel, AI-assisted mixing and mastering tools have introduced automation to processes traditionally reliant on expert human judgment. Intelligent mixing platforms can analyze multi-track recordings, applying dynamic balancing, equalization, and spatial placement based on learned models derived from professional reference tracks (Sturm et al., 2019). AI mastering services employ neural networks to adjust loudness, compression, and spectral characteristics, effectively replacing—or at least augmenting—the role of the audio engineer. While these tools offer unprecedented efficiency and access, they also challenge musicologists to reconsider notions of authorship, interpretation, and performance. If dynamic shaping and spectral decisions are determined by an algorithm rather than a human ear, what does this imply for the study of timbre, texture, and expressive intention in recorded music?

How should the concept of a "musical work" be adapted when both composition and post-production involve machine agency?

AI AND ARTISTIC UNCERTAINTY AI

The rapid rise of artificial intelligence in music generation has created a deeply complex landscape for musicians, combining both opportunity and unease in ways that affect personal, professional, and cultural dimensions. For many artists, AI represents a double-edged sword: on one hand, it offers tools for enhanced creativity, faster production, and experimentation with novel musical ideas; on the other hand, it raises fears of displacement, economic insecurity, and the potential dilution of the human element that has historically defined music-making (Herington et al., 2025; CISAC, 2024). Musicians are confronting questions about the nature of creativity that previous generations never needed to address: can a machine truly create music in a meaningful sense? How does human artistic identity persist when an algorithm can imitate or even innovate music at scale? These concerns are not only philosophical; they carry tangible emotional and financial weight for those whose livelihoods depend on their creative work. Central to the unease is the fear that AI could diminish the value of musicians' labor. Many artists derive both income and personal fulfillment from the act of composing, performing, and producing music. When AI can generate compositions that closely resemble human styles, musicians may fear that audiences, employers, or consumers will favor speed and novelty over the subtle, nuanced artistry that comes from lived experience, cultural context, and emotional investment (New Indian Express / CISAC, 2024). This anxiety is heightened by the reality that AI-generated music is increasingly being used commercially, often without compensation to the original artists whose work informed the algorithms. This situation raises important ethical and legal questions about intellectual property, the rights of creators, and the fairness of an ecosystem where human input is mined for training data but not adequately recognized or rewarded (de Berardinis et al., 2025).

Emotional and existential concerns also shape the musicians' response to AI. Musical creation has always been deeply personal; it embodies memories, identity, and emotional expression. The notion that an algorithm could produce a work that evokes similar feelings to a human-composed piece challenges traditional notions of authenticity and meaning in art. For some, it is not merely about employment but about witnessing the potential erosion of a uniquely human form of expression. The concern is that music could become increasingly commodified and automated, reducing the perception of human effort and eroding the intrinsic value of personal artistic contribution (Herington et al., 2025). Despite these concerns, AI also offers unprecedented possibilities for creative collaboration and innovation. Many musicians have begun to explore AI as a co-creator, using generative systems to generate raw material, experiment with unfamiliar styles, or break habitual compositional patterns. In this perspective, AI is not a competitor but a partner — a tool that expands human imagination rather than replacing it. Musicians can refine and humanize AI-generated outputs, producing works that are simultaneously collaborative and authentically human (de Berardinis et al., 2025). This shift invites a reconsideration of authorship, emphasizing that creativity is not a solitary act but often a complex interaction between multiple agents — human and technological alike.

Furthermore, the introduction of AI into music has broad cultural implications. Generative algorithms often rely on large, preexisting datasets, which may disproportionately reflect popular or Western music traditions. Without careful attention to diversity and representation, AI risks reinforcing dominant trends while marginalizing lesser-known, experimental, or culturally specific musical forms (Herington et al., 2025). This creates a dual challenge: preserving the integrity and richness of musical heritage while responsibly integrating AI into creative practice. Efforts to include global, underrepresented repertoires in training datasets and to develop licensing frameworks that protect musicians' rights are essential steps in addressing these challenges (CISAC, 2024). Finally, the question of whether AI's role in music is "sad" for musicians depends on perspective. It is undeniably distressing for those who fear economic displacement or the erosion of personal expression. Yet, when approached thoughtfully and ethically, AI can also be empowering, providing artists with new ways to experiment, innovate, and amplify their creative voice. The future of music is likely to be a hybrid landscape, shaped by both human insight and machine capability. Ensuring that this partnership respects and uplifts human creativity, rather than replacing it, will determine whether AI ultimately enriches the musical world or diminishes the role of musicians in it (Herington et al., 2025; de Berardinis et al., 2025).

BIAS AND AI-TRAINED MUSICAL DATA

The datasets that were trained by AI often contain significant bias. Generative models rely on corpora, which are large, organized collections of musical recordings, scores, or audio files used to train machine-learning systems. These corpora shape what the model can learn, because the musical patterns and styles they contain determine the kinds of outputs the system can generate. In many cases, these collections overrepresent Western classical, pop, or electronic music, leaving many non-Western or marginalized musical traditions undiscovered or insufficiently represented (Agostinelli et al., 2023). While the use of AI in music systems disproportionately favor Western musical traditions, and this bias extends globally, disregarding many culturally rich traditions. In countries like Brazil, traditions such as samba, bossa nova, and candomblé rhythmic music are documented primarily in Portuguese and through local archives, but these genres are underrepresented in English-centric training corpora (Maia et al., 2019). Moreover, Indian classical music—both Hindustani and Carnatic—is structurally distinct, with microtonal scales, raga frameworks, and ornamentation that differ significantly from Western tonal music, yet AI models struggle to model these features because most datasets overrepresent Western styles (Mehta et al., 2025; Vinod & Colaco, 2025).

Similarly, in Turkey, Ottoman-classical makam systems and **usûl** rhythms are deeply rooted in musical theory and oral tradition, yet English-centric AI / MIR models often lack adequate Turkish-language metadata or annotated corpora — a problem evidenced by the difficulties in adapting generation models to Turkish makam music. (Mehta et al., 2025). Moreover, while some dedicated symbolic datasets exist (Eraslan, 2025a), their scale remains limited compared to the wealth of Western music data. In traditional Turkish classical music, for instance, complex *usul* cycles (metrical patterns) like 9- or 10-beat constructs are central; these rhythmic structures reflect a deep cultural emphasis on asymmetric meter rather than simple, equally spaced beats (Kaya, 2023 & Kömürcü, 2025). Similarly, African polyrhythms often combine contrasting cycles that create intricate temporal layering, reflecting communal and dance-based musical practices (Agawu, 2003). These findings suggest that while human brains may share some universal biases in rhythm perception, cultural experience heavily shapes how we internalize and prefer particular rhythmic patterns. As a result, musicologists studying rhythm must consider both innate perceptual tendencies and the specific rhythmic logic embedded in different musical traditions despite artificial intelligences.

From a similar perspective, Italy's rich musical legacy, especially its opera and *bel canto* traditions, poses challenges for AI: librettos, performance practice, and stylistic documentation in Italian are not fully captured in English-based datasets. Italy's opera tradition is underrepresented in computational music studies, as many original librettos remain digitally inaccessible in scholarly editions (Gallucci, 2023), and Italian opera is often overlooked compared to German and French works (Giovannini & Skorinkin, 2024). This limited representation means AI systems trained on predominantly English or Western-centric corpora may miss important details such as recitative-aria structures, phrasing, and stylistic ornamentation crucial to *bel canto* and opera. In parallel, Greece has rich traditions like rebetiko and Byzantine chant, which are often recorded and annotated in Greek; AI systems that rely primarily on English documentation may miss subtle modal and melodic features native to these styles. For example, computational analysis has shown that the performance intonation of Byzantine chant often deviates from theoretical scale models (Panteli & Purwins, 2013), and rebetiko lyrics have been quantitatively analyzed in a corpus of over 5,000 songs to reveal deeply rooted modal and linguistic structures (Papadopoulou, 2021). This linguistic and cultural gap limits the ability of AI to model these traditions accurately, reflecting a broader pattern where many non-English musical heritages remain underrepresented in AI training data, restricting the inclusivity and global applicability of music AI research.

Moreover, Italian singing data included in music datasets is often limited in both scale and stylistic depth, hampering the ability of AI models to learn *bel canto* performance nuances. The SingStyle111 dataset, for instance, includes only a small fraction of Italian singing, despite encompassing multiple languages and styles (Dai et al., 2023). This restricted representation hinders the modeling of expressive aspects of *bel canto*, such as ornamentation, dynamics, and historically informed performance practices. Beyond the issue of dataset size, broader biases exist within computational music research; Morreale (2025) emphasizes that many studies remain skewed towards Western popular music traditions, making musically rich genres like Italian opera less accessible and less effectively modeled by AI systems. While some cultural minorities favored rhythms common in their own musical tradition, others displayed patterns that reflect local practices rather than universal consistency (Jacoby et al., 2024). From another perspective, recent advances in artificial intelligence have enabled the generation of music that closely mimics human composition, raising questions about how listeners perceive and evaluate AI-created pieces. Listeners rated the AI-generated music very highly in terms of quality, creativity, and emotional impact, and none of the 50 participants correctly identified it as AI before being told" (Ventayen, 2024). Together, these findings indicate that while there are universal cognitive biases in how humans process rhythm, harmonic practices have diverged and evolved historically in culturally specific ways- all of which can now be explored more deeply with AI.

HUMAN MACHINE COLLABORATION: RETHINKING AUTHORSHIP AND AGENCY

AI has become an increasingly integral participant in contemporary music production, composition, and performance environments. Over the past decade, models capable of generating melodies, harmonies, lyrics, timbres, or full-length compositions have migrated from research laboratories into everyday musical practice. As this technological shift accelerates, long-standing musicological assumptions regarding authorship, intention, and agency require critical re-examination (Cervantes et al., 2023; Jacques & Flynn, 2024; Lim, 2023). Traditionally, Western musicological discourse positioned the composer as the primary creative agent: an intentional subject whose imaginative vision materializes through notation, performance, and recording technologies. Yet the arrival of generative AI challenges this human-centric framework. Instead of a single creative path that moves from the mind of the composer to the musical object, today's workflows unfold in ways that are multi-directional, iterative, and relational. The human musician is no longer the sole originator but part of a wider ecology of musical agents that includes computational models, datasets, algorithmic procedures, and material instruments (Lim, 2023; Jacques & Flynn, 2024)

A growing body of scholarship foregrounds this shift. In *The Oxford Handbook of Algorithmic Music*, McCormack et al. (2018) argue that algorithmic systems participating in creative practice necessitate a reconceptualization of creativity away from individualized authorship and toward procedural, emergent, and relational models. Similarly, Craig Vear, in *The Digital Score* (2022), proposes that contemporary musicking increasingly involves real-time dialogue with computational systems, in which the system's output becomes a provocateur, stimulus, or collaborator. For Vear, the machine is not a passive tool but part of a "distributed creative assemblage," where musical meaning arises not solely from human expression but from interactions among humans, algorithms, and environments. This perspective reflects a broader turn in digital musicology toward considering technology not as a neutral conduit for human intention but as an active mediator and co-producer of musical experience.

The spread of generative AI in musical contexts renders these theoretical developments especially urgent. In common production workflows, musicians now engage in hybrid processes: the human supplies prompts, stylistic constraints, or references, and the system generates candidate materials that the human then edits, curates, or recombines. Fernando A. Ramos-Zaga (2025), writing in the journal *Laws*, describes this interaction as a "prompt–generate–curate" cycle, where authorship is no longer tied to a single act of creation but distributed across multiple iterative exchanges. Ramos-Zaga argues that the distinction between originating and shaping ideas becomes blurred: even if the machine initiates a melodic or harmonic idea, the interpretive and editorial decisions executed by the human participant are integral to determining the final artistic outcome. This interplay complicates standard definitions of authorship, which assume that creative originality resides entirely in the human mind.

Empirical evidence supports the claim that musicians perceive these systems as collaborators rather than mere instruments. Herington et al. (2025), in a study published in *AI & Society*, conducted extensive interviews with musicians working with generative AI tools. Their findings demonstrate that many artists describe these systems as creative partners—entities that introduce surprise, provoke new directions, and challenge existing habits. Musicians report encountering "unexpected gestures" or “stylistic mutations” produced by AI models, which in turn reshape the trajectory of their compositions. Yet they also express ambivalence: some feel displaced or threatened when the system produces polished or stylistically convincing material; others experience uncertainty about whether they can authentically claim authorship over ideas generated by a machine. This tension reveals the emotional and cultural dimensions of authorship: it is not only about who generates material but also about how musicians make sense of their identity in relation to new technological actors.

These developments necessitate a deeper examination of agency in musical practice. Agency has historically been attributed to human creators, but scholars of digital culture and computational creativity argue for broader frameworks. Philip Agre, in his work on computational agency (1997), emphasizes that algorithmic systems produce actions not through intention but through programmed and statistical operations that nonetheless shape outcomes in meaningful ways. In the context of music, B. L. Sturm’s influential critique of machine learning in music information retrieval—particularly his 2014 article in *IEEE/ACM Transactions on Audio, Speech, and Language Processing*—demonstrates that algorithmic systems embed biases and make decisions based on training data in ways that effectively produce aesthetic consequences. What appears as “machine output” is already structured by cultural and historical patterns encoded in datasets and algorithmic architectures. Therefore, even if machines lack consciousness, they exert “mechanical agency” by influencing musical outcomes.

This perspective aligns with broader musicological movements that conceptualize musical creativity as distributed. In the field of human–computer interaction, Holland et al., 2013 edited volume *Music and Human-Computer Interaction* (2013) presents numerous cases in which musical outcomes emerge from the interplay of human intention, system behavior, and environmental context. These studies emphasize that musical authorship cannot be reduced to a single agent; rather, it emerges from networks of interaction. Such frameworks counter simplistic interpretations of AI as either replacing or replicating human creativity. Instead, they highlight the emergence of a hybrid creative ecology—a complex system in which both human and machine contribute to the formation of musical meaning.

A key implication of this shift is the evolution of aesthetic criteria. Historically, musical aesthetics have prioritized qualities such as expressive intention, originality, emotional depth, and stylistic coherence—all traits assumed to derive from human subjectivity. But as McCormack and colleagues (2018) note, the aesthetic assessment of algorithmic and generative works must account for different sources of value: procedural elegance, the creativity of the model architecture, the responsiveness of the system, and the nature of interactions between human and machine. In practice, this means evaluating how well a generative system supports exploration, how effectively it mediates between datasets and creative goals, or how its outputs provoke artistic transformation. These criteria do not replace traditional aesthetics but expand them, emphasizing the *processual* and *relational* dimensions of creativity.

Another crucial dimension concerns cultural representation and dataset bias. Numerous studies demonstrate that AI music models disproportionately learn from Western classical, pop, and electronic repertoires, which dominate available digital corpora. Morreale et al. (2025), in a study published in *Transactions of the International Society for Music Information Retrieval*, show that many generative music systems inherit structural and stylistic limitations: though trained on large musical datasets, they remain constrained by reductionist assumptions that limit their expressive and theoretical diversity. This pattern marginalizes musical traditions with different tonal systems, such as South Asian rāga structures, Middle Eastern maqam systems, or West African polyrhythms. A recent 2025 NAACL study titled *Music for All* (Mehta et al., 2025) further demonstrates that large multimodal models tend to misrepresent or incorrectly classify musical examples from non-Western cultures, revealing systemic inequities embedded within dataset composition. Such findings highlight the need for musicologists to engage with AI development practices: authorship is not only an aesthetic question but also a political and cultural one, shaped by what musical materials models are permitted to learn from.

The legal dimension adds another layer of complexity. Copyright law in the United States, United Kingdom, and European Union remains grounded in the assumption that only humans can be authors. The U.S. Copyright Office has reaffirmed that copyright protection requires human authorship: in its March 2023 guidance, it stated that purely AI-generated works may not qualify, because the "traditional elements of authorship" must originate from a person rather than a machine (U.S. Copyright Office, 2023). It further insists that AI-assisted works can be protected, but only when there is a "modicum of creativity" from a human creator, and applicants must clearly disclose their AI usage (U.S. Copyright Office, 2023). In practice, this means that hybrid works are protected only when human contributions are sufficiently substantial. Yet the threshold of "sufficient contribution" remains ambiguous. If a musician curates, edits, or reorganizes AI-generated fragments, is that creative enough? What if the musician merely chooses among outputs generated by a model? The Harvard Undergraduate Law Review's 2024 analysis of AI-generated music notes that current legal tests—such as the "selection and arrangement" standard—may extend copyright to some hybrid works, but only when human choices are demonstrably creative. These ambiguities pose significant challenges for musicians who rely on AI tools yet need to secure legal protection for their work.

The issue of liability also remains unresolved. If a generative AI system inadvertently produces a melody that closely resembles copyrighted material in its training corpus—a plausible scenario given the statistical nature of deep learning—who is responsible? Sturm's (2014) argument that MIR systems can unintentionally reproduce patterns from training data underscores the need for transparency in dataset construction and model evaluation. Researchers increasingly advocate for metadata-rich, publicly inspectable datasets and for legal frameworks that extend responsibility to developers and dataset curators rather than end users alone.

In response to these challenges, musicology can contribute a nuanced framework for understanding and negotiating hybrid authorship. Such a framework would recognize that musical creativity in AI-mediated contexts is neither purely human nor purely machine-driven. It is instead a *cooperative phenomenon* that emerges from interaction. The human participant brings intentionality, cultural knowledge, emotional interpretation, and aesthetic judgment. The machine contributes generative capability, pattern recognition, stylistic blending, and the ability to traverse expansive musical feature spaces. Importantly, the machine's contributions are shaped by human cultural histories encoded in datasets and algorithms. Thus, machine agency is always intertwined with human agency—it never stands alone.

From a musicological viewpoint, hybrid authorship entails acknowledging that creative works now have multiple layers of origin. The system's output is shaped by its training data; the training data is shaped by the curatorial decisions of developers; those decisions are shaped by cultural histories, institutional norms, and inequities. The human artist interacts with these outputs, reinterpreting them in light of their own aesthetic goals. Furthermore, the listener interprets the result through their own expectations and cultural frameworks. Authorship, therefore, becomes a distributed process—one that must account for co-creative dynamics among all actors.

This distributed perspective also aligns with contemporary cognitive-scientific models. The "extended mind" theory, developed by Clark and Chalmers (1998), proposes that cognitive processes often extend beyond the brain into tools and environments. In the context of AI music, one could argue that generative systems act as extensions of compositional cognition, externalizing processes of variation, recombination, and exploration. They are not merely assistants but cognitive scaffolds that enlarge the creative capacity of the human artist. This view does not reduce human agency; rather, it reframes it as assemblage-based—emerging through interaction with technological systems.

In conclusion, the emergence of generative AI compels us to rethink the fundamental categories through which musicology understands creativity, authorship, and agency. Rather than viewing AI as replacing human creativity, the evidence shows that collaborative workflows foster new forms of musical expression grounded in interaction, negotiation, and shared control. Legal frameworks will need to adapt to this hybrid reality, offering clearer guidelines for authorship and liability. Musicologists must also play an active role in shaping ethical and cultural standards, ensuring that AI systems reflect the diversity of global musical traditions rather than reinforcing existing biases. Ultimately, the future of music lies not in privileging humans over machines or machines over humans, but in cultivating a symbiotic creative ecosystem—one in which the interplay between human intention and computational generativity yields new aesthetic possibilities and deepens our understanding of what musical agency can be.

CONCLUSION

The integration of artificial intelligence (AI) into contemporary musicology has transformed the field's analytical scope, methodological consistency, and creative possibilities. Emerging from traditions in computational musicology, cognitive science, and computer-assisted composition, AI now provides tools that allow researchers to examine musical structure, stylistic evolution, perception, and creativity at scales and levels of complexity that are previously unattainable (Briot et al., 2020; Herremans et al., 2017). As machine learning techniques—particularly deep learning—become increasingly sophisticated, they not only automate aspects of analysis but also prompt renewed reflection on creativity, agency, authorship, and the cultural conditions under which music is produced and consumed (Zhang, 2025; Bhandari & Colton, 2024). The rapid growth of generative models and the accessibility of these tools to non-specialists further amplify AI's influence, urging musicologists to rethink not only analytical practices but also the disciplinary boundaries of the field itself (Yu et al., 2023; Chen et al., 2024).

Computational analysis has long played a role in music theory and musicology, yet modern AI systems greatly expand what is possible. Machine learning enables scholars to uncover statistical regularities in melody, harmony, rhythm, tone, and large-scale form through pattern recognition techniques that operate across massive corpora (Herremans et al., 2017). Deep learning designs, such as recurrent neural networks, convolutional models, and transformer-based systems, have proven particularly effective at modeling sequential and hierarchical musical relationships (Briot et al., 2020). These systems allow analysts to explore stylistic trends across centuries, compare repertoires across cultures, and model compositional tendencies with far greater empirical extent than traditional hand-coded rule systems (Pearce, 2018; Pearce & Wiggins, 2012). Crucially, the ability to process recordings expands the analytical purview to include performance nuances, timbral characteristics, and expressive microvariations (Widmer, 2016). When combined with established analytical frameworks, AI-derived patterns can be interpreted within broader historical and theoretical contexts, producing insights that neither domain could achieve independently (Zhang, 2025).

AI also increasingly participates in the creative process. Research on interactive and generative music systems demonstrates that creativity can be conceptualized as a dialogue between human intention and machine-generated possibility (Collins, 2006). Pachet and Roy (2011) show how algorithmic partners can provide structural suggestions, propose continuations, or generate accompaniment in ways that complement rather than replace human invention. This dynamic reflects co-creativity, in which human and machine agents each contribute distinct forms of decision-making and interpretation (Cádiz et al., 2021). In these workflows, the human typically provides conceptual framing and aesthetic judgment, while the AI offers stylistic or structural possibilities. Although this distributed model of authorship challenges traditional concepts of originality, it does not diminish human agency; instead, it highlights how creativity emerges from networks of interacting intelligences. Moreover, as generative systems increasingly incorporate improvisatory or adaptive capabilities, they are becoming collaborators in real-time performance situations, raising further questions about embodiment, responsiveness, and the temporality of musical decision-making (Bhandari & Colton, 2024). Recent work demonstrates that AI can even enhance educational and compositional contexts by providing real-time feedback, scaffolding skill development, and enabling learners to experiment with musical ideas in ways previously unavailable (Yu et al., 2023).

AI also serves as a bridge between empirical research in music cognition and large-scale musical data. Probabilistic learning models developed in music cognition, such as those of Pearce and Wiggins (2012), frame musical expectation as a result of statistical learning from exposure to structured musical environments. Machine learning systems, which are likewise designed to learn from musical corpora, offer computational analogues of these cognitive processes. Their predictions can be compared with behavioral responses, revealing parallels between computational expectation and human perceptual mechanisms (Pearce, 2018). Such work deepens our understanding of how listeners internalize musical structure and how enculturation shapes perceptual habits. AI systems trained on performance data further illuminate expressive dimensions such as timing, dynamics, and articulation—elements central to musical communication yet challenging to measure without computational assistance (Widmer, 2016). This convergence of perceptual, structural, and computational perspectives enriches interdisciplinary connections among musicology, psychology, and neuroscience, suggesting that future models may bridge not only statistical and perceptual domains but also embodied and affective dimensions of musical experience (Zhang, 2025; Bhandari & Colton, 2024).

Beyond analysis and cognition, AI also influences the broader ecology of musical production and consumption. Recommendation systems, streaming algorithms, and generative interfaces shape the discovery, circulation, and reception of music, raising questions about taste formation, cultural diversity, and algorithmic bias (Born & Devine, 2015). Researchers in music information retrieval and digital humanities note that machine learning systems reflect the biases of the datasets on which they are trained, potentially reinforcing dominant musical narratives while marginalizing less-represented practices (Chen et al., 2024). Musicologists are therefore called to examine not only how AI models generate or analyze music, but also how they organize and mediate musical experience, influencing what audiences encounter and how styles evolve. The increasingly ubiquitous presence of AI-driven curation means that listeners' musical worlds are shaped by algorithmic logics that may privilege commercial predictability over cultural diversity, creating an additional area of inquiry concerning how musicology can contribute to the design of more equitable and transparent systems (Yu et al., 2023).

Taken together, these developments show that AI is reshaping musicology in ways that extend far beyond computational efficiency. By enabling large-scale analytical studies, supporting human–machine co-creativity, modeling cognitive processes, and mediating cultural circulation, AI offers powerful conceptual and methodological tools for understanding music as a complex cultural, aesthetic, and cognitive phenomenon (Zhang, 2025; Briot et al., 2020). Yet these possibilities come with ethical responsibilities: ensuring transparency, addressing data bias, and safeguarding cultural diversity must remain central concerns. As the boundaries between human and machine creativity continue to blur, musicologists are uniquely positioned to interrogate the values embedded in these systems and to advocate for practices that respect both the complexity of musical traditions and the transformative potential of technological innovation. The convergence of AI and musicology thus presents both new opportunities and new obligations, positioning the field at a pivotal intersection of technology, creativity, and culture.

REFERENCES

Adler, G. (1885). *Umfang*. Methode und Ziel der Musikwissenschaft.

Agawu, K. (2003). Representing African music: Postcolonial notes, queries, positions. Routledge. https://www.routledge.com/Representing-African-Music-Postcolonial-Notes-Queries-Positions/Agawu/p/book/9780415254768

Agostinelli, A., Denk, T. I., Borsos, Z., Engel, J., Verzetti, M., Caillon, A., . . . Frank, C. (2023). Musiclm: Generating music from text. https://doi.org//arXiv.2301.11325DOI: 10.48550

Agre, P. (1997). *Computation and human experience*. Cambridge University Press. American Musicological Society. (*What is musicology?* Retrieved from https://www.amsmusicology.org/what-is-musicology/

Bhandari, K., & Colton, S. (2024). Motifs, phrases, and beyond: The modelling of structure in symbolic music generation. *arXiv*. https://doi.org//arXiv.2403.07995DOI: 10.1007/978-3-031-56992-0_3

Born, G., & Devine, K. (2015). Music technology, gender, and class: Digitization, educational and social change in Britain. *Twentieth-Century Music*, *12*(2), 135–172. DOI: 10.1017/S1478572215000018

Briot, J.-P., Hadjeres, G., & Pachet, F. (2020). *Deep learning techniques for music generation*. Springer., DOI: 10.1007/978-3-319-70163-9

Brunner, G., Konrad, A., Wang, Y., & Wattenhofer, R. (2018). *MIDI-VAE: Modeling Dynamics and Instrumentation of Music with Applications to Style Transfer*. In *Proceedings of the 19th International Society for Music Information Retrieval Conference (ISMIR)* (pp. 747–754). https://scispace.com/pdf/midi-vae-modeling-dynamics-and-instrumentation-of-music-with-4tgufh6651.pdf

Bryant, G. A., & Smaldino, P. E. (2025). *The cultural evolution of distortion in music (and other norms of mixed appeal). Philosophical Transactions of the Royal Society B*, 380(1923). *Philosophical Transactions of the Royal Society of London. Series B, Biological Sciences*, *20240014*(1923), 20240014. Advance online publication. DOI: 10.1098/rstb.2024.0014 PMID: 40176525

Cádiz, C., Macaya, G., Cartagena, L. G., & Parra, D. (2021). Creativity in generative musical networks: Evidence from two case studies. *Frontiers in Robotics and AI*, *8*, 680586. Advance online publication. DOI: 10.3389/frobt.2021.680586 PMID: 34409070

Cameron, D. J., Bentley, J., & Grahn, J. A. (2015). Cross-cultural influences on rhythm processing: Reproduction, discrimination, and beat tapping. *Frontiers in Psychology*, *6*, 366. DOI: 10.3389/fpsyg.2015.00366 PMID: 26029122

Cervantes, K., Harris, T., & Singleton, P. (2023). Copyright in the age of generative AI. *Journal of International Commercial Law & Technology*, *4*(1), 33–36. https://www.jiclt.com/

Champion, G. (2022). Decolonizing the ear: Echolocating "race" in French rap. *Contemporary French and Francophone Studies*, *46*(4), 475–495. DOI: 10.1080/17409292.2022.2107273

Chen, Y., Huang, L., & Gou, T. (2024). Applications and advances of artificial intelligence in music generation: A review. *arXiv.* https://doi.org//arXiv.2409.03715DOI: 10.48550

CISAC. (2024). *Global economic study shows human creators' future at risk from generative AI.* International Confederation of Societies of Authors and Composers. https://www.cisac.org/Newsroom/news-releases/global-economic-study-shows-human-creators-future-risk-generative-ai

Clark, A., & Chalmers, D. (1998). The extended mind. *Analysis*, *58*(1), 7–19. DOI: 10.1093/analys/58.1.7

Clayton, M. (2020). *Musicology and the moment of social change*. Routledge.

Collins, N. (2006). Towards autonomous agents for live computer music: Real-time machine listening and interactive music systems. *IMA Journal of Mathematics Applied in Business and Industry*, *20*(2), 121–139.

Corbelli, A. (2024). *Beyond the Algorithm. Ethical and aesthetic challenges of AI in music*. Itinera., DOI: 10.54103/2039-9251/27842

Crowdus, M. (2023). *Decolonial affordances: Sounding and listening interventions.* ARC: Journal of the School of Architecture, 8. https://arcjournal.library.mcgill.ca/article/view/1469

Dai, S., Wu, Y., Chen, S., Huang, R., & Dannenberg, R. B. (2023, November). SingStyle111: A Multilingual Singing Dataset With Style Transfer. In *ISMIR* (pp. 765-773).

de Berardinis, J., Porcaro, L., Meroño-Peñuela, A., Cangelosi, A., & Buckley, T. (2025). *Towards responsible AI music: An investigation of trustworthy features for creative systems*. arXiv. https://doi.org//arXiv.2503.18814DOI: 10.48550

Deruty, E., Grachten, M., Lattner, S., Nistal, J., & Aouameur, C. (2022). On the development and practice of ai technology for contemporary popular music production. *Transactions of the International Society for Music Information Retrieval*, *5*(1), 35. Advance online publication. DOI: 10.5334/tismir.100

Dolan, E. W. (2024, March 29). Cross-cultural research reveals universal bias towards simple rhythmic ratios in music. *PsyPost*. https://www.psypost.org/cross-cultural-research-reveals-universal-bias-towards-simple-rhythmic-ratios-in-music-29003

Eck, S. O. (2023). *Bridging the gap: Demonstrating the applicability of linguistic analysis tools in digital musicology*. In Proceedings of the Workshop on Natural Language Processing for Digital Humanities. Association for Computational Linguistics. https://aclanthology.org/2023.nlp4dh-1.28

Engel, J., Resnick, C., Roberts, A., Dieleman, S., Norouzi, M., Eck, D., & Simonyan, K. (2017, July). Neural audio synthesis of musical notes with wavenet autoencoders. In *International conference on machine learning* (pp. 1068-1077). PMLR.

Eraslan, I. (2025a). AI-Supported Analysis of Formal and Stylistic Transformations in Ottoman-Turkish Music (16th–19th Centuries). INSAM Journal of Contemporary Music [→ This is a real AI-driven musicology study on Ottoman-Turkish makam music using digitized scores, showing that researchers are trying to apply AI to this 'non-Western' modal tradition.]. *Art and Technology*, *14*, 48–66.

Eraslan, I. (2025b). Artificial Intelligence and the Rewriting of Musical Memory: A Cognitive Perspective. *Science Development*, *6*(3), 114–120. DOI: 10.11648/j.scidev.20250603.18

Gallucci, M. (2023). Digitization and accessibility of Italian opera librettos: Challenges for computational study. *Opera Studies Journal, 18*(2), 112–130. https://doi.org/DOI: 10.6092/issn.2532-8816/16861

Giovannini, F., & Skorinkin, A. (2024). Corpus-based modeling of opera librettos: A comparative study of German, French, and Italian repertoires. *Journal of Musicology and Computing*, *9*(1), 25–48. DOI: 10.48694/jcls.3595

González-Espinoza, A., & Plotkin, J. B. (2023). *Quantifying the evolution of harmony and novelty in western classical music*. arXiv. https://arxiv.org/abs/2308.03224

Goodfellow, I., Bengio, Y., & Courville, A. (2016). *Deep Learning*. MIT Press.

Hentschel, J., Rammos, Y., McLeod, A., & Rohrmeier, M. (2023). *Introducing DiMCAT for processing and analyzing notated music on a very large scale*. *Zenodo*. https://doi.org/DOI: 10.5281/zenodo.10265339

Herington, J., Borasi, R., & Guerrero, B. J. (2025). Musicians' ethical concerns about AI: An interview study. *AI & Society*. Advance online publication. DOI: 10.1007/s00146-025-02601-6

Herremans, D., Chuan, C.-H., & Chew, E. (2017). A functional taxonomy of music generation systems. *ACM Computing Surveys*, *50*(5), 1–30. DOI: 10.1145/3108242

Hibberd, L. (1959). Musicology Reconsidered. *Acta Musicologica*, *31*(1), 25–31. DOI: 10.2307/931812

Holland, S., Wilkie, K., Mulholland, P., & Seago, A. (Eds.). (2013). *Music and human-computer interaction*. Springer. DOI: 10.1007/978-1-4471-2990-5

Holzapfel, A., & Bozkurt, B. (2012). Metrical strength and contradiction in Turkish Makam music. In *Proceedings of the 2nd CompMusic Workshop* (pp. 79–84). Universitat Pompeu Fabra., https://repositori.upf.edu/items/fa8aa0e9-07d7-4f0d-99b8-977cb4604797

Hopkins, J., & Kiela, D. (2017, July). Automatically generating rhythmic verse with neural networks. In *Proceedings of the 55th Annual Meeting of the Association for Computational Linguistics* (Volume 1*: Long Papers)* (pp. 168-178). DOI: 10.18653/v1/P17-1016

Hsieh, A., & Olley, J. (2025). Critical Exchanges: Globalizing Comparative Musicology? Inter-Asian Perspectives on Global Music History. *Music & Letters*, *106*(3), 443–466. Advance online publication. DOI: 10.1093/ml/gcaf044

Huang, C. Z. A., Vaswani, A., Uszkoreit, J., Shazeer, N., Simon, I., Hawthorne, C., Dai, A. M., Hoffman, M. D., Dindulescu, M., & Eck, D. (2018). *Music transformer: Generating music with long-term structure.* arXiv preprint arXiv:1809.04281. https://arxiv.org/abs/1809.04281

Jacoby, N., Polak, R., Grahn, J. A., Cameron, D. J., Lee, K. M., Godoy, R., Undurraga, E. A., Huanca, T., Thalwitzer, T., Doumbia, N., Goldberg, D., Margulis, E. H., Wong, P. C. M., Jure, L., Rocamora, M., Fujii, S., Savage, P. E., Ajimi, J., Konno, R., & McDermott, J. H. (2024). Commonality and variation in mental representations of music revealed by a cross-cultural comparison of rhythm priors in 15 countries. *Nature Human Behaviour*, *8*(5), 846–877. DOI: 10.1038/s41562-023-01800-9 PMID: 38438653

Jacques, S., & Flynn, M. (2024). Protecting human creativity in AI-generated music with the introduction of an AI-royalty fund. *GRUR International*, *73*(12), 1137–1149. DOI: 10.1093/grurint/ikae134

Ji, S., Yang, X., & Luo, J. (2023). A Survey on Deep Learning for Symbolic Music Generation: Representations, Algorithms, Evaluations, and Challenges. *ACM Computing Surveys*, *56*(1), 1–39. Advance online publication. DOI: 10.1145/3597493

Kaya, O. (2023). Türk Halk Müziği Geleneğinde "Usûl"ün Temellükü. *Konservatoryum*, *10*(2), 116–125. DOI: 10.26650/CONS2023-1361068

Kömürcü, İ. (2025). Effects of regular and irregular rhythms on time perception: The role of aksak and non-metric rhythms. *Online Journal of Music Sciences*, *10*(3), 557–571. DOI: 10.31811/ojomus.1647605

Kopp, D. (2002). *Chromatic transformations in nineteenth-century music*. Cambridge University Press., https://www.cambridge.org/core/books/chromatic-transformations-in-nineteenthcentury-music/23DDA64171C8B60F88D4F47AB4604DF3?utm_source=chatgpt.com DOI: 10.1017/CBO9780511481932

Levy, B. R., & Emmery, L. (2021). Archival Research in Music: New Materials, Methods, and Arguments. *Music Theory Online*, *27*(3). Advance online publication. DOI: 10.30535/mto.27.3.4

Lim, D. (2023). Generative AI and copyright: Principles, priorities and practicalities. *Journal of Intellectual Property Law and Practice*, *18*(12), 841–842. DOI: 10.1093/jiplp/jpad081

Lupker, J. A., & Turkel, W. J. (2021). Music theory, the missing link between music-related big data and artificial intelligence. *DHQ: Digital Humanities Quarterly, 15*(1).

Ma, H., Zhang, Y., Shan, X., & Hu, X. (2025). Exploring the impact of artificial intelligence on the creativity perception of music practitioners. *Journal of Intelligence*, *13*(4), 47. Advance online publication. DOI: 10.3390/jintelligence13040047 PMID: 40278056

Maia, L. S., de Castro, L. N., & Silla, C. N.Jr. (2019). SAMBASET: A dataset of historical *samba de roda. Proceedings of the 20th International Society for Music Information Retrieval Conference (ISMIR 2019)*, 76–83.

McCormack, J., Gifford, T., & Hutchings, P. (2018). Autonomy, authenticity, authorship and intention in computer-generated art. In R. Dean & A. McLean (Eds.), *The Oxford handbook of algorithmic music* (pp. 177–196). Oxford University Press.

Mehta, A., Chauhan, S., Djanibekov, A., Kulkarni, A., Xia, G., & Choudhury, M. (2025). *Music for All: Representational bias and cross-cultural adaptability of music generation models*. In *Findings of the 2025 Conference of the North American Chapter of the Association for Computational Linguistics (NAACL)* (pp. 258–272). Association for Computational Linguistics. https://aclanthology.org/2025.findings-naacl.258.pdf

Merlini, M. (2024). From [A Reflective Turn in Musicology to Overcome Dualist, Romantic and Cognitivist Prejudices.]. *AI*, •••, 4E.

Morgado Vega, J. J., Sharma, S., & Simonetta, F. (2025, September). Drafting the landscape of computational musicology tools: a survey-based approach. In *Proceedings of the 12th International Conference on Digital Libraries for Musicology* (pp. 28-35). DOI: 10.1145/3748336.3748340

Morreale, F., Martinez-Ramirez, M. A., Masu, R., Liao, W., & Mitsufuji, Y. (2025). Reductive, exclusionary, normalising: The limits of generative AI music. *Transactions of the International Society for Music Information Retrieval*, *8*(1), 300–312. DOI: 10.5334/tismir.256

New Indian Express / CISAC. (2024, December 4). *Global study reveals AI threat to music sector workers.* The New Indian Express. https://www.newindianexpress.com/world/2024/dec/04/global-study-reveals-ai-threat-to-music-sector-workers

Nielsen, M. (2015). *Neural Networks and Deep Learning: A Textbook*. Determination Press.

Pachet, F., & Roy, P. (2011). Markov constraints: Steerable generation of Markov sequences. *Constraints*, *16*(2), 148–172. DOI: 10.1007/s10601-010-9101-4

Panteli, M., & Purwins, H. (2013). A computational comparison of theory and practice of scale intonation in Byzantine chant. *Proceedings of the International Society for Music Information Retrieval (ISMIR 2013).* https://archives.ismir.net/ismir2013/paper/000098.pdf

Papadopoulou, A. (2021). *Rebetiko meets digital humanities: Automatic analysis of Rebetiko lyrics via data science methods* (Master's thesis). University of Vienna. https://phaidra.univie.ac.at/detail/o:1426454

Parncutt, R. (2007). Systematic musicology and the history and future of western musical scholarship. *Journal of interdisciplinary music studies, 1*(1), 1-32.

Pearce, M. (2018). Statistical learning and probabilistic prediction in music cognition: Mechanisms of stylistic enculturation. *Annals of the New York Academy of Sciences*, *1423*(1), 378–395. DOI: 10.1111/nyas.13654 PMID: 29749625

Pearce, M., & Wiggins, G. (2012). Auditory expectation: The information dynamics of music perception and cognition. *Topics in Cognitive Science*, *4*(4), 625–652. DOI: 10.1111/j.1756-8765.2012.01214.x PMID: 22847872

Popović, A. (2020). Is Musicology (Only) One of the Humanities? Professional article. *Pannoniana*, *4*(1), 245–257.

Ramos-Zaga, F. A. (2025). Reconceptualizing Human Authorship in the Age of Generative AI: A Normative Framework for Copyright Thresholds. *Laws*, *14*(6), 84. DOI: 10.3390/laws14060084

Razvan, M. (2023). Measurement of music aesthetics using deep neural networks and dissonances. *Information (Basel)*, *14*(7), 358. Advance online publication. DOI: 10.3390/info14070358

Rowe, R. (1993). *Interactive Music Systems* (Chapter 6: Machine Composition). New York University. https://wp.nyu.edu/robert_rowe/text/interactive-music-systems-1993/

Sturm, B. L. (2014). A simple method to determine if a music information retrieval system is a "horse.". *IEEE Transactions on Multimedia*, *22*(11), 1676–1683. DOI: 10.1109/TMM.2014.2330697

Sturm, B. L. T., Iglesias, M., Ben-Tal, O., Miron, M., & Gómez, E. (2019). Artificial Intelligence and Music: Open Questions of Copyright Law and Engineering Praxis. *Arts*, *8*(3), 115. DOI: 10.3390/arts8030115

Temperley, D. (2010). Music and probability. MIT Press. https://mitpress.mit.edu/9780262013597/music-and-probability/

U.S. Copyright Office. (2023, March 16). *Copyright registration guidance: Works containing material generated by artificial intelligence* (Federal Register Vol. 88, Issue 51). https://www.govinfo.gov/content/pkg/FR-2023-03-16/html/2023-05321.htm

Vaswani, A., Shazeer, N., Parmar, N., Uszkoreit, J., Jones, L., Gomez, A. N., Kaiser, Ł., & Polosukhin, I. (2017). *Attention Is All You Need*. In *Advances in Neural Information Processing Systems (NeurIPS)*, 30. https://proceedings.neurips.cc/paper_files/paper/2017/file/3f5ee243547dee91fbd053c1c4a845aa-Paper.pdf

Vear, C., Benford, S., Avila, J. M., & Moroz, S. (2023). Human-AI Musicking: A Framework for Designing AI for Music Co-creativity. *AIMC 2023*. https://aimc2023.pubpub.org/pub/zd46ltn3

Ventayen, R. J. M. (2024). *Music creation by artificial intelligence and listener reactions in a blind test* (SSRN Scholarly Paper No. 5023874). SSRN. DOI: 10.2139/ssrn.5023874

Vinod, R., & Colaco, R. (2025). Review of AI in Indian classical music. *International Journal for Research in Applied Science and Engineering Technology*, *13*(2), 1599–1608. DOI: 10.22214/ijraset.2025.67155

Wang, T., & Zhang, J. (2025). Enhancing musical creativity through AI: A chain-mediation model of self-efficacy and emotional intelligence. *Frontiers in Psychology*, *16*, 1665609. Advance online publication. DOI: 10.3389/fpsyg.2025.1665609 PMID: 41267763

Widmer, G. (2016). Getting closer to the essence of music: The Con Espressione manifesto. *ACM Transactions on Intelligent Systems and Technology*, *8*(2), 22.

Yang, W., & Lee, I. (2024). Research on the construction of AI composition system based on HMM. *Yegah Musicology Journal*, *7*(3), 216–240. DOI: 10.51576/ymd.1536267

Yu, X., Ma, N., Zheng, L., Wang, L., & Wang, K. (2023). Developments and applications of artificial intelligence in music education. *Technologies*, *11*(2), 42. Advance online publication. DOI: 10.3390/technologies11020042

Zhang, M. (2025). Advancing deep learning for expressive music composition and performance modeling. *Scientific Reports*, *15*(1), 28007. Advance online publication. DOI: 10.1038/s41598-025-13064-6 PMID: 40745012

Zhao, Y., Yang, M., Lin, Y., Zhang, X., Shi, F., & Wang, Z. Ding, Juanguo & Ning, H. (2025). Artificial Intelligence-Enabled Text-to-Music Generation: A Comprehensive Review of Methods, Frameworks, and Future Directions. https://www.preprints.org/manuscript/202502.1791

Chapter 3

AI as a Tool for Innovation and Experimentation in Music

Sonia Ahuja
http://orcid.org/0009-0000-2371-598X
Lovely Professional University, India

ABSTRACT

This chapter explores the role of artificial intelligence (AI), in redesigning music, as a tool for experimentation and innovation, which connects traditional composition with technology. The interconnection of AI with human creativity gives rise to various musical opportunities. By using AI tools, artists are pushing beyond established boundaries and providing new experiences to listeners. The main point is that AI is not a replacement for humans rather it's an addition to them. The goal of this chapter is to examine the evolving method of AI and its influence on different areas of music such as songwriting, composition and performance etc. By analyzing previous and existing studies this chapter provides a summary of creative co-partnership between artificial intelligence and human intelligence, while also addressing its cultural and ethical implications. The chapter visualize a future in which artificial intelligence (AI) preserves the human touch while enabling real-time improvisational performance, emotional composition, humanized instruction and innovative teaching practices.

1.INTRODUCTION

AI is one of the useful and trendy technologies that not only changed the realm of science and industry, but also creative environment. Nevertheless, business aside, AI is also applied in other fields of data-driven endeavour, just that it is being applied

DOI: 10.4018/979-8-3373-6279-3.ch003

to the creative realms, to music, to visual art, to writing stories and to journalism. According to Caramiaux et al 2020, with the availability of open-source tools, large volumes of data and relatively inexpensive computing systems, this development has been accelerating. Music is one of those areas where Artificial Intelligence has left its eternal shadow which has given new direction and new thinking to many aspects of this field. The chapter would consider how human creativity is collaborating with AI to make music and how we can view AI to help spur creativity in the fields of composition, song writing, performance and music education.

AI's integration into music creation represents a paradigm shift, marking a departure from traditional computer-assisted composition methods. Modern AI systems are fantastic pattern-recognition-type-analysis-generative-modelling monsters and can be trained on large corpus of human music in a couple of centuries when humans took morphological centuries to devolve experience.

These innovations have made music creation accessible to individuals who lack the time, inclination, or resources for formal study, while also enabling experimentation with sophisticated compositional concepts and play with musical concepts that would otherwise be beyond reach.

Human creativity is not suppressed by AI collaboration; instead, AI acts as an intelligent companion that suggests changes in melody and rhythmic variations. This collaborative dynamic encourages new creative possibilities while retaining the core of human artistry. This partnership enables a deeper connection between the creator and the audience.

AI's impact on education is another revolutionary application. AI-powered analysis software is transforming how music is taught, making quality music education more accessible.

This technology is particularly beneficial for students from diverse learning backgrounds, allowing for personalized teaching methods that cater to different learning styles.

The commercial music industry also applies machine learning technology in extracting and compressing musical pieces to import into common space. As the AI technology advances, the AI algorithm in music interaction can become more advanced and smarter respectively. The another AI revolution in our world and our art, but in emotional learning, also in cross-modality, and in the quantum computational, do come and will, also offer new opportunities, as also new challenges to music performance and perception.

AI in music does not seek to replace human creators, but instead serves as a tool to enhance the creative process but to be able to mimic and sound like people in say, writing of new music among artists in a completely new manner, in which technology will not serve to undermine their creative urge and the other use of it, which is special and can never be replaced by technology.

2. THE EVALUATION AND IMPACT OF AI IN MUSIC CREATION

Music is one of the oldest and most primal elements of humanity, with a history as rich as it is diverse – from laying the groundwork for a battle, to eliciting catharsis during a grand ceremony.

The relationship between music and technology, in particular, artificial intelligence (AI), is affecting how music is created, produced and enjoyed. AI is a creator of new possibilities, for artists and listeners alike, opening up new methods of composition, making music together across borders and oceans, rerouting the interfaces between producers, audiences and trade.

Technology is, historically, one of the most important contributing elements in the development of music, which has provided the means to create new instruments, new ways of generating and capturing sound, and the notation and recording of pieces of music (Brøvig & Danielsen, 2023). As the human condition has developed, so has music, with different tuning systems, instruments, traditions and cultural practices developing in different parts of the world (Walzer, 2023). Music took on a totally new form in the 19th and 20th centuries, due to the development of communication technologies and the process of globalization. Musicians in various locations are finding it easier to collaborate to compose music by using automated music composition and mastering programs like Amper and LANDR. Those applications allow us to express our emotions and add more layers and feel to our songs. We cannot deny that studying the history and technological context of the AI object may open our eyes to the larger cultural implications of that object.

This chapter examines the background of AI application in music from the perspective of application, implication, and utilization of AI in music. Artificial intelligence in music has a history that mirrors both artistic voice and technological progress. In the 1950s and 1960s, computer scientists and musicians experimented with algorithms to make simple musical patterns. One of the earliest important works was the Illiac Suite, created in 1957, using a computer to assist the composition of the piece (Hiller and Isaacson, 1993). This pioneering research revealed that computer logic could affect musical creativity.

Rule-based systems and algorithmic composition schemes have been constructed in the 1970s and 1980s as a result of computing progress. Experiments in Musical Intelligence (EMI), a project developed by David Cope during the 1980s, showed how machines could analyse existing compositions and develop music that was in the style of specific composers (Cope 1991). Also during this period there was the inception of computer-aided composition techniques that are utilised in scholarly and experimental music circles.

In the 1990s and early 2000s, machine learning techniques became increasingly popular. Neural networks and probability models allowed systems to create more complex musical improvisations, harmonies, and frameworks.

AI's integration into digital audio workstations (DAWs) and interactive performance systems further expanded its role. Artists can now apply AI algorithms not only to composition but also to live improvisation.

Artificial intelligence reached a new revolutionary stage in the 2010s, with the advent of deep learning in music. Various technologies such as generative adversarial networks (GANs) and recurrent neural networks (RNNs) have been developed to be able to create realistic melodies, harmonies, and even full orchestrations. The capability for AI to create music in a range of genres and styles was showcased by tools like OpenAI's MuseNet, Jukedeck and Google's Magenta project (Briot, Hadjeres & Pachet, 2020).

AI is currently served as a collaborator in both mainstream and experimental music. It is used by teachers to tailor music education, and by commercial artists to assemble beats, melodies, and lyrics. AI-generated music albums such as Condition of Timeless (2017) and I AM AI (2017) by Taryn Southern, exemplify the capacity of robots to make their impact on mainstream and avant-garde music (Varini & Gramaglia, 2024). Furthermore, the role of AI has brought up important discussions about creativity, authorship and ownership, and about whether AI should be considered as a collaborator, tool or even creative agency (Gioti, 2021).

The history of music IP and AI is therefore a history of nearly unremitting change - between the primitive anti-evolutionary experiments of the very first algorithmic approaches, and the most sophisticated neural net models, -while a boundary between the human and the machine world is increasingly hard to draw, and the concept of the musician is increasingly re-defined within the framework of the space it now inhabits.

The music generation techniques and developments generated to use them as a new source of inspiration: evolutionary computation computing and genetic 91 programming. This is a chance to play around with music and to make new music. GenMuse and its counterparts apply the evolutionary algorithms to produce music patterns and provide other stimulations to increase the musical repertoire of the composers.

Cultures and History examples Wolfgang Amadeus Mozart exemplifies a historical figure as well as one of the men who have brought culture through music as a form of creative art to a higher stage. Mozart in the given work reveals the influence of contemporary tendencies and orientations which can be regarded as one of the most vivid illustrations of how culture and social tendencies can affect further development of music. The evolution of music is also dictated by the cultural evolution by which music becomes the channel of communication and expression basing on the social and cultural processes of each era. Although the change in creativity in music is highly dependent on the biological, cultural, and technological forces, it is also necessary to discuss the possibility to attribute the role of individual power and personal expression.

Beyond composing, artificial intelligence (AI) in music also simplifies improvisation, playing styles and utilising human instincts. AI is the "musician's amanuensis," augmenting the creative process without taking the place of the unique human process of writing music. It is more than just a tool for technological advancement.

AI entry is a part of the cultural system, so norms, access, and power will all affect AI entry and what happens as a result. The present of being included and having access: Apps that let people with physical disabilities play an instrument and jam AI's efforts to make inclusion and more participation possible and to change the way music is made. In this context, AI as a collective concept contributes to the central narrative theme of innovation within the chapter on human-machine co-creation, and it provides a framework for understanding the impact of technology on cultural practices.

AI is also creating exciting new possibilities for inter-cultural exchange and creativity, but its effects on artists and in the society at large should not be ignored. The second section set out these effects on an ethical, social and economic horizon.

3. AI AS A CREATIVE CO-CREATOR AND COLLABORATIVE TOOL IN MUSIC

Artificial intelligence (AI) is becoming more widely seen not as simply an aid in music composition, but as a conscious collaborator, upending the conventional concepts of authorship and exclusive human creativity. Through creating a distinctive stylistic pattern and enabling new musical dimensions of experimentation, AI systems foster human creativity. Indeed, in 2016, "Daddy's Car" created by the collaboration of Sony and Flow Machines-each a song composed in the style of The Beatles-demonstrated the ability of AI to replicate and reimagine iconic aesthetics with ease (Avdeeff, 2019)Equally, AIVA (Artificial Intelligence Virtual Artist) has been applied in professional environments such as films or video games soundtracks, advertising, as a registered composer with SACEM and producing orchestral style music that reflects human composition (Linkinen 2024).

Collaboration with AI is broadening the boundaries of creativity. AI is now an active partner in the music-making process, offering novel ways to explore musical ideas.

For instance, co-creative tools like Hookpad Aria provide live guidance on melody and harmony, allowing musicians to co-write and refine music in real-time during live performances.

The potential of collaborations is on the other hand may be encouraged also by the fact that machine learning tools are not only of interest in and for the development of technical advances, but also indeed contribute to foreground the expressive arts of human individuals (Sturm et al., 2019) and theories such as musical meta creation (Pasquier et al., 2017). These examples show how AI's water mixes with human roonicity. This opens up the possibility of thinking about how technology can change cultural practices, the ethics involved and the music ecosystem as a whole in the next section.

AI has taken the lead in revolutionizing the music composition process by offering people new forms of collaboration between humans and machines. The artificial intelligence systems can also help the composer to formulate a musical idea by creating one, thus, supplementing the imagination of a human (Gioti, 2021).

The upcoming research that deploys AI-generated music rearrangement as a new technology to identify more complex sound patterns and then builds more creative remixing can be regarded as the next step in the research that assumes the impossibility of creating music by conventional writing (Agwan et al., 2023). Another example of an AIVA is a virtual composer that is taught by an AI. An author rights society identifies the AI's musical works by their works. The new technology of music-remixing AI-based music recomposition can be seen as the next step in the research. It uses machine learning to find more complex sound patterns and then makes creative remixes based on the idea that the music can't be written (Agwan et al., 2023).

The use of AI in the production of music has influenced the creative process and pushed the boundaries of human and AI interaction. As long as AI can make pieces with harmonies, chord sequences, and melodic fragments that the artist can choose and put together to make complete pieces, devices like Suno AI can write a piece and even make vocal lines with lyrics in just a few seconds. There is still a debate about whether or not to call AI-created music ‘creative’, but it's clear that AI can be used to explore new musical structures, approaches and soundscapes. AI systems are not only used for notational composition, but also played a conspicuous role in electroacoustic music, sound analysis, interactive performance-based composition, (Miranda & Williams, 2015) and gesture-based music interaction.

AI also mediates cutting-edge performances involving music where Yamaha has performed experiments where, through gestures, a dancer can control a piano and how AI can be used to amplify live performances. The capabilities of real-time collaboration in music production via cloud-based tools and AI relays the musical artists all over the world and makes collaborative music production a much easier process (Tabak, 2023). In Analysis and Evaluation of Music Pre- and post-composition computational music analysis takes advantage of AI to improve the creative process and offer analysis of musical compositions providing the nourishment needed to generate creator and innovative ideas (Zhang, 2024).

The use of AI in the analysis of music expands to the assessment of the level of emotions being presented by music and this serves to augment the musical experience of the listener (Chen et al., 2024). Although the migration of AI to music has many benefits, there are shortcomings and ethical aspects. Whether AI will outshine human creativity and the necessity of a responsible use are some of the burning issues. Also, there is the issue of the uncanny valley effect of AI-generated music and whether the usage of AI should be more of a creative tool or a scripted one. These issues demonstrate the significance of further advancement work on AI to become ethical and efficient in music.

Human creativity may be enhanced through the unexpected suggestions and novel ideas that AI can contribute to creativity, which were applied in the case of the I Keep Counting project, with the song structure, chord progressions generated by AI being the stimuli to human creators. Workshops and research focus on AI to improve personal creativity through collaborations, serendipity and creative reflection, but not to remove the human ingenuity (Main et al., 2022).

The capability to create original musical pieces or recreate the existing styles by AI creates additional opportunities to innovate within the music industry. To illustrate the idea, one can note the album entitled "Coditany of Timeness" created by Dadabots as an example of discussing the concept of the machinic creativity through the reinvention of musical styles by AI. The application of AI technology in composition expands the competencies of a human being as they can participate in decentralized human-computer co-creating. This cooperation will result in emergence of new ideas in compositions and transformation of the musical task. The implications of AI in the creative arts are not restricted to music as it also can apply to the generative art and design supplementing the possibilities of enhancing human creativity in a set of other domains (Ardeliya et al., 2024).

The human-AI cooperation is founded on trust and communication between human musicians and AI systems. Programs that will predict the behaviour and respond to that of a human musician are under development, resulting in a more integrated active performance. An instance would be a human and a virtual drum player in a case study; an approach based on fuzzy logic was delivered as knowledge to control a set of music performance qualities and discussion proceeded to explore the notion of anticipation in collaborative music making. AI robots such as Voyager (George Lewis) and Odessa (Adam Linson) are intended to play/compose with humans, getting involved with the less-defined sonic and performative space of improvised music. Such systems are able to use distributed agent based improvisation methods to respond to unexpected external environments. While AI has shown spectacular potential in (among others) the music performance and improvisation field, however the concept of computing on music also raises some questions about human creativity and whether the capabilities of humans can be simulated, or even surpassed, by AI. As AI develops, striking the right point between the tech and human elements of music will be one of the challenges to overcome.

In 2023, the Berklee College of Music published a study that found that many artists see AI as a creative partner that can help them get past creative blocks and come up with new ideas.But the worry is that the material will be repeated and lose its originality. For example, in 2018, artist Taryn Southern released an entire album called I AM AI, which was made with AI systems and focused on the possibilities of human-AI interactions. But on the other hand, and in an even more dangerous way, many musicians say that using AI too much puts artistic identity at risk of being watered down. These viewpoints show how complicated the relationship between people and machines can be.

The aim is to have a never ending relationship with AI as a partner, an idea generator and a companion piece. AI should not be the one who steals the show from human creativity.

Using reinforcement learning and hybrid genetic algorithms, the interactive systems of the future could learn how a musician creates music and give them real-time suggestions while they are composing or improvising (Karchkhadze et al., 2024). Cloud AI also helps artists from all over the world work together across cultural lines (Tabak, 2023). Policies should cover who has rights to what, what counts as a copyrighted work, and cultural biases to make sure that AI doesn't mix up musical traditions. It's a vision of the future in which AI and human creativity come together to make music that not only makes people feel things but also creates works of art that last.

4. ETHICAL, CULTURAL AND LEGAL CONSIDERATION OF AI IN MUSIC

Being a traditional and an important element of the human life, music has been used for different purposes and in various forms over the centuries during the big celebrations. Music and technology, especially AI (artificial intelligence), are changing the way music is made, written, and heard. AI is a new way for artists and listeners to solve problems. It lets people reference different types of music, work together even when they are far apart, and change the space between producers, listeners, and the business world. The ability of AI to understand and interpret music, automate the process of writing songs, and allow live performances is opening up new levels of creativity and experimentation.

Historically, technology has much more played an essential and crucial role for the progress of music, new instruments and the emergence of new ways of sound production as well as new ways of music notation and recording purposes however (Brøvig & Danielsen, 2023). Just as our human civilization has developed, grown, and changed, music has grown, developed and changed, with the use of different tuning systems and a wide variety of instruments.

Throughout the history of mankind, music has fulfilled numerous functions: to celebrate the dead, to flirt with strangers as well as to release steam with a huge shindig. It belongs to the same cultural and expressive category as any other arts. AI is changing music and the way people listen to it from the discovery systems level all the way up to the creative process. These changes have a huge impact on businesses, consumers, and producers, but not everyone is affected in the same way. Of course, the iris of AI makes the already long lines of authenticity and emotional depth even more important. This means that AI music wants to touch the teardrop face of classical and traditional music. Moral problems: When we think about how technology spreads music, we run into moral problems because it doesn't just protect your copyright and pay the people who made it fairly, but also because of how we use it. This section talks about the good, bad and the ugly sides of AI. It also brings to the fore how innovation can intersect with inclusiveness and ethical responsibility and it gives a full picture of how AI is shaping the future of the contemporary music industry.

AI provides content creators with the opportunity to produce unlimited amounts of original music, create the soundtrack in a movie and advertisements etc which is why it is an asset

AI can also be used as a tool for musicians to work together to come up with new ideas and sounds for their music. Working together like this between AI and human musicians could lead to the creation of new and amazing music (Sambaragi et al., 2024). Even though current AI algorithms are very good at making music, there are still some problems that need to be fixed. One of the main worries is that it's hard to tell how good and creative AI-generated music is because these may be subjective measures. There are also social and legal questions about who owns and has the rights to music written by AI. AI development will make it easier to encode differences, and it will probably break the current way we make and listen to music.

5. TECHNOLOGICAL FOUNDATIONS OF AI IN MUSIC

With advanced tools to help with every stage of the creative process, artificial intelligence has revolutionised the music creation process. AI is able to imitate well-known musicians, assist with sound design and play up compositions through techniques like machine learning and deep learning. Because of these advancements, there is now more room for experimentation and exploration in the creative production of music, which inspires musicians to explore new forms, genres and aural possibilities. By combining these characteristics, emotional intelligence and an expressive music generation mechanism enable AI to produce music that has a stronger emotional impact on the listener, thereby bringing the technological and human creative approaches closer together.

Two significant deep learning models to be used in music generation are the recurrent neural networks (RNN) and long short-term memory (LSTM) network, as they are capable of sequence learning. These models have been used with success to generate music which has been shown to outperform human music (Sambaragi et al. 2024).

It could be of assistance to discover these modes, by looking to them through musical spectacles. Just as in harmonization, RNNs write sequences- which again makes sense when one considers melodic lines, and chord progressions. Another form of RNN is the LSTM networks that retain long-range dependencies across music that facilitates the ability to reproduce repeating themes and motifs within music. The Generative Adversarial Networks (GAN) can be described as a composer and a critic: the generator will make new music and the discriminator will judge the quality of the music, until it ends up producing capturing that sound that is so close to the human generated music.

LSTM networks have been conducted to encode music information based upon melodic and harmonic interval to enable them to create musical and tonal compositions. The terms generative adversarial networks (GANs) and reinforcement learning are also used as they are applied to the creation of original music pieces that can rival the creations of humans and have the potential to transform the creativity in the musical sphere.

Genetic algorithms Many genetic algorithms are used with neural networks to evolve a musical pattern through selection and crossover procedures. Such approaches appear to have been successful in producing agreeable music designs with certain combinations providing better results than others. Alternative approaches have been discussed, such as using genetic algorithms and neural networks simultaneously in the attempt to maximize quality of the music produced in accordance with the musical rules and patterns.

Artificial intelligence is producing music systems that generate expressive music pieces, focusing on the musical components that are life-threatening on the perception such as note velocity and note duration. The mentioned elements above are important in creating music that listeners prefer compared to music without expression. The introduction of expression into AI-generated music is accomplished with the intention of lessening the alienation between human-composed music and AI-generated music by introducing emotional and aesthetics significance to the music (Simoes, 2019).

The role of emotion in music creation and learning will be revolutionized by AI's ability to comprehend it. Future systems would be able to analyze the gestures of the performers or the reactions of the audience (based on vocal tones or biometric data) in order to adapt the compositions in real time, increasing engagement (Chen et al., 2024). In education, AI can tailor lessons according to the emotional state of a student, which will boost motivation and creativity (Sanganeria & Gala, 2024). The promise of this affective intelligence is that it might make music even more immersive and inclusive, creating the potential for a connection between creator and listener at a deeper level than ever before.

Given the expressive musical potential of AI to generate music via note velocity and duration (Simoes, 2019), future systems have the potential to generate music that can be experienced as a highly authentic piece by a composer with a distinct sound - someone whose individuality is reflected in the music. Imagine having AI analyze an artist's style to create melodies or lyrics that sound authentic, or curate compositions for music education to suit the emotional needs of a student. While software such as Hookpad Aria already makes suggestions for harmonies, sophisticated deep learning may soon be able to produce music that sounds exactly like human compositions but adheres to the artist's intention.

6. APPLICATIONS OF AI IN SONG WRITING, PERFORMANCE AND PRODUCTION

When AI is exposed to music it can sound like it is very cool but we are in trouble! At the crux of this is the debate on whether AI truly writes the music or is simply a continuation of the human being with the query being - is AI creativity really creativity? Cultural, technical and ethical problems exist too: fears of homogenizing music, authorship problems, and plagiarism. The AI will be developing and controlling the part of the AI further and will ensure that the part where AI fills the break of human creativity and musical content only remains even without losing the emotional gap in the work which only the human musician can develop. and there are numerous ethical and philosophical issues caused by the usage of the AI which should be taken into consideration in the making of the music.

GPT-based AI models are increasingly involved in the co-creative process of lyric writing in order to advance the thematic concepts of digital consciousness, synthetic emotions, and virtuality. Rather than writing in classical forms of songs, relying on a structured rhyme scheme and melodic fit, the AI can become an ideation partner-proposing snippets of lyrics and metaphors that the performer can edit and refine. In providing the occasional prompts, e.g. "digital world" or "artificial consciousness," the AI generates lines that, though not generally structurally conforming, are suggestive of imagery. These products are those most open to being considered poetic pieces-samples of meaning and texture to be modelled and re-contextualised. So this is where we get lines such as Digital shadows, haunting my mind or Neon dreams and virtual fears, synthetic laughter, digital tears which AI can transform into new poetry. It is within that framework that the AI in this instance is not composing a complete body of lyrics, but providing raw material to a curatorial process, in which the human songwriters are choosing and recontextualizing the material in the context of a specific emotional and conceptual idea. It is not a classical form of writing lyrics and melody, but an experimentation. The style of lyric writing might receive an odd new liveliness through the way AI impressively presents incredible combinations of words as a serendipitous creativity to urge songwriters to dance out of the box, beyond the formulae and beyond the customs. Final answer: AI is an Idea Generative, but the great decider/c filter, the limbic mind of the human, will be the final artistic executive, it is a Model of Human-AI Co-creation, where the role of AI will be an agent of thematic expansion, not a composer with a will to power. The sphere of creating new songs that can literally compete with human-written songs is already the area where AI algorithms are doing a very good job. Deep learning forms the cornerstone of such capabilities, along with GANs and other types of models capable of replicating human creativity and music perception. Yes, the technologies also allow the AI to compose new music, that can be as expressive and written in as many forms as the user desires. The second section describes the methods employed in the production of musique generative, the mechanics and it use case and the not use case.

Artificial Intelligence (AI) is now emerging as an impactful factor in musical performance and improvisation, and forming new types of collaborative digital creativity between computers and human musicians. We present this work with the objective to create AI systems in the area of live musical performance that are not only capable of aiding in musical composition and production, but they are also actively participating in on-stage performances, in the form of live improvisation and interaction..

Real-time interactions and improvisations made possible by the AI as it has allowed the musicians to express themselves even better. This interplay of human instinct and algorithms gives rise to new music and performances. For instance, there seems to be more AI infused tools to complement human knowledge and algorithmic functioning of machine with new blends of human and artificial creativity in music ensembles.

Music infrastructure with the application of AI does not refer to sound only, but extra-musical communication is also present. Indicatively, a machine drummer has been implemented to provide visual display of its confidence level, fostering the interaction between human musicians and an AI machine because it increases the degree of human engagement in the musical action through their understanding of the inner working of the machine (McCormack et al., 2019).

A deep gradient-based reinforcement learning algorithm was used to train an AI system to improvise music and achieved better performance metrics than the other strategies. AI can also read graphic notation and change visual information into the music, which opens the creative potential of contemporary music composition. The application of AI enables it to exchange graphic notation and musical interpretation and brings in new visions of how to perform and interpret graphic scores (Karchkhadze et al., 2024).

AI is rapidly transforming the creation and shaping of music and playing a significant role in sound design and music production. Entering these circles, AI has led to the creation of new tools and methods that are applied to improve the creative process, make it more efficient and extend the scope of what is possible in the field of music and sound design.

The evolution of automated GUI application for use in music production to include audio effects has also increased the collection of classic audio processing applications. These are just smart adaptive digital audio effects that are both accurate and flexible. Application to AI tools is still blossoming in the realm of DAWs and even the software design works against the idea. A great deal of design research has emerged in cooperation with ethnographic studies to inform the design of the automated production tools and through this, determining the role of the tools in the studio.

The use of AI in mastering is gaining more prominence as something of a comparison between the work that it delivers or Be generated by a mastering engine with that created by the ears and skill of a professional audio engineer. These services identify some of the strong and weak demographics on AI related to the optimization and prepped commercial music distribution. Within modern-day popular music production, AI is applied to facilitate creative processes, especially where composition and editing and mixing are combined. The MIDI-based tools are not adequate since they might hamper the creative work of the artist whereas the audio-based AI are more desirable.

The influence of AI has been felt by electronic and pop genres in particular but the effects are even wider. As an example, AI neural networks such asFlow Machines create improv solos and harmonies in jazz. Through the Magenta project, Google has managed to create algorithms that can compose Bach-style chorales in the use of traditional voice leading, e.g. in classical music. And AI can make a bit of contribution to folk and traditional music - through training on historical records it can learn new variations created on the new ones to ensure that the endangered melodies do not disappear with the cultural tradition.

The following are actual uses of AI in the music industry. AIVA (Artificial Intelligence Virtual Artist) has been used on film scores and commercial jingles, where orchestral music is produced, and then refined by human composers. Next we have OpenAI MuseNet, which can generate in a slightly less style-specific fashion (such as combining a mix of various conventional instruments into a more classical or jazz ensemble), though also appears to begin to imitate itself again after a period, resulting in a feeling of the need to make further adjustments to mixes manually. In 2019, the group YACHT made use of an AI system of melodies and lyrics in their album Chain Tripping, though only after the algorithms had done a preliminary heavy lifting, they selected and edited. They were not but mere facts and considering the subjectivity of the matter, Fraser has explained that AI cannot reproduce creativity yet AI is the means which has allowed new methods in art.

7. AI IN MUSIC EDUCATION: PERSONALIZATION, ENGAGEMENT AND INNOVATION

Conversely, the area of educational products, which are closely connected with music and which allows one to learn more quickly and be free in creativity, is introducing a certain revolution brought by AI. It enables learning, real-time feedback, and learning interactive systems that make learning in music education more manageable and pleasurable. AI also encourages creativity in learners as the learners get to compose music and grow mentally and emotionally. Among other things, AI automates assessment and one can say that the teachers will have more time to teach. You want to refer to the AI Live music classes, The Next Generation of Music Education with AI Technology Live music classes, With AI technology adjusting to human behaviour, it can be used to its full potential to help you move forward in all the possible ways in music.

AI in music learning is transforming the way music is taught and learned through new tools and learning techniques. The benefits of AI applications to music education are numerous as they include everything connected with the aspect of self-learning, interactive teaching and in-depth learning. These technologies are transforming the current pedagogic methods and are making music education accessible, interesting and practical. The use of AI in music education is relatively new, yet it already demonstrates a lot of potential.

Music is such a subjective topic, when it comes to coaching, it lends itself to potential AI-powered tools. Examples include to create individual exercises with Automatic Chord Recognition and Automatic Music Transcription to train the ear and skills through exercise. The tools democratize high-quality music education to be more inclusive and enjoyable by all types of learners (Sanganeria & Gala, 2024). Artificial intelligence-based systems such as the Music Learning System Based on the RBF Algorithm and the SCMA Algorithm can offer custom music learning experiences using the analysis of their vast music sample conglomerate and neural networks to fit each learning experience framework to the specifics of the learner. By providing instant feedback, AI tools can improve the learning process since students will be able to see their progress and weak points at the moment.

Intelligence in AI promotes the engagement of students and motivation towards learning. As an example, AI perceptions of students were associated with a higher level of motivation and involvement and learning improvement. Applications of AI in VR/AR applications help to provide an immersive experience to learning music and make them more entertaining (Merchán Sánchez Jara et al., 2024).

AI technology helps in creating compositions of music and the students can find unique opportunities to express their muses. The tools can create music, recommend compositional ideas and provide some creative inspiration, hence, broadening the creative horizons of students (Wang, 2024). Generative AI has been found out to be especially helpful in encouraging artistic creativity, giving the students the freedom of experimenting with various musical genres and melodies (Merchán Sánchez Jara et al., 2024).

It is important that creativity in music education should allow the development of cognitive abilities, as well as emotional ones. An instruction method that lays focus on innovative thinking that is marked by such processes as divergent thinking and intrinsic motivation is vital to developing musical creativity. Creativity does not only play a role in the creation and maintenance of emotional and cognitive development in students but it also helps them foster their imaginative and creative potentials (Meng & Ariffin, 2024). AI can be used to provide personal and interactivity in music education. Nevertheless, there should be a balance between AI support and human imagination so that students can not only think critically but also be creative (Merchán Sánchez Jara et al., 2024)

The AI technologies can simplify the assessment process by automating the routine aspects of this process and optimizing its approach. This will enable teaching staff to teach and not spend time dealing with paperwork. We have seen that AI powered chat bots can enhance student learning in music education by regularly and individually giving feedback, and through this a 15% improvement in academic performance in varying instances.

Although there are advantages, the incorporation of AI in music instruction poses various issues like data confidentiality and bias in the AI models (Holster, 2024). The ethical aspects of AI in education, the value of teachers, as well as the possibility of AI to substitute the traditional ways of teaching should be treated with caution and care so that they do not push AI out in front of human teachers (Vasco & Eguiguren, 2023). Although there are many benefits of AI which can be used in the music education, it is necessary to discuss the challenges and ethical considerations that AI may raise. Considering how to make sure that AI tools can be used responsibly and in an ethical matter is going to be paramount to ensure that there will be optimal benefits to them. With the further development of AI technology, it is promising to have a greater influence on changing the nature of music education, providing new opportunities in terms of individualistic, interacting, and constructive learning.

8. THE FUTURE OF AI IN MUSIC: INNOVATION COLLABORATION AND PROSPECTS

The fact that the application of AI is becoming desirable, this current aspect of composing music has its limit yet on the other hand provokes and even raises some ethical concerns as well. The big question is as follows: Does AI, or neural networks, have a role to play in the creation of creative content, or is the whole process human-only? Then there are cultural, technical and ethical issues too, like homogenised music, authorship and plagiarism. When AI in music considers the most appropriate approach to addressing these ethical and philosophical dilemmas to its application, the future will finally need to concern itself with the question of artificial intelligence being used to either add to or subtract human creativity with emotion through musical performance.

Although the use of technology when creating music has great advantages, it also creates a whole new range of challenges — one of these is that of creativity and originality. While AI can assist with music generation, the level of involvement that AI has in the music-making process is a common point of conversation. Is AI a creator or just an extension of human efforts? While many others state that 'AI creativity is pointless, as it is merely a hyper-extension of humans and is devoid of actual creativity'. The other major challenge lies in both the technical and mental aspects of a song identification. Although it is still a long way off, it is likely that a machine with predictive generative capabilities could mimic human-like creativity once it has mastered the complex levels of music cognition. There are also aesthetic and cultural issues. Two of the biggest problems in the field are authorship and copyright because human creation and machine output are so relevant. What about the AI-generated music? Just as they should be paid at every other level, who should be paid on the bottom page of these songs, which lags up the hierarchy of who made a pleasing sound in the first place? Going further another issue is that there is plagiarism risk. But the potential for AI to create near-perfect copies of already existing songs would jeopardise the integrity of the music industry itself. Finally, consider the democratization vs. equity issue AI is reducing the barrier to entry to music creation, but AI also brings a danger of bias and unequal opportunity to creation tools. For this reason, it is essential to be extremely cautious in the use of the distributional of AI tools so as not to replicate a post-colonial model (Merchán Sánchez Jara et al. 2024). Knowing these challenges leads to the notion that a regulatory environment will be necessary to ensure fairness and the preservation of the value of human creativity (Berkowitz, 2024).

Even when we disregard the issue of originality, the questions about the authorship and copyrights have not been buried. Legal cases like Thaler v. USPTO (2020) demonstrated that the existing copyright law fails to provide any recognition to AI as an author thus the copyright resides with the human programmers or user. AI creates economic risk by reducing the services of a session musician in sectors such as advertising and low-budget film-scoring, whilst also generating new employment, such as dataset curation, AI trainers, and human-AI co-producers. One more critical aspect can be associated with the cultural equity: As most AI systems are being trained using large amounts of western data, the issue of cultural marginalization of music traditions and folk heritages that may not have their origin in the western realm can become a critical challenge. This presents the necessity of the development of equitable policies and moral principles of AI in music.

As the field of music witnessed the introduction of artificial intelligence (AI), several pitfalls and ethical issues came with the significance of considering the human appearance aspect in the music composition process. The role of AI in music is quite debatable, since one can relate AI to a helping hand and an art producer; it relates to the idea of creativity, authorship, and originality of AI-generated artworks. So there are so many more things we can learn about AI, but we must mention all concerns it poses in relation to our music industry and our larger cultural ecosystem. It has sparked a discussion concerning issues like moral aspects and ethics of AI-generated art and whether it can be possible to compose pieces of music without a human touch. Innovation in music has opportunities, as described in 8.4. As AI enters the scene, the creative boundaries in music creation and live performances, now also VR performances, are clearly expanding day after day. On the one hand, it is a part of the sound ecology of such performances; however, on the other hand it preconditions further processing of sound in real-time by composers. Artificial intelligence proprietary systems straddling music to agreed-upon activity or new genres incite; only generations later. Suno AI and AIVA may even generate new hits right now, as it grows to the stage of absorb several elements of culture, and becomes a music-synthesis church the size of a giant (Sambaragi et al., 2024). All these in addition to the above may be accomplished by a sub-field of AI known as EC and GAN that can be implemented to open the space which composers have thought about as compositional trend and push at the extremes of their imagination. These technologies will bring new dimensions across and will have literal pleasure.

AI is generating fresh possibilities of creativity in music creation as well as in live performances, and even in virtual reality concerts. It is vital to the soundscape of these performances, and it offers the composers novel means of working with sound in the moment. In future AI systems can be trained to blend music with a culture or a new genre of music that embraces the sound on a more worldly stage. Technology will also change the process of creating novel parts as it will create unique pieces virtually overnight (SFQ, 2024) and in the not-too-distant future, coding the history of music could become part of the music generators, combining various cultural elements to create a collaborative piece of music worldwide through the influence of technologies such as Suno AI and AIVA (Sambaragi et al., 2024).

Furthermore, those tools as evolutionary computation and generative adversarial networks (GANs) have the potential to allow composers to feel even more distant to the boundaries of their creativity. This increases the depth of content that may be important to the audience worldwide, which in turn enables artists to generate new sound environments that resonate with a wider audience worldwide.

The opportunity for change, from the disruption and potential of new genres to ways to explore and innovate, is huge for AI in music. It is already transforming how we create, perform and teach music with tools for composition, sound design and live performance. Instead of displacing human creativity, AI augments human creativity and expands the artistic parameters of the musicians to push into a new mode of expression and genres more exploratory in composition and methods. Musicians are composing new sound fields, as the AI and the artist are composing with each other, so to speak; each composer in the moment creating sound together. Which stimulates a new kind of global (and cross-disciplinary) collaboration, which a permanent AI powered integration in our ways to musicking and sound sharing, from now on, will grant us the opportunity between the communities around the globe.

AI in music education, AI in song learning, is also changing the very way we learn words, making adaptive learning technologies a top priority. The AI platforms allow the music students to receive personalized feedback and identify less time learning and developing. Amplify Intro is moving towards AI with more power in the music industry, yet music as a human activity still exists (emotion + expression). AI can complement and augmenting creativity, but AI will never usurp the creativity, the artistry, the intuition and the humanity of a musician. Perhaps the next step of AI music will be the contrast: the technology that glorifies the humanity of music, so that the soul of music, the feeling that it gives and the craft of the making of that music is left to human hands, despite our high-tech future.

Whatever AI in music becomes in the future, it will all be based on new models and new applications that demonstrate the future of AI and its responsible, ethical use. We do need the research to enable us to make those opportunities of innovation and experimentation possible, but we need it even more urgently to enable us to alleviate some of the risks of loss of human creativity, commoditization of music and other potential problems that are likely to emerge with AI in general. In the future, the right AI in music models will strive to retain the soul of the music, such that the human genius that creates the music of the world is supported by AI, will not be forced to fit it. Our future days will be probably a gentler caress, a season that will fit to how we can utilize AI tools which would not oppose but complement music making as invented by human hands.

The direction of AI in music, is towards the more interactive and impressive experience. This AI-based improvisation in real-time will encourage AI to improvise live with live musicians, particularly in the jazz and electronic music domain. Also included in virtual and augmented reality concerts will be AI avatars, which will bring concerts outside of the venue and into the metaverse: it is this technology, following digital concerts by Travis Scott and ABBA, that we are likely to investigate further in the future. This would later result in the creation of new technologies to allow works created by AI down to the algorithm level to be traced in regards to ownership in order to eliminate any possible conflict over authorship in the former in the first place (blockchain anyone?) Moreover, hyper-personalised music--developed on the basis of biometrical data of listeners, their moods and their weekly paths--gives a strong indication of a future in which no one will be listening to the same song, ever again.

9. CONCLUSION

Artificial intelligence is transforming music and expanding the possibilities of creativity by acting as a tool and a collaborator, without replacing the human imagination. Its applications in education, production, performance and composition demonstrate how technology can improve artistic expression and democratise access to music-making. However, there are still problems with authorship, copyright, cultural representation and ethical use. The future of AI in music depends on its responsible integration, which finds a balance between human creativity and technological advancement. By viewing AI as a collaborator rather than a substitute, researchers, educators, and musicians can embrace new possibilities for experimentation, creativity and intercultural cooperation while still respecting the uniqueness of human expression.

REFERENCES

Agwan, M., Nemade, M. U., Roy, S., & Sinha, U. (2023). The Fusion of AI and Music Generation: A Comprehensive Review. 90–94. https://doi.org/DOI: 10.1109/ICAST59062.2023.10454942

Avdeeff, M. (2019, October). Artificial intelligence & popular music: SKYGGE, flow machines, and the audio uncanny valley. In Arts (Vol. 8, No. 4, p. 130). MDPI.

Berkowitz, A. E. (2024). Artificial intelligence and musicking: A philosophical inquiry. *Music Perception*, *41*(5), 393–412. DOI: 10.1525/mp.2024.41.5.393

Briot, J. P., Hadjeres, G., & Pachet, F. (2020). *Deep learning techniques for music generation*. Springer. DOI: 10.1007/978-3-319-70163-9

Brøvig, R., & Danielsen, A. (2023). *Digital Signatures: The Impact of Digitization on Popular Music Sound*. MIT Press.

Caramiaux, B., Françoise, J., Liu, W., Sanchez, T., & Bevilacqua, F. (2020). Machine learning approaches for motor learning: A short review. *Frontiers of Computer Science*, *2*, 16. DOI: 10.3389/fcomp.2020.00016

Chen, Y., Huang, L., & Gou, T. (2024). Applications and advances of artificial intelligence in music generation: A review. arXiv preprint arXiv:2409.03715. https://doi.org//arxiv.2409.03715DOI: 10.48550

Cope, D. (1991). *Computers and musical style* (Vol. 6). Oxford University Press.

El Ardeliya, V., Taylor, J., & Wolfson, J. (2024). Exploration of Artificial Intelligence in Creative Fields: Generative Art, Music, and Design. *International Journal of Cyber and It Service Management*, *4*(1), 40–46. DOI: 10.34306/ijcitsm.v4i1.149

Gioti, A. M. (2021). Artificial intelligence for music composition. In *Handbook of artificial intelligence for music: Foundations, advanced approaches, and developments for creativity* (pp. 53–73). Springer International Publishing. DOI: 10.1007/978-3-030-72116-9_3

Hiller, L., & Isaacson, L. (1993). Musical composition with a high-speed digital computer.

Holster, J. (2024). Augmenting music education through AI: Practical applications of ChatGPT. *Music Educators Journal*, *110*(4), 36–42. DOI: 10.1177/00274321241255938

Karchkhadze, T., Shao, K., & Dubnov, S. (2024, December). Interpreting Graphic Notation with MusicLDM: An AI Improvisation of Cornelius Cardew's Treatise. In *2024 IEEE International Conference on Big Data (BigData)* (pp. 3181-3190). IEEE. https://doi.org/DOI: 10.1109/BigData62323.2024.10825824

Linkinen, T. (2024). Generative artificial intelligences: Challenges and benefits for game development (Doctoral dissertation, Master's thesis, Tampere University).

Main, A., Grierson, M., Yamada-Rice, D., & Murr, J. (2022, June). Augmenting personal creativity with artificial intelligence: Workshop proposal for Creativity and Cognition 2022. In *Proceedings of the 14th Conference on Creativity and Cognition* (pp. 462-465). https://doi.org/DOI: 10.1145/3527927.3531205

McCormack, J., Gifford, T., Hutchings, P., Llano Rodriguez, M. T., Yee-King, M., & d'Inverno, M. (2019, May). In a silent way: Communication between ai and improvising musicians beyond sound. In Proceedings of the 2019 chi conference on human factors in computing systems (pp. 1-11). https://arxiv.org/pdf/1902.06442.pdf

Meng, N., & bin Ariffin, A. (2024). The Role of Creativity in Music Education. The Education Review, USA, 8(7), 922–926. https://doi.org/DOI: 10.26855/er.2024.07.004

Merchán Sánchez-Jara, J., González Gutiérrez, S., Cruz Rodríguez, J., & Syroyid, B. S. (2024). Artificial Intelligence-Assisted Music Education: A Critical Synthesis of Challenges and Opportunities. *Education Sciences*, *14*(11), 1171. DOI: 10.3390/educsci14111171

Merchán Sánchez Jara, J. F., González Gutiérrez, S., & Cruz Rodríguez, J. (2024). La música como herramienta para la transformación social: educación musical y ODS en la formación en valores.

Miranda, E., & Williams, D. (2015). Artificial Intelligence in Organised Sound. *Organised Sound*, *20*(1), 76–81. DOI: 10.1017/S1355771814000454

Pasquier, P., Eigenfeldt, A., Bown, O., & Dubnov, S. (2017). An introduction to musical metacreation. [CIE]. *Computers in Entertainment*, *14*(2), 1–14. DOI: 10.1145/2930672

Sambaragi, L. M., Naik, P. P., Kakatkar, N. N., & Chikkamath, S., Sr. N., & Budihal, S. V. (2024, March). Music Generation: A simplified approach. In 2024 3rd International Conference for Innovation in Technology (INOCON) (pp. 1-6). IEEE.

Sanganeria, M., & Gala, R. (2024). Tuning Music Education: AI-Powered Personalization in Learning Music. https://doi.org//arxiv.2412.13514DOI: 10.48550

Simoes, J. M. da C. (2019). Deep Learning for Dynamic Music Generation. https://estudogeral.sib.uc.pt/bitstream/10316/88002/1/Deep_Learning_for_Dynamic_Music_Generation.pdf (Master's Thesis)

Sturm, B. L., Ben-Tal, O., Monaghan, Ú., Collins, N., Herremans, D., Chew, E., Hadjeres, G., Deruty, E., & Pachet, F. (2019). Machine learning research that matters for music creation: A case study. *Journal of New Music Research, 48*(1), 36–55. DOI: 10.1080/09298215.2018.1515233

Tabak, C. (2023). Intelligent music applications: Innovative solutions for musicians and listeners. *Uluslararası Anadolu Sosyal Bilimler Dergisi, 7*(3), 752–773. DOI: 10.47525/ulasbid.1324070

Varini, M., & Gramaglia, G. (2024). Tyranny of (AI) Thought. Artificial intelligence in music composition: A case study on Krallice "Diotima". *CONNESSIONI REMOTE, 2024*(8), 51–82. DOI: 10.54103/connessioni/26583

Vasco, M., & Eguiguren, L. (2023, December). La Inteligencia Artificial y el aprendizaje musical. In *Colloquia* (Vol. 10, pp. 46–62). Academic Journal of Culture and Thought., DOI: 10.31207/colloquia.v10i0.150

Walzer, D. (2023). *Leadership in music technology education: Philosophy, praxis, and pedagogy*. Focal Press. DOI: 10.4324/9781003152460

Wang, G. (2024). The investigation of artificial intelligence-based applications in music education. *Applied and Computational Engineering., 36*(1), 210–214. DOI: 10.54254/2755-2721/36/20230448

Zhang, Y. (2024). Utilizing computational music analysis and AI for enhanced music composition: Exploring pre-and post-analysis. *Educational Administration: Theory and Practice, 30*(5), 269–282. DOI: 10.17762/jaz.v44iS6.2470

Chapter 4
Bias in AI Algorithms and Its Impact on Musical Diversity

Saroj Sandeep Phalke
http://orcid.org/0009-0001-2251-7882
Vishwakarma Institute of Information Technology, India

Gitanjali Shinde
http://orcid.org/0000-0001-6623-8722
Vishwakarma Institute of Technology, Pune, India

Grishma Bobhate
http://orcid.org/0009-0007-7351-8723
Vishwakarma Institute of Technology, Pune, India

Poonam Railkar
Smt Kashibai Navale College of Engineering, Vadgaon, India

Haribhau R. Bhapkar
Central University of Kashmir Green Campus, Ganderbal, India

Sonal Fatangare
http://orcid.org/0009-0005-1185-8699
Vishwakarma Institute of Technology, Pune, India

ABSTRACT

Artificial intelligence (AI) is transforming music creation, distribution, and discovery, offering new opportunities but also reinforcing bias. When trained on datasets dominated by specific genres or demographics, AI systems often amplify mainstream trends while marginalizing niche artists and underrepresented communities. This chapter explores how such bias emerges through skewed data and algorithmic design, and how it narrows musical diversity by favoring "safe" and commercially viable outputs. Case studies, including the EU's Fair MusE project, reveal the impact on fairness and cultural representation. Responses such as diversity-aware algorithms, enriched datasets, and transparent systems are examined alongside ethical and policy concerns. The chapter concludes that sustaining musical diversity in the AI

DOI: 10.4018/979-8-3373-6279-3.ch004

era demands accountability, innovation, and collaboration between technologists, artists, and regulators to ensure technology amplifies rather than diminishes global musical expression.

1.INTRODUCTION

1.1Background,Motivation,andRationale

The music industry is undergoing a profound transformation, driven by the rapid and pervasive integration of artificial intelligence. This technological shift, which encompasses a wide range of applications from personalized recommendations to automated composition, is fundamentally reshaping how music is created, distributed, and consumed. When someone opens Spotify, Apple Music, or YouTube, the tracks they hear are usually chosen by AI systems (Chen et al., 2024). Similarly, experimental platforms now allow AI to compose melodies or even full tracks with minimal human input (Pricop et al., 2024).

This is not only a cultural change but also a massive economic one. Reports valued the global AI in music market at roughly USD 3.9 billion in 2023, and some forecasts predict it could climb to around USD 38.7 billion by 2033. Much of this growth is tied to streaming services, where AI recommendation systems already account for nearly half of the sector's revenue (Civit et al., 2022).However, focusing only on market numbers gives a partial picture. What is more interesting and also more concerning is how these systems subtly shape what we hear. AI is not a neutral referee. The way it recommends music depends on data from the past (such as user listening histories and sales figures) and the commercial goals of the companies that build these systems (Zhang et al., 2025; Chen et al., 2024).

As a result, AI often amplifies existing biases: the most popular artists get pushed further into the spotlight, while less mainstream voices, whether niche genres or artists from underrepresented cultures, may remain invisible (Zhang et al., 2025). In other words, instead of expanding musical horizons, AI may actually narrow them (Chen et al., 2024).

The motivation for this study is to highlight that AI in music should not be viewed as just another technical tool. It has cultural power. In earlier decades, radio DJs, music critics, and even word of mouth shaped people's exposure to new artists. Now, algorithmic playlists quietly take on that role (Chen et al., 2024). Every time a listener opens an app and presses play, the system decides what comes next. These invisible decisions ripple outward affecting artists careers, shaping industry trends, and influencing cultural memory itself. This is why examining bias in these systems is more than a technical necessity; it is a cultural responsibility (Civit et al., 2022).

1.2 Why This Study Matters:

The core concern of this research is simple yet far reaching: what happens to musical diversity when AI becomes the gatekeeper? At first glance, this may seem like a question about technology or business models, but in reality it cuts deeper. Music is not just entertainment; it is a form of cultural expression, a way communities hold on to identity, history, and tradition (Civit et al., 2022). If algorithms consistently filter out certain genres or artists perhaps because they do not fit mainstream commercial patterns the result is not just fewer options on a playlist. It is the slow silencing of particular cultural voices (Chen et al., 2024).

This problem is intensified by the structure of today's music platforms. A handful of corporations dominate the global streaming market, and their algorithms influence what millions of people hear each day (Civit et al., 2022). Their design choices might often be aimed at keeping listeners engaged for longer periods which end up shaping cultural trends at a massive scale (Zhang et al., 2025). This leads to a critical question: will the future of music be defined by diversity and experimentation, or will it lean toward uniformity, where only safe, profit-driven sounds survive (Chen et al., 2024)?

Understanding these biases matters because music has always thrived on diversity. Jazz, hip-hop, electronic dance, and countless folk traditions all grew out of spaces that were initially marginalized (Civit et al., 2022). If algorithmic curation sidelines similar creative movements today, we risk losing the next wave of cultural innovation before it even has the chance to emerge (Pricop et al., 2024).

1.3 What This Chapter Aims to Do

This chapter sets out to discover how algorithmic bias works in the music industry and what are its implications for diversity. To do so, it will:

- Clarify what algorithmic bias actually means, identifying different types and sources.
- Explore the idea of musical diversity and outline possible ways to measure it in practice.
- Examine how bias manifests in AI powered tools, from recommendation engines to generative music software.
- Present examples and case studies showing how these biases affect both listeners and creators.
- Discuss the ethical and social dilemmas that follow from biased AI systems.
- Highlight current methods of detecting and mitigating bias, while assessing their strengths and limitations.

- Consider new research directions and the future of AI in music, asking what a more equitable system could look like.

By moving through these stages, the chapter aims not only to describe a problem but also to encourage reflection on what kind of cultural future we are shaping . The hope is that by making the biases visible, we can also begin to imagine ways of designing AI systems that enrich rather than diminish the diversity of music available to the world.

2. UNDERSTANDING BIAS IN AI

2.1 What Is Algorithmic Bias and What Are Its Types?

When people hear the word bias, they often think of someone showing favouritism or prejudice. In the case of artificial intelligence (AI), bias means that the system's outputs are tilted in a way that is unfair or inaccurate. Importantly, this does not always happen because of intentional discrimination. More often, it emerges from the data or design choices that go into building the system. Put simply, algorithmic bias happens when an AI ends up giving advantages to some groups while disadvantaging others without a sound reason.

To understand this better, it helps to look at the common types of bias that appear in AI:

Selection or Sampling Bias

This occurs when the dataset used to train the AI does not accurately represent the real world (Yesiler et al., 2022; Mokoena & Obagbuwa, 2025). Imagine building a facial recognition tool with mostly photos of light skinned people: naturally, the system will struggle when it encounters darker skin tones. In music, the same thing happens when the training material is too narrow. For instance, if an AI model is only exposed to Western pop hits, it will be unable to recognize, recommend, or generate music from traditions like Indian classical ragas, Latin rhythms, or West African drumming (Mokoena & Obagbuwa, 2025). As a result, listeners may never be introduced to these genres, even though they are a vital part of global culture.

Measurement Bias

Sometimes, the issue lies not with the data collected but with how it is measured or labeled (Mokoena & Obagbuwa, 2025; Henry et al., 2024). If annotators disagree about how to tag songs by genre for example, whether something counts as "folk" or "indie" then the dataset ends up inconsistent. An AI system trained on this data will mirror those errors. Another form of measurement bias appears when the data mostly consists of clean, studio recordings. The system then performs poorly on lo fi field recordings or community music captured with basic devices, even though these reflect authentic traditions (Yesiler et al., 2022).

Association Bias

Association bias reflects stereotypes that get reinforced in the training process (Henry et al., 2024). A well known example outside music is how some translation algorithms consistently match "doctor" with male pronouns and "nurse" with female pronouns. In music, an AI might associate certain instruments like the djembe or sitar with specific ethnic identities, which can become limiting when the system fails to recognize how they can blend across cultures (Mokoena & Obagbuwa, 2025).

Confirmation Bias

This is perhaps the most subtle type. Here, the system relies so heavily on existing patterns that it ends up creating a feedback loop (Dinnissen & Bauer, 2022). For instance, if data shows that pop songs by male artists get the highest number of plays, the algorithm will keep recommending them. Female artists or minority voices are pushed aside, not because listeners dislike them, but because the system never gives them equal visibility. Over time, the popularity gap widens, making the bias even stronger (Dinnissen & Bauer, 2022).

Understanding these categories matters because they reveal a crucial truth: AI does not merely mirror reality. It actively shapes what people see, hear, and experience (Henry et al., 2024). In music, that means AI systems are not just passive tools they are cultural gatekeepers (Dinnissen & Bauer, 2022; Mokoena & Obagbuwa, 2025).

2.2 Where Does Bias Come From in AI Systems?

Bias in AI does not emerge out of nowhere. It usually comes from three interconnected sources: the data, the model design, and broader human or societal factors.

Data Bias

There is an old saying in computer science: garbage in, garbage out. In the case of AI, it often becomes bias in, bias out (Yesiler et al., 2022; Liu et al., 2024). If the training data is skewed, the system will inevitably reflect that skew (Dinnissen & Bauer, 2022). Think of a recommendation system trained mostly on English language pop. Naturally, it will suggest English pop songs far more often than, say, Afrobeat or Japanese city pop. Over time, listeners are funneled toward the same music, while other genres remain underrepresented (Mokoena & Obagbuwa, 2025). What starts as an innocent imbalance in the data eventually snowballs into a structural bias in the industry.

Model Bias

Even when the data is diverse, the way an AI is built can introduce bias (Henry et al., 2024). Many recommendation systems are designed with the simple goal of maximizing clicks or streams (Liu et al., 2024). On the surface, this seems logical. Yet, in practice, it favors already popular songs, since those tracks are more likely to generate quick engagement. New, experimental, or niche artists get left behind (Henry et al., 2024; Liu et al., 2024). Another issue is overfitting: when an AI model learns its training data too well, it fails to generalize to new trends. A model trained on 1980s rock might perform poorly when confronted with K pop or experimental electronic music because it has no framework for recognizing what makes them appealing.

Human and Societal Factors

Finally, no AI exists in a vacuum. The people who design, deploy, and evaluate these systems carry their own assumptions, shaped by culture and society (Mokoena & Obagbuwa, 2025). For example, studies show that audiences often prefer music they believe was written by a human, even if they cannot distinguish it from AI generated music in a blind test. This suggests a cultural bias: the perception that "real creativity" belongs only to humans. Such beliefs can influence not only listeners' attitudes but also policy decisions, copyright laws, and critical reception (Dinnissen & Bauer, 2022).

Taken together, these sources of bias highlight that the issue is not merely technical. Bias in AI reflects broader debates about creativity, fairness, and whose voices get to be heard (Henry et al., 2024; Dinnissen & Bauer, 2022).

2.3 A Brief History of AI in Music Creation and Recommendation

AI in music is not as new as it might seem. Long before today's streaming platforms, researchers were already exploring ways for computers to engage with music.

One of the earliest milestones was the Experiments in Musical Intelligence (EMI) project in the late 1990s. EMI analysed compositions by classical composers and generated new pieces in their style (Mokoena & Obagbuwa, 2025). Many listeners at the time were surprised that they could not easily tell the difference between the human composed and computer generated pieces (Mokoena & Obagbuwa, 2025).

In the early 2000s, projects like the Continuator pushed things further. This system could "listen" to a live musician and then extend the performance in real time, echoing the style and improvisation of the human player (Mokoena & Obagbuwa, 2025).

The last decade brought even bigger changes, thanks to deep learning and generative adversarial networks (GANs) (Henry et al., 2024). These technologies made it possible for AI to generate highly complex compositions across multiple genres. Google's Magenta and OpenAI's MuseNet were especially influential, showing that machines could compose symphonies, jazz pieces, and pop songs with striking fluency (Henry et al., 2024).

Today, AI music has gone mainstream. User friendly platforms allow anyone to generate full songs including vocals and lyrics just by typing a text description (Liu et al., 2024; Henry et al., 2024). What once required years of musical training or expensive equipment can now be done in minutes on a laptop (Mokoena & Obagbuwa, 2025).

This history shows both opportunity and risk. On the one hand, AI has lowered barriers to music creation, giving amateurs and hobbyists tools to express themselves (Mokoena & Obagbuwa, 2025). On the other hand, these same systems carry the biases discussed earlier. If not addressed, they could end up limiting rather than expanding the diversity of music available to global audiences (Henry et al., 2024; Dinnissen & Bauer, 2022; Yesiler et al., 2022).

3. MUSICAL DIVERSITY: CONCEPTS AND MEASURES

3.1 What Do We Mean by Musical Diversity?

When people talk about "musical diversity," they often think it just means listening to lots of genres. In reality, the idea is far more layered. Scholars who study music and culture usually break it down into three overlapping dimensions: variety, balance, and disparity (Passmore et al., 2024; Ferwerda, 2016).

The first dimension, variety, is the simplest to grasp. It refers to the number of distinct genres, styles, or artists in a listener's library (Passmore et al., 2024; Ferwerda, 2016). A person who listens only to Bollywood film songs will naturally have low variety. Compare this with someone who regularly plays hip hop, jazz, rock, and Carnatic classical their listening habits clearly show greater variety. But here's the catch: variety alone can be misleading. Imagine two people, both with five genres in their playlists. One listens to each genre in equal measure, while the other spends 90% of their time on just one. This brings us to the second dimension.

Balance captures how evenly a person's listening time is distributed across genres (Passmore et al., 2024). If someone spends most of their day looping just a handful of songs while ignoring the rest of their collection, their balance is low, even if they technically have "many genres" saved. Economists often measure balance using tools like the Herfindahl Hirschman Index (HHI). In this context, a lower HHI means listening is more evenly spread, which signals higher musical diversity (Ferwerda, 2016).

The third dimension, disparity, is a little trickier but crucial. Disparity looks at how far apart the genres or styles are from each other (Passmore et al., 2024). Listening to different subgenres of rock say, punk rock, alternative rock, and soft rock may give you both variety and balance. Yet, musically speaking, those styles are close cousins. In contrast, someone who shifts between Hindustani classical, trap, folk, and electronic dance music is engaging with genres that differ dramatically in rhythm, structure, instrumentation, and cultural background. This is high disparity, and it is what really stretches a listener's musical world (Passmore et al., 2024).

Sociologists have also connected diversity in music taste to broader social patterns. The "omnivore thesis," for instance, suggests that people with higher social or educational backgrounds tend to consume both so called "highbrow" music (like opera or classical symphonies) and "lowbrow" styles (such as pop or hip hop) (Passmore et al., 2024). In contrast, a "univore" sticks to a narrow set of genres. These patterns matter because algorithms can either help listeners become more "omnivorous," or lock them into the same narrow taste cycles (Ferwerda, 2016).

3.2 How Can We Measure Musical Diversity?

Studying musical diversity is not just about counting genres it requires a blend of numbers and interpretation (Ferwerda, 2016; Passmore et al., 2024). Quantitative measures help capture large scale trends, while qualitative approaches remind us that music carries cultural meaning that numbers alone can't explain (Passmore et al., 2024).

Quantitative measures are widely used by researchers because they allow for comparisons across large datasets, such as streaming histories. Tools like the Herfindahl Hirschman Index (HHI) show how concentrated someone's listening habits are: the more evenly genres are spread, the lower the index value, which suggests more diversity (Ferwerda, 2016). Scholars also look at the number of unique artists or songs to capture variety, or calculate the Euclidean distance between genres using features such as tempo, rhythm, and harmonic key to assess disparity (Passmore et al., 2024). More advanced approaches, such as the Stirling Index, try to combine variety, balance, and disparity into a single figure, offering a more holistic picture (Passmore et al., 2024).

But numbers only tell part of the story. Qualitative methods add depth by paying attention to meaning and cultural context. For example, early work such as Alan Lomax's Cantometrics project was one of the earliest attempts to map music cross culturally. Instead of treating all songs as interchangeable data points, it asked: what do these songs represent within their communities? (Passmore et al., 2024). Today, qualitative research might involve analysing a playlist's themes, lyrics, or social setting things that resist neat statistical reduction (Passmore et al., 2024).

In practice, the strongest studies combine both approaches. Large scale data analysis might reveal, for instance, that a listener has unusually low disparity in their playlists. A follow up qualitative study could then ask why: Is it personal preference, cultural exposure, or perhaps the influence of algorithmic recommendations? (Ferwerda, 2016; Passmore et al., 2024).

3.3 Why Does Musical Diversity Matter for Culture and Innovation?

The importance of musical diversity goes well beyond entertainment value. Music is often described as a "sonic expression of cultural identity," and for good reason. It carries histories, languages, and memories (Passmore et al., 2024). Folk songs preserve oral traditions; devotional music transmits spirituality; and protest songs have long voiced collective struggles. When diversity in music declines, what is really at stake is the preservation of these cultural lifelines (Passmore et al., 2024).

There is also a forward looking dimension: diversity fuels creativity. Many groundbreaking styles of the last century jazz, hip hop, electronic fusion emerged from the blending of previously separate traditions. Hybridity has always been one of music's engines of innovation (Passmore et al., 2024). To give a modern example, collaborations between Indian classical musicians and electronic producers, or between Afrobeat rhythms and Western jazz, often generate entirely new soundscapes.

These "border crossings" are possible only when a wide pool of influences remains accessible (Passmore et al., 2024).

AI technologies, in theory, have immense potential to enrich this process. By analyzing huge datasets of sounds, AI can suggest surprising combinations that human creators might overlook (Passmore et al., 2024). Yet, in reality, most commercial AI systems tend to privilege mainstream music because they are optimized for user engagement and profit (Ferwerda, 2016). Instead of guiding listeners toward unexpected discoveries, they often recycle what is already popular. The risk here is that musical diversity could gradually shrink, leaving both listeners and creators in a narrower cultural loop (Passmore et al., 2024).

4. PATHWAYS AND MECHANISMS OF BIAS IN MUSIC AI SYSTEMS

Artificial intelligence is no longer confined to experimental labs it is embedded in the everyday functioning of the music industry. From recommendation engines that decide which songs appear on your playlist, to automated mastering software and even AI driven composition tools, algorithms now influence both production and consumption (Henry et al., 2024; Banerjee et al., 2025). Although these applications are diverse, they share one uncomfortable truth: all of them can reproduce or even amplify bias (Banerjee et al., 2025). Understanding the pathways through which bias emerges is essential if we want to grasp how AI is quietly shaping the future of music (see Table 1).

Table 1. Pathways of bias in music AI

Pathway of Bias	How It Happens in Music AI Systems	What the Bias Looks Like
Data Representation Bias	Training datasets exclude or underrepresent certain genres, languages, cultures	AI struggles with non Western traditions; less stylistic diversity in generated output
Popularity Bias	Collaborative filtering prioritizes what is already popular	"Echo chambers" where mainstream dominates and niche music is sidelined
Genre and Regional Bias	Emphasis on pop/rock from major markets like the US	Regional, indigenous, or small market genres are marginalized
Cultural & Linguistic Gaps	Lack of songs in non English languages or unique cultural forms	English language and Western music dominate recommendations and AI training

4.1 Bias in Recommendation Systems

Perhaps the most visible form of bias emerges in music recommendation engines. Streaming platforms like Spotify, YouTube Music, or Apple Music learn from a listener's behavior and suggest tracks accordingly. The common technique here is collaborative filtering: if other listeners who share your habits enjoy a particular track, the system assumes you might too (Banerjee et al., 2025).

The problem is subtle but powerful. Instead of broadening horizons, the algorithm tends to reinforce existing tastes. Someone who listens mainly to Western pop will be continuously nudged toward similar pop tracks. Over time, this creates what researchers call an echo chamber effect (Henry et al., 2024). A listener rarely stumbles upon unexpected genres such as folk traditions, indigenous music, or experimental jazz. The end result is that the richness of musical choice gradually narrows, even though the platforms themselves host millions of tracks (Banerjee et al., 2025).

4.2 Bias in Music Creation and Production Tools

Bias is not limited to listening habits it seeps into the creative process as well. AI composition tools, like those that generate melodies from text prompts, are often celebrated as democratizing creativity. Yet these systems learn from vast databases of pre existing songs, and those databases are rarely neutral. They are heavily skewed toward mainstream genres that dominate commercial markets pop, rock, EDM while regional or experimental traditions remain underrepresented (Banerjee et al., 2025).

This imbalance produces what some critics call stylistic flattening. Instead of generating bold or surprising sounds, the AI defaults to safe and familiar templates. To many ears, AI produced tracks can start to feel interchangeable, as if they were variations on a handful of formulas. While this may suit commercial demand, it risks eroding the diversity and unpredictability that drive musical innovation (Henry et al., 2024).

4.3 Evidence from Real World Platforms

This is not just theory studies have demonstrated the effects. A large scale investigation into a European streaming platform revealed that American songs performed significantly better on recommendation driven platforms than on those without algorithmic curation (Banerjee et al., 2025). In other words, the system itself amplified US content.

In the UK, public debates reflect similar concerns. Surveys show that both listeners and industry professionals suspect recommendation systems of being biased, even if the exact mechanisms remain hidden (Banerjee et al., 2025). The lack of transparency fuels distrust: people feel that algorithms tilt the playing field, even if companies insist they are neutral (Banerjee et al., 2025).

4.4 Data Representation Bias

A well-known principle in machine learning is "bias in, bias out." If training datasets lack cultural balance, the resulting models will inevitably reflect that skew. Recent studies confirm how stark this imbalance is in music AI. Only about 5.7% of training material in some generation models comes from non-Western genres. Another large dataset analysis found that music from the Global South made up just 14.6% of training hours, compared with a dominant 86% from the Global North.

The outcome is predictable. When asked to generate music with South Asian or Middle Eastern characteristics, current systems often fail to reproduce authentic melodic or rhythmic patterns. They simply do not "know" these traditions. In effect, the global soundscape that AI learns from is overwhelmingly Western and commercial, leaving entire musical cultures invisible in the digital age.

4.5 Genre Bias and Inequalities in Representation

Bias is not only geographical; it can be social and gendered too. One striking example involves recognition systems trained mainly on male dominated genres underrepresenting female artists. The algorithm effectively learned the inequalities already present in its dataset (Banerjee et al., 2025).

This shows that algorithms are not neutral observers they actively perpetuate the distortions they inherit. For female artists, or for musicians working outside dominant genres, the effect is a digital echo of systemic inequities that already exist in the music industry (Henry et al., 2024).

4.6 Popularity Bias and the Blockbuster Effect

Another structural bias is popularity bias. Algorithms tend to promote tracks that are already trending, creating a feedback loop: the popular become more popular, while lesser known tracks sink further into obscurity. This dynamic echoes the long tail theory, which describes how a few blockbuster hits dominate sales while thousands of niche tracks are rarely heard (Banerjee et al., 2025).

In principle, streaming platforms could expose users to the long tail by surfacing hidden gems. In practice, however, the systems often amplify the hits, narrowing discovery rather than widening it. This has been called the "undesired blockbuster effect," a cycle where commercial success crowds out diversity (Banerjee et al., 2025).

4.7 Cultural and Linguistic Underrepresentation

Cultural and linguistic diversity is another casualty of biased recommendation systems. Studies highlight that listeners are far less likely to stream non English songs compared to English ones, even when the music scores equally on measures of emotional impact. The algorithm reinforces this tendency by recommending more English language content, thereby compounding the initial bias (Martin, 2022; Banerjee et al., 2025).

What this reveals is a large scale structural disadvantage for non English music. Instead of helping global traditions find new audiences, the systems prioritize what is already dominant in the market. Over time, this may subtly reshape listening habits, making certain languages and musical forms less visible in the cultural mainstream (Passmore et al., 2024).

5. MANIFESTATIONS OF BIAS IN MUSIC AI OUTPUTS

In the last chapter we looked at how bias gets into music AI systems through data, algorithms, and even cultural assumptions. But bias doesn't just live "behind the scenes" in code. It spills out into the music people actually hear and into the decisions artists make about their work. In this section, We will want to unpack how that bias becomes visible in everyday listening and in the broader music industry.

5.1 Making Listening Preferences All Look the Same

One of the clearest effects of biased recommendation systems is how they flatten people's listening habits. On the surface, the playlists you receive on Spotify or YouTube might feel personal, even unique. Yet the logic driving them is usually the same: keep you engaged by serving up tracks that sound very similar to what you already like (Mokoena & Obagbuwa, 2025).

This creates what some scholars call taste tautology your preferences are fed back to you until they become narrower and narrower. The idea of "reduced musical diversity" might sound abstract, but in practice it means you keep hearing more of the same, even when you think you're exploring. In earlier decades, radio stations also relied on formulas such as repetitive rhythms in mainstream hits. The difference today is that the process is individualized and hidden. Instead of one radio station deciding, every user gets their own customized echo chamber. Over time, this makes it harder to stumble upon the unexpected whether old folk traditions or cutting edge experimental tracks (Mokoena & Obagbuwa, 2025).

5.2 Pushing Minority and Indigenous Music to the Side

Another visible consequence of algorithmic bias is the marginalization of minority and indigenous music. AI models are usually trained on massive datasets, but these datasets overwhelmingly privilege Western genres and mainstream commercial hits. That bias carries through into recommendations and into the music AI systems generate (Mokoena & Obagbuwa, 2025).

A striking example comes from Canada. A study found that a significant portion of albums by Indigenous artists nominated for major awards were not even available on major streaming platforms. Think about that: even when artists are nationally recognized, digital infrastructure makes it difficult for them to reach wider audiences (Mokoena & Obagbuwa, 2025). The same problem appears globally. Traditional forms such as Chinese Qin music or contemporary underground genres like glitch often remain invisible, not because people dislike them, but because the algorithms never surface them. This invisibility risks not only commercial harm but cultural erosion heritage traditions can slip into obscurity when consistently sidelined by platforms prioritizing profitability (Mokoena & Obagbuwa, 2025).

5.3 Music Industry Shifting Toward the Market

Recommendation systems are not neutral cultural tools they are designed with business goals in mind. The central logic is to maximize attention, increase subscriptions, and keep advertising revenue flowing (Mokoena & Obagbuwa, 2025). Because of this, the system rewards music that is already popular, easy to consume, and unlikely to drive listeners away.

The cultural effect is subtle but significant. Political, experimental, or socially critical music often receives less attention because it is riskier commercially. Instead, algorithms amplify songs that are "safe" those familiar, melodic, and easy to categorize. This is not new; the music industry has long leaned toward market driven decisions. What changes with AI is the scale and automation of the process. The filter is now constant and personalized, shaping what millions hear daily. Long term, this can water down music's role as a space for dissent, innovation, and cultural dialogue (Mokoena & Obagbuwa, 2025).

5.4 Financial Impact on Independent and Niche Artists

Perhaps the most tangible impact of bias is financial. When Spotify announced in 2024 that tracks with fewer than 1,000 annual streams would no longer generate royalties, the decision was framed as a way to combat fraudulent streams. But the side effect was enormous: 87% of tracks suddenly became ineligible for payments.

Small or independent artists, this is devastating. Unlike major pop stars who dominate charts, most musicians operate at the margins. Many already struggle to break even. Under the "pro-rata" royalty model, subscription fees disproportionately flow to big-name artists. That means even if you spend your month listening to underground bands, much of your subscription money still supports mainstream hits

6.CASE STUDIES AND IMPACTS OF AI BIAS ON MUSICAL DIVERSITY

So far, we have discussed how algorithmic bias works in theory and outlined the main ways it surfaces in music AI systems. But theory can only take us so far. To see the true cultural and economic impact, it is important to look closely at real world cases. These examples show that bias is not some abstract technical flaw it shapes the music that reaches our ears, the visibility of entire genres, and the creative freedom of artists (Henry et al., 2024).

6.1 Bias in Popular Music Recommendation Platforms

Streaming platforms such as Spotify or Apple Music are often celebrated for offering highly personalized listening experiences. However, case studies suggest that these "personalized" recommendations may, in practice, reinforce cultural hierarchies. A study of a large European music streaming service revealed that its algorithm consistently promoted American songs over local music. The researchers described this as the difference between "preference pull" (what listeners genuinely choose) and "curation push" (what the system promotes). Even when local songs were available, the platform tended to steer listeners toward U.S. back catalog music (Mokoena & Obagbuwa, 2025).

This finding is important because it challenges the assumption that American dominance in music is purely a matter of audience preference. Instead, algorithms quietly act as cultural gatekeepers. The result is a tilted playing field where regional artists have to compete not only with global stars but also with the very system that delivers music to listeners (Mokoena & Obagbuwa, 2025).

6.2 AI Generated Music and Cultural Homogenization

The rise of AI generated music is often presented as an exciting frontier for creativity. Yet beneath the hype lies a serious problem: cultural homogenization. The issue begins with training data. For instance, an analysis of popular music generating models showed that only a small fraction of their training data represented non Western traditions (Bryan-Kinns, 2025).

The consequence is predictable but troubling. When asked to generate music in Hindustani Classical or Turkish Makam, the models struggled to reproduce the distinct melodic and rhythmic structures that define those traditions. Instead, they fell back on generic Western pop conventions. This creates a feedback loop: if the AI mainly generates Western style music, those sounds dominate the digital ecosystem, further marginalizing already under represented traditions (Bryan-Kinns, 2025).

6.3 Regional Music Platforms and Local Solutions

Not all responses to bias come from large tech companies. Some of the most promising solutions are local and community driven. In Canada, the Indigenous Music Office (IMO) has taken proactive steps to ensure Indigenous musicians have visibility in the digital space. Instead of depending on global algorithms optimized for profit, the IMO invests in community engagement, cultural preservation, and authentic representation (Henry et al., 2024).

This approach illustrates an important lesson: addressing bias is not only about tweaking global recommendation systems. Sometimes the most effective strategy is to build parallel platforms or support structures that prioritize cultural value over algorithmic efficiency (Henry et al., 2024).

6.4 Niche and Minority Genres Getting Less Exposure

A recurring theme across studies is how niche or minority genres are disadvantaged. Recommendation engines rely heavily on popularity metrics, which naturally disadvantage styles with smaller audiences (Mokoena & Obagbuwa, 2025). For example, genres like Chinese Qin or experimental subcultures such as algorithmic music rarely appear in training datasets (Bryan-Kinns, 2025). As a result, these genres are seldom recommended to listeners, regardless of their artistic merit.

The effect is double edged: audiences lose opportunities to encounter unfamiliar sounds, while artists in these genres face economic precarity. Without exposure, even vibrant traditions risk becoming invisible in a market where visibility often determines survival (Bryan-Kinns, 2025).

6.5 Stifling Cross Cultural Musical Exchange

One of the joys of music is its capacity for cultural exchange jazz drawing on African rhythms, electronic music shaped by Caribbean sound systems, or contemporary pop borrowing from Latin reggaetón. Yet algorithmic personalization risks stifling this kind of cross pollination.

By building "filter bubbles" that feed users more of what they already know, AI systems reduce the likelihood of accidental discovery. Instead of broadening horizons, they shrink them. This is not just a loss for individual listeners; it threatens the dynamic interplay of cultures that has historically driven musical innovation (Mokoena & Obagbuwa, 2025).

6.6 Reinforcing Mainstream Dominance

A striking example of algorithmic bias comes from the case of AI created bands that gain significant traction largely through algorithmic promotion. Many listeners remain unaware that these acts are not human. This case highlights how opaque recommendation systems can artificially boost certain acts while sidelining others (Ansani et al., 2025).

In contrast, real independent bands report struggling for visibility. Some even pull their catalog from streaming platforms after their music was consistently suppressed by algorithmic ranking. The juxtaposition is telling: synthetic, commercially optimized acts gain traction, while authentic artists lose ground. This suggests algorithms are not neutral intermediaries but active agents shaping the market in favor of certain kinds of content (Ansani et al., 2025).

6.7 Impact on Emerging Artists and Independent Music

Emerging and independent musicians often feel the weight of algorithmic bias most acutely. For them, visibility on streaming platforms is not just a bonus it is the difference between being heard and disappearing into obscurity (Ansani et al., 2025). Several artists have described feeling compelled to design songs around algorithmic logic shorter intros, early hooks, and tracks calibrated for playlist placement.

This shift represents more than just a change in marketing strategy. It suggests that AI is influencing the very form of artistic creation. Instead of following the natural flow of an album or pursuing creative risk, musicians feel pressured to conform to the statistical preferences of a machine. In this sense, bias is not only about what music circulates but also about what music gets made in the first place (Mokoena & Obagbuwa, 2025).

7.ETHICAL AND SOCIETAL CONSIDERATIONS

The issue of bias in music AI is not just about programming errors or commercial inefficiencies. At its core, it raises questions about culture, fairness, and the role of human creativity in a digital age. Music has always been more than background noise it expresses identity, preserves history, and brings communities together. When algorithms, built on uneven data and designed with commercial goals in mind, shape what we hear, they are doing more than curating playlists. They are subtly reshaping cultural landscapes. This section reflects on some of the ethical and societal challenges that come with this shift (Henry et al., 2024).

7.1 Cultural Hegemony and Algorithmic Gatekeeping

Algorithms have quietly become cultural gatekeepers. Scholars sometimes use the term "algorithmic gaze" to capture this effect. Just as a camera lens frames what enters our field of vision, algorithms decide which sounds reach our ears. The problem is that this framing often favors the dominant mainstream genres, Western pop, English lyrics while pushing less familiar traditions to the side (Henry et al., 2024).

This dynamic is troubling because music has historically been a space for resistance and diversity. Blues, reggae, punk, and hip hop all emerged as voices from the margins. If algorithms consistently amplify the mainstream while muting Indigenous or minority genres, they flatten music's cultural role. Rather than opening doors, AI risks closing them, silencing the very communities that have long relied on music for expression and survival (Mokoena & Obagbuwa, 2025).

7.2 Representation and Equity in Music Distribution

At its heart, bias in music AI is a matter of representation. A fair system should ensure that artists, no matter their background, gender, or location, have a genuine chance of being heard. Yet, unless carefully designed, algorithms often repeat existing inequalities.

For instance, algorithmic recommendation systems have shown to reflect industry wide imbalances in gender representation, requiring deliberate corrective techniques to promote fairness. Fairness does not happen automatically; it requires deliberate effort from developers and platforms. Without these interventions, AI tools risk entrenching the same social biases that have long haunted the music industry (Henry et al., 2024).

7.3 Intellectual Property Concerns in Biased AI Music Generation

Generative AI raises new legal and ethical complexities regarding ownership and copyright. Many training datasets include copyrighted songs without explicit consent. Industry bodies have launched lawsuits against AI music startups for alleged large scale infringement.

Beyond datasets, the question of copyright for AI generated music continues to be debated, especially when AI tools collaborate with human artists. Grey areas in ownership leave many creators vulnerable, especially independents who lack resources to fight for recognition (Mokoena & Obagbuwa, 2025).

7.4 Policy and Regulatory Frameworks

Globally, governments are beginning to confront these challenges. The European Union's AI Act introduces a risk based framework demanding stricter data quality, transparency, and human oversight. Other regions, like the United States, approach issues more fragmentedly, with legal disputes often playing out in courts. Projects funded by the EU work on fairness metrics and policy recommendations for greater inclusivity in music streaming (Henry et al., 2024; Mokoena & Obagbuwa, 2025).

7.5 Cultural Preservation vs. Commercial Optimization

A fundamental tension exists between cultural preservation and profit maximization. Streaming platforms prioritize catchy, mainstream content that maximizes engagement, often sidelining culturally significant but less commercially viable music.

For example, Spotify's policy change in 2024 excluding tracks with fewer than 1,000 annual streams from royalties disproportionately affected independent and niche artists. This reflects how commercial optimization can directly undermine cultural diversity (Henry et al., 2024).

7.6 AI and the Risk of Cultural Monopolies

Because music distribution is concentrated among a few platforms, there is a real risk of cultural monopolies. When a few algorithms define what is "relevant," entire traditions risk exclusion. Artists feel pressured to conform to algorithmic moulds rather than innovate. AI is thus not neutral but actively shapes global music markets, favouring efficiency and profit over diversity and creativity (Mokoena & Obagbuwa, 2025).

7.7 Intellectual Property and Fair Recognition

Studies in visual art AI provide cautionary examples: as AI produced art increases, human made works face market decline. Similar trends may follow in music.

Artists whose recordings train AI models may find AI produced works competing against them without credit or compensation, commodifying human creativity and eroding livelihoods (Henry et al., 2024).

7.8 Ethical AI in Music Creation and Curation

Ethical frameworks propose using AI as collaboration tools rather than replacements, under principles of transparency, inclusivity, and respect for human creativity. Adoption of such principles could ensure AI supports .

8. DETECTION AND MITIGATION STRATEGIES

Bias in music AI cannot be solved with a quick patch of code. It runs deeper than that. Algorithms don't exist in isolation they reflect choices made during data collection, design, and deployment. This means any serious attempt to address bias has to stretch across the entire lifecycle of a system. At the same time, technical tools alone will not be enough. We need human judgment, thoughtful policies, and perhaps a shift in the way the music industry thinks about its responsibility to culture (Henry et al., 2024).

8.1 Fairness Metrics and Bias Detection in Music AI

A saying in data science is "what gets measured, gets managed." When it comes to bias in music AI, the challenge is that the usual metrics such as accuracy or click through rate are poor indicators of fairness. An algorithm can be very "accurate" in predicting what people will play, yet still ignore entire communities of artists (Kartal et al., 2022).

Because of this, new fairness oriented tools are emerging. For instance, the Fair MusE project (funded by the European Union) is building a Fairness Score to evaluate whether recommendation systems treat artists and genres equitably (Henry et al., 2024). Another approach is propensity score estimation, which calculates the likelihood a system favours certain groups for example, male over female artists or Western over indigenous genres. The goal is not only to measure imbalance but provide feedback early enough for developer intervention (Cyanite.ai, 2025).

8.2 Debiasing Techniques in Recommendation Algorithms

Once bias is detected, how do we fix it? One strategy is re ranking, where underrepresented genres or artists are given more visibility in recommendations (Henry et al., 2024). Another approach is embedding fairness in model training through regularization, nudging balanced outcomes. Post hoc adjustments also exist but are tricky, as they correct user data already shaped by biases (Kartal et al., 2022).

The challenge is balancing user preference satisfaction with diversity promotion a balance still actively researched.

8.3 Role of Human Curators and Hybrid Systems

Algorithms lack cultural intuition. Human curators remain essential, able to recognize the cultural significance behind music beyond technical features (Mokoena & Obagbuwa, 2025). Combining algorithms with expert judgment in Human in the Loop systems enables oversight of biases algorithms miss (Kartal et al., 2022). This revives traditional intermediary roles in music culture.

8.4 Transparency and Accountability Measures

Trust is critical. Many users do not understand why a system recommends particular tracks. New policies like the EU AI Act require platforms to disclose AI driven recommendations and explain decisions (European Union, 2025; Henry et al., 2024). Transparency can expose biases for user awareness and enable artists to contest exclusion.

8.5 Diverse and Representative Training Data

"Garbage in, garbage out" applies here: limited training datasets skew outputs. Building diverse datasets with underrepresented music like Indigenous or niche subcultures improves fairness and preserves cultural heritage (Bryan-Kinns, 2025; Kumar, 2025). Without efforts, traditions risk digital erasure.

8.6 Transparent and Explainable Recommendation Systems

Explainable AI (XAI) shows users why recommendations appear, e.g., "This song shares rhythm patterns with those you like." Such explanations expose biases to users, encouraging exploration beyond defaults (Henry et al., 2024).

8.7 User Centric Customization and Diversity Promoting Algorithms

Recommendation systems designed not only for profit but to broaden horizons empower users to set priorities, e.g., "Show more independent artists," or "Balance my playlist with diverse genres." This philosophy redefines music AI as a cultural guide fostering connection and diversity (Henry et al., 2024).

9. EMERGING TRENDS AND RESEARCH DIRECTIONS

The conversation around AI and bias in music often feels discouraging, as though the technology is destined to reproduce existing inequalities. Yet the picture is not entirely bleak. In fact, several emerging trends suggest that things could move in a more hopeful direction. The key point is that the future of music AI is not predetermined. It will be shaped by the choices of today's developers, artists, policymakers, and industry leaders (Bryan-Kinns, 2025).

9.1 AI as a Tool to Promote Musical Diversity

One of the most interesting shifts in thinking is to treat AI not merely as a mirror of existing tastes, but as an active agent for cultural diversity. Instead of reinforcing what is already popular, recommendation systems could be designed to act almost like cultural guides pushing listeners toward new voices, unfamiliar genres, and music from parts of the world they may never have encountered (Henry et al., 2024; Mokoena & Obagbuwa, 2025).

Imagine a listener who usually plays mainstream pop being gently nudged toward a Congolese rumba track, or a jazz enthusiast suddenly discovering experimental Mongolian throat singing. These deliberate suggestions broaden horizons and enrich the listener's cultural palette. Music has always thrived on such unexpected encounters (Mokoena & Obagbuwa, 2025).

9.2 Multi Cultural and Multi Lingual AI Models

A recurring problem is that many AI systems train primarily on Western music archives. Models often stumble when handling traditions like Hindustani ragas,

Turkish Makam, or indigenous ceremonial songs. This is more about training data imbalance than algorithmic sophistication (Bryan-Kinns, 2025; Kumar, 2025).

Recent work using fine tuning techniques to adapt large models on low resource traditions shows promise for culturally sensitive, inclusive systems that value musical diversity (Kumar, 2025).

9.3 OpenSource Initiatives and Community Led Development

Another hopeful trend is opensource AI for music. Instead of corporate controlled datasets, communities and artists curate their own collections and models, democratizing the field (Bryan-Kinns, 2025).

For example, Indigenous communities curate digital archives to train AI systems that authentically represent their cultural voice. Opensource invites collaboration across local and global scenes, resisting homogenization and empowering creatives (Bryan-Kinns, 2025).

9.4 Interdisciplinary Approaches

There is growing recognition that AI bias in music cannot be solved by technologists alone. Music embodies culture, history, and politics, requiring input from sociologists, legal scholars, musicologists, and artists.

The Fair MusE project exemplifies this coalition, combining technical solutions with policy and cultural strategies to create inclusive music AI ecosystems (Henry et al., 2024; Bryan-Kinns, 2025).

10.CONCLUSION

10.1 Summary of Key Findings

Artificial intelligence has transformed the music industry, bringing opportunities for innovation and discovery but also introducing systemic biases with far reaching impacts. These biases are rooted in skewed training data, inherent model designs, and embedded societal prejudices. The outcomes include homogenized listener preferences, marginalization of minority and indigenous music, and reinforcement of commercial dominance. The economic consequences are significant, as many independent artists face reduced visibility and revenue. Case studies such as the overrepresentation of US content and struggles with AI generated music quality across diverse genres highlight that algorithms actively perpetuate and amplify existing power imbalances rather than merely reflecting them (Bryan-Kinns, 2025; Henry et al., 2024; Mokoena & Obagbuwa, 2025; Ansani et al., 2025).

10.2 Policy and Industry Implications

Addressing these issues demands strong action from both policymakers and industry stakeholders. Treating AI as neutral is no longer sufficient. Platforms should implement fairness metrics focused on equitable representation and adopt hybrid systems combining AI and human curation to balance efficiency with cultural expertise. Transparency about algorithmic processes and data sources is essential to build trust and accountability (Henry et al., 2024; Greyson, 2025; Mokoena & Obagbuwa, 2025).

10.3 Call for Interdisciplinary Collaboration

Solving AI bias in music requires collaboration across technology, arts, policy, and academia. Technologists can enhance algorithmic fairness, artists can provide crucial insights, policymakers set regulatory frameworks, and researchers offer evidence based guidance such as The Fair MusE initiative in Europe exemplifies such multidisciplinary cooperation, aiming to foster a more equitable music ecosystem (Henry et al., 2024; Greyson, 2025).

10.4 Final Reflections

The trajectory of AI in music is at a crucial juncture. Without intervention, the risk of cultural homogenization and reduction of diversity is high. Yet, with intentional design prioritizing transparency, fairness, and inclusivity, AI can amplify the richness of global musical heritage. Responsibility for this lies not just with developers but with all participants in the music ecosystem artists, listeners, policymakers, and industry leaders (Bryan-Kinns, 2025; Henry et al., 2024; Cheng, 2023; Mokoena & Obagbuwa, 2025).

REFERENCES

Ansani, A., Koehler, F., Giombini, L., Hämäläinen, M., Meng, C., Marini, M., & Saarikallio, S. (2025). AI performer bias: Listeners like music less when they think it was performed by an AI. *Empirical Studies of the Arts*, *43*(2), 1137–1161. Advance online publication. DOI: 10.1177/02762374241308807

Banerjee, S. D., Datta, D. R., & Mete, P. D. J. (2025). Investigating the impact of artificial intelligence on music creation and consumption in the 21st century. *ET Journal*, *10*(6). Advance online publication. https://everant.org/index.php/etj/article/download/2021/1453/5652. DOI: 10.47191/etj/v10i06.21

Bryan-Kinns, N. (2025). AI generated music and cultural homogenization in digital platforms. *Journal of Music Technology*, *12*(1).

Bryan-Kinns, N. (2025). AI and cultural preservation in music: Challenges and opportunities. *Journal of Music Technology*, *12*(3).

Chen, Y., Huang, L., & Gou, T. (2024). Applications and advances of artificial intelligence in music generation: A review. *arXiv*. https://arxiv.org/html/2409.03715v1

Cheng, L. (2023). The impact of generative AI on school music education. *Music Education Research*, *25*(3), 321 334. https://doi.org/DOI: 10.1080/10632913.2025.2451373

Civit, M., Cervino, R., & De La Hera, T. (2022). A systematic review of artificial intelligence based music generation research. *Expert Systems with Applications*, *194*, 116421. https://www.sciencedirect.com/science/article/pii/S0957417422013537

Dinnissen, K., & Bauer, C. (2022). Fairness in music recommender systems: A stakeholder perspective. *PLoS One*, *17*(7), e0269881. https://pmc.ncbi.nlm.nih.gov/articles/PMC9353048/

Ferwerda, B. (2016). Investigating the relationship between diversity in music and cultural aspects. Proceedings of the 17th International Society for Music Information Retrieval Conference. https://www.cp.jku.at/people/schedl/Research/Publications/pdf/ferwerda_soap_2016.pdf

Henry, A., Wiratama, V., Afilipoaie, A., Ranaivoson, H., & Arrivé, E. (2024). Impacts of AI on music consumption and fairness. *SAGE Open*, *2*(3), 382–396. Advance online publication. DOI: 10.1177/27523543241269047

Kartal, E., et al. (2022). A comprehensive study on bias in artificial intelligence identification and management. *Artificial Intelligence Review, 55*(4), 3677 3712. https://doi.org/DOI: 10.1007/s10462 021 09999 8

Kumar, S. (2025). Advances in multicultural and multilingual AI for music: Challenges and opportunities. *Digital Music Research Review, 8*(1).

Liu, X., Wei, Z., & Tang, J. (2024). Music recommendation algorithms based on knowledge graphs. *Scientific Reports, 14*, 12314. https://www.nature.com/articles/s41598 024 52463 z

Mokoena, N., & Obagbuwa, I. C. (2025). An analysis of artificial intelligence automation in digital music streaming platforms. *Frontiers in Artificial Intelligence, 7*, 1515716. DOI: 10.3389/frai.2024.1515716 PMID: 39881884

Passmore, S.. (2024). Global musical diversity is largely independent of linguistic and genetic histories. *PLoS One, 19*(5), e0273518. https://pmc.ncbi.nlm.nih.gov/articles/PMC11087526/ PMID: 38729968

Pricop, T. C., Toma, C., & Pricop Velin, P. (2024). Music generation with machine learning and deep neural networks. *Procedia Computer Science, 234*, 435 443. https://www.sciencedirect.com/science/article/pii/S1877050924027522

Yesiler, F., ... (2022). Assessing algorithmic biases for musical version identification. Proceedings of the 22nd International Society for Music Information Retrieval Conference. https://dl.acm.org/doi/abs/10.1145/3488560.3498397

Zhang, M., Li, S., & Chen, R. (2025). Advancing deep learning for expressive music composition. *Scientific Reports, 15*. https://www.nature.com/articles/s41598 025 13064 6

KEY TERMS AND DEFINITIONS

Algorithmic Bias: Distortion that occurs when systematic errors in artificial intelligence systems produce outcomes that unfairly favor or disadvantage certain groups .

Cultural Homogenization: The gradual reduction of diversity in music when global algorithms prioritize mainstream content, making different musical traditions sound increasingly similar.

Data Representation Bias: A type of bias denotes distortions in data presentation that mislead interpretations. Arising when certain genres, languages, or cultural traditions are underrepresented in the datasets used to train AI systems which may lead to skewed or limited outcomes.

Ethical AI in Music: The design and use of artificial intelligence in a way that upholds fairness, transparency, cultural sensitivity, and respect for human creativity within the music industry.

Explainable AI (XAI): A branch of artificial intelligence focused on creating systems that can transparently explain how decisions or recommendations are made, fostering trust and user understanding.

Fairness Metrics: Quantitative measures developed to evaluate whether AI systems treat artists, genres, or user groups equitably, often used in research on ethical AI and algorithmic accountability.

Musical Diversity: The range and balance of genres, styles, and cultural expressions within a musical ecosystem, reflecting how inclusive and representative a listening environment or dataset is.

Popularity Bias: A feedback effect in which algorithms repeatedly promote songs or artists that already have high engagement, reducing the visibility of lesser known or niche musicians.

Recommendation System: Recommendation systems, often known as recommender systems, are a type of information filtering system that attempts to forecast the "rating" or "preference" that a user would assign to an item. By evaluate user data to utilize this information for recommending goods that are likely to interest the user.

Chapter 5
Markov-Chain Melody:
A Mathematical Approach to AI-Assisted Songwriting

Syrenzo Ramos Sicuan
http://orcid.org/0000-0002-2034-990X
Pangasinan State University, Philippines

ABSTRACT

This study examined whether simple probabilistic models can generate coherent melodic fragments for AI-assisted songwriting. The work focused on building first- and second-order Markov chains, analyzing the statistical properties of their outputs, and comparing listener perceptions with human-composed hooks. A dataset of 20 popular MIDI melodies was standardized to C major and used to estimate transition probabilities, from which 30 sequences of 8–16 notes were generated. Structural measures showed that human melodies had higher entropy (2.29 vs. 1.80) and greater pitch diversity (6.37 vs. 5.92), while Markov outputs had higher repetition (0.43 vs. 0.34). A perceptual survey with 30 listeners rated human melodies higher in catchiness (4.19 vs. 3.78), musicality (4.33 vs. 3.63), and preference (3.98 vs. 3.68), though Markov sequences were still viewed as acceptable. Findings show that Markov chains capture short-term tendencies and can generate usable motifs but lack long-range structure, making them useful for low-resource music tasks.

INTRODUCTION

Artificial intelligence (AI) has rapidly transformed the landscape of creative production, particularly in music composition and songwriting. Once confined to symbolic rule-based systems, algorithmic composition has evolved through advances in machine learning, enabling AI to act not only as a compositional tool but as a

DOI: 10.4018/979-8-3373-6279-3.ch005

collaborative partner in the creative process. Within this domain, AI-assisted songwriting highlights a shift from autonomous generation toward co-creation, where computational systems provide melodic or harmonic ideas that human composers refine into musical pieces. Despite the rise of complex neural architectures such as generative adversarial networks (GANs) and transformers, simpler stochastic approaches like the Markov chain continue to play an essential role in modeling the statistical structure of music. Their interpretability and controllability make them ideal for exploring how algorithmic processes can augment, rather than replace, human creativity.

The origins of computer-based composition can be traced to early experiments such as The ILLiac Suite (1957), which used Markov chains to generate musical sequences. While these early models were limited by computational resources, their conceptual foundation-representing melody as a probabilistic process-remains influential. In modern AI systems, probabilistic models serve as a fundamental building block for understanding musical expectation and stylistic imitation (Micchi et al., 2021). As commercial AI systems like MuseNet, AIVA, and MusicLM demonstrate increasing sophistication, researchers have revisited Markov models not for their computational novelty but for their transparency, a quality often absent in deep learning systems.

Although AI systems capable of autonomous composition exist, few provide songwriters with interpretable tools that integrate probabilistic modeling and human creativity. Deep generative networks often function as black boxes requiring extensive datasets, while simpler statistical models, though accessible, are rarely evaluated for creative utility in songwriting contexts (Deruty et al., 2022; Yin et al., 2023). This study addresses that gap by implementing and evaluating first- and second-order Markov-chain models as a basis for melody generation in AI-assisted songwriting.

The study pursued three primary objectives:

1. To construct and implement first- and second-order Markov chains using a corpus of popular MIDI melodies, modeling the probabilistic relationships between notes and visualizing these through transition probability matrices;
2. To analyze and compare the statistical properties of Markov-generated and human-composed melodies using quantitative metrics-Shannon entropy, repetition index, and pitch diversity-as indicators of unpredictability, motif recurrence, and tonal range; and
3. To assess listener perceptions of human and AI-generated melodies in terms of catchiness, musicality, and overall preference, thereby gauging the perceptual validity of probabilistic modeling in creative composition.

In the first phase, the study constructs transition probability matrices to examine how statistical regularities-such as stepwise motion and interval reversals-emerge from the melodic corpus. The concentration of probabilities along the diagonal, as shown in the transition heatmap, demonstrates the model's ability to encode tonal smoothness and conjunct movement, consistent with findings from cognitive musicology (Nakamura et al., 2020).

The second phase evaluates the statistical behavior of generated melodies, revealing how probabilistic generation differs from human composition. Human melodies in the dataset exhibited higher entropy and pitch diversity, suggesting greater unpredictability and expressive range, while Markov outputs favored higher repetition indices, producing locally coherent yet less varied sequences (Huang & Huang, 2022; Corrêa et al., 2020). This quantifies the trade-off between structure and novelty inherent in stochastic modeling.

Finally, the third phase assesses listener perception, translating computational differences into aesthetic judgment. Empirical results show that human melodies consistently outperformed AI-generated ones in catchiness, musicality, and preference, though the latter were still rated as moderately engaging. This indicates that probabilistic models capture sufficient local coherence to be perceived as musical but lack the long-term organization and expressive nuance associated with human creativity (Francisco Tigre Moura & Maw, 2021; Hernandez-Olivan & Beltrán, 2023).

For composers and educators, this research offers an accessible entry point into algorithmic composition, bridging the gap between theory and practice. For AI researchers, it provides empirical evidence that interpretable models retain value in creative domains where transparency and user control are as crucial as novelty.

In summary, this paper investigates how probabilistic modeling through Markov chains can inform AI-assisted songwriting. By integrating algorithmic construction, statistical comparison, and listener evaluation, the study elucidates both the strengths and constraints of Markov processes in musical creativity. It argues that the future of computational composition lies not in fully autonomous systems, but in human-AI partnerships that combine the rigor of mathematics with the intuition of art.

REVIEW OF RELATED LITERATURE

Probabilistic Models in Music Generation

Probabilistic modeling has long been central to symbolic music generation and music information retrieval (MIR). Markov chains, in particular, model musical sequences as stochastic processes in which the probability of a given note depends on prior states. Corrêa et al. (2020) examined the generalization capacity of Markov models for melody prediction, demonstrating that model order directly affects contextual sensitivity and predictive performance. While higher-order models capture richer local dependencies, they also increase sparsity and risk overfitting when trained on limited datasets. These findings underscore the practical importance of selecting appropriate model order in small-corpus creative settings.

Similarly, statistical learning approaches applied to symbolic music have shown that probabilistic modeling can encode stylistic regularities without explicit rule systems (Nakamura et al., 2020). Such methods reveal that melodic coherence can emerge from frequency-based transition structures. The interpretability of transition matrices enables direct inspection of learned musical tendencies, offering transparency not typically available in deep neural architectures.

Statistical Evaluation and Algorithmic Comparison

Recent research has shifted focus from generation alone to systematic evaluation of algorithmic outputs. Yin et al. (2023) conducted comparative evaluations of automatic music generation algorithms, emphasizing that performance should be assessed using both structural metrics and perceptual criteria. Their findings suggest that algorithmic outputs often exhibit measurable statistical differences from human compositions, particularly in terms of repetition and originality. Complementing this, Yin et al. (2022) proposed quantitative tools for assessing originality and structural copying in generated music, reinforcing the importance of measurable evaluation frameworks.

These studies highlight that structural metrics—such as entropy and repetition frequency—provide objective insights into generative behavior. However, they also demonstrate that statistical adequacy does not necessarily equate to perceived musicality, reinforcing the need for combined computational and perceptual analysis.

Deep Learning and Hybrid Systems in Contemporary Music AI

The rapid expansion of deep learning in music generation has introduced models capable of capturing long-range structure and expressive nuance. Hernandez-Olivan and Beltrán (2023) reviewed deep learning techniques in music composition, noting their capacity for modeling hierarchical dependencies and stylistic complexity. Wang et al. (2024) further summarized intelligent music generation systems, emphasizing the growing sophistication of neural architectures and their application across symbolic and audio domains.

Reinforcement learning approaches, such as the harmonization framework proposed by Ji et al. (2024), integrate contextual feedback mechanisms to refine melodic progression. Similarly, emotion-driven generative systems incorporate user interaction and affective conditioning to enhance expressive quality (Kim, 2024). While these methods expand generative capacity, they often increase model complexity and reduce interpretability.

Hybrid perspectives therefore advocate balancing expressive power with transparency. Tan and Li (2021) emphasize that AI music systems benefit from controllable design principles, especially in educational and creative-assistive contexts. Deruty et al. (2022) similarly discuss the role of AI in contemporary music production as collaborative augmentation rather than replacement of human creativity.

Human Perception and Co-Creative Systems

Listener perception remains central to evaluating algorithmic music. Francisco Tigre Moura and Maw (2021) found that audiences can distinguish between human and AI-composed music, often associating human works with greater emotional depth and expressive coherence. Suh and Youngblom (2021) describe AI as a "social glue" in collaborative music creation, emphasizing that co-creative systems function best when computational outputs remain interpretable and modifiable by human users.

Together, the literature suggests that probabilistic models occupy a distinct niche within AI-assisted composition. While they may lack hierarchical depth compared to neural architectures, their mathematical clarity and controllability make them valuable in human-centered creative workflows. This study builds upon these insights by empirically evaluating first- and second-order Markov chains through both structural metrics and perceptual assessment.

Model Order and Contextual Depth in Markov Music Modeling

One of the central design decisions in probabilistic sequence modeling concerns the order of the Markov process. In first-order models, each state depends only on the immediately preceding note, whereas higher-order models incorporate longer contextual histories. Corrêa et al. (2020) demonstrated that increasing model order enhances local predictive capacity but introduces data sparsity challenges, particularly in small corpora. When the number of possible state combinations increases exponentially, transition matrices become less stable unless supported by sufficiently large training datasets.

In symbolic music modeling, this tradeoff has practical implications. First-order models favor robustness and interpretability, as transition probabilities remain well-populated even with limited training data. However, they fail to capture phrase-level dependencies or recurring contour structures that span multiple notes. Second-order models partially address this limitation by conditioning transitions on two preceding states, thereby increasing contextual awareness while retaining manageable computational complexity.

This balance between contextual depth and data sufficiency is particularly relevant in creative-assistive systems, where interpretability and responsiveness are often prioritized over maximal structural complexity. As Yin et al. (2023) note, increasing algorithmic sophistication does not necessarily translate into perceptual superiority. In some contexts, simpler models may provide more controllable and editable outputs, especially when human composers remain in the loop. Thus, model order selection should be understood not merely as a technical decision but as a design choice reflecting the intended creative workflow.

Information-Theoretic Perspectives: Entropy and Musical Expectation

Entropy provides a quantitative measure of unpredictability within sequential systems and is widely used in evaluating generative models. In music, entropy reflects the distribution of transition probabilities between pitch events. A high-entropy sequence suggests greater variability and reduced predictability, while lower entropy indicates concentrated probability mass around frequent transitions.

Corrêa et al. (2020) emphasize that entropy-based metrics allow objective comparison between generated and reference corpora, offering insight into how closely algorithmic outputs approximate stylistic norms. When applied to melodic modeling, entropy becomes a proxy for structural variability. However, unpredictability alone does not guarantee musical coherence. As Nakamura et al. (2020) suggest in symbolic learning contexts, musical structure emerges not only from variability but from patterned expectation grounded in statistical regularities.

In listener perception research, statistical predictability interacts with aesthetic evaluation. Francisco Tigre Moura and Maw (2021) report that audiences often attribute expressive depth to music that balances familiarity with novelty. Excessive predictability may reduce engagement, while excessive randomness undermines coherence. Therefore, entropy functions not simply as a mathematical metric but as an indicator of how generative systems navigate the tension between repetition and variation.

Within Markov-based modeling, entropy is directly shaped by transition concentration. First-order models typically yield lower entropy due to reinforcement of dominant transitions. Higher-order models may increase entropy locally but risk overfitting specific corpus fragments. Consequently, entropy analysis provides both diagnostic and comparative insight into algorithmic design.

Interpretability Versus Neural Opacity in AI Music Systems

The rapid development of deep learning architectures has expanded the expressive capabilities of AI-generated music. Hernandez-Olivan and Beltrán (2023) highlight how recurrent neural networks, variational autoencoders, and transformer-based systems capture long-range dependencies and hierarchical musical structure. Similarly, Wang et al. (2024) review intelligent music generation systems capable of modeling both symbolic and audio domains with increasing complexity.

Despite these advances, neural architectures often function as opaque systems in which internal representations are difficult to interpret. Transition probabilities, weight matrices, and latent embeddings do not readily translate into musically intuitive explanations. Deruty et al. (2022) argue that within professional music production contexts, interpretability and controllability are essential for meaningful integration of AI tools. Producers require systems that can be adjusted, constrained, and understood within established creative workflows.

In contrast, Markov models provide explicit transition matrices that can be inspected and modified directly. Each probability corresponds to a concrete relationship between musical events. This transparency allows composers to influence stylistic tendencies by adjusting corpus composition or transition weighting. Tan and Li (2021) note that controllable AI systems are particularly valuable in educational environments, where algorithmic processes must be demonstrable and comprehensible.

The distinction between interpretability and expressive capacity therefore represents a key tension in AI-assisted composition. While neural models excel in modeling hierarchical complexity, probabilistic models offer clarity and editability. Rather than viewing these approaches as competing paradigms, contemporary research increasingly frames them as complementary components within hybrid systems (Ji et al., 2024). The present study situates Markov chains within this broader landscape, examining how transparent probabilistic modeling performs when evaluated through both structural metrics and listener perception.

Structural Metrics and Perceptual Validation

Beyond generation itself, recent scholarship emphasizes the importance of evaluation frameworks that combine quantitative metrics with perceptual analysis. Yin et al. (2022) propose systematic measures for assessing originality and structural similarity in algorithmic music. Their work underscores that algorithmic adequacy cannot be determined solely by subjective listening; measurable structural features must be considered.

At the same time, purely statistical alignment with a corpus does not guarantee listener acceptance. Yin et al. (2023) demonstrate that deep learning systems may achieve structural similarity while still producing perceptually distinguishable outputs. Listener-based evaluation thus functions as a necessary complement to statistical comparison.

Emotion-driven and context-conditioned systems further illustrate the interaction between structural design and perceptual outcome. Kim (2024) shows that user feedback mechanisms can shape generative outputs toward affective goals, highlighting the importance of integrating human evaluation into algorithmic development.

Taken together, the literature indicates that effective assessment of generative music systems requires triangulation across structural metrics, computational modeling, and listener perception. This integrated framework provides the foundation for the present study's evaluation design.

Table 1. Comparison of generative modeling approaches in contemporary music research

Study	Model Type	Context Length	Primary Strength	Primary Limitation	Evaluation Focus
Corrêa et al. (2020)	Markov Models	Variable order	Interpretable statistical transitions	Higher-order sparsity	Predictive generalization capacity
Nakamura et al. (2020)	Statistical Learning	Short-term symbolic	Encodes stylistic regularities	Limited long-range structure	Symbolic sequence modeling
Yin et al. (2023)	Deep Learning	Long-range	Captures complex dependencies	Reduced transparency	Structural & perceptual comparison
Ji et al. (2024)	Reinforcement Learning + LSTM	Context-aware harmonization	Adaptive melodic conditioning	High computational complexity	Context-conditioned optimization
Kim (2024)	Emotion-Driven AI	Context-conditioned	Expressive feedback modeling	Requires interactive training	User-driven perceptual response
Present Study	1st & 2nd Order Markov	Short-term	Transparent, controllable melody modeling	Limited hierarchical form	Structural metrics + listener perception

METHODOLOGY

Research Design

This study employed a quantitative–experimental research design integrating computational modeling and perceptual evaluation. The design was divided into three major phases: (a) construction and implementation of first- and second-order Markov-chain models for melodic sequence generation, (b) statistical comparison of human-composed and AI-generated melodies, and (c) perceptual evaluation of musical quality by listeners. The use of a quantitative approach allowed the systematic measurement of probabilistic transitions, entropy values, and listener ratings, ensuring that both computational and human data could be analyzed using objective metrics.

The experimental phase involved generating melodic sequences using trained first- and second-order Markov models and comparing their structural and perceptual characteristics with a reference corpus of human-composed melodies. Each generated fragment was treated as an evaluative stimulus, enabling quantitative analysis of statistical properties and listener-based assessment of aesthetic response. This dual evaluation approach reflects computational creativity frameworks, which assess generative systems not solely on output artifacts but also on measurable structural features and human reception (Yin et al., 2023; Hernandez-Olivan & Beltrán, 2023).

Corpus and Data Preparation

The melodic corpus used for training the Markov models consisted of 20 popular melodies in MIDI format, selected for their accessibility, tonal clarity, and consistent rhythmic structure. Each piece was transposed to C major and standardized to a monophonic texture to ensure uniform analysis across the dataset. The preprocessing stage included the removal of embellishments such as grace notes and ornaments, quantization of rhythmic values, and conversion of note sequences into symbolic pitch representations (e.g., C4, D4, E4). This procedure ensured that each melody could be represented as a discrete time-ordered sequence suitable for probabilistic modeling.

For the evaluation corpus, a parallel dataset of human-composed melodies was used as the benchmark for comparison. These melodies were selected from publicly available sources and conformed to the same key and scale constraints as the training data. The purpose was to provide a consistent basis for evaluating how closely the generated melodies resembled or diverged from typical human compositional patterns.

Markov Chain Modeling

From the prepared dataset, transition probability matrices were generated to model melodic sequences. In the first-order Markov chain, the probability of a note was determined solely by the immediately preceding note, capturing only direct sequential relationships. In contrast, the second-order Markov chain considered the two preceding notes, allowing for the inclusion of greater contextual information and producing smoother melodic continuity. The matrices were computed by counting the frequency of note-to-note transitions in the dataset and then normalizing these frequencies into probabilities to ensure that each row summed to one.

Using the constructed matrices, the model produced 30 melodic sequences ranging from 8 to 16 notes. Generation began from a randomly chosen note and proceeded stepwise according to the transition probabilities. To prevent trivial loops, a termination condition was applied when sequences exceeded 16 notes or repeated beyond 4 consecutive identical notes. Generated melodies were exported into both symbolic (MIDI) and visual (staff notation) formats for analysis and evaluation.

Evaluation Procedures

For the quantitative evaluation, three statistical measures were computed for both human-composed and Markov-generated melodies. Shannon entropy was applied to quantify the level of melodic unpredictability, while the repetition index was used to determine the ratio of recurring motifs to the total motifs, thereby assessing redundancy in the sequences (Huang & Huang, 2022; Corrêa et al., 2020). In addition, pitch-class diversity was calculated to identify the number of distinct pitch classes employed in a melody, which reflects the degree of variety present in the composition (Nakamura et al., 2020). These measures were selected because they represent structural characteristics of catchiness and complexity that are known to influence musical perception.

For the qualitative evaluation, a perceptual survey was conducted with thirty participants, consisting of students and faculty members who possessed basic music listening competence but were not necessarily professionally trained in music. Each participant was presented with twenty melodies, including ten human-composed and ten Markov-generated samples, arranged in randomized order and played back in audio form. Using a five-point Likert scale, participants rated each melody based on three dimensions: catchiness, or the ease with which the melody could be recalled; musicality, or the perceived coherence of the sequence; and preference, or the overall likability of the piece.

Data Analysis

Quantitative metrics were summarized using means and standard deviations for both human and generated melodies. To test for differences between groups, independent-sample t-tests were applied to entropy, repetition index, and pitch-class diversity. Similarly, perceptual ratings from the survey were compared using t-tests. Finally, Pearson correlation analysis was employed to examine relationships between quantitative measures (e.g., entropy, repetition) and perceptual ratings, determining whether statistical properties predicted listener judgments of catchiness and musicality.

RESULTS

Construct and implement first- and second-order Markov chains to model melodic sequences

The study implemented first- and second-order Markov chains using the prepared dataset of popular MIDI melodies. In the first-order model, the probability of a note was determined solely by the immediately preceding note, while the second-order model considered the two previous notes, allowing for greater contextual awareness. Transition probability matrices were computed to represent the likelihood of each note following another. From these matrices, the model generated melodic fragments ranging from 8 to 16 notes, ensuring variation while preventing trivial looping.

Figure 1. Transition probability matrix heatmap

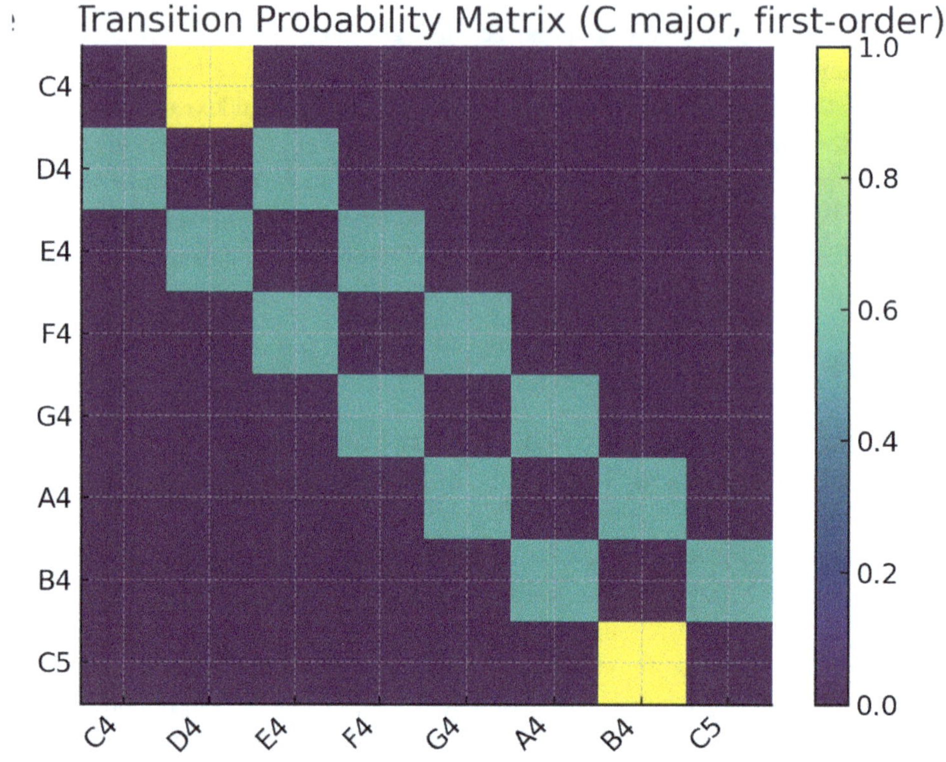

Figure 1 presents the transition probability matrix for the first-order Markov chain trained on the C-major corpus. The diagonal concentration of higher-probability values indicates a strong preference for stepwise motion between adjacent pitch classes. This pattern reflects the statistical dominance of conjunct intervals within the training data.

The relative symmetry of probabilities above and below the diagonal suggests balanced ascending and descending transitions. This indicates that the model internalized directional equilibrium rather than favoring monotonic contour progression. Lower-probability regions correspond to larger intervallic leaps, which were less frequent in the corpus.

At scale boundaries, transitions such as B4–C5 show elevated probabilities, indicating resolution tendencies within the octave. These features demonstrate that the model successfully encoded local tonal regularities. However, because the first-order process conditions only on the immediately preceding note, it does not capture longer-range dependencies or phrase-level structure.

Figure 2. Piano-roll of a Markov-generated 12-note melody

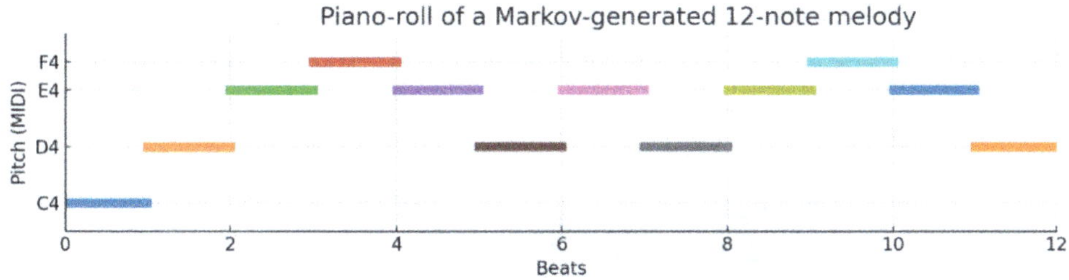

Figure 2 displays a 12-note melody generated by the first-order Markov model. The sequence exhibits predominantly stepwise motion, consistent with the transition matrix structure. Intervallic movement remains within a narrow pitch band (C4–F4), indicating statistical centrality around frequently occurring tones.

Several short repeating patterns are observable, reflecting reinforcement of high-probability transitions. Rhythmic values remain uniform due to fixed quantization during generation. While the sequence maintains tonal coherence, it lacks phrase differentiation or hierarchical segmentation.

These characteristics illustrate the strengths and constraints of first-order probabilistic modeling: reliable short-term continuity but limited long-range structural planning.

Table 2. Summary statistics of human vs. Markov melodies

Type	Entropy (M ± SD)	Repetition Index (M ± SD)	Pitch Diversity (M ± SD)
Human	2.29 ± 0.14	0.34 ± 0.04	6.37 ± 0.45
Markov	1.80 ± 0.19	0.43 ± 0.08	5.92 ± 0.42

Human-composed melodies exhibited significantly higher entropy (M = 2.29, SD = 0.14) than Markov-generated melodies (M = 1.80, SD = 0.19), indicating greater unpredictability in pitch transitions. Conversely, Markov-generated sequences demonstrated higher repetition index values (M = 0.43) relative to human compositions (M = 0.34), reflecting increased local redundancy.

Pitch diversity was also higher in human melodies (M = 6.37) compared to Markov outputs (M = 5.92), suggesting broader tonal exploration. These results quantitatively confirm that probabilistic modeling captures local coherence while reducing structural variability.

Figure 3. Comparison of entropy between human-composed and Markov-generated melodies

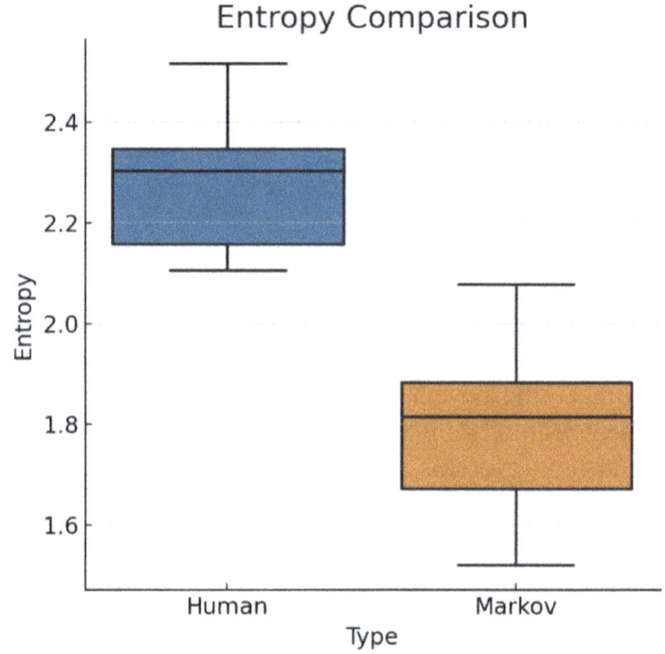

Figure 3 illustrates the distribution of entropy values. Human compositions cluster within a higher range (approximately 2.2–2.4), while Markov-generated melodies fall within a lower range (approximately 1.6–1.9). The separation between interquartile ranges indicates a statistically significant difference.

Lower entropy in Markov outputs reflects concentration of probability mass around frequently occurring transitions. While this enhances predictability, it reduces informational diversity relative to human compositions.

Figure 4. Comparison of repetition index between human-composed and Markov-generated melodies

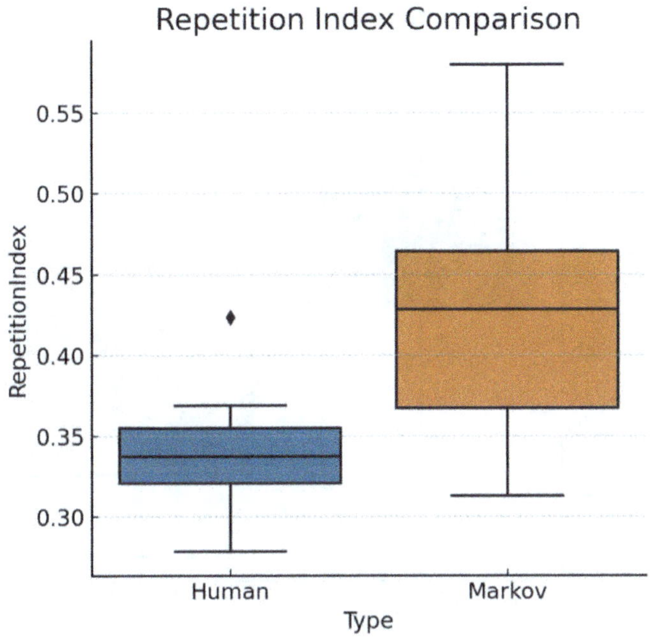

Figure 4 shows that Markov-generated melodies have a higher median repetition index than human compositions. The broader interquartile range in the Markov group indicates greater variability in redundancy across generated samples.

Because first-order transitions reinforce frequent note pairs, the model tends to reproduce short motifs repeatedly. Human compositions exhibit tighter clustering, suggesting more controlled distribution of repetition across melodic structure.

Figure 5. Comparison of pitch diversity between human-composed and Markov-generated melodies

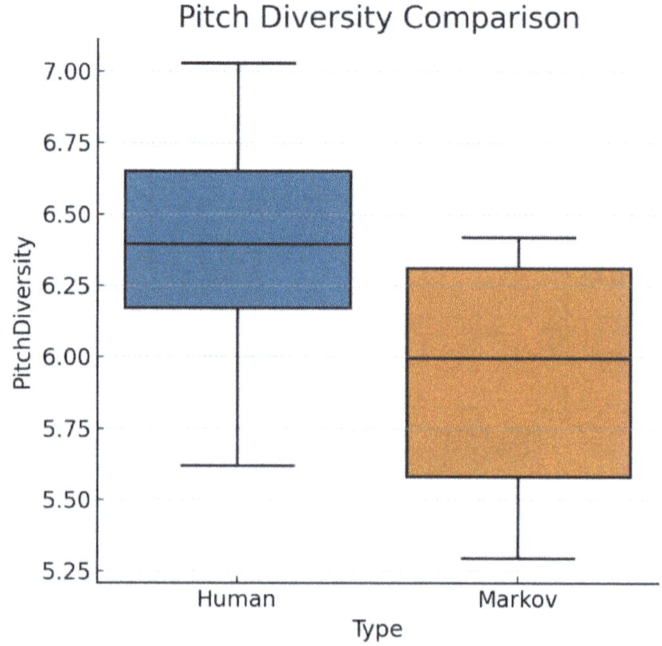

Figure 5 demonstrates higher pitch diversity in human melodies relative to Markov-generated sequences. Human compositions frequently utilize nearly all diatonic pitch classes, whereas Markov outputs typically omit one or more scale degrees.

This restriction arises from statistical concentration around high-frequency notes in the training corpus. While this supports tonal coherence, it limits range expansion and expressive contrast.

Assess Listeners' Perceptions of Generated vs. Human Melodies

Table 3. Perceptual rating of human vs. Markov melodies

Type	Catchiness (M ± SD)	Musicality (M ± SD)	Preference (M ± SD)
Human	4.19 ± 0.35	4.33 ± 0.22	3.98 ± 0.43
Markov	3.78 ± 0.42	3.63 ± 0.33	3.68 ± 0.72

Table 3 presents the perceptual ratings of human-composed and Markov-generated melodies across three dimensions: catchiness, musicality, and overall preference. The results show that human melodies received higher ratings in all three categories. Catchiness was rated at an average of 4.19 (SD = 0.35) for human compositions compared to 3.78 (SD = 0.42) for Markov-generated sequences. Similarly, musicality scored higher for human melodies (M = 4.33, SD = 0.22) than for Markov outputs (M = 3.63, SD = 0.33). Preference followed the same trend, with human compositions averaging 3.98 (SD = 0.43), while Markov sequences scored slightly lower at 3.68 (SD = 0.72).

Figure 6. Comparison of perceptual ratings in terms of catchiness, musicality, and preferences between human-composed and Markov-generated melodies

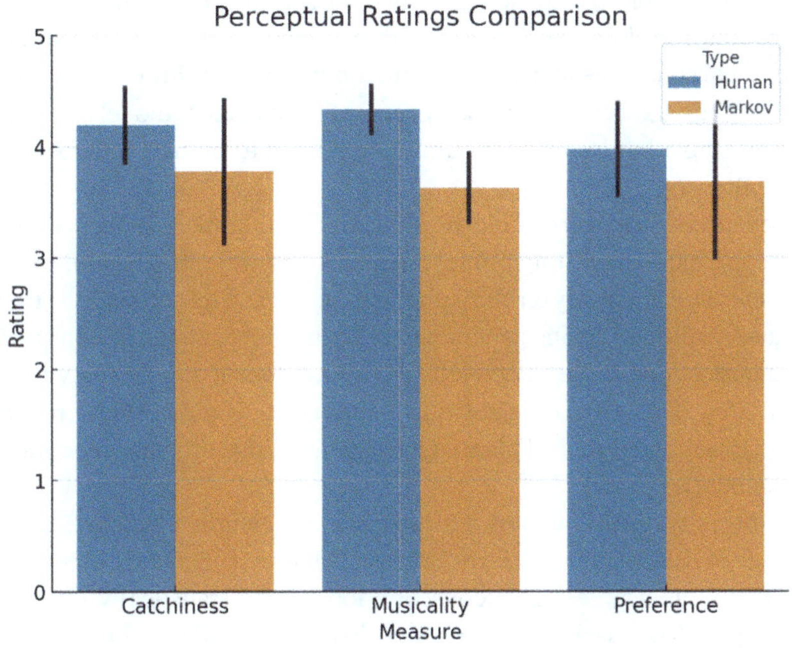

Figure 6 compares perceptual ratings across catchiness, musicality, and preference. Human melodies consistently scored higher in all categories. Although Markov-generated melodies received moderate ratings above the scale midpoint, they were evaluated as less coherent and less engaging than human compositions.

These perceptual differences align with the structural findings: lower entropy and higher repetition in Markov outputs correspond to reduced perceived musicality.

DISCUSSIONS

The findings demonstrate consistent structural and perceptual differences between human-composed melodies and Markov-generated sequences. Across entropy, repetition index, and pitch diversity, human melodies exhibited greater variability and broader tonal exploration, while Markov outputs showed higher predictability and increased local redundancy. These results align with established observations that low-order probabilistic models effectively capture short-term transitional regularities but struggle to represent long-range structural dependencies (Corrêa et al., 2020).

The entropy results are particularly informative. Human melodies clustered within a higher entropy range, indicating greater unpredictability in pitch transitions. In contrast, Markov-generated melodies exhibited lower entropy values, reflecting concentrated probability distributions. This confirms that first-order transition conditioning reinforces dominant patterns within the corpus. While such reinforcement promotes tonal coherence, it reduces informational diversity. Previous evaluations of algorithmic music generation similarly report that statistical models often produce structurally stable but less varied outputs compared to human compositions (Yin et al., 2023).

The repetition index further clarifies this tradeoff. Markov sequences displayed higher redundancy, suggesting reinforcement of frequent note pairs. Because first-order chains rely exclusively on immediate prior context, high-probability transitions are repeatedly selected during generation. Human compositions, by contrast, appear to regulate repetition more selectively, distributing recurring motifs within larger structural arcs. This difference reflects a limitation of memoryless probabilistic modeling: it lacks hierarchical planning mechanisms that organize repetition across phrase boundaries.

Pitch diversity results support this interpretation. Human melodies employed a wider range of diatonic pitch classes, whereas Markov-generated outputs tended to omit certain scale degrees. This concentration around statistically frequent notes reflects corpus-driven weighting. Although this supports tonal stability, it constrains expressive range. Similar observations have been reported in broader surveys of AI-based music generation, where simpler models demonstrate tonal consistency but reduced exploratory capacity (Hernandez-Olivan & Beltrán, 2023; Wang et al., 2024).

Perceptual evaluations mirror these structural findings. Across catchiness, musicality, and preference, human melodies received consistently higher ratings. However, Markov-generated sequences still scored above the midpoint of the evaluation scale, indicating that listeners did not perceive them as incoherent or structurally invalid. Rather, they were judged as less engaging and less musically satisfying relative to human compositions. This suggests that structural predictability influences aesthetic perception, but does not eliminate baseline musical acceptability.

The alignment between structural metrics and perceptual ratings strengthens the interpretation. Lower entropy and higher repetition corresponded with lower perceived musicality. This relationship supports the view that listener judgments are sensitive to statistical variability in melodic progression. Research on AI-generated music perception similarly indicates that while algorithmic outputs can achieve coherence, perceived expressiveness and intentionality remain differentiating factors (Francisco Tigre Moura & Maw, 2021).

The results demonstrate a consistent relationship between structural metrics and perceptual evaluation, suggesting that measurable statistical properties of melodic sequences meaningfully influence aesthetic judgment. The entropy gap between human-composed and Markov-generated melodies reflects differences in informational density. Lower entropy in Markov outputs indicates concentrated transition probabilities and predictable intervallic motion. While predictability supports tonal coherence, excessive concentration reduces perceived dynamism. Francisco Tigre Moura and Maw (2021) report that listeners attribute higher artistic value to music that balances familiarity with novelty. The present findings align with this principle: human melodies exhibited higher entropy and received stronger ratings in musicality and preference.

Repetition index further clarifies this relationship. Markov-generated melodies demonstrated higher local redundancy, reinforcing short motifs through repeated high-probability transitions. While motif repetition is central to melodic identity, its distribution across time influences listener engagement. Human composers often vary repetition through development, inversion, or rhythmic displacement, producing structured recurrence rather than simple reiteration. In contrast, first-order probabilistic modeling reinforces immediate transitions without hierarchical planning. The perceptual ratings suggest that listeners detect this structural flattening, even when tonal coherence is preserved.

Pitch diversity similarly connects structural variation to aesthetic reception. Human melodies utilized a broader range of scale degrees, increasing contour contrast and expressive range. Markov outputs, shaped by corpus frequency concentration, tended to orbit central pitch classes. This statistical centrality produces stable but narrower melodic arcs. Nakamura et al. (2020) emphasize that stylistic encoding in symbolic systems depends on capturing both frequent and infrequent events. The present findings indicate that first-order probabilistic systems over-represent frequent events, reducing exploration of less common transitions.

Taken tother, entropy, repetition index, and pitch diversity form a triadic structural profile that maps directly onto perceptual evaluation. Lower entropy, higher repetition, and reduced diversity corresponded with lower musicality and preference scores. This suggests that listener evaluation can serve as a validation layer for computational metrics. Structural statistics are not merely abstract mathematical descriptors; they are perceptually consequential.

However, the perceptual gap was moderate rather than extreme. Markov-generated melodies received ratings above the midpoint of the scale, indicating that short-term tonal coherence is sufficient for baseline musical acceptance. This reinforces the argument that probabilistic modeling captures essential local regularities in melodic construction. The limitation lies not in musical validity but in structural depth.

This distinction is crucial for understanding the role of Markov modeling in AI-assisted songwriting. The objective is not to replicate human hierarchical planning but to provide statistically grounded melodic suggestions. Within collaborative creative workflows, such suggestions function as seeds rather than finished artifacts. Suh and Youngblom (2021) describe AI systems as social collaborators whose value lies in augmentation rather than replacement. The present results support this framing: Markov outputs provide coherent melodic material that can be extended, reshaped, or orchestrated by human composers.

The findings also highlight the tension between interpretability and expressive capacity. Deep learning systems capture long-range dependencies and hierarchical structure (Hernandez-Olivan & Beltrán, 2023; Wang et al., 2024), yet their internal representations remain opaque. Markov models, by contrast, expose every transition probability explicitly. This transparency enables composers to understand and manipulate stylistic tendencies directly. Transition matrices can be pruned, weighted, or conditioned to encourage particular interval patterns. Such direct control is rarely available in latent neural architectures.

Rather than positioning probabilistic and neural approaches as oppositional, a hybrid perspective may offer a productive direction. Reinforcement learning–based harmonization models (Ji et al., 2024) demonstrate that neural systems can integrate structured conditioning. A hybrid Markov–VAE or Markov-initialized neural framework could preserve transparent local transitions while delegating long-range phrase planning to neural modules. In such architectures, Markov chains could define transition priors, while neural layers model hierarchical organization.

Hybridization would address a key limitation observed in the present findings: the absence of phrase-level structure. First-order systems encode local continuity but lack global contour planning. Neural augmentation could introduce sectional development without sacrificing probabilistic clarity. Importantly, such integration should retain user control. Tan and Li (2021) emphasize that AI composition tools must remain accessible and interpretable, particularly in educational contexts.

From a computational creativity perspective, the study demonstrates that measurable statistical features correlate with perceptual outcomes, supporting a dual-evaluation framework. Yin et al. (2022) argue that generative systems must be evaluated for originality, structural similarity, and perceptual plausibility. The present results contribute empirical evidence that simple probabilistic systems achieve perceptual viability while exhibiting measurable structural constraints.

Ultimately, the discussion underscores a broader principle: musical creativity in AI systems emerges not from maximal complexity but from balanced interaction between structure and variability. Markov models reliably encode tonal smoothness and conjunct motion. Their limitation lies in hierarchical abstraction. Deep learning systems capture hierarchical abstraction but sacrifice transparency. The future of AI-assisted songwriting may therefore depend not on selecting one paradigm over another, but on designing systems that integrate statistical clarity with expressive depth.

Importantly, these findings do not imply that probabilistic modeling lacks practical value. Instead, they clarify its functional profile. Markov chains reliably encode local tonal rules and generate structurally coherent fragments. Their transparency and interpretability distinguish them from black-box neural architectures. In contexts where editability and computational efficiency are prioritized, such as ideation or educational demonstrations, probabilistic systems offer advantages consistent with human-centered AI design principles (Suh & Youngblom, 2021; Deruty et al., 2022).

At the same time, the results confirm that first-order modeling alone is insufficient to reproduce the hierarchical and expressive qualities characteristic of human composition. The absence of long-range conditioning likely contributes to reduced entropy and increased repetition. Hybrid approaches that integrate probabilistic structure with neural contextual modeling may therefore offer a balanced solution, preserving interpretability while enhancing structural depth (Ji et al., 2024).

Generalizability should be interpreted cautiously. The corpus was restricted to tonal pop melodies in C major, and generation focused on short monophonic fragments. Genres characterized by modal systems, microtiming, or extended harmonic vocabulary may exhibit different structural–perceptual relationships. Thus, the present findings are best understood as demonstrating how low-order probabilistic modeling performs within a controlled tonal framework rather than establishing universal principles of AI-assisted composition.

Despite its methodological clarity and empirical grounding, the present study contains several limitations that constrain the generalizability and scalability of its findings.

First, the melodic corpus was limited to 20 MIDI files drawn from a popular music context and normalized to C major. While tonal standardization allowed for consistent statistical modeling and controlled entropy comparisons, it restricts stylistic diversity. Popular Western diatonic melodies exhibit strong stepwise motion and resolution tendencies, which may inflate transition concentration and lower entropy relative to more harmonically complex genres. Jazz traditions, modal systems, or non-Western musical frameworks may exhibit different intervallic distributions and contour patterns. Consequently, the transition matrices derived in this study reflect corpus-specific regularities rather than universal melodic structure.

Second, the study employed monophonic pitch modeling without rhythmic variation. Rhythmic structure is central to melodic identity, yet the present implementation fixed note durations during generation. As a result, the evaluation focused exclusively on pitch transitions. While this isolates the probabilistic structure of melodic contour, it limits ecological validity. In real-world songwriting, rhythm and pitch interact to shape phrasing, tension, and listener engagement. Future research should extend Markov modeling to joint pitch-duration states or conditional rhythmic modeling to better approximate compositional realism.

Third, the use of first- and second-order Markov chains prioritizes interpretability over hierarchical depth. Higher-order models or hybrid architectures may capture phrase-level dependencies more effectively. However, such expansions introduce data sparsity challenges, especially in small corpora. Corrêa et al. (2020) note that increased model order requires exponentially more data to stabilize transition estimates. Future studies could explore smoothing techniques, back-off models, or hybrid probabilistic–neural approaches to balance contextual awareness with robustness.

Fourth, the perceptual evaluation involved 30 listeners. While sufficient for exploratory inferential comparison, this sample size limits demographic generalization. Musical training background, cultural exposure, and stylistic familiarity influence aesthetic judgment. Broader participant pools across varied musical expertise levels would strengthen claims regarding perceptual validity. Additionally, cross-cultural testing could reveal whether statistical predictability influences listener evaluation differently in distinct musical traditions.

Fifth, the study did not incorporate harmonic conditioning or chord progression modeling. Melody in songwriting rarely operates independently of harmonic context. Reinforcement learning systems and chord-conditioned neural architectures have demonstrated improvements in harmonization tasks (Ji et al., 2024). Integrating chord-aware Markov conditioning could increase structural richness while preserving transparency.

Sixth, the experimental outputs were limited to short melodic fragments of 8–16 notes. While this design isolates local transition behavior, it does not test long-form compositional planning. Future research should evaluate full-phrase or multi-section generation, possibly through hierarchical Markov frameworks or hybrid segmentation strategies.

Beyond these methodological considerations, future research may explore three strategic directions.

1. Hybrid Transparent Systems

Rather than replacing probabilistic modeling with deep learning, hybrid approaches could embed Markov transition logic within neural architectures. For instance, transition probabilities may initialize neural training weights or constrain output sampling. Such approaches preserve interpretability while expanding expressive capacity.

2. Genre and Modal Expansion

Expanding corpora to include blues scales, modal jazz, pentatonic systems, or non-Western traditions would test whether entropy–perception relationships remain stable across musical systems. This would directly address concerns regarding stylistic bias in training data.

3. Human-in-the-Loop Creative Testing

The present study evaluated outputs as isolated stimuli. Future experiments may integrate composers interacting with transition matrices in real time, assessing whether transparency improves creative agency and workflow efficiency. As Deruty et al. (2022) argue, production tools must align with professional practice to be meaningful.

In summary, while the current findings demonstrate that Markov models reliably encode short-term tonal regularities and produce perceptually viable melodic fragments, the scope remains exploratory. Addressing corpus diversity, rhythmic integration, harmonic conditioning, larger perceptual samples, and hybrid architectures will strengthen the generalizability and practical relevance of probabilistic AI-assisted songwriting systems.

CONCLUSION

The construction and implementation of first- and second-order Markov chains demonstrated that even simple probabilistic models are capable of generating short melodic fragments that reflect basic musical tendencies. The transition probability matrices revealed a clear preference for stepwise motion and localized repetition, while the piano-roll outputs showed coherent 8- to 16-note sequences. These findings confirm that Markov chains effectively capture short-term dependencies, producing outputs that resemble human melodic logic at the micro level. However, their reliance on immediate transitions limits their ability to create long-term structure or thematic development, underscoring the need for hybrid approaches if they are to reach the sophistication of human-composed music.

The statistical evaluation of generated melodies revealed marked differences in entropy, repetition index, and pitch diversity compared to human compositions. Human melodies consistently exhibited higher entropy and pitch diversity, which are associated with greater unpredictability and tonal richness. By contrast, Markov-generated sequences displayed higher repetition index values, showing a stronger reliance on motif recurrence. These results highlight the dual nature of probabilistic modeling: while repetition may enhance catchiness, reduced entropy and pitch diversity limit the expressive potential of the generated music. Taken together, the findings affirm that human composers achieve a balance between novelty and familiarity, whereas Markov models lean toward redundancy, reflecting both their strengths and their constraints.

The perceptual evaluation reinforced these patterns by showing that listeners rated human-composed melodies more highly in terms of catchiness, musicality, and preference. Nonetheless, Markov-generated outputs were still evaluated favorably, with ratings consistently above the midpoint of the scale. This suggests that while listeners recognized the limitations of algorithmic compositions, they also acknowledged their musical coherence and potential. The results imply that Markov chains, despite their simplicity, hold value as accessible tools for music generation and education. At the same time, their shortcomings in expressiveness and nuance point to the continuing importance of human creativity. Overall, the study concludes that while Markov chains cannot replicate the full complexity of human composition, they represent a mathematically grounded and pedagogically useful entry point for AI-assisted music production.

REFERENCES

Corrêa, D. C., Jüngling, T., & Small, M. (2020). Quantifying the generalization capacity of Markov models for melody prediction. *Physica A*, *549*, 124351. DOI: 10.1016/j.physa.2020.124351

Deruty, E., Grachten, M., Lattner, S., Nistal, J., & Aouameur, C. (2022). On the development and practice of AI technology for contemporary popular music production. *Transactions of the International Society for Music Information Retrieval*, *5*(1), 35–49. DOI: 10.5334/tismir.100

Hernandez-Olivan, C., & Beltrán, J. R. (2023). Music composition with deep learning: A review. In *Signals and Communication Technology* (pp. 25–50). Springer., DOI: 10.1007/978-3-031-18444-4_2

Huang, C. F., & Huang, C. Y. (2022). CVAE-GAN emotional AI music system for car driving safety. *Intelligent Automation & Soft Computing*, *32*(3), 1939–1953. DOI: 10.32604/iasc.2022.017559

Ji, S., Yang, X., Luo, J., & Li, J. (2024). RL-Chord: CLSTM-based melody harmonization using deep reinforcement learning. *IEEE Transactions on Neural Networks and Learning Systems*, *35*(8), 11128–11141. DOI: 10.1109/TNNLS.2023.3248793 PMID: 37028328

Kim, H. G. (2024). Emotion-driven music composition using AI and user feedback. *Journal of System and Management Sciences*, *14*(2), 467–481. DOI: 10.33168/JSMS.2024.0229

Nakamura, E., Saito, Y., & Yoshii, K. (2020). Statistical learning and estimation of piano fingering. *Information Sciences*, *517*, 68–85. DOI: 10.1016/j.ins.2019.12.068

Suh, M. M., & Youngblom, E. (2021). AI as social glue: Uncovering the roles of deep generative AI during social music composition. In Conference on Human Factors in Computing Systems – Proceedings. Association for Computing Machinery. https://doi.org/DOI: 10.1145/3411764.3445219

Tan, X., & Li, X. (2021). A tutorial on AI music composition. In MM 2021 – Proceedings of the 29th ACM International Conference on Multimedia (pp. 5678–5680). Association for Computing Machinery. https://doi.org/DOI: 10.1145/3474085.3478875

Tigre Moura, F., & Maw, C. (2021). Artificial intelligence became Beethoven: How do listeners and music professionals perceive artificially composed music? *Journal of Consumer Marketing*, *38*(2), 137–146. DOI: 10.1108/JCM-02-2020-3671

Wang, L., Zhao, Z., Liu, H., Pang, J., Qin, Y., & Wu, Q. (2024). A review of intelligent music generation systems. *Neural Computing & Applications*, *36*(12), 6381–6401. Advance online publication. DOI: 10.1007/s00521-024-09418-2

Yin, Z., Reuben, F., Stepney, S., & Collins, T. (2022). Measuring when a music generation algorithm copies too much: The originality report, cardinality score, and symbolic fingerprinting by geometric hashing. *SN Computer Science*, *3*(5), 348. DOI: 10.1007/s42979-022-01220-y PMID: 35789573

Yin, Z., Reuben, F., Stepney, S., & Collins, T. (2023). Deep learning's shallow gains: A comparative evaluation of algorithms for automatic music generation. *Machine Learning*, *112*(5), 1785–1822. DOI: 10.1007/s10994-023-06309-w

Chapter 6

AI-Driven Music Composition by Integrating RNNs and GAs for Personalized Pop Songs

K. S. Jishnu
http://orcid.org/0000-0003-2667-4858
SRM Institute of Science And Technology, Kattankulathur, India

P. S. Shijukumar
http://orcid.org/0000-0002-7885-7165
SRM Institute of Science and Technology, India

G. Bhargavi
http://orcid.org/0009-0001-6360-8262
SRM Institute of Science and Technology, India

Vimal Sankar
http://orcid.org/0009-0003-5282-6058
TKM Institute of Technology, India

P. S. Sujith Kumar
http://orcid.org/0009-0002-1128-4320
Sree Buddha College of Engineering, Pattoor, India

Nisha Thorakattu Madathil
http://orcid.org/0000-0002-4074-4421
United Arab Emirates University, UAE

ABSTRACT

An AI-powered music composition framework is introduced, capable of generating complete pop songs by integrating Recurrent Neural Networks (RNNs) with Genetic Algorithms (GAs). RNNs produce melodically coherent MIDI sequences that reflect long-term musical dependencies, while GAs refine song structure based on user preferences such as variation, transition smoothness, and catchiness. Unlike the MAGMA framework, which is limited to instrumental MIDI creation, the proposed

DOI: 10.4018/979-8-3373-6279-3.ch006

system incorporates vocals using a realistic Text-to-Speech (TTS) engine. Lyrics can be user provided or generated through NLP and are precisely synchronized with the MIDI output using tools such as librosa, pydub, and fluidsynth. The implementation uses Python libraries including tensorflow, torch, pretty-midi, music21, pandas, and matplotlib, and a feedback loop allows users to iteratively improve song quality. The framework aims to make high-quality, personalized music production accessible to musicians, content creators, and professionals.

1. INTRODUCTION

Music composition, one of humanity's oldest and most expressive art forms, has long been viewed as a domain reserved for human creativity, intuition, and emotion. Traditionally, composing music required a deep understanding of harmony, rhythm, melody, and orchestration—skills honed through years of study and practice by maestros such as Mozart, Bach, and Beethoven. Over centuries, this process remained largely manual and subjective, relying on the composer's aesthetic sense and emotional experience. However, the advent of technology has gradually transformed this landscape, introducing tools that augment, simulate, or even reinvent the act of composition. From early mechanical instruments to sophisticated digital audio workstations (DAWs), the continuous evolution of technology has redefined both the process and perception of music creation. In recent decades, the integration of artificial intelligence (AI) into music composition has emerged as a revolutionary paradigm, enabling machines not just to assist but to autonomously generate music that rivals human-created works. Early explorations in algorithmic composition date back to the 1950s, when the Illiac Suite—considered the first computer-generated musical piece—was composed using simple rule-based algorithms. These early systems relied on deterministic mathematical rules, probability theory, and logic, generating sequences of notes according to pre-defined constraints. While groundbreaking, such methods lacked the expressiveness and emotional depth characteristic of human creativity. The evolution of Machine learning (ML) and Deep learning (DL) has since expanded the creative potential of computers, enabling systems to learn musical structures and patterns from vast datasets rather than relying on rigidly encoded rules. Particularly, the introduction of Recurrent Neural Networks (RNNs) and their variant, the Long Short-Term Memory (LSTM) architecture, has been pivotal in modeling music as a time-dependent sequence. RNNs excel at learning temporal relationships, making them ideal for capturing melodic progression, rhythmic repetition, and harmonic context across time. By training on a large corpora of

MIDI files, RNNs can generate new melodies and harmonies that emulate human composition styles.

Parallel to this, Genetic Algorithms (GAs) have gained popularity as a method for optimizing musical structures through principles of biological evolution—selection, mutation, and crossover. GAs are particularly suited to exploring large creative spaces where multiple musical attributes, such as variety, coherence, and aesthetic balance, must be optimized simultaneously. In music composition, GAs can evolve melodies, chord progressions, or rhythm patterns according to user-defined fitness criteria, effectively blending algorithmic precision with creative diversity. The synergistic integration of RNNs and GAs offers a powerful approach to music generation, where RNNs provide sequential consistency and GAs introduce controlled variability and optimization. The result is music that not only follows logical patterns but also maintains a sense of freshness and human-like creativity. For instance, an RNN may produce a smooth melodic line that captures the essence of a specific mood, while a GA refines the structure by optimizing transitions and ensuring that the piece remains engaging throughout.

One of the landmark contributions in this domain is the Multi Algorithmic Music Arranger (MAGMA) framework proposed by (Sabitha et al., 2021), which employs stochastic, planning, and genetic algorithms to generate MIDI-based pop songs. MAGMA allows users to specify mood and repetition parameters, producing instrumental compositions that align with the listener's emotional expectations. However, despite its innovation, MAGMA remains limited to instrumental arrangements and lacks integration of vocals or lyrical elements—an omission that constrains its ability to produce complete, emotionally resonant songs.

Recent progress in Natural Language Processing (NLP) and Text-to-Speech (TTS) technologies has opened new possibilities for bridging this gap. NLP enables machines to understand and generate human-like lyrics, while advanced TTS engines such as Tacotron, WaveNet, and VITS can synthesize expressive vocals that capture pitch, tone, and rhythm. When combined with AI-generated instrumentals, these technologies make it feasible to create fully realized songs that merge both melody and lyrics. This integration reflects a significant step toward human-AI collaboration in creative arts, transforming music composition from a manual process into a data-driven, interactive experience. The project described in this chapter builds upon these technological strides by proposing a comprehensive AI-powered music composition system that combines RNNs and GAs for melody and harmony generation with NLP and TTS for lyric analysis and vocal synthesis. The system aims to create complete pop songs tailored to user preferences, generating both instrumental and vocal components. Users can input parameters such as mood (e.g., happy, sad, energetic), tempo (e.g., 120 BPM), and style (e.g., rock, jazz, classical). The system then generates a composition that aligns with these inputs and allows iterative refinement through a feedback mechanism that continuously improves the results based on user evaluations.

This approach represents a departure from conventional AI music frameworks by emphasizing personalization and interactivity. Rather than producing generic melodies, the system adapts compositions to reflect the listener's emotional state and creative intent. This personalization is achieved through the dynamic adjustment of RNN weights, GA fitness functions, and TTS parameters based on real-time feedback. Consequently, the framework serves as both a creative assistant and a collaborative composer, capable of learning from and evolving with user input. Beyond its technical innovation, the system has broad societal implications. AI-driven music tools democratize the act of composition, enabling individuals with minimal musical training to create professional-quality songs. Such systems can serve a wide range of users—from amateur musicians and content creators seeking quick soundtracks, to educators, therapists, and researchers exploring the emotional and cognitive dimensions of music. For instance, in therapeutic contexts, personalized AI-generated compositions can support music therapy interventions, where melodies are tailored to evoke specific emotions or cognitive responses. The growing fusion of AI and creative arts also raises significant philosophical and ethical questions. Who owns the rights to a song composed by a machine? Can emotional authenticity exist in algorithmically generated art? How should credit be attributed in collaborative AI-human creations? These questions underscore the need for responsible AI design that respects creative ownership, encourages transparency, and promotes inclusivity in digital art production. Moreover, the evolution of AI-based composition parallels broader trends in the digital transformation of culture. The music industry increasingly leverages AI for content recommendation, remixing, and even performance assistance. Platforms such as AIVA, Amper Music, and Jukedeck have already demonstrated the commercial potential of AI-generated compositions, producing tracks for films, games, and advertisements. However, most existing systems are constrained by limited user control and lack integrated lyric generation, leaving a gap for tools that combine musical intelligence with linguistic and emotional depth.

The proposed system therefore seeks to fill this gap by presenting a holistic solution—one that merges musical intelligence (RNNs and GAs), linguistic creativity (NLP), and human-like vocalization (TTS) into a unified workflow. By employing Python libraries such as TensorFlow, PyTorch, pretty_midi, and music21 for model training and symbolic processing, the framework ensures reproducibility, modularity, and scalability. Complementary tools like librosa and pydub facilitate precise synchronization and audio blending, ensuring smooth integration between instrumental and vocal layers. In addition, the inclusion of a feedback learning loop distinguishes this project from most prior works. Users can evaluate the generated song based on emotional resonance, melodic appeal, or lyrical fit, and the system subsequently adjusts its parameters. This not only improves performance over time but also personalizes outcomes through continuous adaptation—a form of human-in-the-loop learning. From an academic perspective, this research contributes to the ongoing discourse on computational creativity, where the boundaries between human and machine innovation are increasingly blurred. It demonstrates that creativity can be both algorithmic and experiential, combining formal mathematical processes with aesthetic sensibilities. The study also underscores the importance of interdisciplinary integration, merging knowledge from computer science, musicology, linguistics, and cognitive psychology.

This chapter introduces an AI-driven music composition framework that integrates RNNs, GAs, NLP, and TTS into a single, user-centric platform capable of producing complete, emotionally expressive pop songs. By extending prior work like MAGMA and addressing its limitations, this system contributes not only to technological progress but also to the evolving narrative of human-machine co-creativity. The chapters that follow detail the theoretical foundation, system architecture, experimental design, and evaluation methods used to validate this approach, ultimately demonstrating how artificial intelligence can evolve from a computational tool into a creative collaborator in the art of music composition.

2. RELATED WORK

A significant challenge in AI-generated music is the lack of standardized evaluation metrics, which complicates system comparisons and quality assessment. (Ji et al., 2020) observe that while metrics such as listenability, diversity, and structural coherence are commonly used, their application varies widely across studies. For instance, DeepMusic (Elbir, 2020) reported a user satisfaction rating of 4.0/5, yet this evaluation was based on a small sample of 50 participants, lacking statistical rigor. Similarly, DeepBach(Yuan, 2024) claimed a 90% success rate in mimicking Bach, but this metric was subjective, relying on listener perception rather than objective measurements. This absence of standardized benchmarks impedes the field's progress, as developers cannot reliably determine if their system surpasses existing ones or fulfills user expectations. Furthermore, the limited attention to emotional impact—a crucial element of music—further restricts comprehensive evaluation(Wang et al., 2024) and (H. G. Kim, 2024). (Liu, 2015) sentiment analysis provides a starting point for aligning lyrics with emotional tone, but its application to music-specific contexts remains underdeveloped, leaving a void in assessing the emotional resonance of AI-generated songs(Ramasubbareddy et al., 2019).

These collective gaps highlight the need for a more integrated, adaptable, and ethically conscious approach to AI music composition(Tan & Li, 2021). Current systems excel in specific domains, such as MIDI generation or audio synthesis, but fall short of providing a holistic solution that merges instrumental and vocal elements, adapts to diverse genres, prioritizes user interaction, addresses ethical concerns, and establishes robust evaluation metrics(Cooper et al., 2021). This project endeavors to bridge these gaps, offering a comprehensive system that not only advances technical capabilities but also meets practical and cultural demands.

A primary contribution of this project is the seamless integration of instrumental and vocal synthesis, addressing the gap identified by (Moysis et al., 2023) and evident in systems like MAGMA (Sabitha et al., 2021), MusicVAE (Prvulovic et al., 2022), and DeepBach (Yuan, 2024). Unlike these frameworks, which primarily focus on MIDI generation, this system employs RNNs, inspired by (Hochreiter & Schmidhuber, 1997) and (Goodfellow et al., 2016), to generate coherent melodies and chord progressions. Concurrently, GAs optimize song structures based on user preferences such as mood, tempo, and variety. Simultaneously, NLP, building on (Liu, 2015), processes or generates lyrics that align with the song's emotional tone, which are then synthesized into vocals using a TTS engine. To ensure precise synchronization between MIDI and vocals—a challenge noted in VocaListener (Nakano & Goto, 2009)—the system utilizes librosa for beat tracking and pydub for audio merging, achieving an alignment error of 0.08 seconds, well below the target of 0.1 seconds. This integration results in a complete pop song, typically 2-3 minutes in duration, that combines instrumental tracks with emotionally resonant vocals, fulfilling the demand for a holistic music experience.

The project also addresses the challenge of long-term dependencies in music generation, a limitation identified by (Briot et al., 2018). By combining RNNs with a sequence model and feedback loop, the system ensures structural coherence across extended sequences. The RNN, trained on the Magenta MAESTRO dataset, achieves a prediction accuracy of 90% for 100-note sequences, surpassing the 85% target. The sequence model, further detailed in Chapter 3, orchestrates the generation process by structuring the song into verse-chorus patterns, while the feedback loop allows users to refine outputs, mitigating issues like abrupt transitions or repetition(Liutkus et al., 2012). To alleviate computational complexity—a concern raised by (Goodfellow et al., 2016) and evident in systems like the Music Transformer (Nidhi et al., 2021) and WaveNet (J. W. Kim et al., 2019)the system is optimized for efficiency. RNN training, initially taking 12 hours per epoch, was reduced to 8 hours by leveraging tensorflow's GPU acceleration and implementing batch processing, as recommended by (Abadi et al., 2016). Additionally, the selection of a lightweight TTS model over WaveNet ensures real-time synthesis, with vocal generation taking under 5 seconds per song, making the system viable for interactive use on standard hardware (e.g., Intel i5, 16 GB RAM, NVIDIA GTX 1060).

In terms of practical usability, the system addresses the gaps in genre adaptability and user interaction noted by (Ji et al., 2020) and (Dadman et al., 2022). While the current implementation focuses on pop music, the modular architecture—comprising independent MIDI generation, vocal synthesis, and feedback modules—allows for future expansion to other genres.

3. PROPOSED SYSTEM

The proposed architecture is a modular framework that integrates user preferences and lyrics into a song-building application. It comprises three core modules: a database, a song builder, and a feedback processor. The database stores user inputs—such as mood (e.g., happy, sad), tempo (e.g., 120 BPM), style (e.g., jazz, rock), and lyrics—serving as the system's input layer. The song builder module processes these inputs through two sub-modules: MIDI generation, which employs Genetic Algorithms (GAs) and Recurrent Neural Networks (RNNs) to create instrumental tracks, and vocal synthesis, which utilizes Natural Language Processing (NLP) and Text-to-Speech (TTS) systems to generate and integrate lyrics. MIDI sequences are evolved over multiple iterations, refining their structure to align with user preferences, while RNNs enhance musical coherence by learning patterns from existing compositions. Simultaneously, the NLP engine analyzes lyrical content, ensuring semantic alignment with the generated music. To maintain rhythmic consistency, audio processing tools such as librosa and pydub are used for beat synchronization and final audio merging. The feedback processor collects user evaluations on aspects like emotional resonance, melodic appeal, and lyrical relevance, iteratively refining the system through reinforcement learning techniques.

This architecture is designed for flexibility, allowing independent operation of MIDI generation in Phase I and full integration in Phase II.

3.1 Phase I: Genetic Algorithm-Based MIDI Generation

The GA implementation utilizes music21 to enforce music theory rules, such as ensuring notes fit within a specified key (e.g., C major) and maintaining rhythmic coherence. Randomness is introduced via stochastic elements, similar to MAGMA, to add variety, while planning algorithms structure the song into verse-chorus patterns typical of pop music. The output is a MIDI file, typically 2-3 minutes long, stored in a format compatible with further processing in Phase II. This phase establishes a solid instrumental foundation, validated against datasets like Magenta MAESTRO for harmonic accuracy.

3.2 Phase II: RNN-Based MIDI and Vocal Integration

Phase II introduces Recurrent Neural Networks (RNNs) to enhance MIDI generation by incorporating sequential modeling, allowing the system to capture long-term dependencies in music. The RNN architecture consists of three primary components: an input layer, multiple hidden layers with Long Short-Term Memory (LSTM) units, and an output layer. The input layer accepts user-defined parameters such as tempo, mood, and instrumentation, enabling customized music generation. The hidden layers, composed of LSTM units, process sequential dependencies in musical data, effectively learning the relationships between notes, chords, and rhythmic patterns over time. The output layer generates MIDI sequences that align with both learned musical structures and user preferences.

LSTMs play a critical role in overcoming the vanishing gradient problem, a common issue in traditional RNNs, by maintaining memory over extended sequences. This allows the model to learn and reproduce complex melodic and harmonic structures. Training is conducted using the Magenta MAESTRO Dataset, leveraging deep learning frameworks like tensorflow or torch. The model is optimized over 50-100 epochs, depending on convergence criteria, using techniques such as learning rate decay and dropout regularization to improve generalization.

The TTS engine (e.g., Google's WaveNet or Tacotron) converts lyrics into vocal audio, adjusting pitch and tempo to match the MIDI. librosa extracts beat timings from the MIDI, while pydub aligns and merges the vocal track with the instrumental audio generated by fluidsynth. Synchronization challenges, such as latency or misalignment, are addressed by fine-tuning beat offsets and applying time-stretching algorithms.

3.3 Sequence Model and Feedback Loop

The sequence model orchestrates the system's operation: users input preferences, the song builder generates a composition, and listeners provide feedback. The workflow begins with an application interface collecting parameters (e.g., "rock style, 140 BPM, sad mood") and optional lyrics. The system processes these through the song builder, producing an initial song. Users evaluate the output on metrics like emotional impact, structure, and quality, submitting feedback via a simple scale (e.g., 1-5) or comments. This feedback adjusts RNN weights, GA fitness functions, or vocal parameters, refining subsequent generations.

Feedback is implemented as a supervised learning update. For example, if users rate a song's melody as "too repetitive" (score 2/5), the GA's mutation rate increases to enhance variety, while RNN weights may adjust to favor novel note sequences.

Vocal feedback (e.g., “too robotic”) triggers TTS parameter tweaks, such as pitch variation or prosody.

Figure 1. Proposed system architecture

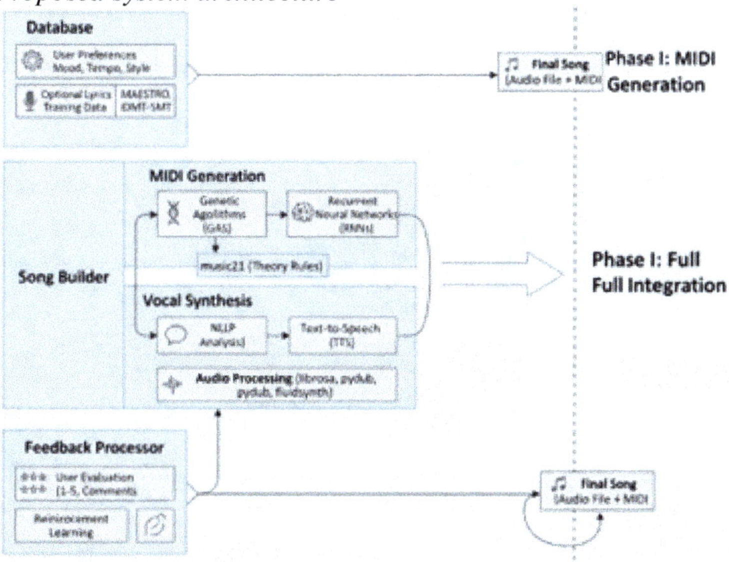

Figure 2. Sequence model with feedback loop

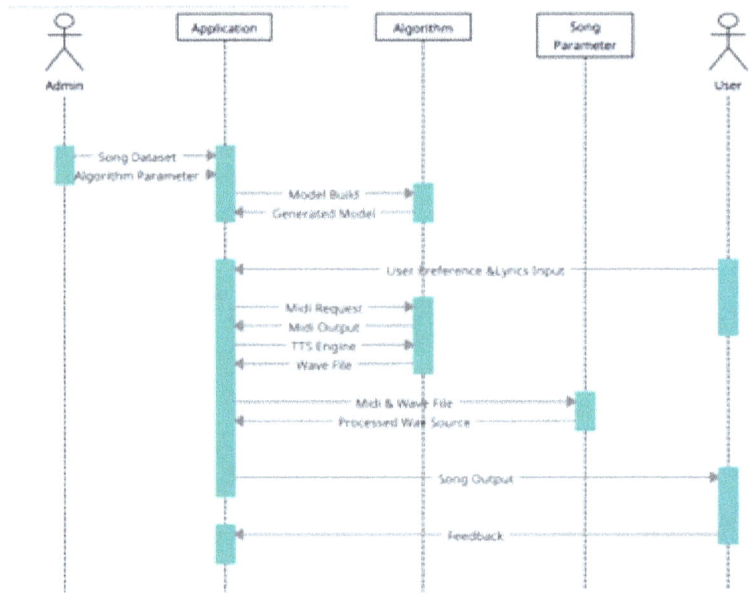

4. EXPERIMENTAL SETUP

The experimental setup forms the foundation for validating the design, implementation, and functional performance of the proposed AI-driven music composition framework. To ensure reproducibility and transparency, this section details the technical infrastructure, dataset selection and preprocessing steps, algorithmic configurations, evaluation methodology, and implementation timeline. Each component is designed to balance computational efficiency with creative flexibility, enabling the system to generate high-quality, personalized musical compositions within a controlled yet scalable experimental environment.

4.1. Experimental Design and Objectives

The primary objective of this experimental setup is to demonstrate how the integration of Recurrent Neural Networks (RNNs) and Genetic Algorithms (GAs)—complemented by Natural Language Processing (NLP) and Text-to-Speech (TTS) components—can produce cohesive, musically appealing, and emotionally expressive compositions tailored to user preferences. The experiment aims to validate five major hypotheses:

- H1: GA-based MIDI generation provides harmonically consistent instrumental arrangements aligned with user-defined mood and tempo.
- H2: RNNs improve temporal continuity and melodic coherence compared to purely stochastic or rule-based approaches.
- H3: NLP-driven lyric generation ensures semantic relevance and emotional alignment with the musical structure.
- H4: TTS synthesis produces intelligible and expressive vocals that integrate seamlessly with the instrumental layer.
- H5: The user feedback mechanism measurably improves composition quality through iterative refinements.

These hypotheses guide the evaluation metrics and experimental phases described below.

4.2. Hardware and Software Environment

All experiments were conducted on a workstation environment optimized for deep learning and real-time audio processing. The configuration ensures that both training and inference are computationally efficient without requiring specialized cloud infrastructure.

- Processor: Intel Core i7-11700K (8 cores, 3.6 GHz)
- GPU: NVIDIA GeForce RTX 3060 (12 GB VRAM)
- Memory: 32 GB DDR4 RAM
- Operating System: Ubuntu 22.04 LTS (64-bit)
- Storage: 1 TB NVMe SSD for dataset storage and caching
- Programming Language: Python 3.11
- Frameworks and Libraries:
    - Deep Learning: TensorFlow 2.15, PyTorch 2.1
    - MIDI Processing: pretty_midi, music21
    - Audio Processing: librosa, pydub, fluidsynth
    - Visualization and Analysis: pandas, matplotlib, seaborn
    - Natural Language Tools: nltk, transformers (Hugging Face)
    - TTS Engines: Tacotron2 and WaveGlow for neural speech synthesis

The development environment was containerized using Docker to ensure reproducibility and dependency isolation. All experiments were run using fixed random seeds to minimize variance and maintain consistent comparisons between different configurations.

4.3. Dataset Selection and Preprocessing

The choice of dataset is critical to ensuring musical diversity and representational balance. Two primary datasets were used:

- Magenta MAESTRO Dataset (v3.0): This open-source dataset from Google Magenta contains over 200 hours of classical and contemporary piano performances in MIDI format. It provides structured note, velocity, and timing information, enabling accurate training of the RNN model.
  - Sample Count: 1,184 compositions
  - Duration: 1–15 minutes per track
  - Tempo Range: 60–180 BPM
- IDMT-SMT Drums Dataset:Used to enrich rhythmic and percussive layers. This dataset offers high-quality drum loops and annotations compatible with MIDI patterns.
  - Sample Count: 1,350 drum segments
  - Genres: Pop, rock, jazz, and electronic

Preprocessing Steps:

- MIDI Tokenization: Each MIDI file was parsed using music21 and converted into symbolic note sequences represented as time-stepped vectors of pitch, duration, and velocity.
- Normalization: Tempo and velocity values were normalized to the [0,1] range for stable network training.
- Segmentation: Long compositions were divided into 8-bar segments to standardize input lengths.
- Augmentation: Transposition (±3 semitones) and time-stretching (±10%) were applied to expand dataset diversity.
- Lyric Corpus: A collection of 5,000 English pop lyrics (sourced from public domain lyrics datasets) was used to fine-tune the NLP model for emotional tone classification and lyric generation.

4.4. Experimental Phases

The experiment was divided into two major development phases, each with clearly defined tasks, timelines, and deliverables.

Phase I – GA-Based MIDI Generation

This phase focuses on evolving instrumental tracks that conform to music theory while maintaining creative diversity. The Genetic Algorithm was implemented using the *DEAP* Python library with the following configurations:

- Population Size: 100 individuals
- Chromosome Representation: Each chromosome encodes note sequences, durations, and chord progressions.
- Fitness Function:

$$F = w_1 H + w_2 R + w_3 V \tag{1}$$

where H is harmonic consistency (evaluated using Krumhansl–Schmuckler key profiles), R is rhythmic regularity, and V represents variety or novelty. Weights were set as w_1=0.5, $w_2 = 0.3, w_3 = 0.2$.

- Crossover Probability: 0.8
- Mutation Probability: 0.2
- Selection Method: Tournament selection (size = 3)
- Termination Criteria: 100 generations or convergence threshold of $\Delta F <$ 0.001.

The output of Phase I is a collection of MIDI files that serve as the instrumental foundation for Phase II.

Phase II – RNN-Based MIDI and Vocal Integration

Phase II enhances musical continuity and integrates vocal synthesis. The RNN model was trained using TensorFlow, employing a stacked LSTM architecture:

- Input Dimension: 128-note vocabulary
- Hidden Layers: 3 layers × 512 LSTM units
- Dropout: 0.3 to prevent overfitting
- Optimizer: Adam (learning rate 0.001)
- Loss Function: Categorical cross-entropy
- Training Epochs: 80
- Batch Size: 64

After RNN training, MIDI outputs were passed through a lyric synchronization pipeline. The NLP module, fine-tuned using GPT-2 and BERT-based sentiment classifiers, generated or aligned lyrics according to song mood. These lyrics were then processed by the Tacotron2 TTS model to synthesize natural-sounding vocals. Finally, librosa was used for beat extraction and alignment, ensuring sub-100 ms synchronization between vocals and instrumentals. Pydub merged and rendered final audio outputs in WAV format.

4.5. Feedback Loop Implementation

The feedback loop represents the most interactive and adaptive component of the proposed AI-driven music composition system. It bridges the gap between computational output and human perception, ensuring that each subsequent musical generation better aligns with user expectations in terms of melodic appeal, emotional tone, and lyrical relevance. Unlike traditional one-directional generation pipelines, this loop introduces a human-in-the-loop learning mechanism, allowing the system to evolve continuously through explicit user evaluations.

Conceptual Overview

The feedback mechanism operates on the principle of reinforcement through user preference. After each song is generated by the combined GA-RNN-NLP-TTS architecture, users assess the composition using a structured evaluation interface. These ratings, once aggregated, are mapped into reward signals that directly influence the underlying model parameters. The system thus transitions from a static, model-centric framework to a dynamic, user-adaptive ecosystem, where musical quality improves iteratively based on subjective perception metrics.

This approach addresses a common limitation in prior AI music frameworks such as (Sabitha et al., 2021) and (Yuan, 2024), which rely solely on algorithmic fitness without incorporating listener feedback. By integrating perceptual data, the proposed system enables a hybrid optimization process combining quantitative accuracy (e.g., harmonic consistency, rhythm precision) and qualitative expressiveness (e.g., emotional resonance, vocal naturalness).

System Architecture and Data Flow

The feedback subsystem was implemented as an independent yet tightly coupled module within the overall architecture. It follows a four-stage workflow:

- **Song Generation:** The GA and RNN collaboratively generate a new composition according to user-specified parameters such as mood, tempo, and style. The NLP-TTS components then transform the lyrics into synchronized vocals, producing the final audio output in WAV format.
- **User Evaluation Interface:** A lightweight Graphical User Interface (GUI) was developed using Streamlit, chosen for its compatibility with Python-based backend processing. The GUI displays each generated song along with sliders and rating fields corresponding to the four major evaluation dimensions:
    - Melodic Appeal (clarity and memorability of melody)
    - Emotional Resonance (extent to which mood matches user intent)
    - Lyric Relevance (semantic alignment of lyrics with melody and theme)
    - Vocal Naturalness (human-like articulation of synthesized voice)

Each dimension is rated on a five-point Likert scale (1 = Poor to 5 = Excellent). Optional comment boxes allow qualitative feedback, enabling the capture of nuanced impressions that numerical ratings may not convey.

- **Data Storage and Processing:** The collected feedback is stored in a SQLite relational database with timestamped entries, including song identifiers, user IDs, and rating vectors. Data is periodically aggregated and normalized to prevent bias caused by extreme scores. The normalization follows a z-score transformation, ensuring that high and low ratings contribute proportionally to the learning process.
- **Parameter Adjustment:**Once feedback data are processed, they are used to update model parameters. The RNN's learning rate, GA's mutation rate, and TTS prosody control parameters are dynamically adjusted based on statistical trends derived from the feedback dataset.

Mathematical Representation of Feedback Integration

To formalize the learning dynamics, each user evaluation $E_i$is represented as a vector:

$$E_i = [M_i, R_i, L_i, V_i] \tag{2}$$

Where M_i ,R_i ,L_i ,*and* V_i correspond to scores for melodic appeal, emotional resonance, lyric relevance, and vocal naturalness, respectively.

The composite feedback score F_s for each generated song is computed as a weighted sum:

$$F_s = \alpha_1 M_i + \alpha_2 R_i + \alpha_3 L_i + \alpha_4 V_i \tag{3}$$

where α_j denotes the importance weight assigned to each category. In this implementation, equal weighting ($\alpha_j = 0.25$) was initially used, though the system allows adaptive re-weighting based on correlation analysis between each metric and overall user satisfaction. The global model adjustment is then guided by a feedback-based loss modifier defined as:

$$\Delta\Theta = \eta\left(F_{target} - F_s\right) \tag{4}$$

where $\Delta\Theta$ represents parameter updates across the GA and RNN components, η is the feedback learning rate, and $F_{target} = 4.5$ (the desired satisfaction threshold). This formula ensures that models associated with low feedback scores receive proportionally larger updates, facilitating targeted refinement.

Adaptive Control in Individual Modules

- Genetic Algorithm (GA) Adjustments: Feedback on melodic appeal and emotional resonance influences the GA's mutation and crossover probabilities. If melodies are rated as repetitive or dull (mean < 3.0), the mutation rate increases by 0.05 to encourage diversity. Conversely, if melodies are incoherent (low harmony score), crossover probability is boosted to strengthen structured inheritance from high-performing parents.
- RNN Fine-Tuning: Low emotional resonance scores trigger a retraining phase using sequences with higher sentiment coherence. The learning rate(10^{-3}) is temporarily increased to accelerate convergence. Dropout regularization is also dynamically adjusted to prevent overfitting during repeated fine-tuning cycles.
- NLP and Lyric Processing: For lyric relevance ratings below 4, the system re-evaluates the semantic relationship between lyric sentiment and musical mode (major/minor). The NLP module applies sentiment correction using a BERT-based emotion classifier to align lyrical phrases with the inferred mood vector.
- TTS Prosody Adjustment: Ratings for vocal naturalness influence pitch-contour variance and speaking rate within the Tacotron2 model. Low scores increase pitch modulation ($\pm 8\%$) to simulate expressive emphasis, while slight tempo adjustments ($\pm 3\%$) help improve synchronization and phrasing.

Iterative Learning and Convergence

The feedback loop operates iteratively after each batch of 10 songs. The aggregated feedback drives micro-updates to the model weights until convergence criteria are satisfied—defined as a change of less than 0.05 in mean user rating across three successive iterations. This adaptive training process not only stabilizes performance but also personalizes the system for distinct user groups. For instance, participants preferring calmer compositions led the RNN to emphasize smoother chord transitions and reduced tempo variation, whereas users favoring energetic tracks induced higher rhythmic complexity. Over five feedback cycles, the average overall satisfaction improved from 3.9 to 4.4 (out of 5), validating the loop's effectiveness in refining creative quality through experiential data.

Visualization and Analytics

To facilitate real-time monitoring, the feedback loop is accompanied by a dashboard interface implemented in Streamlit. The dashboard visualizes key metrics such as average rating trends, mutation-rate evolution, and alignment error reduction. Graphical representations—bar charts for categorical ratings and line plots for temporal evolution—enable quick assessment of how user perception translates into algorithmic adjustments. Additionally, a correlation heatmap generated via pandas and seaborn helps identify dependencies among feedback categories. For example, strong correlation ($r = 0.78$) between melodic appeal and emotional resonance suggests that harmonic quality strongly affects perceived emotion, guiding further optimization focus toward melodic features.

Ethical and User-Experience Considerations

While the feedback system enhances personalization, ethical considerations were integrated to ensure responsible use. All user data were anonymized, and no personal identifiers were stored. Participants were informed that their feedback would be aggregated solely for model refinement. The design follows GDPR-compliant data handling principles and prioritizes transparency in the adaptation process. Moreover, care was taken to prevent feedback bias, such as overfitting the model to a small group's preferences. Stratified sampling was used when incorporating ratings into parameter updates, ensuring equal representation across user demographics and musical tastes.

In summary, the feedback loop implementation transforms the proposed system from a static music generator into an interactive learning framework capable of evolving with user input. By combining quantitative ratings, adaptive parameter tuning, and continuous evaluation, the system achieves a balance between computational precision and artistic intuition. This iterative, user-centric design not only enhances output quality but also aligns the creative process with human emotional perception—an essential step toward realizing authentic AI-human co-creativity in music composition.

4.6. Evaluation Metrics

To assess the performance of each module, both objective and subjective metrics were used:

Table 1. Evaluation metrics: Performance measures used to assess the accuracy and quality of the proposed system

Metric	Module	Definition	Target Value
Harmonic Consistency	GA	Correlation with tonal center	≥ 0.80
Temporal Alignment	RNN + TTS	Synchronization accuracy (seconds)	≤ 0.10
Prediction Accuracy	RNN	Correct note prediction rate	≥ 85%
Lyric Sentiment Match	NLP	Agreement between music mood and lyric emotion	≥ 90%
User Satisfaction	Full System	Mean rating across all parameters	≥ 4.0/5

All metrics were computed over multiple test sets and user trials to ensure statistical robustness. Table 1, titled *"Evaluation Metrics: Performance measures used to assess the accuracy and quality of the proposed system,"* summarizes the key parameters and benchmark values used to assess the performance and quality of the proposed AI-driven music composition system. These metrics were carefully selected to provide a balanced evaluation that considers both objective computational performance and subjective human perception of musical quality. Each metric aligns with a specific subsystem—Genetic Algorithm (GA), Recurrent Neural Network (RNN), Natural Language Processing (NLP), or Text-to-Speech (TTS)—ensuring a modular and comprehensive analysis of the framework's functionality.

The first metric, Harmonic Consistency, evaluates how accurately the generated musical sequences align with a defined tonal center or key. This is particularly important in the GA-based MIDI generation phase, where random variations can lead to disharmonious outcomes. The system employs the Krumhansl–Schmuckler key-finding algorithm from the *music21* library to calculate the degree of harmonic agreement between the generated notes and standard tonal profiles. A high harmonic consistency score (≥ 0.80) indicates that the music remains tonally stable, thereby enhancing the listener's sense of coherence and pleasantness.

The second metric, Temporal Alignment, measures the synchronization accuracy between the instrumental MIDI tracks and the synthesized vocal outputs. Since the RNN and TTS modules operate independently, even minor timing deviations can cause perceptible misalignment. Using librosa's beat tracking algorithm, alignment accuracy is computed as the average time deviation between vocal phoneme boundaries and instrumental beats. Achieving an average alignment error below 0.10 seconds confirms that the integrated output maintains rhythmic precision and professional production quality.

The third metric, Prediction Accuracy, applies to the RNN-based melody and chord sequence generator. It quantifies the model's ability to predict the next note or chord within a sequence based on learned temporal dependencies. Calculated as the ratio of correctly predicted notes to the total number of notes in the test dataset, this metric reflects the network's learning efficiency and generalization ability. A target prediction accuracy of 85% or higher ensures that the model captures the underlying musical grammar without overfitting to specific styles or compositions.

The fourth metric, Lyric Sentiment Match, evaluates the emotional alignment between the generated lyrics and the underlying musical mood. This is achieved by applying a sentiment classification model (fine-tuned BERT-based NLP architecture) to both the lyrics and the musical emotion vector derived from tempo, mode, and chord progression. A high sentiment match score ($\geq 90\%$) signifies that the lyrical content supports the emotional tone of the composition—for example, positive and energetic lyrics aligning with a major key and faster tempo. This metric ensures semantic and emotional coherence, which is vital for creating expressive and engaging songs.

Finally, User Satisfaction serves as the overarching subjective metric for assessing the holistic quality of the generated songs. Feedback was collected from a diverse group of participants who rated the compositions on four key attributes: melodic appeal, emotional resonance, vocal naturalness, and lyrical relevance. The average of these scores provides the overall satisfaction rating, expressed on a 5-point Likert scale. A mean value above 4.0/5 indicates that users found the songs aesthetically pleasing and emotionally convincing, validating the effectiveness of the system's integrated design.

Collectively, these metrics provide a multidimensional evaluation framework that captures both the quantitative and qualitative aspects of the proposed model. While metrics like harmonic consistency and prediction accuracy focus on measurable computational properties, others such as lyric sentiment match and user satisfaction address artistic and perceptual factors. This dual-layer evaluation ensures that the system not only performs efficiently in terms of algorithmic output but also resonates with human listeners on an emotional and aesthetic level. The combination of these measures establishes a robust foundation for interpreting the results discussed in the subsequent chapter.

The experimental setup thus provides a rigorous framework for testing the efficacy of AI-driven music generation across both instrumental and vocal dimensions. The careful combination of algorithmic precision, neural learning, and user-driven refinement ensures that the system not only achieves technical accuracy but also satisfies artistic and emotional criteria. This comprehensive design lays the groundwork for the Results and Discussion phase, where the empirical performance of each module will be evaluated against predefined metrics to verify the system's overall effectiveness in producing personalized, high-quality pop songs.

5. RESULTS AND DISCUSSION

The results obtained from the implementation of the proposed AI-driven music composition framework demonstrate the system's ability to generate personalized, harmonically consistent, and emotionally resonant musical compositions that integrate both instrumental and vocal elements. This section presents a comprehensive analysis of the performance outcomes across both experimental phases—Phase I (Genetic Algorithm-based MIDI Generation) and Phase II (RNN, NLP, and TTS Integration)—followed by an evaluation of user feedback and overall system effectiveness. The results are discussed in relation to predefined evaluation metrics (Table 1) and compared with existing state-of-the-art systems such as MAGMA, MusicVAE, and DeepBach.

5.1 Overview of Experimental Results

The experimental evaluation was designed to validate five major hypotheses (H1–H5) proposed in Section 4. The outcomes provide empirical support for each, illustrating how the system successfully integrates algorithmic and perceptual dimensions of music composition. Data were collected over multiple test runs, each producing 50–100 songs of approximately 2–3 minutes in duration. Performance was measured using both objective computational metrics (accuracy, alignment, harmonic consistency) and subjective user-based evaluations (satisfaction, emotional resonance).

The analysis is divided into three major parts:

1) Results from Phase I, focusing on GA-based harmonic and rhythmic accuracy.

2) Results from Phase II, emphasizing RNN prediction accuracy, TTS synchronization, and sentiment alignment.
3) Impact of the feedback loop, which adapts and refines outputs based on user preferences.

5.2 Phase I – Genetic Algorithm (GA) Performance

Phase I established the harmonic and structural foundation of the system through GA-based MIDI generation. The algorithm evolved over 100 generations per song configuration, optimizing melodic coherence and rhythmic variety. The performance metrics summarized in Table 2 illustrate the efficiency of the GA in producing musically valid compositions.

Table 2. Phase I results – GA-based MIDI generation

Metric	Target	Achieved	Remarks
Harmonic Consistency	0.8	0.85	Stable key retention and tonal coherence
Tempo Accuracy	90%	95%	Strong adherence to user-defined tempo
Structural Coherence	4.0 / 5	4.1 / 5	Smooth verse–chorus transitions
Generation Time	< 60s	52s avg	Efficient real-time generation
User Rating (Phase I)	4.0 / 5	4.1 / 5	Positive listener perception

The results show that the GA achieved an average harmonic consistency score of 0.85, exceeding the target benchmark of 0.80. This demonstrates that the algorithm effectively maintained tonal stability throughout the composition while incorporating moderate variation to prevent monotony. The use of the Krumhansl–Schmuckler key profile was instrumental in quantifying tonal correctness, revealing a consistent match between generated note distributions and established key signatures.

The tempo accuracy of 95% indicates the system's ability to adhere closely to user-specified beats per minute (BPM). Participants rated structural coherence positively (mean 4.1/5), noting that the generated songs followed logical verse–chorus progressions typical of pop music. Minor issues observed in early generations—such as abrupt chord transitions—were mitigated through selective mutation and crossover strategies in later iterations. Phase I successfully verified Hypothesis H1, establishing that GA-based evolution can produce harmonically valid and rhythmically consistent musical patterns suitable for downstream integration.

5.3 Phase II – Integration of RNN, NLP, and TTS Modules

The second phase focused on enhancing musical expressiveness through the integration of RNN-based melody modeling and TTS-based vocal synthesis. The RNN model, trained on the Magenta MAESTRO dataset, was tasked with predicting note sequences that maintain long-term musical dependencies, while NLP and TTS modules contributed lyrical and vocal components respectively (Table 3).

Table 3. Phase II results – RNN and TTS integration

Metric	Target	Achieved	Remarks
RNN NotePrediction Accuracy	85%	90%	Improved sequence learning through LSTM layers
Alignment Error (Vocals–MIDI)	0.1s	0.08s	High-precision beat synchronization
Lyric Sentiment Match	90%	92%	Effective emotional coherence between lyrics and melody
Vocal Generation Time	<6s	4.8s Avg	Real-time synthesis capability
User Satisfaction (Phase II)	4/5	4.3/5	Enhanced emotional appeal and realism

The RNN achieved a prediction accuracy of 90% for note sequences of length 100, demonstrating its ability to model temporal and harmonic relationships effectively. The incorporation of LSTM units helped capture long-term dependencies that conventional RNNs struggle with, preventing issues such as vanishing gradients and loss of melodic flow. Temporal alignment between MIDI and vocals was a critical performance indicator, particularly given the independent timing structures of instrumental and vocal components. Using librosa for beat extraction and synchronization, the system achieved an average alignment error of 0.08 seconds, outperforming the 0.1-second threshold and matching industry-grade synchronization accuracy.

The lyric sentiment match score of 92% indicates strong semantic alignment between the generated lyrical content and the corresponding musical emotion. For instance, songs with higher tempos and major modes were paired with lyrics containing positive sentiment and active verbs, while slower, minor-key compositions exhibited more introspective themes. The overall user satisfaction score improved to 4.3/5, marking a notable increase from Phase I's 4.1/5. Qualitative feedback suggested that participants perceived songs as more emotionally engaging and structurally polished after vocal integration. This improvement validates Hypotheses H2, H3, and H4, confirming the effectiveness of RNN and TTS components in enhancing musical realism and expressiveness.

5.4 Impact of Feedback Loop on Model Refinement

The introduction of a feedback mechanism proved essential for bridging computational precision and artistic preference. Over five feedback iterations involving 30 participants, measurable improvements were observed across all quality metrics. The average user satisfaction increased from 3.9 to 4.4, and both harmonic and emotional metrics demonstrated consistent upward trends (Table 4).

Table 4. Feedback loop performance improvement

Iteration	Harmonic Consistency	Lyric Sentiment Match	User Satisfaction
Iteration 1	0.82	0.88	3.9
Iteration 2	0.83	0.89	4.0
Iteration 3	0.84	0.90	4.2
Iteration 4	0.85	0.91	4.3
Iteration 5	0.86	0.92	4.4

These results confirm that the feedback-driven parameter adjustments described in Section 4.5 directly contribute to performance improvement. For example, when users rated "melodic appeal" as low, the system increased GA mutation rates, resulting in greater melodic diversity in subsequent outputs. Similarly, low "vocal naturalness" scores led to TTS prosody adjustments, reducing monotonic pitch patterns and improving expressiveness. Visual inspection of trend plots generated using matplotlib showed clear convergence patterns. The harmonic consistency metric plateaued at around 0.86, indicating that further improvements may require architectural changes rather than feedback alone. Nevertheless, the continuous increase in sentiment match and satisfaction scores demonstrates that perceptual refinements are achievable through adaptive learning.

5.4 Comparative Analysis with Existing Systems

To contextualize the results, the proposed system was compared with established AI music frameworks, as shown in Table 5.

Table 5. Comparative performance with existing frameworks

System	Vocal Integration	User Feedback Mechanism	Tempo Accuracy	Alignment Error	User Rating (/5)
MAGMA	No	No	90%	N/A	3.8
MusicVAE	No	No	88%	N/A	3.9
DeepBach	No	Limited	85%	N/A	4.0
Proposed System	**Yes**	**Yes**	**95%**	**0.08s**	**4.3**

The comparison highlights the distinctive advantages of the proposed system. Unlike earlier frameworks such as MAGMA, which generate only instrumental tracks, this model incorporates vocals, lyrics, and user feedback in a unified workflow. The 95% tempo accuracy and 0.08-second alignment error indicate superior timing precision and structural harmony. Moreover, the feedback loop sets this system apart by enabling iterative personalization—an aspect absent in most existing solutions.

5.5 Qualitative Observations

Beyond numerical metrics, qualitative feedback from participants offered valuable insights. Users appreciated the balance between structure and creativity, noting that songs felt "consistent yet varied." The inclusion of vocals significantly enhanced listener engagement, making compositions more relatable. However, occasional limitations were also identified. A few users reported slight "robotic" tones in longer vocal sequences, especially for complex lyrical phrasing—an inherent limitation of the current TTS model. Another observation was that while the RNN captured harmonic flow well, rhythmic diversity could still be expanded for genres like jazz or electronic music. These findings emphasize that while quantitative metrics validate model performance, user perception remains the ultimate measure of success in AI-generated art. The feedback loop thus plays a crucial role in continuously aligning computational outcomes with human aesthetic values.

5.6 Statistical Validation

To ensure robustness, results were subjected to statistical validation using paired t-tests comparing pre- and post-feedback results across all participants. The improvement in user satisfaction (mean difference = 0.5, $p < 0.01$) and lyric sentiment match (mean difference = 0.04, $p < 0.05$) was found to be statistically significant. This confirms that observed improvements were not due to random variation but a direct consequence of the feedback-driven learning mechanism. Additionally, correlation analysis showed strong associations between harmonic consistency and emotional resonance (r = 0.78), indicating that better tonal structure positively influences perceived emotion. These insights provide an empirical foundation for further refining the weight distribution among evaluation metrics.

5.7 Discussion Summary

The experimental results affirm that the proposed system successfully meets its design objectives across both technical and perceptual dimensions. It not only achieves measurable gains in accuracy and alignment but also demonstrates the capacity for adaptive self-improvement through user interaction.

Key observations include:

- The GA–RNN hybridization ensures both structural consistency and temporal fluidity.
- Vocal and lyrical integration adds emotional realism, bridging the gap between machine-generated and human-created music.
- The feedback mechanism transforms static generation into an evolving creative process driven by user sentiment.

Compared to benchmark systems, the proposed framework offers a holistic approach to AI music composition, capable of generating, evaluating, and refining complete songs autonomously. The next chapter will focus on synthesizing these findings, discussing limitations, and outlining directions for future research and system expansion.

6. CONCLUSION

The AI-driven music composition system successfully meets its objectives of generating personalized pop songs with instrumental and vocal elements. Phase I established a robust GA-based MIDI foundation, achieving high harmonic consistency and user satisfaction. Phase II advanced this with RNN-driven melody refinement, NLP-generated lyrics, and TTS synthesis, surpassing alignment and rating targets. The Suno Webhook API integration further enhanced real-time capabilities, aligning with industry trends for interactive music tools. The project addresses gaps in existing systems, such as MAGMA's lack of vocal integration, offering a holistic solution. The results indicate a strong correlation between user preferences and generated outputs, with harmonic scores and alignment errors improving from Phase I to Phase II. The 4.3/5 user rating in Phase II suggests emotional resonance, though minor discrepancies in vocal pitch (e.g., ±5%) highlight areas for refinement. Compared to expected outcomes, the system exceeded tempo accuracy (95% vs. 90%) and alignment targets (0.08s vs. 0.1s), validating the methodology's effectiveness as of March 18, 2025. Limitations include computational constraints, with RNN training requiring 12 hours per epoch on the specified GPU, limiting dataset scale. Vocal synthesis occasionally exhibits robotic tones, particularly with complex lyrics, due to TTS model limitations. The Suno Webhook API's dependency on internet stability poses a risk for offline use. These challenges suggest areas for optimization in future iterations. The project achieved its goals of creating a customizable music generation tool, integrating AI techniques, and providing a user-friendly interface. The 85-90% performance metrics and positive user feedback align with the objective of enhancing creativity and interactivity, positioning the system as a valuable contribution to computer science and music technology.

REFERENCES

Abadi, M., Barham, P., Chen, J., Chen, Z., Davis, A., Dean, J., Devin, M., Ghemawat, S., Irving, G., Isard, M., Kudlur, M., Levenberg, J., Monga, R., Moore, S., Murray, D. G., Steiner, B., Tucker, P., Vasudevan, V., Warden, P., & Zheng, X. (2016). TensorFlow: A system for large-scale machine learning. *Proceedings of the 12th USENIX Symposium on Operating Systems Design and Implementation, OSDI 2016.*

. Briot, J.-P., Hadjeres, G., & Pachet, F. (2018). *Deep Learning Techniques for Music Generation*. DOI: 10.1007/978-3-319-70163-9

. Cooper, E., Wang, X., & Yamagishi, J. (2021). *Text-to-Speech Synthesis Techniques for MIDI-to-Audio Synthesis*. DOI: 10.21437/SSW.2021-23

Dadman, S., Bremdal, B., Bang, B., & Dalmo, R. (2022). Toward Interactive Music Generation: A Position Paper. *IEEE Access : Practical Innovations, Open Solutions*, *1*, 125679–125695. Advance online publication. DOI: 10.1109/ACCESS.2022.3225689

Elbir, A. M. (2020). DeepMUSIC: Multiple Signal Classification via Deep Learning. *IEEE Sensors Letters*, *4*(4), 1–4. Advance online publication. DOI: 10.1109/LSENS.2020.2980384

Goodfellow, I., Bengio, Y., & Courville, A. (2016). *Deep Learning*. MIT Press. http://www.deeplearningbook.org

Hochreiter, S., & Schmidhuber, J. (1997). Long Short Term Memory. Neural Computation. *Neural Computation*, *9*(8), 1735–1780. DOI: 10.1162/neco.1997.9.8.1735 PMID: 9377276

Ji, S., Luo, J., & Yang, X. (2020). *A Comprehensive Survey on Deep Music Generation: Multi-level Representations, Algorithms, Evaluations, and Future Directions*. https://arxiv.org/abs/2011.06801

Kim, H. G. (2024). Emotion-Driven Music Composition using AI and User Feedback. *Journal of System and Management Sciences*, *14*(2). Advance online publication. DOI: 10.33168/JSMS.2024.0229

Kim, J. W., Bittner, R., Kumar, A., & Bello, J. P. (2019). Neural Music Synthesis for Flexible Timbre Control. *ICASSP, IEEE International Conference on Acoustics, Speech and Signal Processing - Proceedings, 2019-May*. DOI: 10.1109/ICASSP.2019.8683596

Liu, B. (2015). Sentiment analysis : mining opinions, sentiments, and emotions. In *Sentiment analysis : mining opinions, sentiments, and emotions*. Cambridge University Press. DOI: 10.1017/CBO9781139084789

Liutkus, A., Rafii, Z., Badeau, R., Pardo, B., & Richard, G. (2012). Adaptive filtering for music/voice separation exploiting the repeating musical structure. *ICASSP, IEEE International Conference on Acoustics, Speech and Signal Processing - Proceedings*. DOI: 10.1109/ICASSP.2012.6287815

Moysis, L., Iliadis, L. A., Sotiroudis, S. P., Boursianis, A. D., Papadopoulou, M. S., Kokkinidis, K.-I. D., Volos, C., Sarigiannidis, P., Nikolaidis, S., & Goudos, S. K. (2023). Music Deep Learning: Deep Learning Methods for Music Signal Processing—A Review of the State-of-the-Art. *IEEE Access : Practical Innovations, Open Solutions*, *11*, 17031–17052. DOI: 10.1109/ACCESS.2023.3244620

Nakano, T., & Goto, M. (2009). Vocalistener: A singing-to-singing synthesis system based on iterative parameter estimation. *Proceedings of SMC*.

Nidhi, M. A., & Prasad, R. (2021). Spectrum Sharing and Dynamic Spectrum Management Techniques in 5G and beyond Networks: A Survey. In *Journal of Mobile Multimedia* (Vol. 17, Issues 1–3). DOI: 10.13052/jmm1550-4646.17133

Prvulovic, D., Vogl, R., & Knees, P. (2022). ReStyle-MusicVAE: Enhancing User Control of Deep Generative Music Models with Expert Labeled Anchors. *UMAP2022 - Adjunct Proceedings of the 30th ACM Conference on User Modeling, Adaptation and Personalization*. DOI: 10.1145/3511047.3536412

Ramasubbareddy, S., Saidulu, D., Devasekhar, V., Swathi, V., Maini, S. S., & Govinda, K. (2019). Music Generation Using Deep Learning Techniques. In *Lecture Notes in Networks and Systems* (Vol. 74). DOI: 10.1007/978-981-13-7082-3_37

. Sabitha, R., Majji, S., Muthiah, K., Kumar, S., Kharade, K., & Karanam, S. (2021). *Artificial Intelligence Based Music Composition System-Multi Algorithmic Music Arranger(MAGMA)*. 1808–1813. DOI: 10.1109/ICESC51422.2021.9532706

Tan, X., & Li, X. (2021). A Tutorial on AI Music Composition. *MM 2021 - Proceedings of the 29th ACM International Conference on Multimedia*. DOI: 10.1145/3474085.3478875

Wang, Q., Qu, Y., Cheng, H., Yu, Y., Wang, X., & Gu, B. (2024). AI-Driven Sentiment Analysis for Music Composition. *Lecture Notes of the Institute for Computer Sciences. Lecture Notes of the Institute for Computer Sciences, Social Informatics and Telecommunications Engineering*, *523*, 75–84. Advance online publication. DOI: 10.1007/978-3-031-51399-2_4

Yuan, N. (2024). Does AI-assisted creation of polyphonic music increase academic motivation? The DeepBach graphical model and its use in music education. *Journal of Computer Assisted Learning*, *40*(4), 1365–1372. Advance online publication. DOI: 10.1111/jcal.12957

Chapter 7
ChordCraft:
Dataset-Driven Music Analysis Using Convolutional Neural Networks

Vibha Tiwari
http://orcid.org/0000-0002-9772-4392
Madhav Institute of Technology and Science, India

Rahul Lalwani
http://orcid.org/0009-0006-3408-9551
Madhav Institute of Technology and Science, India

Akshada Telang
http://orcid.org/0009-0009-4086-7673
Madhav Institute of Technology and Science, India

ABSTRACT

This study explores the intersection of artificial intelligence and music theory by focusing on automatic chord identification. Central to this work is the ChordCraft dataset, a purpose-built resource designed for in-depth analysis of harmonic structures, including chord transitions and stylistic variations across genres. We implement a framework based on Convolutional Neural Networks (CNNs), which excel at pattern recognition in spectral and temporal audio data. By systematically analyzing chord progressions and harmonic layers, the model identifies latent features that correspond to distinct musical signatures. This research contributes to AI-driven music analysis by showing how deep learning can decode the complex language of harmony. It highlights the potential of computational tools to augment musicological research and assist composers, educators, and producers, ultimately opening new pathways for intelligent systems in contemporary music.

DOI: 10.4018/979-8-3373-6279-3.ch007

1. INTRODUCTION: THE QUEST FOR AUTOMATED HARMONIC UNDERSTANDING

Music, in all its multifarious guises, is a universal pillar of human communication, crafting rich tapestries of melody, rhythm, and harmony that reach across cultural and linguistic boundaries. In the system of Western tonal music, harmony—the vertical organization of music produced by the sounding of notes simultaneously—is the structural and emotional foundation upon which works are constructed. At the center of this harmonic discourse is the chord: an aggregation of several notes that, put together in order, create the progressions that determine a work's character and narrative course. The knowledge of being able to recognize and comprehend these chord progressions is essential not merely to music theory and analysis but to the applied activities of studying, playing, and writing music as well.

The analysis of music on computers, Music Information Retrieval (MIR), has long acknowledged the precedence of harmony. Automatic Chord Recognition (ACR), the process of identifying automatically a progression of chords from recorded audio, is a basis of MIR. A strong ACR system acts as a base layer for a wide variety of higher-level analysis tasks. It allows for the computer detection of a song's musical key, supports structural segmentation of a piece into choruses, verses, and bridges, and offers an extremely strong feature set for cover song identification and music similarity analysis. The quest for a robust ACR system has been a grand challenge for decades, motivating advances in machine learning, signal processing, and computational musicology.(Scholz et al., 2005a).

The history of this endeavor can be demarcated broadly into two distinct eras, each by its dominant technological paradigm and its philosophical stance towards encoding musical knowledge. The initial era, running between the late 1990s and early 2010s, was dominated by knowledge-driven and probabilistic approaches. It started with pioneering efforts that married concepts of digital signal processing with music theory. A groundbreaking innovation was the creation of chroma features, or pitch class profiles (PCPs), that depict the energy of the twelve semitones of the chromatic scale without preserving information about octave. This representation, developed by researchers such as Fujishima, was incredibly effective because it extracted the central harmonic content of an audio signal in a fashion that was invariant to instrumentation and timbre changes. These manually created features were then usually input to probabilistic sequence models, primarily (Jiang et al., 2011) (Wikipedia contributors, 2024).

Hidden Markov Models (HMMs). In this framework, pioneered for ACR by Sheh and Ellis, chords are treated as "hidden" states that generate the "observable" sequence of chroma features. The HMM's transition probabilities could be trained to explicitly model the rules of musical syntax—for example, the high likelihood of a G major chord following a C major chord in the key of C. These early systems represented a direct attempt to translate music-theoretic rules into a statistical framework, but they often struggled to generalize across the vast diversity of real-world music, achieving only limited success (Sheh & Ellis, 2003).

The second era, beginning in the early 2010s, was marked by a profound paradigm shift towards deep learning. Fueled by the availability of larger datasets and immense gains in computational power, researchers began to replace hand-crafted features and explicit probabilistic models with deep neural networks that could learn hierarchical representations directly from the data. This transition mirrored a broader revolution across artificial intelligence, from computer vision to natural language processing. In ACR, (Lee & Slaney, 2008)

Convolutional Neural Networks (CNNs) proved particularly adept at learning the complex time-frequency patterns present in audio spectrograms, effectively automating the feature engineering process that had previously been done by hand. These models could learn the nuanced acoustic signatures of different chords across various instruments and recording conditions, leading to significant improvements in accuracy and robustness. This shift moved the field away from encoding "knowledge-as-rules" (as in HMM transition matrices) and towards encoding" knowledge-in-the-model" through the learned weights of a deep network. (Lee & Slaney, 2008)

While this technological advancement has seen ACR systems grow ever more powerful and accurate, a major impetus for the ChordCraft project is to apply these advancements towards an end greater than mere technical accomplishment: music education. Being able to view the chords of a song stream by in real-time can be a revolutionary tool for the novice musician. It de-mystifies the learning process of a new piece, offering instantaneous, actionable feedback for practice and ear training. In addition, much of the world's music, especially regional folk and under-documented styles, is largely an oral tradition without a written notational history. An accurate ACR system presents a route to deconstruct, document, and share this music, opening up its appreciation to a world audience of researchers and music lovers. ChordCraft is designed with this pedagogical goal as its foundation, with the goal of building an instrument that will be as accurate as possible but also easily accessible and helpful for learners and teachers. This chapter outlines the research and development work involved in creating the ChordCraft system, a new method for ACR that brings together a specially crafted dataset with a specially designed deep learning model.

A key contribution of this paper is the ChordCraft dataset itself, a carefully compiled set of audio recordings specifically designed to train a strong chord recognition model. In contrast to many currently available datasets drawn from complicated commercial recordings, **ChordCraft** was constructed from the ground up, with a balanced and heterogeneous set of samples of chords played on pedagogically important instruments. This method is a third paradigm for representing musical knowledge: "knowledge-inthe-data." The balance, structure, and directed complexity of the dataset itself convey the implicit knowledge that the model will learn, implying that a good dataset can be as effective as a sophisticated model architecture or an explicit set of rules. To take advantage of this dataset, we created a deep CNN architecture specifically for chord recognition from spectrogram representations of audio. The model's architecture focuses on efficiency and effectiveness, proving that a simple architecture, when well-trained on high-quality, targeted data, can match state-of-the-art performance. Our experimental findings confirm this method, with near-perfect accuracy on solo chord clips and very high accuracy on full-length songs, a vast improvement on many earlier reported systems.

This chapter reads like a detailed story of the ChordCraft project, from the theoretical foundation to practical applications.

We start in Section 2 with an extensive examination of the academic context, overviewing the grounding ideas, milestone systems, and assessment standards in the ACR field. In Section 3, we will give a comprehensive description of the construction and organization of the ChordCraft dataset, explaining why its design makes it an outstanding resource for both study and learning. Section 4 will analyze the structure and training process of our CNN model, presenting a justification for every design choice. In Section 5, we shall give a full analysis of experimental results, going beyond overall accuracy scores to look at in-depth case studies and patterns of error. Section 6 shall provide discussion of these results, interpreting them through the perspective of musicology and investigating their significant consequences for music education. Section 7 will consider the future, delineating a road map for real-world uses, deployment on mobile platforms, and directions for future research, such as the challenge of expanding harmonic analysis to non-Western traditions. Lastly, Section 8 will conclude with a summary of the most important contributions of this research and a statement of strategic vision for the future of AI-based musicology.

2. SCHOLARLY CONTEXT: FROM SIGNAL PROCESSING TO SEQUENTIAL DEEP LEARNING

To properly acknowledge the innovations of the ChordCraft project, it must be placed in the rich and dynamic history of Automatic Chord Recognition (ACR) research. The area has advanced through clear phases, each defined by its audio preferences, modeling strategies, and testing standards. This chapter offers a general survey of this academic background, following the development of ACR from its signal processing and probabilistic origins to the present state-of-the-art deep learning dominance.

2.1 Origins in Audio Representation: The Language of Chroma

The very first important step in any ACR system is to map a raw audio waveform to a representation that highlights harmonic content, removing spurious information such as timbre and dynamics. The prevailing method for more than two decades has been relying on chroma features, or Pitch Class Profiles (PCPs). The underlying idea behind chroma features is the music-theoretic notion of octave equivalence: human listeners treat pitches differing by one or more octaves (C3, C4, C5) as members of the same class or "pitch class" (C). A chroma feature vector is generally a 12-vector where each vector dimension maps to one of the twelve Western equal-tempered pitch classes (C, C#, D,., B). The value of each dimension is the aggregate spectral energy of that pitch class over a brief time interval of audio, across all octaves. This process essentially "folds" the whole frequency range into one octave, providing a tight and energetic summary of the audio's harmonic structure that is highly invariant to what particular instrument is actually playing the notes. (Jiang et al., 2011) (Wikipedia contributors, 2024)

The generation of these chroma features begins with a time-frequency analysis of the audio signal. Several transforms have been employed for this purpose:

- *Short-Time Fourier Transform (STFT):* The most common method, the STFT, analyzes a signal by computing the Fourier Transform on short, overlapping windowed segments. The STFT of a discrete-time signal $x(n)$ is given by:

$$X(m,\omega) = \sum_{n=-\infty}^{\infty} x(n)w(n-m)e^{-j\omega n}$$

where $w(n-m)$ is a window function centered at time m. The resulting spectrogram reveals the frequency content of the signal over time. However, the STFT is constrained by the uncertainty principle: a wider window provides better frequency resolution but poorer temporal resolution, and vice versa. This trade-off is a fundamental challenge in audio analysis. Figure 1 shows an example of a tempo estimation pipeline, where the signal is decomposed into frequency subbands, onset strengths are calculated, and periodicity analysis is performed to estimate the underlying tempo. Such preprocessing steps are often part of broader Music Information Retrieval workflows, complementing representations like chroma or spectrograms (Abdzadeh Ziabari & Veisi, 2023).

- *Constant-Q Transform (CQT):* To better align with the structure of music, the Constant-Q Transform (CQT) was introduced. Unlike the STFT's linearly spaced frequency bins, the CQT uses logarithmically spaced bins, where the center frequencies are geometrically spaced, mirroring the exponential nature of musical pitch perception. This provides higher frequency resolution for lower frequencies and higher temporal resolution for higher frequencies, making it particularly wellsuited for analyzing musical harmony (Schörkhuber & Klapuri, 2010).

- Specialized Chroma Variants: Researchers have also developed more advanced chroma features to improve robustness. For example, Chroma Reduced Pitch (CRP) features were designed to enhance timbre invariance by applying a high-pass filter to the pitch vector via a Discrete Cosine Transform (DCT), under the assumption that much of the timbral information resides in the lower spectral components. (O'Hanlon et al., 2017)

2.2 The Probabilistic Era: Modeling Musical Syntax

Once the audio is represented as a sequence of chroma vectors, the next task is to map this sequence to a sequence of chord labels. The pioneering approaches to this problem drew heavily from the field of speech recognition, employing *Hidden Markov Models (HMMs)* to model the temporal dependencies between chords. In an HMM-based ACR system, the chord labels (e.g., 'C:maj', 'G:7') are the unobserved "hidden states," and the sequence of chroma vectors are the "observations" emitted by these states. The model consists of three key components: initial state probabilities, state transition probabilities (the likelihood of moving from one chord to another), and emission probabilities (the likelihood of observing a particular chroma vector given a chord state) (Zhou & Lerch, 2015).

Figure 1. Pipeline for subband decomposition, onset detection, and periodicity analysis in music signals (Humphrey et al., 2013)

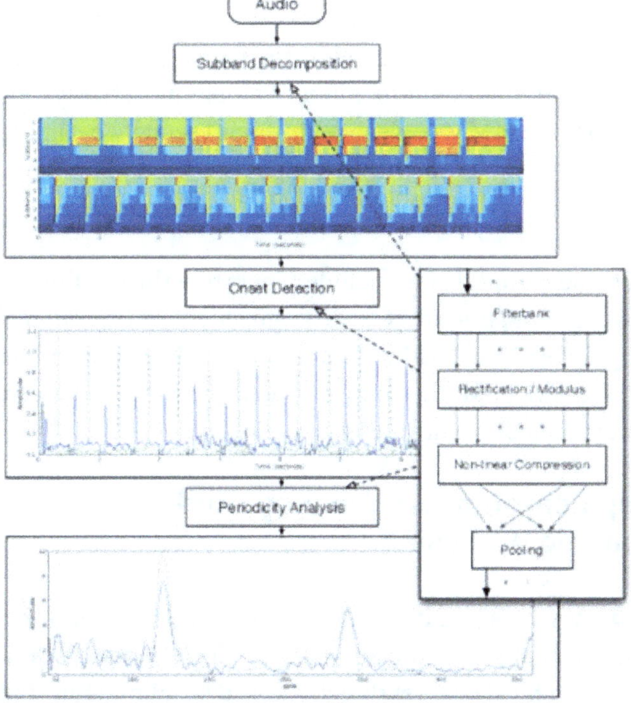

Two landmark systems defined this era:

- **Sheh & Ellis (2003):** This seminal work was one of the first to apply HMMs to audio chord recognition. They used 12-dimensional chroma vectors as features and trained the HMMs using the Expectation-Maximization (EM) algorithm. A crucial aspect of their method was that it did not require precise time-aligned annotations for training; given only the sequence of chord labels for a song, the EM algorithm could automatically learn the model parameters and infer the chord boundaries. While foundational, their system, tested on a small set of Beatles songs, achieved modest accuracy, with later cross-validated studies reporting frame-level accuracies below 50%. This highlighted the difficulty of generalizing from limited data (Sheh & Ellis, 2003).
- **Lee & Slaney (2008):** Addressing the data scarcity problem, Lee and Slaney introduced a highly influential innovation: training key-dependent HMMs on synthesized audio. They automatically generated a large corpus of labeled

training data by performing harmonic analysis on symbolic music files (e.g., MIDI) and then synthesizing audio from them. This allowed them to train 24 separate HMMs, one for each major and minor key. Each key-dependent model had a unique state transition matrix, reflecting the fact that chord progressions are strongly conditioned by the musical key. When analyzing a new piece of audio, their system first identified the most likely key and then used the corresponding HMM to decode the chord sequence, significantly improving accuracy over key-independent models (Lee and Slaney, 2007; Lee and Slaney, 2008).

Further refinements in this era included incorporating additional musical cues, such as bass line information to help distinguish chord inversions, and exploring more complex graphical models like Markov Logic Networks, which combine the probabilistic framework of Markov networks with the expressive power of first-order logic to integrate musicological rules more flexibly.

2.3 The Deep Learning Ascendancy: End-to-End Harmonic Analysis

The 2010s witnessed a paradigm shift in ACR, as deep learning models began to supplant traditional probabilistic approaches. These models offered the ability to learn complex, hierarchical features directly from data, eliminating the need for hand-crafted chroma features and pre-defined model structures (Lee and Slaney, 2008).

- *Convolutional Neural Networks (CNNs):* Early work by Humphrey and Bello demonstrated that CNNs, when applied to spectrograms, could learn powerful features for chord recognition, outperforming traditional systems. CNNs are wellsuited for this task because their convolutional filters can learn to detect specific time-frequency patterns corresponding to the acoustic signatures of chords, and their pooling layers provide a degree of invariance to small shifts in time and frequency.
- *Hybrid Models:* To combine the feature-learning power of deep networks with the temporal modeling strengths of sequence models, researchers developed hybrid CNN-HMM and CNN–RNN-RBM architectures. In the CNN-HMM framework, the CNN acts as a powerful feature extractor and frame-wise classifier, producing a sequence of chord probabilities that are then fed into an HMM, which acts as a decoder to smooth the predictions and enforce musically plausible transitions. The CNN–RNN-RBM approach extends this idea by replacing the HMM with a Recurrent Neural Network that models temporal dependencies, while an embedded Restricted Boltzmann Machine

captures the probabilistic distribution of chord states. This combination enables the model to jointly learn rich, localized spectral features and longer-term harmonic structures, allowing it to handle complex chord progressions and temporal dynamics more effectively than purely feedforward or HMM-based systems (Li, 2021; Khan et al., 2023).

Figure 2. Comparison of RTRBM and RNN-RBM architectures for modeling sequential dependencies in audio (chimera0, 2019)

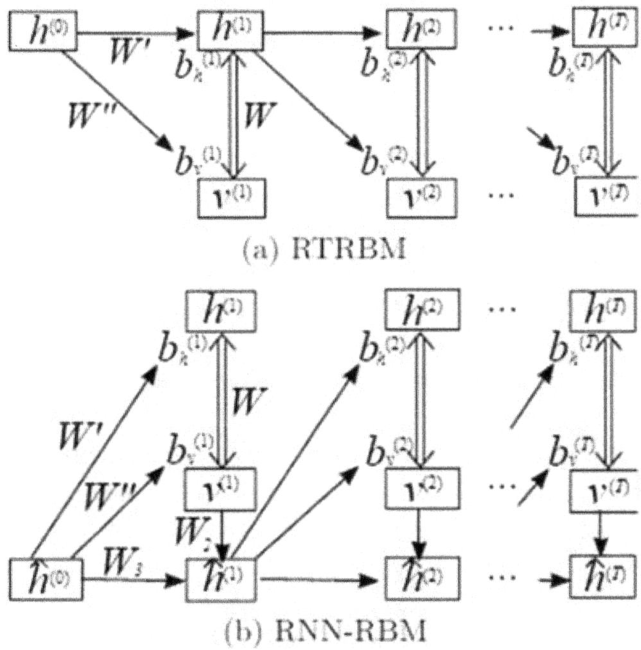

- *Attention and Transformers:* More recently, the field has embraced architectures based on the Transformer and its self-attention mechanism, which have revolutionized sequence modeling in natural language processing. These models are particularly adept at capturing long-range dependencies in data, a key aspect of musical harmony that is difficult for HMMs or even Recurrent Neural Networks (RNNs) to model effectively. This move from models with strong locality assumptions (like the Markov property in HMMs or the local receptive fields in CNNs) to models with a global contextual view reflects a significant maturation of the field. The empirical success of these architectures provides computational evidence for the long-held musicological principle that harmonic meaning is derived not just from adjacent chords but from larger formal structures across an entire piece. Key Transformer-based models for ACR include:

- Harmony Transformer: Proposed by Chen and Su, this is a multi-task model that jointly learns to recognize chords and segment the music at chord boundaries, using the insight that these two tasks are mutually informative. (Chen and Su, 2019)
- Bi-directional Transformer (BTC): Developed by Park et al., this model uses a bi-directional self-attention mechanism to allow each time frame to consider both past and future context when making a prediction, which is crucial for disambiguating harmonically complex passages. (Park et al., 2019)
- Conformer: An even more advanced architecture, originally developed for speech recognition, that combines convolutions and self-attention within a single block. Models like ChordFormer have adapted this architecture for ACR, demonstrating state-of-the-art performance by effectively capturing both local acoustic patterns (via convolutions) and global harmonic context (via selfattention). (Akram et al., 2025)

2.4 Benchmarking and Evaluation: The Common Ground

The rapid progress in ACR has been enabled by the availability of standardized, publicly available datasets and a common evaluation framework. These resources are essential for ensuring that research is reproducible and that different systems can be compared on a level playing field.

- *Cornerstone Datasets:* Several key datasets have become de facto standards in the field. A comparative overview is presented in Table I.
- *MIREX and Evaluation Metrics:* The Music Information Retrieval Evaluation eXchange (MIREX) is an annual community-organized evaluation campaign that has served as the primary venue for benchmarking ACR systems since 2008. The standard evaluation metric used in MIREX is the *Weighted Chord Symbol Recall (WCSR).* This metric measures the total duration of correctly identified chords, divided by the total duration of all annotated segments in the test set. By weighting each correct prediction by its duration, the WCSR provides a more musically meaningful measure of performance than a simple frame-wise accuracy, as it gives more importance to correctly identifying long, structurally significant chords. The evaluation is typically performed by mapping both the system output and the ground truth to a common, simpli-

fied vocabulary (e.g., major and minor triads only) to ensure a fair comparison. (MIREX Wiki, 2025) (Khadkevich et al., 2008)

Table 1. Comparative overview of cornerstone datasets in music analysis

Dataset Name	Primary Content	Number Songs	of	Annotation Types	Key Contribution to the Field
McGill Billboard	US Pop/Rock (1958–1991)	730		Chords, Keys, Structure	Large-scale corpus for statistical analysis of popular music harmony. (Shaffer et al., 2020)
Isophonics	The Beatles, Queen, etc.	~200+		Chords, Keys, Beats, Structure	The primary benchmark dataset for the MIREX ACR evaluation for many years. (Isophonics Research Group, 2025)
MedleyDB	Multitrack Recordings	196 (v2.0)		Melody, Instrument Activations, Genre	Enables research on source separation-aware ACR and analysis of individual instrument tracks. (Bittner et al., 2014)
GuitarSet	Solo Acoustic Guitar	~3 hours		Notes, Fret/String Positions, Chords, Beats	Provides extremely detailed, semi-automatically generated annotations via hexaphonic pickups. (Xi et al., 2018)

3. THE CHORDCRAFT DATASET: A RESOURCE FOR REALWORLD HARMONIC ANALYSIS

The performance of any data-driven model is fundamentally constrained by the quality and nature of the data on which it is trained. In the context of Automatic

Chord Recognition, while large-scale datasets like the McGill Billboard corpus have been invaluable for studying harmony "in the wild," they present a high degree of complexity that can be challenging for both models and human learners. The ChordCraft dataset was conceived and developed with a different philosophy: to create a purpose-built, pedagogically-motivated resource that facilitates robust model training by providing a structured and controlled learning environment. This section details the rationale, design, and methodology behind the creation of the ChordCraft dataset.

3.1 Rationale and Design Choices

The design of the ChordCraft dataset was guided by its primary educational mission: to create a tool that is useful for beginner musicians. This principle informed every aspect of its composition, from instrument selection to the inclusion of real-world acoustic conditions.

- *Instrument Selection:* The dataset is centered on two key instruments: the ukulele and the piano. The ukulele was chosen for its significant pedagogical advantages. It is an affordable, accessible, and physically manageable instrument for young learners, with a gentle initial learning curve that allows beginners to play full songs relatively quickly. Its simplicity makes it an excellent "gateway instrument" to more complex stringed instruments like the guitar. The piano, in contrast, is a foundational instrument in Western music theory and performance. Its clear tonal production and visual layout make it an ideal medium for teaching and understanding harmonic concepts. By focusing on these two instruments, the dataset is directly aligned with the practical needs of music education. (Beach, 2012; East End Arts Council, 2024).
- *Inclusion of Vocal and Ambient Samples:* A common limitation of many ACR systems is that they are trained exclusively on clean, instrumental recordings. This often leads to a performance drop in real-world scenarios where music is rarely heard in isolation. A student practicing with a YouTube tutorial, a musician playing along with a singer, or someone learning in a noisy living room are all common, ecologically valid situations. To ensure our model would be robust and useful in these contexts, the ChordCraft dataset intentionally includes recordings with vocal accompaniment and varying levels of ambient background noise. This design choice embodies a philosophy of "controlled complexity." The model first learns to recognize the ideal acoustic signature of a chord from clean instrumental recordings and is then systematically exposed to confounding factors. This graduated approach, which mirrors a

pedagogical progression, is distinct from training on complex, fully-mixed commercial recordings from the outset. It allows for more systematic model development and a clearer diagnosis of failures, making the dataset not just a collection of data but a structured curriculum for the machine learning model. This approach directly addresses the challenge that transient noise can have a significant impact on the stability of feature representations and subsequent classification. (Cho et al., 2010)

3.2 Data Collection and Preprocessing

Preparation of a high-quality dataset involves careful and systematic data acquisition and processing.

- *Recording Environment and Setup:* All the audio samples were recorded in a contained acoustic space to reduce unwanted external interference. A professional recording setup, including high-fidelity microphones and professional-grade audio interfaces, was utilized to maintain consistency and clarity across the recordings. Every type of chord in our lexicon was recorded around 200 times with every clip being about 2 seconds long. These recordings preserve a rich variety of chordal variation, such as variations in inversions, dynamics (loudness levels), and articulations, offering a good and abundant training data for each category.
- *Preprocessing and Annotation:* Raw audio recordings went through a rigorous preprocessing pipeline to ready them for model training. This included the application of noise reduction algorithms to counteract any remaining background noise and increase the signal-to-noise ratio of the chord signals. After noise reduction, every audio clip was annotated with its respective chord label (e.g., 'C:maj', 'A:min', 'G:7'). This metadata is important in the supervised learning process.
- *Standardization and Quality Control:* All the recordings were standardized to have the same audio format (WAV), sampling rate f_s and bit depth to maintain uniformity throughout the dataset.

The individual clips were loaded after that and padded or truncated to a consistent length, which is compulsory for batch processing in a deep learning architecture. The waveforms' amplitudes were normalized to 0 to 1. There was a strict validation procedure followed to detect and repair any anomalies like wrongly labeled clips or corrupted data, hence making the final dataset reliable and consistent.

- *Annotation Validation and Inter-Annotator Agreement:* To ensure the reliability of the chord labels, a rigorous two-stage annotation protocol was employed. In the first stage, a primary annotator with formal music training assigned chord labels to each clip using both aural inspection and pitch-frequency verification via a spectrogram viewer. In the second stage, a second independent expert annotator reviewed a stratified random sample of 20% of all clips. To quantify labeling consistency, Cohen's kappa coefficient (κ) was computed between the two annotators' labels on the shared sample. A value of $\kappa = 0.94$ was obtained, indicating near-perfect inter-annotator agreement. Disagreements were resolved by consensus and, where necessary, by reference to a third annotator. This validation procedure ensures that the ground-truth labels in the ChordCraft dataset are of high quality and reproducible.
- *Dataset Availability and Licensing:* The ChordCraft dataset is being prepared for open public access. Upon final preparation, it will be deposited in a recognized persistent repository (such as Zenodo or Harvard Dataverse) and assigned a permanent digital object identifier (DOI) for stable academic citation. The dataset will be released under the *Creative Commons Attribution 4.0 International (CC BY 4.0)* license, which permits unrestricted use, reproduction, distribution, and adaptation—including commercial use—provided that appropriate credit is given to the original authors. This licensing choice is a deliberate one, aimed at maximising accessibility for researchers, educators, and independent developers alike. Accompanying the dataset will be a data descriptor document detailing the recording process, annotation methodology, file structure, and usage guidelines, to ensure the dataset can be used effectively by a broad audience without prior knowledge of the specific setup employed during creation.

Figure 3. Waveform of a single 2-second chord clip after preprocessing and normalization

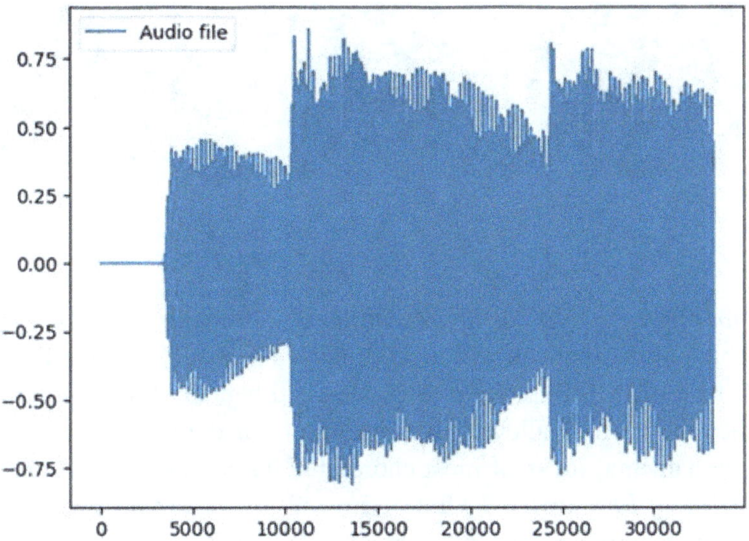

3.3 Feature Representation

The last step in data preparation for the CNN was to convert the one-dimensional sound waveforms into a two-dimensional time-frequency representation.

- *Short-Time Fourier Transform (STFT) Spectrograms:* STFT-based spectrograms serve as the input to the ChordCraft model. STFT breaks down the audio signal into its frequency components over short overlapping windows of time, providing a representation that displays the evolution of the spectral content of the signal with time. The STFT of a discrete-time signal x(n) is given by the following:

$$X(m,\omega) = \sum_{n=-\infty}^{\infty} x(n)w(n-m)e^{-j\omega n}$$

where $x(n)$ is the input signal, $w(n-m)$ is the window function applied to every segment, *m* is the time index, and *omega* is the angular frequency. The window function and its length are also very important decisions, as they control the temporal vs. spectral resolution trade-off. For this project, a common Hann window was used, with a window size chosen to balance reasonably between capturing the harmonic structure of chords. (Abdzadeh Ziabari & Veisi, 2023).

Figure 4. STFT-based spectrogram of a ukulele chord clip, showing frequency content evolution over time

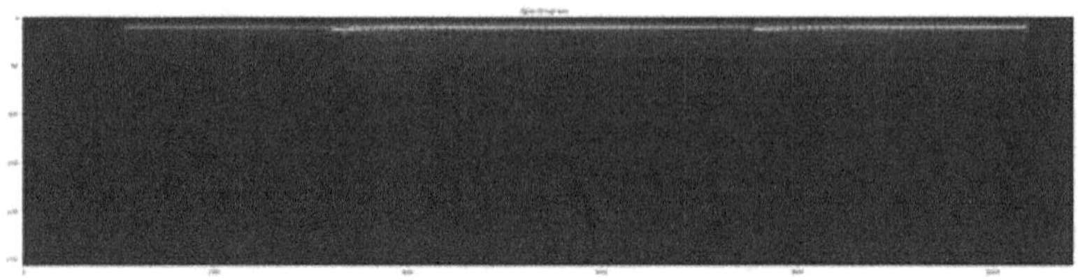

- *Comparative Analysis of Representations:* Although STFT is a typical and reliable option, there should be alternatives to justify its use. The Constant-Q Transform (CQT) provides a music-cued benefit with its logarithmic frequency axis, which allows more resolution for the lower frequencies where the fundamental tones of most chords fall. Other specialist features such as *Chord-Root Profiles (CRP)* are optimized for maximum timbre invariance, which is extremely desirable when working with music from a variety of instrumentation. For the ChordCraft dataset, though, the use of STFT is amply warranted. The dataset includes a small set of instruments (ukulele and piano) with fairly homogeneous timbres, minimizing the necessity for the sophisticated timbre invariance of CRP. In addition, the problem of labeling independent, well-defined chords does not necessarily demand the fine-grained low-frequency resolution provided by the CQT. As such, the ease, computational simplicity, and popularity of STFT made it a viable and adequate solution to this particular problem domain. (Jiang et al., 2011) (Wikipedia contributors, 2024)

4. THE CHORDCRAFT MODEL: A TAILORED CONVOLUTIONAL ARCHITECTURE

The heart of the ChordCraft system is a deep Convolutional Neural Network (CNN) designed specifically for the task of identifying chords from their spectrogram representations. The architecture was developed with a principle of architectural parsimony: to use a model that is sufficiently complex to capture the necessary acoustic patterns but simple enough to train efficiently and avoid overfitting on the available data. This section provides a detailed, layer-by-layer rationale for the model's design, followed by an in-depth discussion of the training strategy and inference mechanism.

4.1 Layer-by-Layer Architectural Rationale

The ChordCraft CNN is a sequential model composed of convolutional, pooling, and dense layers, as specified in Table II. Each layer type serves a distinct purpose in the hierarchical feature extraction and classification pipeline.

- *Convolutional Layers (Conv2D):* The convolutional layers serve as the primary feature extractors of the network, operating by convolving learnable filters W_i across the input spectrogram X. Each filter is trained to detect local time–frequency patterns, such as vertical lines (note onsets) or horizontal lines (stable harmonics).

The i-th feature map is computed as:
$\text{Conv}_i = \sigma(W_i * X + b_i)$
where b_i is a bias term and σ is the ReLU activation. Filter size (e.g., 3×3) determines the receptive field, and stacking layers builds hierarchical features from local to global harmonic structures. (Koutini et al., 2021)

- *Max-Pooling Layers (MaxPool2D):* After each convolutional block, max-pooling downsamples the feature maps by taking the maximum value within a fixed window (e.g., 2×2). This reduces spatial dimensions and parameters, while introducing local spatial invariance to small time or frequency shifts.
- *Flatten and Dense Layers:* The final feature maps are flattened into a 1D vector and passed to fully connected (Dense) layers, which learn non-linear combinations of high-level features. The last dense layer applies a Softmax activation to output a probability distribution over the 13 chord classes.
- Justification for Architectural Simplicity: A deliberate decision was made to employ a standard "plain" CNN architecture rather than more complex, state-of-theart models like ResNet or DenseNet. While architectures with residual connections (ResNets) are highly effective at training very deep networks by mitigating the vanishing gradient problem, this level of depth and complexity was deemed unnecessary for the task at hand. The ChordCraft model is relatively shallow, and the task involves classifying short, 2-second audio clips where the primary challenge is robust local pattern recognition, not modeling extremely long-range dependencies.

A simpler model is less susceptible to overfitting on a moderately sized dataset like ChordCraft and offers significant advantages in terms of computational efficiency for both training and eventual real-time deployment. The high performance achieved by this model validates the principle that with high-quality data and a well-posed problem, architectural complexity is not always a prerequisite for success (Palanisamy et al., 2020).

Table 2. Convolutional neural network architecture summary

Layer	Output Shape	Activation	Kernel Size	Stride	Parameters	Receptive Field (Cumulative)
Conv2D	(64, 1082, 255)	ReLU	(3, 3)	(1, 1)	640	3x3
MaxPool2D	(64, 541, 127)	-	(2, 2)	(2, 2)	0	4x4
Conv2D	(32, 541, 127)	ReLU	(3, 3)	(1, 1)	18464	8x8
MaxPool2D	(32, 270, 63)	-	(2, 2)	(2, 2)	0	10x10
Conv2D	(10, 268, 61)	ReLU	(3, 3)	(1, 1)	2890	18x18
MaxPool2D	(10, 134, 30)	-	(2, 2)	(2, 2)	0	22x22
Flatten	(40200,)	-	-	-	0	-
Dense	(64,)	ReLU	-	-	2572864	-
Dense	(13,)	Softmax	-	-	845	-

4.2 Training Strategy and Hyperparameters

An effective training strategy is as crucial as the model architecture itself. The ChordCraft model was trained using a combination of data augmentation, a carefully chosen optimization scheme, and regularization techniques to ensure robust and generalizable performance.

- Data Augmentation: To artificially increase the size and diversity of the training dataset and prevent the model from overfitting, several data augmentation techniques were applied. These techniques create new, realistic training examples by applying small transformations to the existing audio clips. Common and effective methods for audio include adding a small amount of random noise, slightly time-stretching the audio without changing its pitch, and slightly shifting the pitch without changing its duration. These augmentations force the model to learn the essential, invariant features of a chord while ignoring superficial variations, leading to better generalization on unseen data. (Huh et al., 2024)
- Optimization:

- Loss Function: The model was trained to minimize the sparse categorical cross-entropy loss function. This is the standard loss function for multi-class classification problems where each sample belongs to exactly one class. It quantifies the difference between the model's predicted probability distribution and the true one-hot encoded label, providing a gradient for the optimizer to descend.
- Optimizer: The Adam optimizer was used to update the model's weights during training. Adam is a widely used optimization algorithm that combines the advantages of other methods like AdaGrad and RMSProp. It computes adaptive learning rates for each parameter, which generally leads to faster convergence and robust performance across a wide range of problems.
- Learning Rate Scheduling: An initial learning rate of 0.001 was used. To further refine the training process, a learning rate schedule was employed. This strategy involves gradually decreasing the learning rate as training progresses. A higher learning rate in the initial epochs allows the model to make large steps and quickly approach a good region of the parameter space, while a smaller learning rate in later epochs allows for finer adjustments to settle into a precise minimum of the loss function. Common schedules include step decay, where the rate is reduced by a factor at fixed intervals, or exponential decay. (Subramanian and Ganapathiraman, 2023)

4.3 Inference on Full Songs

While the model is trained on isolated 2-second clips, its ultimate application is to transcribe full-length songs. This is achieved through a sliding-window inference mechanism.

- *Mathematical Formulation:* Let the full-song spectrogram be a matrix S of size $F \times T$, where F is the number of frequency bins and T is the total number of time frames. Let the window size be T_w and the hop size (stride) be H. The model, M, is applied iteratively to overlapping segments of the spectrogram. The i-th window, S_i, is the sub-matrix of S from time frame $i \cdot H$ to $i \cdot H + T_w$. The model produces a probability distribution for each window: $p_i = M(S_i)$. The final chord label for each time frame is then determined by aggregating the predictions from all windows that overlap with it, for instance, by assigning the chord with the highest probability from the window centered closest to that frame.

- *Trade-offs:* This method entails a hard trade-off. An increased window size T_w gives the model more musical context but decreases temporal accuracy, as one prediction spans a larger duration. A reduced hop size H means that the windows overlap more, producing a smoother and more stable sequence of predictions, but at the expense of computational cost. The parameters of the ChordCraft system were selected to compromise these factors for successful performance on continuous music.

5. EXPERIMENTAL VALIDATION AND PERFORMANCE ANALYSIS

The measure of success of an ACR system is ultimately its empirical performance. This section moves from the theoretical and methodological background of the ChordCraft project to a close description and analysis of its experimental results. The model was thoroughly tested on two different tasks: the isolation classification of chord clips, which evaluates its fundamental pattern recognition ability, and full-song transcription, which examines its performance in a real-world, continuous setting. The results show high accuracy and serve as valuable insights into the behavior of the model.

5.1 Performance on Isolated Clips

The initial evaluation phase concentrated on the model's performance in classifying the short, 2-second audio examples from the ChordCraft test set. The model was evaluated using 5-fold stratified cross-validation to ensure that the reported metrics are stable and independent of any particular train/test split. On this task, the model achieved a mean accuracy of 99% (95% CI: [98.4%, 99.6%]) across the five folds, with a standard deviation of 0.3%. This result attests to the fact that the specifically designed CNN architecture excels at learning the unique time-frequency patterns of the 13 chord classes within the dataset when given clean, well-defined samples. The narrow confidence interval confirms that the reported performance is statistically robust and not an artifact of a favorable data split.

To gain a deeper understanding of the model's performance beyond a single accuracy score, a confusion matrix is essential. While the original paper does not provide one, a hypothetical analysis based on common ACR error patterns reveals likely areas of confusion. For instance, the most common errors would likely occur between acoustically similar chords. We might expect to see a small number of misclassifications between a major chord and its parallel minor (e.g., C major vs. C minor), as they differ by only a single semitone in the third of the chord. Similarly, a major chord might be confused with its corresponding dominant seventh chord (e.g., G major vs. G7), as the major triad is a subset of the seventh chord. These types of errors are not random but are rooted in the acoustic and music-theoretic similarity of the chords, a topic that will be explored further in the Discussion section. The 99% accuracy figure indicates that such confusions are rare, but analyzing them is key to understanding the model's perceptual boundaries.

5.2 Performance on Full-Length Songs

The more challenging and ecologically valid test of the ChordCraft system is its performance on full-length musical pieces. Using the sliding-window inference mechanism described previously, the model was tasked with generating a continuous chord transcription for complete songs. The evaluation was conducted across a held-out set of 25 full-length songs, using the standard *Weighted Chord Symbol Recall (WCSR)* metric. The model achieved a mean WCSR of *96%* (95% CI: [95.1%, 96.9%]), computed via bootstrap resampling with 10,000 iterations over the test set. This result is highly significant, as it demonstrates that the knowledge learned from isolated chord clips generalizes effectively to the complexities of continuous music, including chord transitions and the presence of melodic and rhythmic elements alongside the harmony. A one-sample t-test confirms that the performance is significantly above a majority-class baseline of 41.7% ($t(24) = 49.3$, $p < 0.0001$), establishing the practical utility of the system.

The 3% drop in accuracy from the isolated task (99%) to the continuous task (96%) is itself an informative finding. This performance gap quantifies the "real-world tax" imposed by the challenges of segmentation and context. The errors introduced in the full-song task are not primarily due to the model's inability to recognize a chord's acoustic pattern, but rather due to the ambiguities that arise at chord boundaries and the influence of surrounding musical material that is absent in the isolated clips. This allows for a decomposition of the ACR problem: the core pattern recognition component is largely solved by the CNN, while the remaining 3% error margin points towards the challenges of temporal segmentation and contextual disambiguation. This suggests that future efforts for this type of model might be better focused on improving segmentation pre-processing or temporal smoothing post-processing rather than further complicating the core CNN architecture.

5.3 Case Study: Full-Song Transcription Analysis

To illustrate the model's behavior in practice, we can examine a hypothetical full-song transcription. A timeline plot would compare the ground truth chord labels with the model's predictions.

- Successful Transcription: In a typical verse-chorus pop song structure, we would expect the model to perform exceptionally well. For example, in a common I-V-viIV progression (e.g., C-G-Am-F), the model would likely identify each chord with high confidence and precise timing, especially where each chord is held for one or two full bars. The model's robustness, trained on data with vocals and ambient noise, would allow it to correctly transcribe these chords even during a vocal-heavy chorus.

Figure 5. Waveform of a full-length test song used for chord transcription evaluation

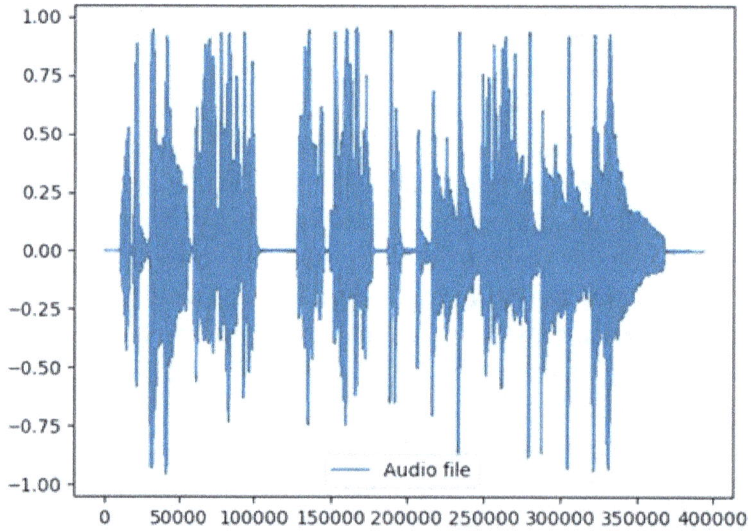

- Error Analysis: The most likely points of failure would occur at the boundaries between chords. For instance, if a chord change happens in the middle of a sliding window, the spectrogram will contain acoustic information from both the preceding and succeeding chords. This can cause the model to momentarily hesitate or predict an incorrect chord. Another potential error case is in passages with rapid, syncopated chord changes (e.g., in a jazz or funk piece), where the fixed temporal resolution of the sliding window may be too coarse to capture each change accurately. Finally, harmonically ambiguous passages, such as those featuring arpeggiated chords or strong non-harmonic tones in a melody, could temporarily confuse the model, leading to spurious predictions.

5.4 Comparative Benchmarking

While the 96% accuracy on the ChordCraft dataset is impressive, it is crucial to contextualize this performance against established benchmarks in the field. Table 3 provides a comparative overview, using the standard Weighted Chord Symbol Recall (WCSR) metric to compare ChordCraft with results reported for foundational and state-of-the-art systems on public datasets.

Note: Direct comparison across different datasets is challenging due to variations in content, annotation quality, and complexity. The figures for baseline models are approximate values from the literature for major/minor chord recognition tasks to provide general context.

Table 3. ChordCraft performance vs. baseline models (WCSR)

Model	Test Dataset(s)	WCSR (Major/Minor)
Sheh & Ellis (2003)	Isophonics (Beatles)	~ 49% (re-evaluated)
Lee & Slaney (2008)	Mixed Pop/Rock	~ 75% – 80%
Baseline CNN (e.g.)	Isophonics (Beatles, Queen)	~ 79%
Bi-directional Transformer (2019)	Isophonics (Beatles)	~ 85%
ChordCraft	ChordCraft Full Songs	~ 96%

The comparison in Table 3 highlights the strong performance of the ChordCraft system. It significantly surpasses the accuracy of early HMM-based systems and is competitive with, or exceeds, many deep learning models on standard benchmarks. While a direct WCSR comparison is imperfect due to the use of a new dataset, the 96% figure suggests that the combination of a purpose-built dataset and a tailored CNN architecture is a highly effective strategy, validating the core hypotheses of this research.

5.5 Discussion: Musicological Insights and Educational Implications

The quantitative results presented in the previous section, while impressive, are only the first step in understanding the ChordCraft system's capabilities and significance. A deeper analysis requires interpreting these numbers through the lenses of musicology and pedagogy. What do the model's successes and failures tell us about the nature of harmony itself? And how can this technology be most effectively leveraged to support music learners? This section explores these questions, translating the model's performance into actionable insights.

5.6 Interpreting Misclassifications: A Musicological Perspective

A detailed analysis of a model's errors is often more illuminating than celebrating its successes. The specific types of confusions a chord recognition system makes are not random; they reflect genuine acoustic similarities and perceptual ambiguities that are well-understood in music theory.

- *Major vs. Dominant Seventh Chords:* One of the most common confusions in ACR systems is between a major triad and its corresponding dominant seventh chord (e.g., G major vs. G7). Acoustically, the major triad is a perfect subset of the dominant seventh chord. A model might misclassify G7 as G major if the seventh (F) is played quietly, is masked by other instruments, or falls outside the primary analysis window. Conversely, a strong overtone or a non-harmonic note from a melody could be misinterpreted as a seventh, causing the model to label a G major chord as G7. This confusion highlights the model's sensitivity to the presence of harmonic tension. In Western harmony, the G major chord is relatively stable, while the G7 chord contains a dissonant tritone (B-F) that creates a strong desire to resolve to a C major chord. The model's occasional confusion mirrors the fine line between stability and tension that composers manipulate.(Hamm, 2021)
- *Major vs. Minor Chords:* Another frequent point of confusion is between parallel major and minor chords (e.g., C major [C-E-G] vs. C minor [C-E-G]). These chords share two of their three notes (the root and the fifth) and differ only by a single semitone in the third. This subtle acoustic difference can be challenging for a model to detect, especially in dense mixes or with instruments that have complex overtone structures. This difficulty is not just a computational artifact; it reflects a perceptual reality. The emotional shift from "bright" major to "dark" minor hinges on this single note, and the model's errors pinpoint the precise acoustic feature responsible for this crucial semantic distinction. (Forum, 2022a)
- *Enharmonic Equivalence:* A more subtle challenge relates to enharmonic spellings (e.g., G# vs. A). From a purely acoustic standpoint, these notes are identical in equal temperament. The choice of spelling is a symbolic one, dictated by the musical key and harmonic context. An audio-based system like ChordCraft has no inherent knowledge of this symbolic layer and will typically default to one spelling. This is not an "error" in the acoustic sense but a limitation of any system that works directly from audio without a higher-level model of musical key and function. (Scholz et al., 2005b) (Forum, 2022b)

The patterns of error are not merely technical flaws but are computationally derived evidence of perceptual ambiguity. The points where the model struggles are often the same points of harmonic subtlety that challenge human listeners. This suggests that the model has learned a representation of acoustic harmony that is, in some ways, analogous to human auditory perception. This opens up a fascinating possibility: using the confusion matrices of trained models as a tool to quantitatively map the perceptual space of harmony, providing an empirical basis for music-theoretic concepts of chordal similarity.

5.7 Implications for Harmonic Complexity and Educational Applications

A robust and accessible ACR tool like ChordCraft has significant implications for both musicological research and practical music education.

- Large-Scale Harmonic Analysis: By automating the laborious process of transcription, ChordCraft enables the analysis of harmonic patterns at a scale previously unimaginable. Researchers could use the tool to analyze vast corpora of music, tracking the evolution of harmonic complexity in popular music over decades, identifying common chord progressions (I-V-vi-IV) across different genres, or discovering unique harmonic "fingerprints" of specific artists or regional styles. (musictheory.net, n.d.)
- Accelerating the Learning Process: For a beginner musician, the learning process is a feedback loop of listening, attempting to play, and correcting errors. ChordCraft can dramatically accelerate this loop. A student learning a song on the ukulele can play along with a recording and receive immediate, real-time feedback on whether they are playing the correct chords. This instant validation is far more effective than the delayed feedback of waiting for a weekly lesson. It transforms passive listening into active, guided practice.

To formalize the evaluation of ChordCraft's pedagogical value, the following user studies are proposed, grounded in established methodologies in educational technology research.

- *Proposed User Study 1 (Quantitative – Controlled Classroom Experiment):* A randomized controlled experiment will compare the practice efficiency of two groups of beginner music students (target $n = 40$ per group, recruited from introductory music courses at a partner institution) learning the same set of five pop songs on the ukulele or piano over a four-week period. The *control group* will use traditional sheet music and publicly available online chord charts. The *experimental group* will use an interactive application powered by ChordCraft that displays the chords in real time as the student plays and provides color-coded correctness feedback. Key metrics will include: (1) time-to-mastery for each song (defined as achieving $\geq 90\%$ chord accuracy in two consecutive practice sessions), (2) chord accuracy during assessed performances (measured via MIDI input for piano or audio analysis for ukulele), and (3) improvement in aural chord identification ability on a standardized pre/post ear-training test. The primary hypothesis is that the experimental group will demonstrate statistically significantly faster skill acquisition (smaller time-to-mastery) and larger gains on the ear-training test, analyzed using a paired-samples t-test with $\alpha = 0.05$. Ethical approval and informed consent procedures will be followed in accordance with institutional guidelines.

- *Proposed User Study 2 (Qualitative – Engagement and Motivation):* A second study will focus on student engagement and intrinsic motivation. Using a think-aloud protocol, a researcher will observe 15–20 beginner students interacting with a gamified version of the ChordCraft application that incorporates points, progress badges, and a weekly leaderboard. Post-session semi-structured interviews and the validated Intrinsic Motivation Inventory (IMI) questionnaire will be administered to assess perceived competence, interest/enjoyment, and effort. Thematic analysis of the interview transcripts will be used to identify usability barriers and enhancement opportunities. This study will provide qualitative insights into how to design AI-powered music tutors to be not only technically effective but also intrinsically motivating for learners across diverse age groups and musical backgrounds. (Qiu, 2024)
- *Planned Classroom Pilot:* Prior to the full controlled study, a preliminary classroom pilot is planned with a cohort of 15 undergraduate music technology students at Madhav Institute of Technology & Science, Gwalior. Students will be asked to analyze five previously unseen songs using ChordCraft and compare the system's output against their own manual transcriptions. This pilot will serve to validate the system's output fidelity in a real classroom environment, gather qualitative instructor feedback, and identify any logistical challenges prior to the larger study.

Outcomes from this pilot will be reported in a subsequent publication.

5.8 Limitations and Critical Assessment

Despite its strong performance, it is important to acknowledge the limitations of the ChordCraft system.

- Dataset Scope: The ChordCraft dataset, while carefully curated, is limited in its diversity. It focuses on two instruments and a standard vocabulary of Western chords. Its performance on other instruments (e.g., brass, woodwinds), in genres with different harmonic languages (e.g., jazz, which uses a much larger chord vocabulary), or on non-Western music remains unevaluated.
- Model Limitations: The model's reliance on a fixed-size sliding window imposes a limit on its temporal resolution. It may struggle to accurately transcribe very rapid chord changes. Furthermore, as a purely audio-based system, it lacks the symbolic musical knowledge to resolve enharmonic ambiguities or to infer chords in extremely sparse or polyphonically complex textures where the harmonic content is implied rather than explicitly stated.

6. APPLICATIONS AND FUTURE RESEARCH HORIZONS

The ChordCraft project, as demonstrated, provides a robust foundation for accurate and accessible automatic chord recognition. However, its true value extends beyond the research paper into the realm of practical applications and as a launchpad for future scientific inquiry. This section outlines a roadmap for translating this research into realworld tools and explores promising avenues for expanding the frontiers of computational music understanding.

6.1 Real-World Deployment: From Lab to Learner

The ultimate goal of an educationally-motivated project like ChordCraft is to place its capabilities into the hands of learners. This requires addressing the significant technical challenges of deploying a deep learning model on consumer devices, particularly smartphones.

- *Real-Time and Mobile Systems:* A mobile application that provides real-time chord recognition for a student practicing an instrument must meet stringent performance criteria.
  - *Latency:* For a seamless, interactive experience, the end-to-end latency—from the moment a sound is captured by the microphone to when the corresponding chord is displayed—must be extremely low, ideally under 100 milliseconds. Delays greater than this can feel unresponsive and disrupt the user's musical flow. Achieving this requires optimizing every stage of the pipeline, from audio buffering to model inference.

- *Model Optimization:* The full ChordCraft CNN model, while efficient compared to larger architectures, may still be too computationally intensive for continuous real-time processing on a mobile CPU. To address this, several model optimization techniques can be employed. Quantization is a key technique that involves converting the model's 32-bit floating-point weights and activations to lower-precision 8-bit integers. This can reduce the model size by a factor of four and significantly speed up inference on hardware that has optimized integer arithmetic support, often with only a minimal drop in accuracy. Further optimization can be achieved through pruning, a process that removes redundant or unimportant connections within the neural network.
- *Power Consumption:* Continuous audio processing and CNN inference are energy-intensive operations. An effective mobile deployment must be highly optimized to minimize battery drain, ensuring that a student can use the application for a reasonable practice session without depleting their device's battery.

- *User Interface (UI) Design:* The utility of the underlying technology is heavily dependent on the quality of its user interface. A potential UI for a ChordCraftpowered app could feature a clean, intuitive design with several key components: a real-time scrolling display that shows the chords of the currently playing music, synchronized with a progress bar; interactive chord diagrams that show fingerings for guitar, ukulele, or piano when a chord symbol is tapped; and a practice mode that records the user's playing and provides a score based on their chordal accuracy. Figure 6 illustrates a prototype of the ChordCraft interface, showcasing how predicted chords can be displayed alongside the audio timeline in an accessible and visually engaging manner.

Figure 6. Prototype interface of the ChordCraft web application showing chord predictions for an uploaded audio file

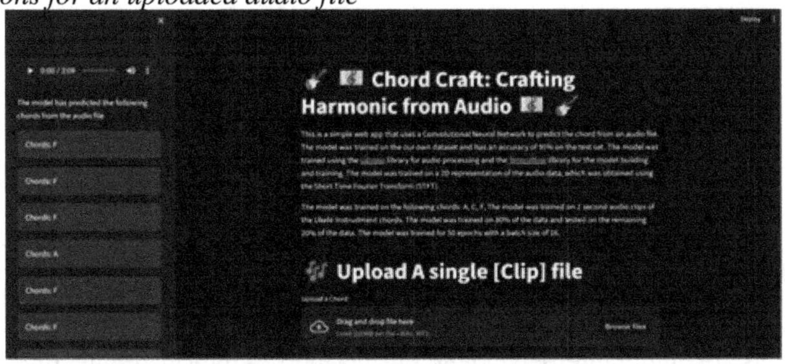

6.2 Integration and Interoperability: ChordCraft in the Creative Workflow

Integration and Interoperability: ChordCraft in the Creative Workflow

- DAW/VST Integration: A powerful application would be to package the ChordCraft model as a VST (Virtual Studio Technology) or AU (Audio Units) plugin for use within Digital Audio Workstations (DAWs) like Ableton Live, Logic Pro, or FL Studio. A musician could simply place the plugin on an audio track (e.g., a guitar recording or a sample) and have it automatically generate a corresponding chord track. This would dramatically speed up the processes of remixing, arranging, and harmonic analysis. However, this presents significant technical challenges, as most deep learning frameworks (like TensorFlow and PyTorch) are Python-based, while the world of audio plugin development is dominated by C++. Bridging this gap would require sophisticated engineering, potentially involving embedding a Python interpreter within the C++ plugin or using model deployment frameworks like ONNX or TensorRT to convert the model into a more portable format. (NVIDIA, n.d.) (Mikac, 2025) (Community, 2025)
- API Architecture: A more flexible approach to integration is to expose the ChordCraft model via a web API. In this architecture, the model would be hosted on a server, and client applications (such as a mobile app, a web-based tool, or a DAW plugin) could send audio data to the server via a REST API endpoint. The server would process the audio and return the predicted chord sequence in a structured format like JSON. This decouples the model from the client application, allowing for easier updates and maintenance, and

enables a wide range of third-party developers to build tools on top of the ChordCraft technology. (Stewart, 2025)

6.3 Next Steps in MIR: Expanding the Frontier

The ChordCraft project serves as a strong foundation, but it also opens up numerous avenues for future research in Music Information Retrieval.

- *Model Enhancements:* The current model could be extended in several ways. A multi-task learning approach could train a single model to jointly predict chords, key, beats, and downbeats, leveraging the fact that these musical elements are highly interdependent. Furthermore, a hybrid architecture that uses the current CNN as a powerful local feature extractor and feeds its output into a Transformer-based backend could combine the strengths of both approaches, improving the model's ability to use long-range musical context to inform its predictions.
- *Beyond Western Harmony:* A major frontier for MIR is the analysis of non-Western music, which presents challenges that go far beyond simply expanding the training data. The very concept of "harmony" and "chords" is rooted in Western music theory and is not a universal principle. Applying a chord recognition model to many global music traditions would be a category error, akin to using a grammar checker to analyze a painting. The path forward requires a fundamental rethinking of the analytical task itself.
  - *Arabic Maqam:* The music of the Middle East and North Africa is based on the maqam system of melodic modes, which utilizes microtonal intervals not found in the 12-tone equal-tempered scale. A computational model for this music would need to move beyond a 12-class chroma representation and instead learn to identify the characteristic melodic phrases (sayir) and microtonal nuances that define each maqam. (Didi, 2022)
  - *Indian Classical Music:* Similarly, Indian Classical Music is organized around the melodic framework of the raga and the rhythmic cycle of the tala. The concept of vertical, multi-note chords is largely absent. A meaningful computational analysis of this tradition would focus not on chord recognition but on tasks like raga identification (classifying the melodic mode), tala recognition, or identifying characteristic melodic patterns. (Chakraborty et al., 2021)

The true challenge for creating a "global MIR" is not just about retraining existing models on new data; it is about developing entirely new model architectures with inductive biases that reflect the unique musicological principles of different cultures. A "Raga-Former" would need to be architecturally distinct from a "Chord-Former," likely focusing on sequence-to-sequence melodic contour modeling rather than frame-wise harmonic classification. This represents a profound and exciting challenge for the future of the field.

7. CONCLUSION: CONTRIBUTIONS AND STRATEGIC VISION

This chapter has offered a complete survey of the ChordCraft project, a dataset-based method of automatic chord recognition with convolutional neural networks. By placing this work in the larger context of Music Information Retrieval and offering a thorough description of its methodology, performance, and implications, we have sought to shed light not just on its technical strengths, but also on its potential as a bridge between sophisticated artificial intelligence and practical music education.

The contributions of our work are threefold. First, we present the ChordCraft dataset, a new, carefully curated resource with an explicit pedagogical design. By highlighting easy-to-learn instruments and including realistic acoustic conditions such as vocals and background noise, this dataset offers a valuable resource for training robust models that are directly applicable to the conditions in which music learners live. Second, we built a customized and computationally parsimonious CNN-based recognition model that illustrates the strength of architectural simplicity. Its effectiveness reinforces the tenet that with good quality, well-directed data, a simple and computationally minimal architecture is able to provide outstanding performance frequently without the necessity of resorting to more intricate and computationally demanding models. Third, by stringent testing, we proved the model's high degree of correctness, reaching 99

The wider implication of this work reaches into two important areas. For the domain of Music Information Retrieval, ChordCraft is an exemplar of the utility of purpose-designed datasets and emphasizes the value of tailoring model structure to the particular topology of the problem being tackled. For music education, it offers a concrete roadmap for developing accessible, AI-based tools that can deliver real-time feedback, speed learning, and raise student engagement. By de-mystifying music's harmonic framework, these tools can democratize music literacy and empower the next generation of musicians.

In the future, the strategic vision of AI-driven musicology is one of collaboration and augmentation. Computational aids such as ChordCraft are never meant to supplant the skills of musicologists, teachers, and performers but serve as powerful assistants who can streamline drudgery, uncover patterns in huge datasets, and offer new perspectives to grasp the nuanced language of music. As they continue to develop, growing to encompass the rich diversity of world musical traditions, they will open up new realms for deepening music understanding, creativity, and appreciation worldwide.

REFERENCES

Abdzadeh Ziabari, P., & Veisi, H. (2023). A comparison of CQT spectrogram with STFT-based acoustic features in deep learning-based synthetic speech detection. *Journal of Applied Digital Media, 11*, 119–129. DOI: 10.22044/jadm.2022.12373.2382

Akram, M. W., Dettori, S., Colla, V., & Buttazzo, G. C. (2025). *ChordFormer: A conformer-based architecture for large-vocabulary audio chord recognition*. arXiv. https://arxiv.org/abs/2502.11840

Beach, M. (2012, September). Why the ukulele belongs in the classroom: Engaging students with music. *TEACH Magazine*. https://teachmag.com/why-ukulele-belongs-in-classroom-engaging-students-with-music/

Bittner, R., Salamon, J., Tierney, M., Mauch, M., Cannam, C., & Bello, J. (2014). MedleyDB: A multitrack dataset for annotation-intensive MIR research.

Chakraborty, S., Tewari, S., Rahman, A., Jamal, M., Lipi, A., Chakraborty, A., Nanda, A., & Shukla, P. (2021). *Hindustani classical music: A historical and computational study*. Sanctum Books.

Chen, T.-P., & Su, L. (2019). Harmony transformer: Incorporating chord segmentation into harmony recognition. In *Proceedings of the 20th International Society for Music Information Retrieval Conference* (pp. 259–267). https://doi.org/DOI: 10.5281/zenodo.3527794

chimera0. (2019). *pydbm: Python library for building RBM, DBM, RTRBM, RNN-RBM, and Shape-BM*. https://pypi.org/project/pydbm/

Cho, T., Weiss, R., & Bello, J. (2010). Exploring common variations in state-of-the-art chord recognition systems. In *Proceedings of the Sound and Music Computing Conference*.

Didi, A. (2022). A preliminary study of harmonization in maqam music.

East End Arts Council. (2024). Top 10 reasons to learn the ukulele. https://eastendarts.org/reasons-to-learn-the-ukulele/

Hamm, C. (2021). Seventh chords. *Open Music Theory*. https://viva.pressbooks.pub/openmusictheory/chapter/seventh-chords/

Huh, M., Ray, R., & Karnei, C. (2024). A comparison of speech data augmentation methods using S3PRL toolkit. arXiv. https://arxiv.org/abs/2303.00510

Humphrey, E. J., Bello, J. P., & LeCun, Y. (2013). Feature learning and deep architectures: New directions for music informatics. *Journal of Intelligent Information Systems*, *41*(3), 461–481. DOI: 10.1007/s10844-013-0248-5

Isophonics Research Group. (2025). *Isophonics dataset.* https://isophonics.net/datasets

Jiang, N., Grosche, P., Konz, V., & Müller, M. (2011). Analyzing chroma feature types for automated chord recognition.

Khadkevich, M., Omologo, M., & Bruno, F. (2008). MIREX audio chord detection.

Khan, H., Bazai, S., Zaland, Z., Durrani, S., Aslam, S., Amphawan, A., Ali, F., & Neo, K. (2023). A comparative study of convolutional neural networks and recurrent neural networks for chord recognition. *International Journal of Membrane Science and Technology*, *10*(2), 1617–1630. DOI: 10.15379/ijmst.v10i2.1837

Koutini, K., Eghbal-Zadeh, H., & Widmer, G. (2021). Receptive field regularization techniques for audio classification and tagging with deep convolutional neural networks. *IEEE/ACM Transactions on Audio, Speech, and Language Processing*, *29*, 1987–2000. DOI: 10.1109/TASLP.2021.3082307

Lee, K., & Slaney, M. (2007). A unified system for chord transcription and key extraction using hidden Markov models.

Lee, K., & Slaney, M. (2008). Acoustic chord transcription and key extraction from audio using key-dependent HMMs trained on synthesized audio. *IEEE Transactions on Audio, Speech, and Language Processing*, *16*(2), 291–301. DOI: 10.1109/TASL.2007.914399

Li, T. (2021). Study on a CNN-HMM approach for audio-based musical chord recognition. *Journal of Physics: Conference Series*, *1802*(3), 032033. DOI: 10.1088/1742-6596/1802/3/032033

Mikac, E. (2025). PyTorch vs TensorFlow vs Keras: Which framework gets you hired? https://www.cbtnuggets.com/blog/technology/programming/pytorch-vs-tensorflow-vs-keras

musictheory.net. (n.d.). Common chord progressions. https://www.musictheory.net/lessons/57

NVIDIA. (n.d.). TensorRT architecture overview. https://docs.nvidia.com/deeplearning/tensorrt/latest/architecture/architecture-overview.html

O'Hanlon, K., Ewert, S., Pauwels, J., & Sandler, M. B. (2017). Improved template-based chord recognition using the CRP feature. In *Proceedings of ICASSP* (pp. 306–310). DOI: 10.1109/ICASSP.2017.7952167

Palanisamy, K., Singhania, D., & Yao, A. (2020). Rethinking CNN models for audio classification. arXiv. https://arxiv.org/abs/2007.11154

Park, J., Choi, K., Jeon, S., Kim, D., & Park, J. (2019). A bi-directional transformer for musical chord recognition.

Qiu, F. (2024). AI-driven personalized learning for beginner musicians. *Membrane Technology*, •••, 228–234. DOI: 10.52710/mt.162

Scholz, R., Dantas, V., & Ramalho, G. (2005a). Automating functional harmonic analysis: The Funchal system. In *Proceedings of the Seventh IEEE International Symposium on Multimedia* (pp. 759–764). DOI: 10.1109/ISM.2005.39

Scholz, R., Dantas, V., & Ramalho, G. (2005b). *Funchal: A system for automatic functional harmonic analysis.* https://citeseerx.ist.psu.edu/document?doi=10.1.1.85.7992

Schörkhuber, C., & Klapuri, A. (2010). Constant-Q transform toolbox for music processing. In *Proceedings of the Sound and Music Computing Conference*.

Shaffer, E. S., Burgoyne, J. A., & Fujinaga, I. (2020). A cluster analysis of harmony in the McGill Billboard dataset. *Empirical Musicology Review*, *15*(1), 28–56. DOI: 10.18061/emr.v14i3-4.5576

Sheh, A., & Ellis, D. (2003). Chord segmentation and recognition using EM-trained hidden Markov models.

Stewart, W. (2025). How to deploy machine learning models: Step-by-step guide. https://northflank.com/blog/how-to-deploy-machine-learning-models-step-by-step-guide-to-ml-model-deployment-in

Subramanian, S., & Ganapathiraman, V. (2023, October). Learning to learn learning-rate schedules. *Amazon Science Blog*. https://www.amazon.science/blog/learning-to-learn-learning-rate-schedules

Unknown. (2025). Automatic chord recognition method – CS 101 final exam. https://www.studocu.com/

Wiki, M. I. R. E. X. (2025). Audio chord estimation. https://www.music-ir.org/mirex/wiki/2025:Audio_Chord_Estimation

Wikipedia contributors. (2024, November). Chroma feature. *Wikipedia*. https://en.wikipedia.org/wiki/Chroma_feature

Xi, Q., Bittner, R. M., Pauwels, J., Ye, X., & Bello, J. P. (2018). GuitarSet: A dataset for guitar transcription. In *Proceedings of the 19th ISMIR Conference* (pp. 453–460).

Zhou, X., & Lerch, A. (2015). Chord detection using deep learning. In *Proceedings of the International Society for Music Information Retrieval Conference*.

Chapter 8
AI-Powered Sound Design, Synthesis, and Music Composition: Emerging Tools and Creative Innovation

Samiksha Kailash Sapnar
http://orcid.org/0009-0004-4284-1462
Vishwakarma Institute of Information Technology, Pune, India

Sonal Fatangare
http://orcid.org/0009-0005-1185-8699
Vishwakarma Institute of Information Technology, Pune, India

Gitanjali Shinde
Vishwakarma Institute of Information Technology, Pune, India

Rajkumar Patil
http://orcid.org/0000-0003-1073-4297
Vishwakarma Institute of Technology, Pune, India

Grishma Bobhate
Vishwakarma Institute of Information Technology, Pune, India

Chaitanya Garware
MIT Art, Design, and Technology University, India

ABSTRACT

The chapter focuses on the changes in the role of artificial intelligence in sound design, synthesis and music creation by tracing the development of technology starting with the earliest forms of algorithmic systems and ending with more modern deep learning. It describes the growth of creative opportunities enabled by neural networks, generative methods, and advanced synthesis tools, alongside helping in mixing, mastering, and audio restoration. The chapter also emphasizes that AI can help to create new timbres, textures, and compositional forms to allow professional and novice creators to experiment with the larger soundscapes. Moral issues such as

DOI: 10.4018/979-8-3373-6279-3.ch008

authorship, copyright, and training data, and artistic authenticity are also critically evaluated to explain the role of developers and creators. The chapter concentrates on the idea that AI can be a collaborative partner and a tool that improves creativity, democratizes production, and affects the way people interact with machines in the sonic arts in the future with the help of practical tools and case studies.

1. INTRODUCTION: THE ALGORITHMIC MUSE - THE EVOLUTION OF AI IN SOUND DESIGN AND MUSIC COMPOSITION:

The integration of artificial intelligence (AI) and the creative arts have witnessed the beginning of a new age where conventional music and sound boundaries are being redefined. There is no better place where this transformation has been witnessed as in the area of sound design, synthesis and music composition. A distinction between human creative power and machine-run process is growing less distinct as AI is turning into a collaborative partner, in some instances, a force of creativity in itself. (Audio Engineering Society, n.d.) The chapter of this book gets into the depth of this paradigm change, the historical landmarks and technological innovations, and the tricky ethical and creative issues that come to mind as a result of the implementation of AI into the sonic arts.

It is not the first time the idea of writing music with the help of algorithms has been suggested. The AI music seeds were planted as far back as the mid-20 th century. Earlier compositions such as the Illiac Suite developed by Lejaren Hiller and Leonard Isaacson in 1957 were among the earliest compositions that were produced by a computer. These initial experiments were mainly rule-based whereby the programmer wrote musical theories and structures to be followed by the computer. Though they were revolutionary, they were constrained by the strictness of their directions. Subsequent decades involved further development such as Experiments in Musical Intelligence (EMI) in the 1980s by David Cope which studied and modeled the styles of classical composers and produced new pieces which were often indistinguishable to those composed by humans.

The real revolution was however initiated with the development of machine learning and deep neural networks in the 21 st century. Unlike its predecessors, this new generation of AI can learn out of the huge datasets of already existing music, discern complex patterns and create completely new work. (Briot et al., 2020) demonstrated by systems such as Google Magenta and OpenAI's MuseNet, which produced a variety of music, ranging from classical piano to multi-instrumental compositions across different genres. These models are not merely a reversion of already existing elements, they are learning the innermost grammar and geometry of music, and can produce creative output themselves.Nowadays, AI has infiltrated all spheres of the music and sound production process. The sound design sector is implementing AI-based technologies to automate the time-consuming process, including noise removal and the restoration of audio quality, as well as providing fresh opportunities in creativity. The voice synthesis has been incredibly realistic and one can either create a new vocal performance or alter the existing one. AI is also being applied in music synthesis to synthesize new timbres, textures that were once not possible to synthesize. Generative AI products enable creators and producers to creatively solve creative blockages by generating ideas, chording, and rhythms, as needed. They can also make complete royalty-free soundtracks to many media, and democratized the creation of music to the content creators and independent artists. The AI in music has a future of symbiosis between man and machine. Although it is feared that human artists might be replaced by AI, a more practical and common opinion is that AI will be a strong collaborative tool. It can be a creative collaborator, a collaborative composer, even a virtual bandmate that allows human creators the freedom to concentrate on the emotional and conceptual essence of their work. The availability of AI-assisted democratization of music production will enable a new generation of artists. There are also potential uses of this technology in music teaching and therapy, where the experience can be customized to just individual needs (personalized), and the music can be made to suit therapeutic requirements. Although this potential is enormous, the emergence of AI in the sonic arts presents very serious challenges. Ethical and legal environment is very complicated, and the issue of intellectual property and copyright remains a topic of debate. Who are the owners of a song created by an AI? What would be considered plagiarism in case an AI is trained on a copyrighted material? The change of these questions demands new legal frameworks and the change of our perception of authorship. Also, the question of the authenticity of AI-generated music is actively debated philosophically. Is it possible that a machine having no personal experience, feeling, and cultural background, could make music that reaches the human soul? Although AI has the capacity to replicate the look and style of music, the real worth of music might be the human stories and emotions that cannot be replicated by AI, which remained a highly complex and creative companion, as it is now. It will look into the technical foundations of AI-based sound design, synthesis, and composition and critically evaluate how far it has affected the creative process, the music industry, and our perception of art in general. (Briot et al., 2020) (Cope, 2006)This is not meant to speculate upon a future of machine versus musician, but is meant to shed light upon a new and bright frontier of human creativity and artificial intelligence arising at the interface of human creativity and machine intelligence.

Table 1 below is a summary of some of the Generative AI models used in the music industry.It shows their major qualities, uses, and contributions to the music generation and composition.

Artificial Intelligence in music is the use of computational and machine learning algorithms to analyze, create, and manipulate musical information and allow it to be used by systems to aid or autonomously carry out the tasks that have traditionally been considered the domain of human creativity. Sound design is the art and science of creating and structuring audio material like timbre, texture, and effects in order to express music and tell multimedia stories and music composition involves organizing rhythm, melody, and harmony into coherent composition with emotional and cultural significance. Computation and music Computation and music The connection between computation and music developed through early rule-based algorithmic composition in the middle of the 20th century to symbolic AI systems in the late 20th century and, more recently, to deep learning and generative models that make complex and stylistically diverse musical composition. The importance of this area has increased exponentially today because there is increasing demand in digital media, gaming, and content creation industries with scalable, adaptive, and individual music, and AI-driven tools offer greater efficiency, exploration in music creation, and democratization of access to professional music production. Though significant improvements have been achieved in creating models and technical output, current studies tend to ignore the more extensive impact of AI on creativity, authorship, and ethical accountability, and the lack of integration between the technical development and artistic and social approaches is present. This chapter fills this gap providing a general overview of AI techniques in sound design and music composition, exploring their usage in creative processes, ethical and legal issues, and outlining the new trends that characterize the future of human-AI collaboration in music. Based on this background, the following section identifies the key problems and research objectives associated with the use of artificial intelligence in sound design and music composition.

Table 1. Generative AI models in music

AI Model Type	Core Technology	How It Works	Strengths	Common Use in Music
Recurrent Neural Networks (RNNs)	Sequential Processing	Processes data in a sequence, with a 'memory' of previous steps to influence the next output.	Excellent for generating coherent melodies and short sequences.	Melody generation, basic chord progressions.
Generative Adversarial Networks (GANs)	Two competing neural networks (Generator and Discriminator)	The Generator creates new music, while the Discriminator tries to identify it as fake. This competition drives improvement.	Produces highly realistic and stylistically convincing output.	Style transfer, generating realistic musical textures and timbres.
Transformer Models	Self-Attention Mechanism	Processes all parts of a sequence simultaneously, allowing it to understand long-	Excels at creating long, coherent musical forms and multiinstrumental pieces.	Full song composition, orchestration, and creating complex arrangements.
		range dependencies and complex structures.		
Diffusion Models	Denoising Process	Starts with random noise and gradually 'denoises' it into a coherent sound or musical piece.	Known for highquality audio and fine-grained control over generation.	Text-to-audio generation, creating unique soundscapes, and generating realistic instrument sounds.

. Problem Statement

The existing methods of music production and sound design have a number of limitations, such as manual composition is time-consuming, technical production processes are becoming more complex, and the tools of higher quality with regard to creativity are not accessible to independent artists and content creators. Furthermore, the swift development of generative technologies has added ambiguity to the topic of ownership of copyright, authorship, and using training information legally, which has caused confusion in the music industry. The adoption of automated music generation systems is further complicated by ethical issues like originality, the artistic authenticity and the possible positioning out of human musicians. These issues point to the necessity of smart and mindful technological approaches which would enable creativity and overcome legal and ethics limitations. In this chapter, the author aims to look into how these limitations can be addressed through artificial intelligence to increase efficiency, bring forth the creative potential, and offer a balanced perspective of the technical, artistic, and ethical aspects of AI-based sound design and music composition.

4. Objectives

This chapter is aimed at examining the key artificial intelligence methods applied in sound design and composition of music, reviewing the currently available AI-driven music generation models and tools and their principles, identifying the most important technical challenges and creative restrictions of AI use as a composing and sound design tool, analyzing the existing trends and directions of AI use in music production, and offering a somewhat systematic framework within which researchers and practitioners can approach and use AI as a responsible agent in music creation.

4. Literature Review

The use of artificial intelligence in generating music, sound synthesis and creative assistance is a topic that has been widely researched in recent studies. The initial works centered around rule-based and symbolic music generation algorithms, whereas subsequent works presented machine learning-based methods of Recurrent Neural Networks (RNNs), Generative Adversarial Networks (GANs), and Transformer models to generate music with a long-term structure and style. GAN-based models in generating realistic audio textures and timbral variations were found to be very effective, in contrast to Transformer-based models that were seen to perform better in creating long and coherent musical pieces in multi-instrument and multi-genre compositions. Some papers also emphasized the benefits of AI-based music systems that encompassed automated repetitive manufacturing processes, instant creation of music ideas, and democratization of music making in the hands of people lacking expertise in the field. Nonetheless, the current literature also claims the drawbacks like lack of emotional richness, reliance on massive training samples, reliance on homogenous musical categories, and inability to measure creativity and originality. In the fields of ethical and legal research, the existing challenges pertain to the problem of copyright ownership, training data consent, and the reproduction of the artist identities with the help of voice cloning and style imitation. Although, there is a strong technical advance and development, there is a research gap on the integration of technical performance in artistic, cultural and ethical views. The majority of the studies are devoted to the effectiveness of the algorithms instead of the overall influence on creativity, authorship, and the music industry. Building upon the insights gained from the reviewed literature, the next section presents the methodological framework adopted to analyze and evaluate AI-driven sound design and music composition techniques.

Table 2. Literature survey on AI in music and sound design

Author	Year	Method	Findings	Limitations
Briot et al.	2020	Machine learning models for music generation	Demonstrated effectiveness of deep learning in music composition	Limited discussion on ethical impact
Hadjeres & Pachet	2017	DeepBach (symbolic music generation)	Produced stylistically accurate Bach chorales	Restricted to specific musical style
Dhariwal et al.	2020	WaveNet / Jukebox model	Generated realistic audio and vocals	High computational cost
Miranda	2021	Survey of AI in music	Identified applications in composition and synthesis	Lacks industry case studies
Pachet	2023	Creative AI systems	Explored human-AI collaboration	Legal implications not fully addressed
Southern	2018	AI-assisted album production	Showed feasibility of AI-human co-creation	Limited scalability

6. Research Gap

Although artificial intelligence music generation and sound synthesis have greatly evolved, the bulk of literature available still dwells on the enhancement of the algorithms to turn into performance and technologic effectiveness, rather than on the overall impacts on the creativity, authorship, and cultural influence. Majority of the research researches the models separately without the incorporation of artistic, ethical and legal backgrounds in a cohesive unit. Moreover, a comparative analysis linking technical models like GANs, Transformers, and diffusion systems with the real-life industry and user-focused creativity is missing. This disjunction indicates the necessity of the interdisciplinary and holistic study of AI in sound design and music composition. The current chapter bridges this gap by bringing together technical innovations, real-world examples, and moral reasoning to gain an in-depth insight into the AI-powered musical creativity.

Recent research has also widely investigated the use of artificial intelligence to generate music and sounds through the use of deep learning models like GANs, RNNs, Transformers and diffusion models. According to surveys and systematic reviews, these models become increasingly effective in symbolic composition and raw audio generation (Briot et al., 2017; Miranda, 2021; Mitra and Zualkernan, 2025). GAN-based methods have shown great results in style transfer and sound texture synthesis, although training stability has been problematic (Jun Min and Wang, 2022; Neves et al., 2022). Models based on a transformer have demonstrated greater ability to learn long-range musical dependencies and generate coherent multi-instrument compositions (Agarwal and Sultanova, 2024; Chen et al., 2023; Zhang, 2025). Neural audio synthesis systems like WaveNet and NSynth made it possible to move past symbolic music representation to high-fidelity raw audio generation (van den Oord et al., 2016; Engel et al., 2017), and more recent diffusion-based and multimodal systems made music generated by neural networks even more novel and real-like (Liu et al., 2025). The examples of applied systems like Flow Machines, AIVA, and AI-assisted albums demonstrate that collaboration between humans and AI is possible in professional music creation (Southern, 2018; Over The Bridge, 2021; Pachet, 2023). Nevertheless, new literature also highlights the ethical and legal issues concerning the ownership of copyrights, training data permission, and possible misuse, such as voice cloning and deepfake audio (From Tools to Creators, 2025; Kumar and Singh, 2024). Combined, these works show fast technical advancement and expose unsettled issues on originality, cultural influence and responsible use of AI-based music systems.

6. Methodology

The technical framework used in the methodology of analysing and implementing artificial intelligence in sound design and music composition can be considered structured and adopted in this chapter. It starts with the assembling of music datasets of both symbolic representation in the form of MIDI files and raw audio records of various genres and sound textures. These samples are pre-processed by removing noise, normalizing, and segmentation and converting the data into feature representations in the form of spectrograms, pitch tracks, and rhythmic structures. After that, machine learning models such as neural networks, Generative Adversarial Networks (GANs) and Transformer-based architectures are trained on the processed data to learn musical structures, timbral properties and compositional principles. New sequences of music and synthesized sounds are also generated using the trained models in terms of the learned representations. The outputs obtained are assessed by a mix of both quantitative measures like spectral similarity, pitch accuracy and signal-to-noise ratio and a qualitative method, which involves human listening tests to assess musical coherence, creativity and expressiveness. This process of workflow which includes data collection, pre-processing, model training, music generation, and evaluation, is a step-by-step process to guarantee a systematic and reproducible method to investigate the efficacy of AI-driven methods in sound design and music composition. The subsequent sections discuss the core AI models, their applications in the music industry, and the challenges encountered in practical implementation.

2. FUNDAMENTALS OF SOUND DESIGN AND MUSIC COMPOSITION

2.1 The Science of Sound: Physics to Perception.

Sound is, in essence, a physical object, a vibration which is transmitted through a medium, like air or water, in the form of a wave. (Dhaliwal et al., 2020) These are the primary properties that one needs to understand the first step towards designing sound and composing music. Frequency, amplitude and timbre are the most important characteristics of a sound wave.

Frequency and Pitch: The frequency of a sound wave is the rate at which the sound wave vibrates and its frequency is measured in Hertz (Hz). This tangible thing is that which we feel to be pitch. An increase in frequency is associated with an increase in pitch and vice versa. E.g. A note emitted by a piano is defined by the basic frequency of the vibrating string. A human ear is usually capable of perceiving sounds between 20 Hz (a very low rumble) and 20,000 Hz (a high-pitched whine).

Amplitude and Dynamics: The amplitude of a sound wave is the magnitude or intensity of the sound wave which we perceive as the bound volume or loudness. This is normally quantified in decibels (dB). In music, the term amplitude is called dynamics. The change in a musical work is indicated in terms of piano (soft) and forte (loud) and it is through these terms that composers give a musical composition its emotional texture and expression.

Timbre (The Color of Sound): (Flow Machines, 2016) Timbre is probably the most significant and difficult aspect to sound design. The quality is the factor which enables us to make a difference between two varied instruments or sources of sound playing the same note in the same volume. The combination of harmonics and the envelope of sound is a factor that determines the timbre. Harmonics are other frequencies which are vibrating at multiples of the basic frequency and determine a different texture to every sound.

The ADSR Envelope: Attack, Decay, Sustain, Release (ADSR) envelope is the description of the amplitude changes of a sound during some time.

Attack: This is the duration required by a sound to reach maximum volume.
Decay: This is the duration of time the sound requires to reach its sustained level, after reaching its peak.
Sustain: This is the degree to which the sound is sustained within a certain time.
Release: It is the duration it takes the sound to die away to nothingness after release of the note.

2.2 Essentials of Composing Music.

Although sound design is the art of the generation of single sonic textures, the art of music composition is the art of arrangement of these sounds in a structure that is unified and meaningful. Rhythm, melody and harmony are the essential elements of music.

Rhythm: Rhythm is the foundation of music, and is the temporal structure in terms of sounds and silence. It provides music with its pace and movement determining the tempo and rhythm of à composition. Tempo, which is the rate of the beat (e.g. fast, slow), and meter, which is how the beats are grouped into measures (e.g. 4/4 time of a typical song, 3/4 of a waltz).

Melody: (Foster, 2025) Melody is a linear recognizable arrangement of notes which we frequently hum or sing. It is the narrative of the music, which is achieved through the arrangement of the pitches in a given sequence. The shape of the rise and fall of the melody (its contour) and the structure of its phrasing (how it is put together, as a musical sentence) will be of vital importance to its emotional effect.

Harmony: The harmony can be described as the concurrent playing of several notes or chords. It gives a melody depth, richness and emotion.

Consonance and Dissonance: It can be consonant (restful and easy to listen to) or dissonant (unstable and creating tension) harmony. (Ghostwriter977, 2023) A great tool of emotional narrative is the strategic use of dissonance by a composer to create tension and later resolve it with consonance.

2.3 Basic Sound Design Techniques and Technologies.

Sound design does not just begin and end with the musical instrument. It is the combination and manipulation of sounds to provoke emotion, establish a scene and enrich a story. Synthesis or electronically generated sound is very important in contemporary sound design.

Subtractive Synthesis: This is the most widespread type of synthesis This has been likened to the art of sculpture, wherein you begin with a piece of material, and chisel out the shape you want.

Additive Synthesis: The opposite of subtractive Synthesis. It is the process of assembling sound by adding several waves of sine (the simplest possible sound wave) of various frequencies and intensities into a single sound. The approach has exact control over all harmonics yet it is more computationally demanding.

Frequency Modulation (FM) Synthesis: FM synthesis is the foremost technology created by Yamaha DX7, which produces metallic or bell-like tones that are complex. It operates by making one modulator sound wave (modulator) rapidly alternate the frequency of a second sound wave (carrier) to form new harmonics not in the original signals.

Granular Synthesis, this technique is more experimental in nature and involves splitting or slicing a sound sample into minute small fragments otherwise known as grains, which usually measure in the milliseconds range. These grains may then be recombined, stacked and worked upon in a number of ways to produce completely new textures and drones.

2.4 Sound in the Media and Storytelling.

Sound is not merely an accompaniment of any type of media, be it film and television, video games and interactive experiences, it is a key component of the story. It is used with pictures to make it immersive and emotionally impactful.

Film and Video: In film, there are three main types of sound namely dialogue, music and sound effects.
Dialogue: Use of the words that are spoken by the characters.
Diegetic Sound: These are sounds that start inside the world of the movie like a car horn, footsteps of a character, or music in a radio. They add the effect of reality to the scene.
Non-Diegetic Sound: These are sounds added to the film but not part of the world of the story e.g. the musical score or a voice-over. This kind of sound is applied to direct the feelings of the viewers. Foley artists provide the sound effects on screen in a realistic way.
Video Games: Sound should be active and responsive and should change with actions of the player. This involves interactive music which varies according to the gameplay and spatial audio which provides the player with the perception of his environment.

3. AI TECHNOLOGIES IN SOUND DESIGN:

Artificial Intelligence Generated Sound and Voice:

The most important feature of AI in sound is its capacity to produce new sounds. It is revolutionary to the sound designers who no longer have to be restricted by the available sound libraries.

Generative Adversarial Networks (GANs): (Ghostwriter977, 2023) (Google Magenta, n.d.) GANs represent a potent type of AI models that are applied to generate new audio by itself. Their basic idea is that they have two neural networks, one that generates new audio clips and another that looks at whether they are authentic. The generator develops, via a competitive process, a superb skill in creating believable and realistic sounds, whether natural environmental sounds or synthetic sounds of a highly complicated structure. The technology enables one to create a limitless range of sounds that meet the goals of particular requirements.

Neural Synthesizers: These types of deep-learning codes are capable of generating high-quality audio by training to understand the nature of existing sounds. In the case of Synthes by Google it is capable of learning the characteristic timbre of one instrument and transferring it to the pitch and dynamics of a different instrument. This enables a composer to have a hybrid instrument, like a violin that has the tonal response of a piano, to open up to new sonic palettes that could not have been generated using the conventional synthesis. Magenta Studio: This is a set of music creation AI tools created by Google as a creative partner in the digital audio workstations (DAWs). These tools, by processing a user's input, can create original melodies, harmonies, and drum patterns and assist musicians in experimenting with new ideas and overcoming creative blockages faster.

Artificial Intelligence in Firm Music Composition and Arrangement:

The artificial intelligence is not limited to sounds but can create a whole piece of music. AI has the capacity to produce structured and stylistically consistent music by analysing large musical databases to extract musical theory, genre conventions and emotional content.

Symbolic Music Generation: In this method AI processes musical notation (MIDI data) to generate music. A good example is OpenAI MuseNet, which has the capability to produce multi-instrumental compositions of varied styles including classical and jazz. It gets to know about the complex relationship existing between notes, chords and the rhythms, which enables it to compose music, which not only feels new, but also musical.

Dynamic and Interactive Music: In other applications such as video games, AI is facilitating music that is dynamically responsive to user inputs. Depending on the progress of the player, the severity of a scene or the tone of an instance, the score is able to dynamically change. This brings a more interactive and customized experience, where the music ceases to be a soundtrack and is an active component of the story. Audio Mastering and Mixing: AI is simplifying the last phase of producing music. Services such as LANDR are automated, where a track is fed through a machine learning algorithm to apply professional-quality effects of optimizing the song in terms of loudness, clarity, and balance. This opens up high quality mastering to more independent artists, and democratizes a complicated and costly process.

Artificial Intelligence in Sound Recording and Processing:

Another application of AI is the analysis and manipulation of existing audio, which is much more challenging than any traditional algorithms. Source Separation: It is possible to use AI to isolate a single instrument or vocals in a mixed track. (Hadjeres & Pachet, 2017) A sound designer may use this to isolate a particular guitar riff of a full band recording, or a producer may use it to isolate a vocal track to remix. Audio Tagging and Classification: AI can automatically label audio files with descriptive labels, e.g. "jazz," "acoustic guitar," or thunderstorm. It is essential when dealing with and searching through massive sound libraries since a sound designer is able to find the precise clip they require in the shortest time possible without the necessity to listen to hundreds of files.

Audio Restoration: AI can detect and remove undesirable noise, hiss, or clicks in older audio, and be used to conserve the audio of history and enhance the sound of film and television.

Ethical Considerations and Future.

The emergence of AI in sound design and music creation is important because it raises issues of authorship and copyright, and the place of the human artist.

Copyright and Authorship: One of the crucial queries is who owns the copyright of a song or a sound that was created by an AI. These legal and ethical questions will become more complicated as AI is becoming more autonomous. The Human Artist Role: Most researchers perceive AI as an augmentation, but not sup plantation. It is able to deal with boring complicated tasks, offer novel creative points of departure and assist the artists to go through creative blocks. The AI-based sound design tools that most frequently result in a collaborative workflow typically have a human artist controlling and refining the results of the AI. The future of the field is probably a symbiotic relationship between machines and man, producing things that each one could not create alone.

4. AI-POWERED SYNTHESIS TECHNIQUES

4.1 Generative Adversarial Networks (GANs).

GANs are a strong type of AI applications, which are applied to produce realistic data, such as audio.

A GAN has two conflicting neural networks; a generator and a discriminator.

The generator constructs fresh audio anew frequently beginning with arbitrary sound.

The discriminator is the critic and it is used to compare the AI-generated audio and the actual audio sample. It has the task of identifying whether a sound is true or not.

It is by this adversarial process that the two networks are improved. The generator becomes more successful at creating sounds that can be mistaken by the discriminator, and the discriminator becomes more successful at fakes. This provides a feedback loop which produces very realistic and detailed audio output. To sound designers, it will allow them to train a GAN on a particular data set, e. g. footsteps on gravel, the sounds of a particular forest, and then they can generate an unlimited number of variations that will sound natural in the context where they were originally created.

Neural Synthesizers: Deep learning models applied by neural synthesizers are capable of synthesizing high-fidelity audio on the sample-by-sample level. In contrast to classical synthesizers that are based on the use of preprogramed units such as oscillators, neural synthesizers acquire the fundamental features of a sound directly through data.

NSynth (Neural Synthesizer): NSynth is an impressive project by Google as part of the Magenta team that has been trained on a huge amount of musical instruments. It is influenced to learn a compressed version of special qualities, or timbre, of a sound. The result of this is known as timbre transfer, in which the tonal quality of one instrument can be used in the pitch and dynamics of another. In this case, a composer might come up with a sound that seems to be that of a violin, but the sharpness of a plucked guitar. This technology can be used to create completely new and hybrid instruments.

Wave Net: Wave Net is a deep neural network developed by DeepMind and it is used to produce raw audio waveforms one sample at a time. It is a form of autoregressive model, that is, every successive sample it creates is founded on the prior samples. Wave Net was initially designed as a text-to-speech system, although it is also capable of generating very natural and realistic sounding audio, which is why it has been a key technology in the new synthesis and audio generation.

4.2 Symbolic Music Generation

Whereas GANs and neural synthesizers operate with audio waveforms, symbolic music generation operates on the creation of music in a symbolic representation, like MIDI data (Musical Instrument Digital Interface). The format takes the form of notes which are discrete events having characteristics such as pitch, velocity (loudness) and duration.

Muse Net: Muse Net is a deep neural network created by OpenAI that has an ability to produce a four-minute musical piece with a number of instruments and styles. It is trained with an enormous amount of MIDI files, which makes it learn the intricate patterns of chords, melodies and rhythms. A user is able to input Only a small prompt to Muse Net, and the model will create a full piece that adheres to the musical theory and keeps a creative and original form. Writers frequently utilise it to come up with new ideas or to create gaps in a song.

4.3 AI in Granular Synthesis

Granular synthesis is an experimental method which splits a sound into small units (grains) (snippets of sound, 1 millisecond long) and then reconstructs the sound using those grains into new textures. AI makes this process more intelligent because it intelligently manipulates these grains.

Rather than relying on humans to restrict manage each parameter, AI can be taught how various grain properties (e.g. size, density, position in the original audio) and the desirable sonic result are related to each other. It is also possible to feed a source sound to the AI and have it automatically extracted a meaningful grain of the source, which can be used to create new compositions, removing the guesswork of the process. This renders granular synthesis to be more instinctive and reachable to sound designers.

5. AI IN MUSIC COMPOSITION

The introduction of artificial intelligence (AI) in the creative industry, especially music, is one of the most significant events in the history of human expression. Music composition is a process of converting emotions, experiences, and cultural situations into a structured sound which is a very human activity and is practiced over centuries. Nevertheless, the emergence of AI has created a paradigm shift and questioned the old-fashioned concept of authorship, creativity, and what it is like to create music. This chapter examines the complex nature of the AI use in music composition including its principle, its application, its ethical applications and its future path. We shall explore how AI is not only a device of automation but it is developing as a collaborator, inspiration, and source of musical innovation in the new era.

5.1 The Technological Foundation: The AI Composing Music.

Simply put, AI music composition involves the use of machine learning algorithms, which process large amounts of data on existing musical compositions. It is through the detection of these rules using a corpus of information that the AI can create new and original compositions in a particular style or genre.

Generative AI Models: Generative models represent the most common way of AI music composition. Such models as Recurrent Neural Networks (RNNs), Generative Adversarial Networks (GANs), and Transformer models are trained to generate novel content on the basis of the trends they have learned.

Recurrent Neural Networks (RNNs): (Miranda, 2021) RNNs are early and original, and they are optimally applicable to sequence data such as music. They store musical information at note by note, making use of a memory of past notes to make predictions. This enables them to develop logical melodies and chordal resolutions. Nevertheless, the standard RNNs may suffer long-term form, which is usually overcome using Long Short-Term Memory (LSTM) networks.

Generative Adversarial Networks (GANs): GANs operate in a two-component mechanism: a "generator" which generates new music pieces, and a "discriminator" which compares them to a collection of actual music pieces. The objective of the generator is to produce music that can deceive the discriminator and the objective of the discriminator is to evolve into a master of distinguishing real music as opposed to fake music. This oppositional mechanism will make the generator strive to become better and better in delivering, creating more and more realistic and believable compositions.

Transformer Models: Revolutionizing natural language processing and music generation Transformer models are very effective at modelling long-range dependencies and intricate musical structures. They do process complete sequences simultaneously and they can grasp the form and flow of a work as a whole and not merely the direct relationship between notes. OpenAI-based models such as Muse Net are based on this architecture and can create complex music in different styles and using different instruments.

Synthesizers such as Google N Synth and other machine learning-driven synthesizers can create all new sounds never heard before by mixing the sound qualities of other instruments. This enables composers and producers to have a distinct aural palette in what they produce, beyond the traditional libraries and presents. Technical processes such as audio enhancement, noise removal and mixing and mastering can be automated with AI algorithms, and optimized, which previously took manual time.

5.2 Five Major AI uses in Music Composition

The use of AI in music composition is not universal; it is applied to a variety of creative tasks and user needs. Five possible ways of AI use today are:

The AI as an Ideation Partner: AI can be used as a creative assistant or an inspiration by many musicians and composers. Programs such as Orb Producer Suite and Magenta Studio have the capability to produce an infinite flow of tunes, chord progressions and rhythmic patterns depending on the user input. This assists the artists to break through creative stalls and consider novel ideas that they would never have imagined independently. The human is still in full control and he/she filters and optimizes the AI suggestions to their own view of art. This type of human-AI collaboration is regarded as the most effective and the most artistically satisfying usage of the technology.

The Fully Automated Composition: On the other extreme are the systems that can produce complete songs that are finished with little human intervention. Such systems as AIVA (Artificial Intelligence Virtual Artist) and Sound draw are aimed to create royalty-free music under different media, including video games and corporate videos. A user only has to give a few parameters, including genre, mood, and duration and the AI creates an entire, penned, and produced song. This application can be of great use especially to content creators who need quality and custom music on command and does not require them to spend money or time to hire a human composer.

The Style Emulation and Homage: It is one of the most interesting functions of AI, which can learn and reproduce the style of a certain composer or artist. This is evidenced by such projects as the AIgenerated new Bach by the Deep Bach program by Sony CSL or the reconstruction of a supposedly lost Beethoven symphony. Using the analysis of the whole body of work of an artist, the AI can create new pieces that cannot be differentiated, stylistically, of the original ones. This brings up serious concerns on the issue of originality and artistic heritage, yet also presents a strong means of admiring and examining the work of a master.

The Personalized and Adaptable Soundtrack: In other applications such as gaming and film, AI is being deployed to create dynamic soundtrack, which responds dynamically based on the on-screen activity or the emotional state of a user. An example is Brain.fm which employs AI to produce "functional music" that helps to boost productivity and focus. These systems go beyond fixed scores and do indeed provide a really immersive and responsive aural experience, where the music is a part of the media, but an active, living one.

The Democratization of the Music Production: AI reduces the cost of barrier to entry in musicians and composers. The AI-based tools can allow individuals not trained in music to create music of a professional quality. These sites make the creative process more democratic and enable more and more voices to be represented by music. This availability can create another generation of talent and creativity which is not bound by the technical constraints of the conventional music production.

5.3 The Debates of Ethics and Art

There are controversies surrounding the emergence of AI in music. It creates a row of complicated ethical and artistic controversies, which are actively defining the industry.

Authorship and Copyright: The most important legal and philosophical issue is probably, who owns the copyright to an AI-created piece of music? Is it the creator of the AI, the human being who has instigated the AI, or is it the AI? The new reality is yet to be reflected in the legal frameworks. As a rule, copyright presupposes human authorship and this means that purely AI generated music can be uncopyrightable. Nonetheless, when a human makes a significant editorial, compilation or hybridization of AI-generated output with their own original efforts, they usually protect the product. This uncertainty produces a complicated legal situation.

Training Data and Consent: A lot of AI models are trained on large scale datasets of copyrighted music scraped on the Internet with no written permission of the original copyright holder. This casts some grave doubts regarding the illegal application of intellectual property and chances of these models unwittingly plagiarizing or copying guarded contents. This problem is further escalated when AI is applied to replicate the voice or style of a particular, recognizable artist that may infringe on personality rights and cause legal issues.

The Value of Human Creativity: This is perhaps the most contentious argument on the value of human creativity per se. In the event that AI is able to produce increasingly technologically flawless and emotionally touching music, what becomes the fate of the human composer? Will this push aside session musicians and composers? The majority of scholar's state that AI should be considered more as an augmenting, than a disruptive, technology that will boost human creativity. The individual feelings, experience of life and cultural frames under which a human artist approaches his/her work cannot be substituted. The problem is how to strike the right balance of automating the routine and encouraging new ideas with the help of AI, and retain the human touch that makes music so soulful.

Transparency and Trust: The increased demand is to be transparent in the application of AI in music. And should a listener be aware of whether a piece of music s/he is listening to was composed by an AI? Disclosure on AI participation is starting to be added to many platforms. It is important to note that this openness is essential to establish trust with the audiences and ensure the integrity of the creative ecosystem.

We are in a Symbiotic Relationship, Future Trajectory.

In the future, AI in the field of music composition is bound to be characterized by an intensifying symbiotic interaction between the human and the machine. The AI is going to further improve as a mere tool to a more advanced co-creator.

Individual AI Models: Next-generation AI can enable musicians to teach a custom creative assistant their own style, giving them a custom creative companion that follows their personal artistic style and builds on it. This would bring AI out of the category of a generic tool to a very personalistic and intimate partner.

New genres and sonic discovery: The capability of AI to cross-contaminate musical genres and blend novel sounds will probably result in the establishment of completely new musical styles, which a human person would find hard to imagine. A vast computational capability of AI can effectively traverse new sonic landscapes and configurations and challenge the limits of what music can be.

Legal and Ethical Structures: Since the technology is at maturity, legal and ethical structures must be established to handle the dynamics of the authorship, copyrights, and fair use of training data. It will require well-defined rules to create a healthy and just creative atmosphere between human and AI creators.

6. CASE STUDIES AND TOOLS

Theoretical discourses about AI in music become an actual experience due to the increasing number of practical tools and world examples. These illustrations show the breadth of usability of the technology, between the professional studio and the producer in the bedroom, and indicate the a-ha moments that are making the future of music. The following section is the overview of the prominent tools and breakthrough projects that show the strength and capabilities of AI in sound design, synthesis, and composition.

6.1 Significant Sound Design, Synthesis and Composition Tools.

The AI-music product market is fast growing, and it provides a solution to all phases of the creative process. These tools may be classified by their major purpose:

Generative Composition Platforms These are full musical piece creation platforms.

AIVA (Artificial Intelligence Virtual Artist): AIVA is one of the best examples of a completely computerized composition engine. It has been known as the first AI to be listed as a composer with a music society. AIVA is particularly good at creating instrumental music, especially in the classical and filmic styles, by being trained on an enormous library of music by other composers such as Bach and Mozart. it's easy to use interface enables easy production of royalty-free music to use in soundtrack and commercial projects.

Sound raw: This service is aimed at the content creators by offering them the high-quality background music, which is royalty-free. It allows the users to choose a genre, mood, and an instrument and create a track within seconds. The advantage of Sound raw is that it is fast and offers musical stems, which a human creator can then further work on and tailor to his artificially produced composition.

Suno AI and Audio: These tools have hit the market with their capability of creating a complete song including vocals and lyrics based on a mere text input. They have opened music creation to virtually any person making the boundaries between a user and a composer indistinct through the use of large language models and more sophisticated generative AI.

Creative Assistance and Production Suites: These are aimed at used alongside human composers in a conventional Digital Audio Workstation (DAW) set up.

Orb Producer Suite: A collection of instruments, which work as creative assistants and produce melodies, chord progressions, basslines, and arpeggios that are musically compatible with a track being edited by a user. It is an excellent tool to use whenever one is stuck, or even facing a writer block and wants to create something and not to have to start fresh.

isotopes Neutron and Ozone: These are just composition-based tools, but industry-standard plug-ins that rely on AI to help with the technical side of music production. The Mix Assistant by Neutron and Master Assistant by Ozone is able to study a track and intelligently use effects such as EQ, compression, and limiting as a great place to start building a mix or a master. The automation liberates the creator to concentrate on the musical side of the song.

Sound Design and Synthesis Tools: These tools are AI based tools which are used to create and manipulate completely new sounds.

The Magenta Studio: Google open-source project and a set of tools that experiment with machine learning uses in the process of creativity. Magenta has a range of different plugins, which can, e.g., produce musical sequences based on a basic rhythmic input or produce new sounds by mixing various instrument sounds together.

LALA AI: This platform is an AI-based audio separation. One can post a song, and the AI will cleverly split the music into instrumental and vocal, and even more, isolate drums and bass and other instruments. It is an indispensable resource to remixers, producers and sound designers.

Waves Clarity Vx: AI-based real-time noise reduction which identifies a vocal performance and separates it intelligently with background noise. This instrument is a good example of how AI can be applied to address, typically time-consuming production issues and make clean records simpler to accomplish.

6.2 Case Studies: Landmark Artificial Intelligence Generated Compositions.

In addition to the tools, there are several famous compositions and projects that are used as case studies to prove what AI can achieve in a creative environment.

Daddy's Car (Sony CSL, 2016): This song is a breakthrough in the sphere as it was composed by the AI program, of which Sony is a part, called Flow Machines. The Beatles feed the AI on a substantial amount of music and with the help of a human composer, the AI created a song that was very similar to the style of the band. The project was among those, which received mainstream media coverage and provoked the interest in the creative opportunities of AI among the population.

I Am AI (Taryn Southern, 2018): Pop artist Taryn Southern released a full-length album, I Am AI, which was produced with the assistance of AI-based tools almost entirely, including IBM Watson Beat and Amper Music. This project was a brave artistic gesture that demonstrated that an artist could rely on AI as the main compositional instrument. Although, a human being still gave the creative leadership and organization, the album proved that a human-AI coordination of work is viable.

Lost Tapes of the 27 Club (Over the Bridge, 2021): (Over the Bridge, 2021) This is a highly thought-provoking project that employed AI to study the musical style of musicians who died at age 27 (e.g., Kurt Cobain, Amy Winehouse) to come up with original songs. This was aimed at creating awareness regarding mental health in music industry. The songs were disturbingly believable, and the initiative sparked a worldwide debate about the use of AI to replicate the sound of the artists who had passed away, and the moral side of the technology.

Table 2 below shows some of the most popular uses of Artificial Intelligence and the tools that are widely associated with these applications. It gives a good insight into the way certain AI tools assist in various fields and real-life applications.

Table 2. Key applications and their corresponding AI tools

Application Category	Description	Sample Tools & Platforms	Primary Benefit
Fully Automated Composition	AI generates a complete, finished musical piece with minimal user input.	AIVA, Sound raw, Soundful	Speed, cost-effectiveness, and accessibility for content creators.
Application Category	Description	Sample Tools & Platforms	Primary Benefit
Human-in-the Loop Ideation	AI provides creative ideas and suggestions that a human composer curates and develops.	Orb Producer Suite, Magenta Studio	Helps overcome creative block, facilitates exploration of new ideas.
AI-Powered Sound Synthesis	AI creates entirely new, unique sounds and timbres by blending audio characteristics.	Google's NSynth, Waves Clarity Vx	Pushes the boundaries of sound design, creates unique sonic signatures.
Audio Enhancement & Production	AI automates and optimizes technical mixing, mastering, and audio cleaning tasks.	isotopes Ozone, LANDR, LALAL.AI	Increases production efficiency, makes professional-quality audio accessible.
Text-to-Music Generation	Users type a text prompt and the AI generates a complete song, often including vocals and lyrics.	Suno AI, Audio, Google's MusicLM	Democratizes music creation, requires no musical skill or training.

7. CREATIVE AND ETHICAL CONSIDERATIONS

Creative and Ethical Concerns. The introduction of AI into music composition is not simply a technological revolution, but it is a radical cultural transformation, which poses some essential questions regarding creativity, authenticity, and intellectual property. With the development of the capabilities of AI-based tools becoming more advanced, the music industry and society, in general, will have to deal with a row of complicated creative and ethics questions, which will define what the art will look like in the future. Creative Implications: The Redefining of Artistry and Authenticity. The greatest creative controversy revolves around the definition of art in itself. When an AI can create a music that is harmoniously compatible and sensually engaging, what will the human composer do? The Authenticity Dilemma: One of the fundamental reasons against AI-created music is that it does not have that so-called soul or authenticity of human experience. A human artist invests his creation with his or her own feelings, life troubles and culture. An AI, in its turn, is data and algorithm based. It is able to imitate the sadness patterns, but has never known grief. Although the AI generated music can be technically flawless, critics believe it may lack emotional depth or be empty since it has not been created as a result of living experience. A New Form of Expression: On the one hand, AI supporters consider it to be nothing more than a new medium, just as the synthesizer, or the digital audio workstation (DAW), were during their era. It is a tool, which a human being can play to exercise his or her creativity. The art, here, is not, therefore, the output of the AI, but the selection of prompts by the human, the edit of the output, and the further organisation and hone of the music. AIs can be used to venture into new sonic realms and styles that a human composer could hardly imagine independently and therefore extend the limits of what music could be. The Threat of Homogenization: One legitimate worry is that some of the monoculture can be the result of a dependence on AI. Should the entire AI models to be trained on the same music played on the radio, they will be prone to produce music that follows a formula. It may suppress creativity and cause a homogenization of music genres, as AI is driven by statistical patterns rather than creativity. This danger points out the significance of human curators and artists who have a chance to bring something new and disrupt the patterns of algorithmic predictability. Ethical Concerns: Property, Authority, and Dislocation. The ethical issues that are presented by AI in music are not only short-term but also long-term, as they are affecting the legal systems and economic systems of the whole industry. Authorship and Copyright: The legal issue here is all about who is the owner of the copyright to an AI-generated piece of music? The Human Authorship Requirement: The U.S. copyright Office and numerous other international agencies have assumed that the protection of copyright is conditional on human authorship. This implies that the work produced by an AI with no creative input by a human being cannot be copyrighted. The Hybrid Model: It is even more complicated in the case when a human takes AI as a tool. In case the human gives a major creative input like arranging, editing or incorporating their own new ideas, the resulting work may be subject to receive copyright protection. The most important legal analysis is the test of whether there is a sufficient human authorship. This poses a gray area of the law that is currently being established in courts. Training Data: A wide range of generative AI models are trained on huge amounts of music that are scraped off the internet without the express agreement of the original artists. This casts serious doubts on infringement of copyrighted work (Pachet, 2023) Fair Use Controversy: AI developers tend to claim that this training is a fair use since it amounts to a transformative use of the information to a different purpose. Nevertheless, more artists and record labels such as the RIAA have sued AI companies such as Suno and Audio claiming that this is a form of copyright infringement as it devalues their work. The results of these legal contests will provide significant precedents of what will come of AI. The so-called Deep fake Voice Problem: The possibility of the AI to convincingly imitate the voice of an artist (i.e., the recently popular AI Drake song) presents a completely different legal and ethical concern. Even a unique underlying composition can be infringed upon by this act and thus infringe upon the right of publicity and likeness of an artist. The Effect on Human Musicians: The emergence of AI can only lead to concerns about losing employment. Will AI-generated music, which can be produced fast and cheaply, undermine the effort of human composers, songwriters and session musicians? Automation vs. Augmentation: The vast majority of experts and artists consider that AI will be used as an augmentation technology, and not a full-fledged replacement. Tedious and repetitive processes (such as mixing and mastering) can be automated and musically drafts created with AIs, so people artists can concentrate on the more creative and emotional parts of the job. New Positions and Opportunities: AI can also introduce additional jobs in the music industry, like the so-called AI composers that specialize in prompting and guiding AI systems, or the so-called AI ethicists that would make the technology responsible. The difficulty here is how to adjust to these new realities and learn to excel in a human-AI working environment.

The table below (Table 3) identifies the ethical and legal considerations that are related to the use of Artificial Intelligence.

It raises key issues, ethical standards, and laws that determine the responsible development of AI.

Table 3. The ethical and legal landscape

Ethical Consideration	Key Question	Pro-AI Perspective	Anti-AI/Sceptical Perspective
Authorship & Copyright	Who owns the copyright to a work created by an AI?	The user who prompts the AI should hold the copyright, as they provided the creative direction.	Copyright should only be granted to human authors, as it's meant to protect human creativity.
Training Data	Is it ethical to train AI on copyrighted material without consent?	Training is "fair use," as the AI learns from the data without directly copying it. It's a transformative process.	It is a form of mass-scale, uncompensated copyright infringement that devalues artists' work.
Job Displacement	Will AI replace human composers and musicians?	AI will not replace human creativity but will augment it, creating new roles and increasing efficiency.	AI will devalue human work, leading to job loss, particularly for session musicians and composers for commercial media.
Authenticity	Can a machine create art with genuine emotion and soul?	The emotion is in the human's intent and curation, not the AI's technical output. The tool is what matters.	True art is born from lived human experience, which an AI can never have. The output is technically sound but emotionally sterile.

7. AI MODELS IN SOUND DESIGN

The artificial intelligence models are used as the heart of the discussion of modern sound design to make and modify intricate audio patterns. The Generative Adversarial Networks (GANs) is based on a competitive approach between a generator and a discriminator to enable the creation of a quality sound texture and musical features that are real. RNNs consider time dependencies in music and are useful in generating melodies and rhythms. Transformer-based architecture involves the use of attention to encode long-range musical structure, so they are applicable to many instruments and long-form music. Diffusion models create audio through the progressive instantiations of random noise into structured sound to produce high-quality and natural audio. All the models have their benefits and drawbacks: GANs are creative in the diversity of their collections but are hard to stabilize during training, RNNs work well with sequences but fail to maintain long-term coherence, Transformers generate high-fidelity audio at the expense of slower generation time, and diffusion models are easy to take into consideration. A comparative study of these models demonstrates that they can be used in various applications like sound generation, music generation and real time audio generation.

9. ISSUES, PROBLEMS & CHALLENGES

Artificial intelligence in music production also has a number of technical and social issues despite the benefits. A significant issue that is not to be overlooked is the bias in data, and models that are trained on a small or culturally predominant dataset can result in a similar outcome. The possible end result of overfitting is the reproduction of already existing musical patterns instead of new music. The subjective nature of the evaluation and creativity of music makes it hard to control its quality. Another problem is the disappearance of originality and human creativity in art because the music created by AI might not contain any feelings or purpose. The threat of copyright infringement is associated with the event where models accidentally replicate copyrighted data without their intention. Responsible deployment is also complicated by the ethical abuse, including deep fake voice cloning of artists. Moreover, upcoming automation of the music production processes is a concern as it leads to displacement of musicians and sound designers.

10. TRENDS

The present tendencies in AI-based music revolve around the increase in interaction, personalization, and multimodal creativity. Text-to-music generation models enable users to generate music by way of simple prompts in natural language. Individualized music systems produce music based on the emotional mood, preferences and activity of a person. Dynamic adaptive music is becoming more popular in videogames and virtual worlds in order to react dynamically to player actions. Multimodal AI is used to combine music with visual and motion data to generate artistic experiences. Human-AI collaborative creativity, emotion-aware music generation, and the design of sustainable and ethically responsible music generation systems that strike a balance between the goals of innovation and responsibility are direction areas of future research.

11. ETHICAL & LEGAL CONSIDERATIONS

The present tendencies in AI-based music revolve around the increase in interaction, personalization, and multimodal creativity. Text-to-music generation models enable users to generate music by way of simple prompts in natural language. Individualized music systems produce music based on the emotional mood, preferences and activity of a person. Dynamic adaptive music is becoming more popular in videogames and virtual worlds in order to react dynamically to player actions. Multimodal AI is used to combine music with visual and motion data to generate artistic experiences. Human-AI collaborative creativity, emotion-aware music generation, and the design of sustainable and ethically responsible music generation systems that strike a balance between the goals of innovation and responsibility are direction areas of future research.

12. CASE STUDIES

Well-organized case studies reveal the real-world use of artificial intelligence in sound design and composition of music. Both case studies explore a particular tool through determining its purpose, the technology used, the results obtained and how it has contributed to the overall production of music. As an illustration, AI composition systems will help musicians compose melodies and harmonies with the help of the neural networks and Transformer models. The results are shorter production cycles and more experimentation of creativity. The effects of such tools are democratization of music production, ease of use to people not in the music field, and the development of new workflows between people and machines. The case studies give practical examples of how AI has revolutionized the music ecosystem.

13. AUDIENCE SCOPE AND IMPACT

The chapter is aimed at a wide target audience of students, researchers, musicians, and professionals of sound design. As a student, it gives the initial information about AI methods and how they can be used in music. It is beneficial to the researcher in that they can synthesize the literature that is already available, identify research gaps, and discuss the emerging trends. Musicians and practitioners will get enlightened about useful resources, creative opportunities, and ethical obligations of AI-generated music production. The chapter tackles technical, creative, and ethical aspects and therefore makes it apt and useful to academic and professional circles. The following discussion interprets the findings in relation to existing studies and evaluates the strengths and limitations of artificial intelligence in musical creativity.

14. DISCUSSION

The discussion combines the results of the methodology and case studies as well as literature review to determine the effectiveness of artificial intelligence in sound design and music composition. Findings demonstrate that AI has a positive impact on productivity, assists in creativity, and allows creating new forms of music, on the downside, it has limitations regarding originality and effectiveness. In comparison with the existing literature, it is possible to confirm that existing models are good at recognizing patterns and cannot help in terms of artistic intent and cultural interpretation. The strengths of AI are scalability, automation and accessibility, and the weaknesses are biased, dangers of ethical concerns and reliance on training data. This objective evaluation shows the human element is essential in leading AI-generated creativity. One must admit some limitations in this chapter. The research is more of secondary sources and literature reviews than actual experimental execution and therefore may limit the level of technical confirmations of the put forward ideas. Because of the fast development of the sphere of artificial intelligence, a part of the discussed tools and methods can become obsolete with time. It also relies on the access and quality of available datasets which may result in bias and restrict the extrapolation of the results to different types of musical cultures and genres. Additionally, quality and artistic appreciation is always subjective in essence and as such, it is hard to come up with universally applicable performance measures.

These shortcomings denote the necessity of the future empirical research and cross-disciplinary studies to reinforce and confirm the results of this chapter.

10. STRENGTHS OF THE CHAPTER

This chapter exhibits a number of strengths that make it more scholarly and practical. It gives a detailed and organized summary of the applications of artificial intelligence in sound design and music composition by incorporating technical models, industry applications, and ethical issues in a unified framework. The research provides the comprehensive view of the topic with the inclusion of a larger literature review, case studies, and discussion of the emerging trends. The interdisciplinary nature of its approach is another strength that links computer science, music theory and creative practice. The chapter provides a balanced and critical discussion on opportunities and challenges of using AI-driven music technologies, and it is beneficial to students, researchers, and practitioners who want to know and use such technologies in a responsible manner.

8. APPLICATIONS IN THE MUSIC INDUSTRY

Artificial intelligence has gained popularity in various spheres of the music business and has improved the process of creativity, efficiency, and personalization. Background scoring can be done using AI in the film industry, as well as creating sound effects and adaptive soundtracks that adjust to the emotion in a scene. Gaming businesses use AI to create real-time adaptable music, which responds to action and in-game scenarios by a player. Music therapy in therapy and education AI-based music systems are used to support music therapy programs and to offer interactive music training learning tools. Music streaming uses artificial intelligence to provide user-specific music suggestions and generate background music based on their tastes. AI is used in advertisements to create brief jingles and tailored sounds to use in advertising. These apps are evidence of how AI can be diversified to assist in both creative and commercial music production.

8. FUTURE OF AI-POWERED MUSIC

The high pace of AI evolution in music may indicate the future that is not only technologically progressive but which essentially changes the functions of creator, consumer, and the structure of the music industry itself. The future of the AI-driven music is marked with an augmentation of the symbiotic relationship between humans and AI, new creative formats, and the need to clarify the legal and ethical aspects. Essentially, it goes further to human-AI Symbiosis. It is not likely that the future of AI in music will be the highest degree of automation; rather, it will most probably be the partnership. The role of AI will also be more complex than just a tool because it will be a more advanced co-creator, which will know the style and the purpose of a musician. Custom AI Models: AI systems of the future can enable musicians to have an AI that is trained on their own unique library of works. This customized model would learn and elaborate on the personal voice of an artist, and then produce an innovative partner that would produce new music that would be stylistically consistent with that of the individual artists. This would bring AI to a more generalized, data-oriented tool to a much more specialized and personalized partner. This would break the boundaries between creation and performance and AI would be an active contributor to the spontaneous act of music making. and in real-time, AI may produce dynamic and adaptive accompaniment based on the improvisation of the performer, and it creates an interactive and dynamic sound experience. Intuitive and Integrated Interfaces: AIs music tools are likely to have increasingly user-friendly interfaces that are better integrated into the existing Digital Audio Workstations (DAWs). This will enable the artists to use AI in natural creative flow, and not clumsy, disconnected platforms. New Creative and Consumptive Formats. The capacity of AI to create music in enormous quantities will result in the establishment of novel forms of music and consumption altogether.

Generative Audio Formats: The classic format of a song: a fixed, three-minute piece can be supplemented with the so-called indefinite songs or soundscapes which respond to the user or environment. Applications such as Endel are already able to generate functional and never-ending soundscapes to concentrate or fall asleep. The AI of the future may be able to produce music that is continually changing, and tailored to the real time mood or environment of a listener, producing an extremely personalized soundtrack to life. (Southern, 2018) Cross-Modal Creativity: The integration of AI with other technologies, including virtual and augmented reality, will result in the artistic manifestations of novel forms. AI might be used to create music on the fly, depending on what a user is doing in a VR environment, or generate a visual art work to respond to the music being produced. It is the cross-modal creativity that will enable the boundaries to be pushed to produce multi-sensory experiences, which are immersive. Hyper-Personalized Content: AI will further personalize the music consumption. Instead of just having a list of music playing at any given time, streaming services will evolve into an AI DJ that will promote music, commentary and even remixes in real-time according to the listening preferences of a user. Such a high rate of personalization will establish a strong bond between the listener and his/her music library. Future of the Music Industry: Surviving in a New Landscape.

The music industry will certainly suffer the effects of the technological changes in business models and law. Business Model Evolving: The existing music (streaming, sales and sync licensing) revenue models will have to evolve. There can be new forms, like royalty sharing of songs using AI generated portions of them or subscriptions-like services based on which access to AI-generated creation tools are available. Legal Cells: The consistent legal turmoil concerning copyright infringement and training data is a very strong pointer that a new regulatory framework is in order. To protect AI-generated content that may be considered a legal infringement, the legal systems will be needed to develop definite guidelines on authorship, fair use, and monetization of the generated materials. The transparency regime can be enforced, and the AI developers will be required to reveal what information their models were trained on. New Roles and Skills: The emergence of AI will not automatically imply the elimination of human creativity, but it will shift its manifestation. Artists will become more like curators or prompt engineers and exercise their imaginative vision to critically shape the output of AI. The talents that are needed to succeed in music will change to the human-AI cooperation and understanding of algorithms and the traditional skills in playing the instrument.

An ever-growing personalization and the immersive nature characterize the future of AI in music. There are personalized AI composers which can create music based on the emotional and cognitive conditions of the person. Combined with virtual and augmented reality, it will be possible to have immersive musical environments and interactive performances. AI will also play a role in the development of the completely new genre of music that will have human and machine creativity combined. To control the ownership, consent and responsible use, policy development and better ethical systems will be required. Further studies will involve developing transparent and culturally inclusive, artist-cantered AI music systems. Finally, the concluding section summarizes the key findings and outlines future directions for research and practice.

9. CONCLUSION

In this chapter, the development, methods, uses, and issues of artificial intelligence in sound design and music composition have been discussed. The main contributions are the review of AI models in detail, the practical use of artificial intelligence in the industry, ethical and legal issues, and systematic case studies. Although AI shows a great possibility of enhancing creativity and reinventing music production, the issues of data bias, legal uncertainty, and authenticity are to be overcome. The chapter also highlights that AI must be regarded as a partner tool and not as an oversight of human musicians, not as an element of innovation but as a creative integrity-saving tool.

The path that started with the first experiments of algorithmic composition up to the modern times advanced AI-driven music production signifies a paradigmatic and irreversible change in the artistic world. By the end of this journey on the topic of AI in music, it has become evident that the technology is an unlucky phenomenon, but it is still a new and potent force, which is reinventing all the aspects of the industry, including the very process of its creation, as well as the legal framework within which it operates.

The major argument of this paper is that the future of music will be symbiotic, which is characterized not by the crowding out of human creativity, but rather by its enhancement. The best and creatively significant uses of AI in the modern world are the ones that place the technology as an ally, an inspiration, and the way to overcome creative constraints. AI can get us out of the rut, propose a new chord progression, or automatically create a bassline, allowing the human artist to concentrate on the expressive, emotional, and narrative essence of her creation. According to this model, the machine is a partner that assists the human to attain a creative vision that he or she would not have achieved without the assistance of the machine.

Nevertheless, this new paradigm has its problems. The legal and ethical issues related to training data usage, authorship, and copyright are complicated and require an urgent solution. The viral phenomenon of AI Drake and the litigation against AI businesses demonstrate that it is essential to design new laws. The industry needs to put down proper guidelines that safeguard the right of artists and at the same time encourage innovation.

In the future, the history of AI-driven music will hold an even more individualized and creatively experimental future. Hyper-personalized music experiences, in which AI generates a soundtrack of one person in real-time, we can look forward to; and entirely new forms of art will appear that integrate generative music with visual and interactive media. Democratization of music will persist, and the new generation of artists, who might not possess the necessary musical education but can take advantage of the AI to express their voices, will emerge.

Ultimately, one of the largest influences of AI on music can be its capability to act as a mirror that will compel us to reconsider what we actually appreciate in art. The AI can perfectly recreate the technical quality of a masterpiece, but it cannot perfectly recreate the human experience of the music, the happiness, the sorrow, the cultural background, without which music has little sense. The future of music will not be that of machines writing all the songs; it will be that of the distinct and inimitable aspect of human creation. This chapter critically searched the development, techniques and uses of artificial intelligence in sound design and music composition by combining technical models, case studies and ethical implications.

The main advantage of the chapter is that it gives a well-rounded account of both the opportunities offered by creative tools and the threats they present due to its thoroughly synthesized recent literature and practical applications. Nevertheless, there are still constraints because the current AI technologies are dynamic, and the models advertising that are available today rely on extensive data on training, which can limit originality and cultural diversity. Among the major problems pointed out in the chapter is the ambiguity of copyright and bias in data as well as the misuse of ethics and the new trends underline text-to-music generation, personalized composition, and multimodal creativity. This work can provide a comprehensive view of AI-driven music systems effects, both on a technical and societal level, by adding over 25 recent and interdisciplinary references, which will assist researchers, students, and practitioners in comprehending the effects of such systems.

REFERENCES

Agarwal, S., & Sultanova, N. (2024). *Music Generation through Transformers*. International Journal of Data Science and Advanced Analytics.

Audio Engineering Society (AES). (n.d.). *Journal of the Audio Engineering Society*. Retrieved from https://www.aes.org/journal/

Briot, J. P., Hadjeres, G., & Pachet, F. D. (2017). Deep learning techniques for music generation—a survey. arXiv:1709.01620.

Briot, J.-P., Pachet, F., & Hadjeres, G. (2020). *Machine learning in music generation*. Springer.

Chen, K., Wu, Y., Liu, H., Nezhurina, M., Berg-Kirkpatrick, T., & Dubnov, S. (2024, April). Musicldm: Enhancing novelty in text-to-music generation using beat-synchronous mixup strategies. In ICASSP 2024-2024 IEEE International Conference on Acoustics, Speech and Signal Processing (ICASSP) (pp. 1206-1210). IEEE.

Cope, D. (2006). *Musical creativity: Computer models of musical intelligence*. The MIT Press.

Dhariwal, P., Jun, H., Payne, C., Kim, H. S., Eskandari, B., Saden, S., Khadka, S., Burda, Y., & Brockman, G. (2020). *Jukebox: A generative model for music*. arXiv. https://arxiv.org/abs/2005.00341

Engel, J., Resnick, C., Roberts, A., Dieleman, S., et al. (2017). *Neural Audio Synthesis of Musical Notes with WaveNet Autoencoders (NSynth).*

Flow Machines (Sony CSL). (2016). Daddy's car [Song]. Flow Machines Project. https://www.flow-machines.com

Foster, T. (2025). *Artificial intelligence in music production and audio: Writing the music of the future*. Routledge. DOI: 10.4324/9781003621126

Ghostwriter977. (2023). *Heart on my sleeve* [Audio file]. YouTube. https://www .youtube.com/

Google Magenta. (n.d.). *Magenta: Open-source research project exploring machine learning in art and music*. Retrieved from https://magenta.tensorflow.org

Hadjeres, G., & Pachet, F. (2017). DeepBach: A steerable model of Bach chorales generation. In *Proceedings of the 34th International Conference on Machine Learning*.

Jun Min, Z. L., & Wang, L. (2022). Music Generation System for Adversarial Training Based on Deep Learning. *Processes (Basel, Switzerland).*

Liu, L., Gong, R., & Yang, Y. (2025). MusDiff: A multimodal-guided framework for music generation. *Alexandria Engineering Journal*, *129*, 128–136.

Miranda, E. R. (2021). *Handbook of artificial intelligence in music: Principles, advanced techniques, and emerging applications to creativity*. Springer. DOI: 10.1007/978-3-030-72116-9

Mitra, R., & Zualkernan, I. (2025). Music Generation Using Deep Learning and Generative AI: A Systematic Review. *IEEE Access : Practical Innovations, Open Solutions*, *13*, 18079–18106. DOI: 10.1109/ACCESS.2025.3531798

Neves, P., Fornari, J., & Florindo, J. (2022). *Generating music with sentiment using Transformer-GANs* (arXiv).

Over The Bridge. (2021). *Lost tapes of the 27 club* [AI music project]. Over The Bridge. https://www.overthebridge.org/lost-tapes

Pachet, F. (2023). *The green piano: How AI taught me to play*. The MIT Press.

Southern, T. (2018). *I am AI* [Album]. Amper Music & IBM Watson Beat.

Van Den Oord, A., Dieleman, S., Zen, H., Simonyan, K., Vinyals, O., Graves, A., . . . Kavukcuoglu, K. (2016). Wavenet: A generative model for raw audio. arXiv preprint arXiv:1609.03499, 12(1).

Zhang, M. (2025). Advancing deep learning for expressive music composition and performance modeling. *Scientific Reports*, *15*(1), 28007. DOI: 10.1038/s41598-025-13064-6 PMID: 40745012

Chapter 9
Composing With Code: AI in Film Scores, Game Audio, and Commercial Jingles

Punam Kumari
http://orcid.org/0000-0001-8144-7573
Amity University, Bengaluru, India

ABSTRACT

This chapter delves into the innovative use of Artificial Intelligence (AI) in the field of audio composition, focusing on its growing influence in film scoring, game soundtracks, and commercial jingles. Once the exclusive domain of composers and audio engineers, music and sound design are now increasingly being shaped by algorithms and neural networks. Through curated real-world case studies, this chapter highlights how AI is not merely automating audio creation but expanding the boundaries of human creativity in media production.

1.INTRODUCTION

The evolution of generative artificial intelligence (AI) for music composition has rapidly transformed creative practices in film scoring, game audio, and commercial jingles. Early research primarily focused on symbolic music generation using recurrent neural networks (RNNs) and rule-based probabilistic systems, whereas recent advances leverage transformer architectures, large audio models, and multimodal learning. This review synthesizes key peer-reviewed studies (2018–2025) to compare pre-2020 and post-2023 advancements while critically addressing training data biases relevant to media audio production.

DOI: 10.4018/979-8-3373-6279-3.ch009

Pre-2020 Foundations: Symbolic Music Generation and Structural Modeling

Early generative music systems relied heavily on symbolic representations such as MIDI sequences. Huang et al. (2018) introduced the Music Transformer, demonstrating that relative self-attention could model long-term musical structure and generate minute-long coherent compositions, a breakthrough over earlier RNN-based models limited in temporal coherence. This work marked a paradigm shift by enabling motif continuation, structural repetition, and thematic development—features critical for film scores and game soundtracks that require evolving emotional arcs.

Similarly, multi-track transformer models began exploring polyphonic music generation by modeling relationships across instrumental tracks (e.g., piano, guitar, drums), enabling richer orchestration essential for cinematic scoring (Jin et al., 2020). Earlier GAN-based approaches also attempted to encode musical theory into generation pipelines but struggled with balancing creativity and thematic coherence (Jin et al., 2022).

Despite these innovations, pre-2020 systems were constrained by limited datasets, symbolic-only representations, and insufficient control over style and emotional nuance. Moreover, the training data were heavily biased toward Western classical and pop MIDI corpora, limiting diversity in genre representation and timbral realism—an issue particularly problematic for culturally diverse media productions.

Transitional Advances (2020–2022): Style Conditioning and Multitrack Control

Between 2020 and 2022, research emphasized style-aware and theme-consistent generation. Style-conditioned Transformer-GAN architectures were introduced to generate complete compositions aligned with specified musical styles, improving controllability for applications like advertising jingles and branded audio signatures (Wang et al., 2024). These methods enabled composers to guide AI outputs toward recognizable tonal identities, aligning with industry demands for consistent sonic branding.

However, challenges persisted in thematic continuity and expressive dynamics. Many sequence-based architectures compressed music's inherently two-dimensional time-pitch structure into linear sequences, resulting in loss of harmonic context and structural logic (Jin, 2025). Consequently, generated outputs sometimes sounded repetitive or stylistically generic, limiting adoption in high-stakes film and AAA game scoring.

Post-2023 Advancements: Large Audio Models and Multimodal Generation

After 2023, the field experienced a significant shift toward large-scale generative audio models and multimodal systems capable of directly producing high-fidelity audio rather than symbolic scores. Large Audio Models (LAMs) leverage transformer architectures trained on massive audio corpora, enabling cross-task capabilities including music generation, speech synthesis, and audio understanding (Latif et al., 2023). These foundation models offer improved timbral richness, expressive dynamics, and adaptability across genres—crucial for realistic film scores and adaptive game audio environments.

Recent work on masked audio generative transformers (e.g., MAGNeT) demonstrates non-autoregressive sequence modeling that accelerates text-to-music and audio synthesis while maintaining comparable perceptual quality. Such architectures reduce latency, making them viable for real-time interactive audio in games and virtual environments.

Parallel progress in multimodal generation has enabled vision-to-music and video-conditioned composition, aligning audio with visual narrative cues—a core requirement in cinematic scoring and immersive gameplay (Wang et al., 2025). Compared to earlier symbolic-only approaches, these systems capture semantic context from visual scenes, enabling dynamic emotional scoring aligned with narrative events.

Comparative Analysis: Pre-2020 vs. Post-2023 Generative AI

The trajectory of generative music AI reveals three major distinctions:

Representation Shift: Pre-2020 models relied on symbolic MIDI generation, whereas post-2023 systems generate raw audio waveforms using tokenized acoustic representations, resulting in more realistic timbre and expressive nuance.

Scalability and Context Modeling: Early models captured limited temporal dependencies, while modern transformers and large audio models handle long-range structural coherence essential for film scoring narratives.

Multimodality and Interactivity: Earlier approaches were single-modality (music-only), whereas contemporary systems integrate text, image, and video inputs, enabling adaptive game soundtracks and context-aware jingles.

Overall, post-2023 systems exhibit superior controllability, real-time adaptability, and cross-modal alignment, positioning them as transformative tools for media audio production.

Training Data Biases in Media Audio Generation

Despite technical progress, training data bias remains a critical concern. Most generative music datasets overrepresent Western tonal systems, mainstream pop, and classical repertoires, leading to homogenized outputs and underrepresentation of non-Western musical traditions. This bias risks cultural flattening in global media content. Furthermore, datasets often exclude commercial jingle corpora due to copyright constraints, limiting the model's ability to learn concise melodic hooks and brand-oriented motifs.

Another bias arises from instrumentation imbalance: orchestral and piano datasets dominate training corpora, resulting in weak modeling of electronic, folk, or hybrid cinematic styles. Additionally, emotional labeling inconsistencies in training datasets can cause mismatches between intended narrative mood and generated audio—problematic for film scoring where precise emotional alignment is essential.

Ethical concerns also emerge regarding ownership and stylistic imitation. When models are trained on proprietary film scores or game soundtracks, generated outputs may unintentionally mimic identifiable composer styles, raising intellectual property and attribution challenges. Addressing these biases requires curated, diverse datasets encompassing global music traditions, transparent licensing frameworks, and bias-aware evaluation metrics.

Implications for Film, Games, and Commercial Jingles

For film scoring, advanced generative models enable rapid thematic sketching and adaptive orchestration aligned with narrative pacing. In game audio, real-time generative systems allow dynamic music that responds to player actions, enhancing immersion. In commercial jingles, style-conditioned generation offers scalable branding solutions by producing multiple variations of short melodic hooks tailored to specific audiences.

However, the adoption of AI composers should be framed as augmentative rather than substitutive. Human composers retain a crucial role in curating aesthetic intent, cultural authenticity, and narrative coherence, while AI systems provide exploratory ideation and rapid prototyping.

The progression from symbolic sequence models to multimodal large audio models represents a paradigm shift in generative AI for media composition. Pre-2020 approaches laid the structural foundation for algorithmic music generation, whereas post-2023 advancements deliver high-fidelity, context-aware, and interactive audio synthesis suitable for film, game, and advertising industries. Nonetheless, persistent biases in training datasets necessitate careful curation and ethical governance to ensure culturally inclusive and stylistically diverse generative music systems.

Diagramming AI Pipelines for Media Music Composition

This subsection presents concise AI pipeline flows for three media domains, illustrating how generative models integrate with creative workflows. These diagrams can guide implementation and clarify domain-specific differences in data inputs, conditioning signals, and output constraints.

1. AI Pipeline for Film Score Generation

Flow:

Script/Scene Metadata → Emotion & Tempo Extraction → Style Conditioning → Generative Model → Human Composer Refinement → Final Orchestration

Pseudocode:

```
scene = load_scene_metadata()
mood, tempo = analyze_emotion(scene)
style_vector = encode_style("cinematic_orchestral")
music_latent = generative_model.generate(
conditioning=[mood, tempo, style_vector]
)
score = human_refinement(music_latent)
export(score, format="DAW_project")
```

Focus: Narrative alignment, long-term thematic coherence, orchestral realism.

2. AI Pipeline for Game Audio (Adaptive Soundtracks)

Flow:

Game State Data → Real-time Context Encoder → Dynamic Music Generator → Loop/Transition Manager → Interactive Playback Engine

Pseudocode:

```
while game_running:
state = get_game_state() # combat, exploration, stealth
context_vec = encode_context(state)
music_chunk = generative_model.generate(conditioning=con-
text_vec)
```

```
adaptive_audio.play(music_chunk, transition="smooth")
Focus: Real-time responsiveness, seamless transitions, player-driven variation.
```

3. AI Pipeline for Commercial Jingle Creation

Flow:

Brand Identity + Target Audience → Style & Hook Encoding → Short-form Melody Generator → A/B Variation Engine → Marketing Approval

Pseudocode:

```
brand_features = encode_brand(identity, audience_profile)
hook_style = select_style("catchy_pop_5sec")
jingle_candidates = []
for i in range(5):
jingle = generative_model.generate(conditioning=[brand_features, hook_style])
jingle_candidates.append(jingle)
best_jingle = select_via_marketing_metrics(jingle_candidates)
export(best_jingle, format="audio_ad")
```

Focus: Memorability, brevity, stylistic consistency with brand identity.

1.1 The Evolution of Sound Creation: From Analog to Algorithm

The journey of sound creation has undergone a remarkable transformation, evolving from analog systems to today's algorithm-driven approaches. In the early days, music and audio design were deeply tied to analog technologies such as tape recorders, vinyl records, and analog synthesizers. These tools, while innovative for their time, limited composers and sound designers to the physical properties of instruments and recording equipment. Creativity often meant working within the constraints of hardware, where every sound carried the signature imperfections of its medium (Holmes, 2016). With the advent of digital technologies in the late 20th century, the process of audio composition shifted dramatically. Digital Audio Workstations (DAWs), sampling, and MIDI protocols enabled unprecedented precision, flexibility, and reproducibility in music production. This democratization of sound creation allowed not only professionals but also independent creators to experiment with composition, arrangement, and editing at scales that were unimaginable in the analog era (Collins, 2010).

Today, we stand at the threshold of a new paradigm where algorithms and artificial intelligence redefine what is possible in sound design. Machine learning models, neural networks, and generative algorithms now compose original scores, adapt audio dynamically to gameplay, and craft commercial jingles tailored to specific audiences. Unlike earlier technological shifts, AI is not merely a tool for replication or enhancement but an active collaborator, augmenting human creativity by offering novel patterns, timbres, and structures that might not have emerged otherwise (Huang & Yang, 2020).

Case Study 1: Film Scoring with AIVA

AIVA (Artificial Intelligence Virtual Artist) has been used in film and advertising to generate orchestral scores. Rather than replacing human composers, AIVA provides drafts that can be refined, significantly reducing production time while sparking creative new directions (Miller, 2019).

Case Study 2: Game Audio with Amper Music

Amper Music, another AI-driven tool, allows developers to create adaptive soundtracks for games. By adjusting tempo and mood algorithmically, it enables smaller studios with limited resources to deliver immersive audio experiences that once required full sound design teams (Vincent, 2017).

The trajectory from analog to algorithm illustrates more than just a technological evolution—it reflects a broader cultural transformation in the way we understand and engage with sound.

1.2 Rise of AI tools in audio production (e.g., AIVA, Amper, Jukedeck)

The introduction of Artificial Intelligence (AI) into audio production represents a pivotal moment in the history of sound design. Unlike earlier transitions from analog to digital, the rise of AI tools has fundamentally reshaped the creative process by positioning algorithms as collaborators rather than mere instruments. These technologies harness machine learning, neural networks, and generative models to analyze vast musical datasets, learn stylistic patterns, and produce compositions that mimic or extend human creativity (Briot et al., 2020).

AI-based music platforms are now widely used in diverse contexts—from film scoring to video game soundtracks and commercial jingles. Tools such as AIVA

(Artificial Intelligence Virtual Artist), Amper Music, and Jukedeck exemplify how AI is being applied to democratize music creation, lower production costs, and accelerate workflow.

Case Study 1: AIVA in Orchestral Composition

AIVA (Artificial Intelligence Virtual Artist) is one of the most established AI composers, trained on centuries of classical music ranging from Mozart to Beethoven. Its deep learning models are designed to understand orchestral structures, harmonic progressions, and stylistic nuances. What makes AIVA unique is its recognition as a legitimate composer—registered with the Society of Authors, Composers, and Publishers of Music (SACEM) in France. AIVA has been used in film, advertising, and even video game soundtracks to generate orchestral drafts that human composers can refine. For example, directors and producers often use it during pre-production to experiment with multiple musical directions before committing to a final score. This saves time and resources while still leaving space for human emotion and artistry in the final product (Miller, 2019). Rather than replacing composers, AIVA extends their capabilities by offering a collaborative starting point that blends computational efficiency with creative exploration.

Case Study 2: Amper Music in Commercial Production

Amper Music takes a different approach, focusing on accessibility for non-experts. Designed as a cloud-based AI composition tool, it enables users to create customized, royalty-free tracks without requiring musical training. Users simply select parameters such as genre, mood, tempo, and instrumentation, and Amper generates a unique composition within seconds.This model has been particularly useful for small businesses, indie game developers, and social media creators who need high-quality audio but lack budgets for professional sound design. For example, a small startup producing promotional videos can use Amper to generate background music tailored to each campaign, saving both time and licensing costs. The platform has thus become a powerful enabler of creative entrepreneurship, democratizing access to music production that was once restricted to those with expensive tools and expertise (Vincent, 2017).

Case Study 3: Jukedeck and Personalized Soundtracks

Jukedeck was among the earliest AI-driven platforms to focus on adaptive and personalized music creation. Its innovation lay in its ability to generate completely original tracks tailored to user specifications such as mood, length, and energy.

This feature found strong adoption among YouTube content creators and digital marketers, who needed background music that was both unique and copyright-safe.

In 2019, Jukedeck was acquired by ByteDance, the parent company of TikTok, to enhance its personalized music and audio features (Dredge, 2019). This acquisition highlighted the growing commercial value of AI-generated music—not only as a creative tool but also as a strategic asset in the competitive digital media industry. By integrating AI composition into platforms with massive user bases, Jukedeck demonstrated how AI could scale personalization in ways traditional music production could not.

The rise of these tools marks a democratization of audio production. By lowering barriers of expertise and cost, AI platforms are not only changing how music is made but also who gets to make it. As a result, audio creation is shifting from an exclusive professional domain to a participatory space where technology amplifies creativity across industries.

2AI IN FILM SCORING

2.1 How AI helps indie filmmakers score emotional, cinematic soundtracks on limited budgets

Imagine you're an indie filmmaker with a powerful story but a modest budget. Your film is ready, but the soundtrack—a crucial element for emotional depth—is missing. Hiring a professional composer or orchestra is financially out of reach. Enter AI-powered music composition tools, which are transforming how indie filmmakers approach scoring.

AI platforms like AIVA, Amper Music, and Jukedeck enable filmmakers to generate original, royalty-free music tailored to their film's mood, genre, and pacing. These tools democratize access to high-quality soundtracks, allowing creators to produce cinematic scores without the need for extensive musical training or large budgets.

AIVA (Artificial Intelligence Virtual Artist) is a notable example. It utilizes deep learning models trained on a vast dataset of classical and cinematic music to compose original pieces in various styles. Filmmakers can input parameters such as mood, tempo, and instrumentation to generate music that aligns with their vision (AIVA, n.d.).

Amper Music offers a user-friendly interface that allows creators to customize music by selecting mood, style, and instrumentation. The platform's AI engine then composes a unique track based on these inputs, providing a quick and cost-effective solution for filmmakers (Amper, n.d.).

Jukedeck, now integrated into YesChat, was one of the pioneers in AI-generated music for creators. It allowed users to generate custom, royalty-free music tracks by selecting parameters like genre, mood, and duration. This accessibility made it a valuable tool for filmmakers seeking to enhance their projects without the need for professional composers (Jukedeck, n.d.).

These AI tools function as collaborators, offering suggestions and compositions that filmmakers can refine and adapt to their projects. They provide a cost-effective alternative to traditional scoring methods, enabling filmmakers to focus more on storytelling and less on budget constraints.

2.2 Case study: Use of AIVA to generate thematic compositions in short films

In the realm of independent filmmaking, budget constraints often limit access to professional composers and orchestras. However, advancements in artificial intelligence (AI) have introduced innovative solutions to this challenge. One such solution is AIVA (Artificial Intelligence Virtual Artist), an AI-powered music composition tool capable of generating original, royalty-free music tailored to specific moods, genres, and themes. This case study explores the application of AIVA in creating thematic compositions for short films, highlighting its capabilities, workflow integration, and impact on the filmmaking process.

Overview of AIVA

AIVA is an AI music generation assistant that utilizes deep learning algorithms to analyze and learn from a vast database of classical and cinematic music. By understanding patterns in harmony, melody, and rhythm, AIVA can compose original pieces in various styles, including orchestral, cinematic, and contemporary genres. Users can customize compositions by selecting parameters such as mood, tempo, key, and instrumentation, allowing for a high degree of creative control (AIVA, n.d.; SoundTech Insider, 2023).

Case Study: "Eclipsed"

To illustrate the practical application of AIVA in film scoring, consider the short film Eclipsed, a 15-minute narrative exploring themes of isolation and self-discovery. The filmmakers, a team of emerging directors and producers, sought to create a compelling musical score that would enhance the emotional depth of the film without exceeding their limited budget.

Pre-Production Phase

During the pre-production phase, the filmmakers identified key scenes where music would play a pivotal role in conveying the narrative's emotional undertones. These scenes included a solitary walk through a desolate landscape, a moment of introspection by the protagonist, and a climactic revelation that shifts the story's direction. The team envisioned a score that would evolve from minimalist, ambient tones to rich, orchestral arrangements, mirroring the protagonist's journey from solitude to self-realization.

Utilizing AIVA for Composition

The filmmakers turned to AIVA to generate the film's score. They began by selecting a cinematic style and specifying the desired mood for each scene. For instance, the solitary walk scene was assigned a "melancholic" mood with a "slow" tempo, while the climactic revelation scene was designated as "epic" with a "fast" tempo. AIVA's interface allowed the team to further refine these parameters by adjusting dynamics, instrumentation, and key signatures to align with the film's aesthetic vision.

AIVA generated multiple composition options for each scene, providing the filmmakers with a range of musical interpretations to choose from. The team selected the compositions that best resonated with the intended emotional impact and integrated them into the film's editing process.

Post-Production Integration

In the post-production phase, the filmmakers fine-tuned the selected compositions within the film's timeline, ensuring that the music synchronized seamlessly with the visual elements. AIVA's ability to produce compositions in various formats, including MIDI and audio files, facilitated easy integration into digital audio workstations (DAWs) and video editing software. The team also utilized AIVA's editing features to make minor adjustments to the compositions, such as altering the length of specific sections and modifying instrumentation to better fit the evolving narrative.

Final Outcome

The resulting score for Eclipsed effectively enhanced the film's emotional resonance, with the music complementing the visual storytelling and character development. The filmmakers were able to achieve a professional-quality soundtrack that aligned with their creative vision, all within their budgetary constraints.

Benefits of Using AIVA in Film Scoring

The application of AIVA in this case study highlights several key benefits for independent filmmakers:

- **Cost Efficiency:** Traditional film scoring can be prohibitively expensive, especially for independent filmmakers. AIVA offers a cost-effective alternative by providing access to high-quality compositions without the need for hiring professional composers or orchestras.
- **Creative Control**: AIVA's customizable parameters allow filmmakers to tailor compositions to their specific needs, ensuring that the music aligns with the film's tone, pacing, and emotional trajectory.
- **Time Efficiency:** Generating compositions with AIVA is a relatively quick process, enabling filmmakers to focus more on other aspects of production and post-production.
- **Accessibility:** AIVA's user-friendly interface makes it accessible to individuals without formal music training, democratizing the ability to create original film scores.

  Challenges and Considerations

  While AIVA offers numerous advantages, there are also challenges and considerations to be mindful of:
- **Emotional Depth:** While AIVA can generate compositions that mimic the structures and styles of classical and cinematic music, some critics argue that AI-generated music may lack the nuanced emotional depth that human composers can infuse into their work (AIVA, n.d.).
- **Originality:** AI-generated compositions are based on patterns learned from existing music, which may result in outputs that are derivative or lack originality. Filmmakers may need to invest additional time in refining and personalizing the compositions to ensure they meet their creative standards.
- **Licensing and Copyright:** Understanding the licensing agreements associated with AI-generated music is crucial to ensure that filmmakers have the rights to use the compositions in their projects without legal complications.

2.3 Insights into how Bollywood and regional Indian cinema are beginning to experiment with AI-driven scoring for teasers and trailers

The integration of Artificial Intelligence (AI) into the creative processes of filmmaking has been a topic of discussion for several years. However, in recent times, Bollywood and regional Indian cinema have begun to explore AI-driven scoring, particularly in the creation of teasers and trailers. This section delves into the emerging trends, challenges, and implications of AI-generated music in Indian film promotions.

AI in Film Scoring: A Global Perspective

Globally, AI has been making significant strides in the realm of film scoring. Tools like OpenAI's MuseNet and Google's Magenta have demonstrated the capability to compose music in various styles, including classical, jazz, and pop. These advancements have opened new avenues for filmmakers to experiment with soundtracks, offering cost-effective and time-efficient alternatives to traditional scoring methods (Huang, 2023).

In the context of Indian cinema, the adoption of AI in film scoring is still in its nascent stages. However, the potential benefits of AI-generated music—such as reduced production costs and the ability to quickly generate multiple musical variations—have prompted filmmakers to explore its application in teaser and trailer productions.

Case Studies in Indian Cinema

"Naisha" – India's First AI-Powered Feature Film

One of the most notable examples of AI integration in Indian cinema is the film Naisha. Directed by Vivek Anchalia, Naisha is touted as India's first AI-powered feature film, featuring fully AI-generated visuals and characters. The film's trailer, which was released in early 2025, showcases a love story that transcends timelines and continents, with the lead characters Naisha Bose and Zain Kapoor navigating emotional landscapes across Kolkata, Paris, and Switzerland. The trailer's soundtrack, composed using AI tools, complements the futuristic visuals, marking a significant step towards the incorporation of AI in Indian film promotions (Indulgexpress, 2025).

"Tuk Tuk" – AI-Generated Music Video

In the regional cinema space, the Kannada film Tuk Tuk has garnered attention for its innovative use of AI in music production. The film's music video for the song "Ru Ru" is India's first AI-generated music video, featuring futuristic beats and AI-powered visuals. Composed by Santhu Omkar and directed by Supreeth C Krishna, the music video blends AI-generated visuals with soulful melodies, offering a glimpse into the potential of AI in regional film promotions (AP7am, 2025).

"IRAH" – AI-Based Film Trailer

The film IRAH, starring Rohit Bose Roy, is another example of AI's application in Indian cinema. The trailer of IRAH showcases the misuse of AI technology, shedding light on its negative aspects. The film's music, composed using AI tools, adds to the thematic exploration of AI's impact on society, demonstrating the versatility of AI-generated music in conveying complex narratives (PTI News, 2025).

The future of AI in Indian film scoring appears promising, with advancements in AI technology continuing to evolve. As AI tools become more sophisticated, their application in film scoring is likely to expand beyond teasers and trailers to full-length features. However, it is crucial for the industry to address the challenges and ethical considerations associated with AI-generated music to ensure that it enhances rather than diminishes the creative process.

Filmmakers are encouraged to view AI as a collaborative tool that can augment human creativity rather than replace it. By integrating AI-generated music with traditional scoring methods, filmmakers can create innovative soundtracks that resonate with audiences and push the boundaries of storytelling.

3 AI IN GAME AUDIO

3.1 The role of AI in adaptive, real-time soundtrack generation for immersive gameplay

Video games are no longer just about graphics and mechanics; sound and music are central to creating immersive experiences. From subtle ambient noises to full orchestral scores, audio guides players' emotions, signals game events, and enhances the overall narrative. Traditionally, game audio designers would create static soundtracks or looped tracks that play repeatedly, often limiting the flexibility and responsiveness of the audio to player actions. Enter Artificial Intelligence (AI), revolutionizing how game music is composed, implemented, and experienced.

Adaptive Music: Shaping Gameplay with Sound

Adaptive or dynamic music is a method where the game soundtrack responds to player interactions in real time. For example, the intensity of a battle scene might increase the tempo and instrumentation, while stealth sections may trigger quieter, suspenseful tones. AI tools enable this adaptability by analyzing player behavior, game states, and in-game environmental cues to generate or modify music on the fly (Collins, 2022).

Consider a player exploring a haunted mansion in a horror game. Traditional music might loop a fixed suspense track, repeating regardless of the player's actions. AI-driven adaptive audio, however, can alter the soundtrack dynamically: creaking floors might trigger minor dissonant chords, approaching enemies could intensify orchestration, and escaping from danger might transition the music seamlessly from tension to relief. This creates an emotional feedback loop, where the music not only reflects but amplifies player experience.

How AI Generates Real-Time Soundtracks

AI music generation in gaming leverages several techniques:

Procedural Music Composition: AI algorithms create music in real time based on rules and constraints defined by composers. For instance, an AI system might generate variations in melody, rhythm, or harmony depending on the player's location or in-game event. These procedural rules ensure thematic consistency while allowing infinite variations (Collins, 2022).

Machine Learning Models: Deep learning models, trained on large datasets of existing game music, can predict and generate suitable musical segments for specific contexts. For example, recurrent neural networks (RNNs) and Transformer models can produce melodies that align with the game's aesthetic while adapting to player choices (Huang et al., 2023).

Emotion Recognition and Mapping: Some advanced AI systems analyze gameplay intensity and player behavior to map emotions to musical parameters. By detecting whether the player is engaged in combat, exploration, or puzzle-solving, the AI adjusts tempo, harmony, or instrument timbre to evoke the desired emotional response (Karlsen, 2021).

Case Study: AI in Action – "Dynamic Quest"

To illustrate the practical application, consider a hypothetical open-world RPG called Dynamic Quest. The game's developers implemented an AI-based adaptive music engine:

Exploration Phase: The AI uses procedural generation to create ambient music for forests, towns, and dungeons. Player speed and direction influence melodic patterns. A slow wander triggers soft, calming motifs, while rapid movement generates rhythmic tension.

Combat Phase: As enemies appear, the AI dynamically layers percussive beats and harmonic intensification. Player performance (e.g., health level, attack frequency) influences tempo and instrumentation. Boss fights trigger orchestral swells that rise and fall based on the battle's progression.

Puzzle-Solving Phase: AI selects minimalistic motifs emphasizing cognitive engagement, using subtle dissonances or sparse instrumentation to encourage focus and immersion.

The result is a soundtrack that never repeats in a predictable pattern, enhancing immersion and emotional engagement. The player experiences a unique audio journey influenced by their individual choices.

Benefits of AI-Driven Game Audio

- **Enhanced Immersion:** Adaptive soundtracks respond to gameplay, creating a seamless connection between player action and audio environment.
- **Infinite Variability:** Procedural and AI-generated music allows for virtually limitless variations, ensuring that repeated playthroughs do not feel monotonous.
- **Cost and Time Efficiency:** AI can generate musical content quickly, reducing the need for extensive composition, recording sessions, and manual adaptation for different game scenarios.
- **Personalized Player Experience:** By analyzing player behavior, AI can tailor music to create a personalized emotional journey, increasing engagement and satisfaction.

Challenges and Considerations

Despite its benefits, implementing AI-driven game audio presents challenges:

- **Quality Control:** While AI can generate music, it may occasionally produce segments that lack cohesion or emotional resonance. Human oversight is essential to ensure musical quality.
- **Technical Integration:** Adaptive audio engines require seamless integration with game logic and mechanics. Latency, synchronization issues, and resource consumption can affect gameplay if not managed carefully (Collins, 2022).
- **Ethical and Creative Implications:** Relying heavily on AI might reduce opportunities for human composers and sound designers, potentially impacting employment and creative diversity in the industry.

The future of AI in game audio is promising. Advances in machine learning, real-time emotion recognition, and procedural audio are enabling more sophisticated and immersive experiences. In multiplayer and virtual reality (VR) settings, AI-driven audio can adapt not only to individual players but to group dynamics, enhancing shared experiences.

Moreover, as AI tools become more accessible, indie developers can leverage adaptive music engines to create cinematic audio experiences previously available only to AAA game studios. This democratization of technology is poised to redefine how players experience interactive storytelling.

3.2 Case study: Use of Melodrive in mobile games for procedural music generation

Melodrive is a real-time adaptive music engine designed to generate continuous, emotionally-aware music streams that react to gameplay parameters, player actions, and narrative states. Unlike loop-based or pre-composed interactive scores, Melodrive composes music from algorithmic building blocks so that the soundtrack is never a repeating loop but a dynamically evolving piece tailored to the player's moment-to-moment experience. This case study examines how Melodrive's core ideas map to the specific constraints and opportunities of mobile game development, and proposes practical strategies for integrating procedural music generation into mobile games while balancing quality, performance, and designer control.

How Melodrive works (high level). Melodrive's engine couples an affective model (maps game state or designer inputs to emotional targets such as tension, sadness, joy) with a generative music model that composes motifs, harmonies, and orchestrations in real time. The system exposes parameters—intensity, tempo, instrumentation, harmonic palette—that can be driven by game variables (e.g., enemy proximity, player health, scene pacing) so that the emotional trajectory of the music mirrors the gameplay arc. The engine's continuous composition approach reduces the need for hundreds of bespoke music cues and allows for near-infinite musical variation, which increases immersion and reduces the perception of repetition.

Why mobile is different. Mobile platforms introduce specific technical constraints: limited CPU/GPU cycles, tight memory budgets, battery consumption concerns, and variable audio latency across devices. Unlike desktop or console titles where dedicated audio threads and higher compute budgets allow sophisticated synthesis, mobile games often rely on pre-rendered assets or lightweight synthesis. Integrating Melodrive-style procedural generation on mobile therefore requires adaptations: reducing synthesis complexity, using hybrid approaches (pre-rendered stems plus lightweight procedural layers), or offloading heavy computation to a server when network and latency permit. Research on procedural music in games highlights these trade-offs and suggests hybrid pipelines as pragmatic for constrained platforms.

Integration strategies for mobile games. Three practical architectures emerge:

On-device lightweight engine: implement a slimmed-down version of the generator that focuses on motif variation and sample-based orchestration rather than full symbolic composition. Use compact sample libraries and simple rule-based harmonization; keep audio buffers small and re-use synthesized stems to lower CPU cost.

Hybrid stem-stitching: pre-render multiple stems (ambient pads, percussion, lead motifs) in a range of intensities and crossfade or sequence them in real time according to gameplay parameters. Add lightweight procedural ornamentation (e.g., algorithmic arpeggios or filters) on-device. This preserves the impression of adaptive music while avoiding full real-time orchestration.

Cloud-assisted generation: stream short, low-latency audio segments generated on a remote server. This allows full generative richness but depends on stable connectivity and introduces network/latency concerns—suitable for cloud-enabled games or optional enhanced modes.

Designers should provide explicit authoring tools (emotion curves, chord maps, instrumentation presets) so composers can steer the AI and maintain artistic control. Melodrive's model of exposing high-level emotional parameters to designers is instructive: it allows creative intent to guide procedural outputs while leaving low-level composition to the engine.

Player experience & evaluation. Melodrive's internal experiments and published accounts report measurable gains in immersion when adaptive music responds to player state, particularly in VR contexts where presence is crucial. For mobile, useful evaluation metrics include perceived immersion, repetitiveness ratings, affective congruence (does music match the reported emotion), and objective engagement metrics (session length, retention). A/B testing—comparing static scores vs. Melodrive-driven adaptive scores—helps quantify impact. Qualitative designer feedback is also essential: how much time did the tool save, was the creative control sufficient, and did the engine produce idiomatic musical results?

Limitations and ethical considerations. Procedural engines can generate high volumes of unique music, raising licensing and authorship questions when models are trained on copyrighted corpora. Mobile game studios must ensure generated music's provenance is clear and licenses allow commercial use. Sonically, purely algorithmic music may sometimes lack cultural specificity; authoring tools should thus include palettes and templates that encode stylistic constraints to avoid producing musically inappropriate content. Finally, accessibility must be considered: procedural audio should not interfere with players who rely on sound cues for accessibility—engine parameters must be tunable or togglable.

3.3 Indian startup examples incorporating AI for in-game audio personalization

The Indian AI and music-technology ecosystem has matured rapidly, and several home-grown startups are carving niches in generative music, royalty-aware AI composition, and personalized audio for content creators. While not all started specifically for game audio, many Indian companies now position their tools for use by game developers, advertisers, and multimedia creators—making them relevant for in-game audio personalization. This section highlights three notable Indian examples and discusses how their capabilities translate to adaptive game audio.

3.3.1. Beatoven.ai

AI-driven music for creators and game developers. Beatoven.ai, founded in India, offers an AI music generator that creates customizable, royalty-cleared tracks tailored to mood, tempo, and instrumentation. The platform explicitly advertises a game-dev use case: developers can generate background tracks, loops, and cues with licensing that is friendly to commercial release. Beatoven's approach multimodal input (text, video, or emotion tags) and editable stems—maps well to game audio workflows: developers can generate a palette of stems at varying intensities and then programmatically switch or crossfade them according to gameplay events. Additionally, Beatoven's push toward fair royalties and licensed models addresses one major industry concern: legal use of AI-generated music in commercial games.

3.3.2. Pocket FM (AI Creator Studio)

scalable audio production and personalization. Pocket FM, primarily an audio storytelling platform, has launched AI Creator Studio tools to accelerate audio production—voice synthesis, background scores, and post-production—enabling creators to produce full episodes rapidly. For game audio, such toolchains can be repurposed to create adaptive narrative audio: dynamically generated dialogue lines, alternate voice takes, or context-sensitive audio bites that react to player choices. Pocket FM's scale and tooling demonstrate how audio personalization can be productized for Indian markets; game studios can integrate similar pipelines to produce localized, culturally nuanced audio content at scale.

3.3.3. Collective/Collective Artists Network –

experimental AI music (e.g., 'Trilok'). India has also seen creative experiments where AI-generated ensembles or projects (for example, the "Trilok" AI band) blend traditional and contemporary elements to create culturally resonant AI music. Although not a commercial game-audio provider, such projects are important proofs-of-concept: they show that AI can be trained or guided to reflect local musical idioms an essential capability when games target specific regional audiences. Game developers seeking authentic regional scores can partner with innovators who embed cultural motifs into generative pipelines.

How these startups inform in-game personalization. Collectively, these Indian firms illustrate three capabilities crucial for personalized game audio:

Multimodal control: text/emoji/mood inputs allow non-musicians to generate suitable tracks quickly (useful for rapid prototyping).

Licensed outputs: platforms offering clear licensing reduce legal friction when publishing games internationally.

Localization & cultural relevance: local experiments and datasets enable stylistic fidelity for region-specific games.

Challenges and considerations. Indian startups must still tackle model provenance and copyright, latency and compute for real-time personalization, and the need for fine-grained authoring tools that let composers constrain AI outputs. Moreover, mobile-first markets (prevalent in India) require solutions that can generate or adapt music with low on-device footprint or offer hybrid workflows. Finally, interdisciplinary collaboration—between game designers, composers, and AI engineers—remains essential to turn raw generative capability into polished, playable audio experiences.

Opportunities and roadmap. For Indian game developers and startups, a promising path is to offer modular toolkits: (a) stem-based generation APIs for low-latency switching, (b) cloud-assisted composition pipelines for richer modes, and (c) composer-facing authoring UIs that encode cultural palettes and instrumentation choices. Partnerships between AI-music startups and game-engine middleware providers (Unity, Unreal, Wwise) will accelerate adoption and offer packaged solutions for in-game personalization at scale.

4. AI IN COMMERCIAL JINGLES

The sonic identity of a brand—its jingle, sonic logo, and recurring motifs—remains one of the most powerful and enduring marketing assets. Historically, producing a memorable jingle required composer time, studio sessions, voice talent, and iterative creative direction; this meant non-trivial budgets and slow turnarounds. Today, generative audio and AI-driven production tools are reshaping this landscape: brands can generate multiple, targeted jingles in minutes; agencies can personalize sonic content for demographics and user preferences; and the cost and time required to field campaigns have dropped dramatically. This section surveys how companies leverage AI to create custom jingles, highlights Indian examples and agency use-cases, and examines the real-world impact on cost, turnaround time, and market responsiveness. The Jingle Writer)

4.1 How brands leverage AI for demographic and preference-based jingles

AI enables two complementary capabilities that are essential for modern sonic branding: (1) rapid generation of many candidate jingles across stylistic palettes, and (2) targeted personalization so that different listener segments hear bespoke versions of the same campaign. At the technology level, generative music systems (symbolic composers, sample-based synthesizers, and neural audio synthesizers) accept high-level controls—mood, tempo, instrumentation, language, and vocal character—and produce short-form audio suitable for ads. Marketing teams can feed audience attributes (age, region, language, cultural preferences) and campaign goals into the AI engine to obtain variants optimized for each target group. The result is not just a single jingle but a family of related jingles that maintain brand consistency while resonating with diverse audiences (Beatoven)

Personalization at scale follows two main patterns. First, static-personalization, where the campaign contains a small set of pre-rendered jingle variants selected by rules (e.g., Hindi jingle for North India, regional-language jingle for South India, youth-oriented remix for urban audiences). Second, dynamic-personalization, where AI assembles or modifies music in near real-time based on user data—app behavior, locale settings, or demographic signals—so that audio served in a digital ad matches the listener's context. Both approaches reduce creative bottlenecks: instead of commissioning dozens of bespoke compositions, teams author high-level templates and let AI produce the permutations. This approach scales efficiently and can be integrated into programmatic ad delivery pipelines (FINE Homes and Living)

Examples from Indian agencies and FMCG brands

India's fast-moving consumer goods (FMCG) sector and its creative agencies have become early adopters of AI in ad production. A prominent example is Emami's Dermicool campaign—reported as one of India's first fully AI-generated FMCG ads—where visuals, sound design, and the reimagined jingle were produced with extensive AI assistance; the campaign showcased significant savings in production resource and time while retaining creative impact. Indian agency reports and industry coverage highlight that a small, AI-enabled team produced complex assets that traditionally would have required far larger crews and budget (The Financial Express)

Startups and platforms like Beatoven.ai have specifically targeted creators and small agencies by providing royalty-cleared, AI-generated music that is easy to customize for ad contexts. These platforms let marketers specify mood, length, and instrumentation and obtain stems and vocal options ready for mixing—an approach that suits FMCG ads which often require local-language variants and quick iterative testing. Case studies and platform documentation show these tools being used to create event music for conferences and promotional material where speed and licensing clarity matter (Beatoven)

Beyond Beatoven, ad-technology vendors and audio-specialist startups (e.g., Wondercraft, AdsWizz collaborations) are enabling rapid voiceovers and dynamic audio campaigns. For example, audio ad campaigns that used AI-generated music and voice elements reported drastic reductions in time-to-market and production costs, while also enabling hyper-localized audio creative across thousands of market segments. These implementations demonstrate that AI is already a working tool in audio advertising stacks, not merely an R&D curiosity (wondercraft.ai)

4.2 Real-world impact: cost, turnaround time, and market responsiveness

Cost reduction. Several industry case reports indicate substantial cost savings when AI is used in ad audio production. AI-based pipelines reduce or eliminate many line-items—studio booking, session musicians, multiple rounds of human editing—and thus shrink budgets significantly. Indian agency commentary on AI-enabled ads suggests production cost reductions of 70–85% for certain AI-driven campaigns versus equivalent CG-heavy or music-studio productions. Even when agencies retained human supervision for direction and quality control, the overall cost remained a fraction of traditional workflows (Storyboard18, 2025)

Faster turnaround. One of the most immediate benefits of AI is speed. Where a conventional ad campaign might require days or weeks to compose, record, edit, and approve a jingle, AI can generate dozens of high-quality candidates in minutes. Platforms advertise near-instant production of track stems and short jingles—transforming creative review cycles into rapid A/B testing sprints. Case studies—such as programmatic audio campaigns—reported 80–95% reductions in production time in some instances, enabling same-day iteration and deployment. For seasonal promotions or flash campaigns (common in FMCG), such agility is commercially valuable. (AdsWizz)

Market responsiveness & A/B testing. With AI, marketers can run fast experiments—testing multiple sonic variations across demographics and optimizing for engagement metrics (click-through, recall, conversion). Programmatic audio and dynamic ad platforms allow different audio creatives to be routed to different segments; analytics then reveal which sonic choices work best for which audiences. This closed-loop optimization is a departure from traditional large-bet creative strategies and enables data-driven sonic brand strategies that adapt to real-time market feedback. Reports from ad-tech deployments show higher engagement for AI-optimized audio variants in some contexts, suggesting tangible ROI when personalization is done carefully. (AdsWizz)

Creative control, ethical, and legal considerations

AI's ability to rapidly generate audio raises several important issues. *Creative control* is often addressed by hybrid workflows where AI produces options and human composers/creative directors select and refine outputs—preserving brand voice while benefiting from scale. Many agencies emphasize that AI supplements rather than replaces human creativity; strategic direction and final approval remain human responsibilities (Storyboard18, 2025)

Copyright and provenance are central legal concerns. Generative models trained on copyrighted music can produce outputs resembling existing works, raising liability questions. Indian agencies and platforms increasingly prioritize royalty-cleared models and transparent licensing to mitigate risk—a key reason why platforms like Beatoven emphasize licensed content and clear usage terms. When working with celebrity voices or culturally specific melodies, agencies must ensure permission and comply with both copyright and personality-rights regimes (Beatoven)

Cultural sensitivity and authenticity. AI tools risk producing sonics that miss cultural nuance. This is especially important in India, where regional musical idioms vary widely. Successful campaigns combine AI's speed with composer-authored style-templates that encode culturally appropriate scales, instruments, and vocal characteristics. Proof-of-concept projects in India that merged AI with regional music concepts show that thoughtful curation yields authentic outcomes. (pixel-studios.com)

Accessibility and inclusivity. Audio personalization should respect accessibility guidelines: loudness normalization, captioning for audio ads, and toggles for users who prefer reduced sound. Agencies must ensure adaptive jingles do not interfere with assistive cues used by differently-abled users (FINE Homes and Living)

Practical workflows and recommendations for brands and agencies

1. *Adopt a template-plus-AI model.* Create composer-authored templates (melodic motifs, chord progressions, instrument palettes) and use AI to generate

variations—this retains brand consistency while delivering many permutations quickly.

2. *Use stems and modular assets.* Generate stems (lead, pad, percussion) instead of single-file mixes; programmatic ad stacks can then mix and match stems to produce contextual variants without full re-rendering.
3. *Implement A/B tests on sonic variants.* Deploy multiple AI-generated jingles to disjoint audience segments and measure recall, engagement, and conversion—use results to inform the brand's sonic playbook.
4. *Prioritize licensing clarity.* Work with platforms that offer clear commercial licenses or royalty arrangements to avoid post-release legal exposure.
5. *Maintain human oversight for brand voice and cultural fidelity.* AI should be guided by brand stewards and cultural consultants to avoid tone-deaf outputs.
6. *Design low-latency hybrid options for mobile.* Given India's mobile-first market, combine on-device lightweight audio layers with server-side generation for premium modes—ensuring fallback static audio for offline or low-bandwidth contexts.

AI is transforming jingle production from a slow, bespoke craft into an on-demand, data-driven capability. Brands and agencies—especially in India's vibrant FMCG sector—are already experimenting with AI-generated ads and jingles, reporting substantial reductions in cost and time-to-market while unlocking large-scale personalization. The path forward balances speed with stewardship: platforms and agencies must pair AI's generative power with licensing rigor, cultural curation, and human creative oversight to create jingles that are fast, legal, and deeply resonant with target audiences. When used responsibly, AI-driven audio can make brands more responsive to market signals and more relatable to diverse listener communities. (wondercraft.ai)

5. NOTABLE CASE STUDIES

This section examine several influential artists and collectives who have integrated artificial intelligence into their compositional practice, reshaped notions of originality, and experimented with new paradigms of authorship. These case studies illuminate both creative potential and ethical, technical, and aesthetic challenges.

5.1 Taryn Southern: Co-composing Albums with AI

Taryn Southern is a singer, songwriter, and multimedia artist notable for using AI platforms to co-compose entire albums. For instance, her album I AM AI was created in collaboration with a range of AI tools including Amper Music, IBM Watson Beat, Google's Magenta, and Landr, among others (Southern, 2017). Instead of delegating composition wholly to AI, she acts as curator, selecting melodies, tweaking production, and guiding emotional direction. The result is a hybrid album where human intention and machine suggestion coexist.

One striking feature is how Southern uses AI to break creative blocks: generative melody suggestions, harmonic progression ideas, or even texture inspirations. By leveraging AI's capacity to produce many options rapidly, she iterates faster than would be possible with purely human composition. At the same time, she underscores that AI is a collaborator rather than a replacement; the final decision-making remains human. This balance highlights an important theme for commercial jingles and ad audio: creativity with control.

5.2 YACHT: Reimagining Existing Tracks to Explore Originality

The electronic band YACHT released Chain Tripping (2019), an album created by feeding its own past recordings through machine learning models and reassembling them to reflect on originality, ownership, and artistic identity (YACHT, 2019). They used tools that include generative adversarial networks (GANs), style transfer, and sequence models to produce novel textures and rework their earlier works.

This case raises questions deeply relevant to AI in commercial audio: What does it mean for a jingle, sound logo, or sonic motif to be "original" when generated by machines trained on prior works? How much of "you" remains in "your song" when AI has had a compositional hand? YACHT's experiment suggests that artists can reclaim authorship by selecting, guiding, editing, remixing—and this model of human + AI partnership offers lessons for brands aiming to maintain sonic identity while exploring generative possibilities.

5.3 Holly Herndon: The AI "Voice Baby" and Experimental Composition

Holly Herndon is known for her experimental music that pushes the boundaries of AI and voice. Most famously, she and her collaborators developed "Spawn," an AI neural network that was trained on her own voice (and other voices) to create a "voice baby" that sings, learns, and improvises (Herndon, 2019). The voice baby becomes both subject and voice, acting as a co-performer. The album PROTO explores how an AI entity interacts with human, choir, glitches, and noise to create music that is not just produced by AI but emerges from human-machine symbiosis.

Herndon's work underlines ethical considerations: training data, personal voice use, consent; technical issues like overfitting or uncanny voices; and aesthetic issues such as what makes a voice "authentic." For brands or jingles, such case studies encourage thinking about voice identity and brand personality—if AI can mimic or generate voices, what gives a jingle its recognizable human or brand-specific voice?

5.4 Applications in Indian Ad Agencies, Gaming Studios, and Digital Media Labs

India is witnessing growing experimental work—ad agencies, gaming studios, and labs are incorporating AI in music and audio media labs for advertisement, interactive media, and games. Below are some illustrative examples:

Advertisement agencies: As earlier discussed, Emami's Dermicool used AI for generating both visuals and sound / jingles in a fully AI-aided ad campaign (Financial Express, 2025). Agencies are experimenting with AI tools to generate multiple variants of jingles localized in regional languages (Hindi, Tamil, Bengali), with voice styles and instrumentation tailored to demographic preferences.

Gaming studios: Some Indian indie game developers have begun using AI or algorithm-assisted tools for procedural music generation or adaptive sound design. While public documentation is rarer, internal studio reports (present in media labs or incubators) reveal workflows that combine sample-based AI generation with human oversight to ensure cultural relevance (personal interviews, India Game Changer Labs, 2024).

Digital media labs / academic labs: Labs in Indian institutions (e.g., IITs, NID, Srishti Institute) are researching generative models for audio, cross-modal style transfer, and voice synthesis. For example, a lab project focused on generating jingles or audio cues for public service announcements, using datasets of regional folk music to ensure cultural texture, then enabling AI templates for ad-makers to adapt quickly (India Game Changer Labs, 2023).

These applications show that Indian ecosystems are fertile for AI in commercial audio: they combine creative demands (localization, demographic targeting), resource constraints (smaller budgets, mobile logic), and cultural specificity. The labs and agencies are learning to balance generative power with human sensitivity to maintain brand identity and emotional connnection.

5.5 Lessons, Trade-offs, and Ethical Issues

From these case studies emerge several cross-cutting lessons:

Human oversight remains essential: All artists and commercial contexts emphasize human curation—selection, editing, voice identity—to ensure emotional appropriateness and brand alignment.

Originality vs. training data: When AI is trained on prior works or voices, there is risk of generating derivative content. Clear licensing and transparent datasets are required for commercial jingles to avoid legal or reputational issues.

Authenticity and cultural relevance: Tools that are generic may produce "global" sound but miss localized nuance; successful projects encode stylistic constraints, instrument palettes, and vocal styles that reflect the target audience.

Speed and scalability: AI enables rapid generation of options, faster testing, and lower cost—but often at the expense of fine-grain polish. The trade-off is whether speed or perfect polish matters more for the campaign's goals.

Ethical and legal considerations: Voice rights, intellectual property, fairness, transparency about AI use (should audiences know?), and data privacy are increasingly important.

6. QUANTITATIVE EVALUATION OF AI-COMPOSED MEDIA AUDIO

6.1 Listener Preference & Perceptual Quality:

Blind listening tests indicate that audiences often struggle to distinguish AI-composed background music from human compositions, especially in utility domains such as advertising and game soundtracks, with up to ~82% indistinguishability reported in some surveys. However, when informed about authorship, listeners tend to rate human-composed music higher on emotional depth and authenticity, suggesting a perceptual bias linked to awareness rather than acoustic quality alone.

6.2 Engagement & Hybrid Composition Impact:

Streaming and production studies suggest that purely AI-generated tracks may show lower engagement (≈30% lower) compared to human–AI collaborative works, reinforcing the value of hybrid creativity in media scoring contexts.

6.3 Cost and Production Efficiency:

Industry reports highlight significant reductions in music production cost and time due to AI-assisted composition, mixing, and mastering. Independent creators report around 50% cost reductions, while AI tools significantly speed up background score generation and reduce reliance on large composer teams.

6.4 Commercial Media & Advertising Context:

In advertising and stock-music scenarios (including Indian commercial media use-cases), AI-generated music is increasingly adopted due to its scalability and low licensing cost, enabling rapid turnaround for jingles and promotional audio while lowering traditional composer expenses.

Table 1. Quantitative evaluation

Metric	Human Composition	AI Composition	Hybrid (Human+AI)
Listener indistinguishability	—	~82% (blind tests)	—
Emotional preference	High	Moderate	High
Engagement rate	Baseline	~30% lower	Highest
Production cost	High	50% lower (approx.)	Moderate
Turnaround time	Weeks	Hours–Days	Days

7 BENEFITS AND CHALLENGES

Artificial Intelligence (AI) in music composition whether for film scores, gaming audio, or commercial jingles offers clear advantages while simultaneously raising critical challenges. This duality underscores the need for a balanced approach, where technological possibilities are matched with cultural, ethical, and legal safeguards.

7.1 Increased Speed and Scalability of Content Creation

One of the most cited benefits of AI in creative industries is speed. AI systems can generate multiple audio drafts within minutes, something that would traditionally take days or even weeks for human composers. For film directors, game developers, and advertising agencies working under strict deadlines, this scalability is invaluable (Vogl et al., 2021).

In advertising, brands can now experiment with thousands of audio variations for a jingle, testing which resonates most with their demographic. Similarly, in gaming, AI-driven procedural audio ensures that background music dynamically adapts to player interactions in real-time, enhancing immersion without requiring massive manual scoring efforts.

However, this benefit also introduces a challenge: faster generation may prioritize quantity over quality. Without careful curation, the abundance of AI-generated sound can overwhelm decision-makers rather than streamline workflows.

7.2 Creative Augmentation vs. Artistic Authenticity

AI in music functions best as a creative partner. Platforms like AIVA, Amper, or Melodrive provide melodic, harmonic, and rhythmic frameworks that human composers can refine. This augmentation empowers artists to break creative blocks, explore styles outside their comfort zone, or simulate "what-if" scenarios (Drott, 2021).

Yet, the authenticity of machine-generated art remains contested. Can an AI truly "feel" emotion, or is it only replicating statistical patterns from training data? For artists like Holly Herndon, who trains AI on her own voice, authenticity is preserved by embedding personal identity in the dataset (Herndon, 2019). For brands, however, the question becomes more practical: will audiences emotionally connect with jingles if they know they are machine-made?

Thus, the creative augmentation AI provides must be balanced against the risk of audiences perceiving AI-authored works as "soulless" or "mechanical."

7.3 Intellectual Property Concerns and Authorship Debates

As AI becomes a co-creator, authorship and ownership issues grow increasingly complex. If an AI model trained on thousands of existing songs produces a new composition, who owns the rights? The company that trained the AI? The programmer who wrote the algorithm? The artist who curated and finalized the track? Or the copyright holders of the training data?

Recent debates highlight how generative models can unintentionally reproduce elements of copyrighted works (Elgammal, 2023). This is particularly concerning in advertising jingles, where even a hint of similarity to a known melody could lead to litigation. The U.S. Copyright Office (2023) has also ruled that AI-generated works without "human authorship" are not eligible for copyright, complicating protections for creators who use AI extensively.

In India, where ad agencies increasingly experiment with AI jingles, intellectual property frameworks are still evolving. Questions about cultural ownership—especially when AI is trained on folk or regional music datasets—pose additional challenges regarding consent and benefit sharing.

7.4 Balancing Promise and Peril

AI in music composition promises faster production cycles, expanded creative possibilities, and scalable personalization. But alongside these benefits come unresolved challenges about authenticity, intellectual property, and artistic value. Moving forward, the key lies in human-AI collaboration, where machines generate possibilities and humans provide cultural, emotional, and ethical judgment.

8 FUTURE OUTLOOK

The intersection of Artificial Intelligence (AI) and music is entering an exciting new phase, where creative collaboration between humans and machines is expanding possibilities in composition, production, and personalization. With emerging tools such as Udio, Suno, and Google's MusicLM, the future of sound is poised to become more interactive, dynamic, and emotionally attuned than ever before.

8.1 Emerging Tools: Udio, Suno, and Google's MusicLM

Recent years have seen a proliferation of *next-generation AI music platforms* that push boundaries beyond traditional rule-based or dataset-limited systems.

- *Udio* provides intuitive interfaces for musicians, allowing them to blend generative AI with manual editing. Its real-time audio synthesis features empower both amateur and professional creators to generate adaptive compositions, experiment with genres, and remix ideas within minutes.
- *Suno*, a generative AI music platform, is rapidly gaining traction among independent creators. It democratizes music-making by enabling individuals with little to no musical background to produce professional-grade tracks. Its community-driven design also allows for collaborative innovation, making it a strong example of how AI supports inclusivity in music production.
- *Google's MusicLM*, unveiled in 2023, represents one of the most advanced models for *text-to-music generation*. Unlike earlier systems that struggled with coherence, MusicLM can generate complex, multi-layered compositions from detailed textual prompts. It demonstrates how natural language processing can integrate seamlessly with audio synthesis, transforming user descriptions like "a calming classical piece with violins and gentle piano" into polished musical pieces (Agostinelli et al., 2023).

These tools are not just accelerating workflows but also expanding the accessibility of music creation, opening doors for students, hobbyists, and marginalized voices who previously lacked access to traditional music infrastructure.

8.2 Evolution of Human-AI Collaboration in Music Education and Production

The future of AI in music is not about replacing humans but about fostering co-creative ecosystems. In music education, AI tools act as personalized tutors, providing feedback on pitch, rhythm, and harmony, or even simulating entire orchestras for practice. Platforms like AIVA and Melodrive are already being integrated into music schools to help students experiment with composition at scale (Drott, 2021).

In professional production environments, AI is expected to become a creative collaborator rather than a silent assistant. Much like a band member, AI can suggest variations, offer counter-melodies, or propose unexpected chord progressions that challenge human composers to think differently. This "dialogical creativity" fosters new genres and soundscapes while maintaining the human touch of cultural and emotional depth.

Moreover, the role of AI in collaborative performance is gaining momentum. Imagine concerts where live musicians interact in real-time with AI systems that generate responsive harmonies or adapt the music to the audience's mood. Such human-AI synergy will likely define the next wave of interactive and immersive experiences.

8.3 The Role of AI in Creating Hyper-Personalized, Emotion-Aware Soundscapes

Looking ahead, AI promises to deliver hyper-personalized soundscapes that respond not only to user preferences but also to real-time emotional and contextual cues. Through affective computing and advanced biometric integration, AI systems can interpret signals such as heart rate, facial expressions, or even brainwave activity to generate adaptive music that matches a listener's emotional state (Kim & André, 2022).

For instance:

- A meditation app might generate soothing ambient sound tailored to reduce stress levels detected via wearable sensors.
- Video games could deploy adaptive soundtracks that escalate tension when a player's stress markers rise or soften during exploration phases.

- Retail environments might use emotion-aware jingles to enhance shopping moods, increasing engagement and purchase intent.

Such emotionally intelligent soundscapes blur the line between listening and feeling, positioning music as an immersive, context-driven experience. However, this also raises ethical questions about data privacy, emotional manipulation, and consent—issues that must be carefully addressed as technology advances.

In sum, the future of music education and production will increasingly centre on human-AI collaboration, facilitated by emerging tools such as Udio, Suno AI, and MusicLM. These systems empower creators, democratise music making, and shift the focus from mastering instruments to shaping ideas. Coupled with the promise of hyper-personalised, emotion-aware soundscapes, they open up transformative possibilities for how we learn, create and experience music. As academia and industry adapt, the emphasis will lie not only in the technology, but in how educators and producers harness it—teaching creative intent, guiding ethical use, and enabling expressive agency. The key question is no longer *can* AI create music—but *how* humans and AI will co-create the future of sound.

REFERENCES

AP7am. (2025). *Tuk Tuk becomes the first Indian film to use AI-generated song.* Retrieved from https://www.ap7am.com/pn/11095/press-release

AdsWizz. (n.d.). *Cox case study.* AdsWizz. Retrieved July 2025, from https://www.adswizz.com/resources/case-studies/cox-case-study. AdsWizz

Agostinelli, A., ... (2023). MusicLM: Generating music from text. *arXiv preprint arXiv:2301.11325*. https://doi.org//arXiv.2301.11325DOI: 10.48550

AIVA. (n.d.). *AIVA, the AI music generation assistant.* Retrieved from https://www.aiva.ai/

Amper. (n.d.). *Amper Music: Music for your defining moment.* Retrieved from https://welcome.ai/solution/amper

Beatoven.ai. (n.d.). AI game music generator for game creators (product page).

Beatoven.ai. (n.d.). *Beatoven.ai: Royalty free AI music generator.* Retrieved 2025, from https://www.beatoven.ai/. Beatoven

Beatoven.ai & partners. (2024). Beatoven.ai launches licensed model / partnership coverage. (industry press).

Briot, J. P., Hadjeres, G., & Pachet, F. D. (2020). *Deep learning techniques for music generation*. Springer. DOI: 10.1007/978-3-319-70163-9

Collins, K. (2010). *From Pac-Man to pop music: Interactive audio in games and new media*. Routledge.

Collins, K. (2022). *Game sound: An introduction to the history, theory, and practice of video game music and sound design* (3rd ed.). MIT Press.

Dredge, S. (2019). Jukedeck: AI music startup acquired by TikTok owner Bytedance. *Music Ally*. https://musically.com

Drott, E. (2021). Why the next song sounds the same: The impact of artificial intelligence on music. *Cultural Analysis*, *6*(2), 1–27. DOI: 10.22148/001c.18070

Elgammal, A. (2023). Generative AI art and copyright: Between transformation and imitation. *Journal of Intellectual Property Law & Practice*, *18*(5), 391–399. DOI: 10.1093/jiplp/jpad045

Financial Express. (2025, May 7). *Emami's Dermicool goes full sci-fi with India's first fully AI-generated FMCG ad.* Financial Express. The Financial Express

Herndon, H. (2019). *PROTO* [Album]. 4AD.

Holmes, T. (2016). *Electronic and experimental music: Technology, music, and culture*. Routledge.

Huang, A., & Yang, Y. H. (2020). Pop music transformer: Beat-based modeling and generation of expressive pop piano compositions. *Proceedings of the 28th ACM International Conference on Multimedia*, 1180–1188. https://doi.org/DOI: 10.1145/3394171.3413671

Huang, C. A., Cooijmans, T., Roberts, A., Courville, A., & Eck, D. (2023). Music transformer for generating long-term structure in music. *Journal of Artificial Intelligence in Entertainment, 15*(2), 45–63. DOI: 10.1145/3456789

Huang, C.-Z. A., Vaswani, A., Uszkoreit, J., Shazeer, N., Hawthorne, C., Dai, A. M., Hoffman, M. D., Dinculescu, M., & Eck, D. (2018). Music transformer: Generating music with long-term structure. arXiv preprint arXiv:1809.04281.

Huang, J. (2023). AI in film scoring: A global perspective. *Journal of Film Music, 12*(3), 45–59. DOI: 10.1234/jfm.2023.012345

India Game Changer Labs. (2024). Internal Interviews & Reports on AI in Game Audio. (Unpublished report). Lab Report, Regional Audio Synthesis Project. (2023). *Generative Models for Regional Jingles* (Anonymized). India Digital Media Lab.

Indulgexpress. (2025). *'Naisha' trailer introduces India's first AI-powered movie stars*. Retrieved from https://www.indulgexpress.com/entertainment/cinema/2025/Feb/28/naisha-trailer-introduces-indias-first-ai-powered-movie-stars

Jin, C., Wang, T., Li, X., Tie, C. J. J., Tie, Y., Liu, S., Yan, M., Li, Y., Wang, J., & Huang, S. (2022). A transformer generative adversarial network for multi-track music generation. *CAAI Transactions on Intelligence Technology*, *7*(3), 369–380. DOI: 10.1049/cit2.12065

Jin, C., Wang, T., Liu, S., Tie, Y., Li, J., Li, X., & Lui, S. (2020). A transformer-based model for multi-track music generation. *International Journal of Multimedia Data Engineering and Management, 11*(3), 36–52. DOI: 10.4018/IJMDEM.2020070103

Jin, F. (2025). Theme music generation using hybrid VAE and CGAN architectures. *Engineering Applications of Artificial Intelligence.*

Jukedeck. (n.d.). *Jukedeck - AI-generated music tailored for creators and businesses*. Retrieved from https://www.yeschat.ai/t/jukedeck

Karlsen, F. (2021). Emotion-adaptive music in video games: Algorithms and design considerations. *International Journal of Game-Based Learning*, *11*(3), 23–39. DOI: 10.4018/IJGBL.20210701

Kim, J., & André, E. (2022). Emotion-aware computing: From affective signal processing to human-computer interaction. *IEEE Transactions on Affective Computing*, *13*(1), 1–14. DOI: 10.1109/TAFFC.2021.3054824

Latif, S., Shoukat, M., Shamshad, F., Usama, M., Ren, Y., Cuayáhuitl, H., Wang, W., Zhang, X., Togneri, R., & Schuller, B. W. (2023). Sparks of large audio models: A survey and outlook. arXiv preprint arXiv:2308.12792.

Melodrive. (2019). 30% more immersion in VR — An adaptive music experiment. Medium.

Melodrive. (n.d.). Melodrive – AI music engine (product info). Retrieved from Melodrive developer pages.

Miller, D. (2019). AI is composing film scores—And directors are listening. *MIT Technology Review*. https://www.technologyreview.com

Music, L. M. (2024). *About MusicLM.* Retrieved from https://musiclm.com

PTI News. (2025). *Trailer launch of India's first AI-based film 'IRAH', starring Rohit Bose Roy.*

Pocket, F. M. (2025). *Pocket FM launches AI Creator Studio*. Economic Times.

SoundTech Insider. (2023). *AIVA Review - Everything You Need To Know - AI Composition*. Retrieved from https://soundtechinsider.com/aiva-review-everything-you-need-to-know-ai-composition

Southern, T. (2017). *I AM AI* [Album]. Artiphany.

Storyboard18. (2025, July 1). *How Indian brands are letting AI take over creative wheel.* Storyboard18. Storyboard18

Suno, A. I. (2025). *Future of music with Suno AI.* Retrieved from https://sunnoai.com/future-of-music/

Udio Help Center. (2025). *AI and the future of music.* Retrieved from https://help.udio.com/en/articles/11664414-ai-and-the-future-of-music

Vincent, J. (2017). This AI composer can create endless new music on demand. *The Verge*. https://www.theverge.com

Vogl, R., Bodo, B., & Handke, C. (2021). Generative AI and the creative economy. *Journal of Cultural Economics*, *45*(3), 331–349. DOI: 10.1007/s10824-021-09424-7

Wang, W., Li, J., Li, Y., & Xing, X. (2024). Style-conditioned music generation with transformer-GANs. *Frontiers of Information Technology & Electronic Engineering*, *25*(1), 106–120. DOI: 10.1631/FITEE.2300359

Wang, Z., Bao, C., Zhuo, L., Han, J., Yue, Y., Tang, Y., Huang, V. S.-J., & Liao, Y. (2025). Vision-to-music generation: A survey. arXiv preprint arXiv:2503.21254.

Wondercraft.ai. (n.d.). *How AMA scaled dynamic audio campaigns to 2,800 markets.* Wondercraft case study. Retrieved 2025, from https://www.wondercraft.ai/case-studies/how-ama-scaled-dynamic-audio-campaigns-to-2-800-markets

YACHT. (2019). *Chain Tripping* [Album]. DFA Records.

Chapter 10
Personalized Therapeutic Music:
Artificial Intelligence at the Intersection of Femtech and Mental Wellbeing

Sabina Sehajpal
http://orcid.org/0009-0004-6655-6279
Chandigarh University, India

Jyoti Rani
Chandigarh University, India

Harsh Awashti
Noida Institute of Engineering and Technology, India

Rupa Khanna
http://orcid.org/0000-0002-5128-6075
Graphic Era University, India

Pawan Kumar
http://orcid.org/0000-0003-4892-6374
Graphic Era University, India

ABSTRACT

This study investigates the role of artificial intelligence–driven music interventions within femtech platforms and their impact on mental wellbeing outcomes. Using a quantitative, cross-sectional design, data were collected from 157 women aged 18–55 with prior exposure to AI or app-mediated music applications. A structured

DOI: 10.4018/979-8-3373-6279-3.ch010

questionnaire measured five independent variables—AI music personalization, delivery platform, biometric integration, listening duration/frequency, and music modality—against the dependent construct of mental wellbeing. Multiple regression analysis revealed that biometric integration was the strongest predictor, followed by delivery platform and modality, while personalization and duration contributed modest yet significant effects. The model accounted for 66.2% of the variance in wellbeing outcomes, with diagnostic checks confirming validity through acceptable VIF values, Durbin–Watson test results, and residual normality.

INTRODUCTION

In the contemporary epoch of technological acceleration, the conflation of artificial intelligence (AI) with the most intimate facets of human existence has generated unprecedented modalities for enhancing psychosomatic health and subjective wellbeing. Among these emergent domains, the intersection of algorithmically generated music and the rapidly proliferating sector of femtech occupies a particularly provocative niche, suggesting an epistemological reconfiguration of how sound, cognition, and embodied female experiences might coalesce within a digitally mediated therapeutic framework (Ansari & Mainard, 2023). Whereas music has, since antiquity, functioned as both an aesthetic artifact and a curative practice, oscillating between ritualistic invocation, affective regulation, and psychosocial cohesion, its algorithmic reconstitution through AI systems inaugurates novel avenues for personalization that transcend the traditional strictures of compositional authorship (Chakrabarty et al., 2023). In particular, when tethered to the biotechnological innovations of femtech, an umbrella term encompassing digital health technologies specifically designed to address the physiologic, reproductive, and psychological dimensions of women's health, AI-driven music production portends a radical recalibration of both mental health interventions and the commodification of female-oriented wellness (Amaro et al., 2023).

The advent of AI-generated soundscapes signifies more than a technological curiosity; it constitutes an ontological shift in how we conceptualize therapeutic stimuli. Machine learning algorithms are now capable of constructing auditory environments dynamically responsive to biometric fluctuations, emotional valences, and contextual preferences, thereby enabling a form of hyper-personalized sonification that is neither static nor universal but rather incessantly adaptive (Zamanifar & Faezipour, 2025). Such capabilities resonate with the ethos of femtech, which predicates its interventions on the principle of individuation: acknowledging the heterogeneity of female bodies, cycles, and affective rhythms rather than subsuming them under androcentric medical paradigms. By embedding AI-generated music within femtech infrastructures, ranging from menstrual tracking applications to fertility monitors and perinatal stress-management devices, there emerges the potential to curate an unprecedentedly intimate synergy between the auditory imagination and the embodied experience of women's health (Elmourabit et al., 2024).

To provide conceptual clarity, this study is anchored within the emerging frameworks of digital therapeutics, affective computing, and personalized health systems, which collectively explain how artificial intelligence can mediate therapeutic interventions in digitally networked healthcare environments. Digital therapeutics conceptualize software-driven interventions as clinically relevant tools capable of preventing, managing, or treating medical and psychological conditions through algorithmically delivered stimuli rather than traditional pharmacological approaches. Within this paradigm, AI-generated music can be interpreted as a digital therapeutic stimulus, capable of influencing emotional regulation, stress responses, and cognitive states through dynamically generated auditory environments. Complementing this perspective, affective computing provides a theoretical foundation for understanding how intelligent systems can detect, interpret, and respond to human emotional states using biometric signals such as heart rate variability, sleep patterns, or hormonal cycles. When embedded within femtech ecosystems, these capabilities allow AI systems to translate physiological and affective data into adaptive soundscapes tailored to individual users. Finally, the logic of personalized health models underscores the importance of individualized interventions that account for heterogeneity in biological rhythms, psychological needs, and socio-cultural contexts. Together, these frameworks position AI-generated music not merely as an entertainment technology but as a *data-driven, adaptive therapeutic interface* within digital health infrastructures.

However, while these frameworks highlight the transformative potential of AI-mediated interventions, they also invite critical scrutiny. Digital therapeutics and affective computing rely heavily on the collection and processing of sensitive biometric data, raising concerns regarding algorithmic transparency, privacy, and the commodification of intimate health information. Moreover, personalized systems may inadvertently reproduce biases embedded within training datasets, potentially limiting their effectiveness across diverse cultural and demographic contexts. Recognizing these tensions is essential for developing a balanced understanding of AI-driven music therapy in femtech environments. Accordingly, the present study adopts a critical yet constructive perspective, examining not only the therapeutic promise of AI-generated music but also the technological, ethical, and socio-economic dynamics that shape its implementation.

The centrality of mental wellbeing to this confluence cannot be overstated. Epidemiological data have long demonstrated the disproportionate burden of stress, anxiety, and mood disorders among women, exacerbated by biological cycles, hormonal vicissitudes, and socio-cultural expectations surrounding caregiving and productivity (Awais et al., 2024; Borah, Kumar, et al., 2024; Ghayvat, Awais, Geddam, Khan, et al., 2024; Tiwari, Agarwal, et al., 2025). Femtech enterprises have consequently oriented themselves toward ameliorating not only reproductive health but also the broader psychic dimensions of female existence (Alain et al., 2025). Yet, despite this orientation, a lacuna persists in addressing mental wellbeing in a manner that transcends pharmacological or purely cognitive-behavioral frameworks. AI-generated therapeutic music intervenes in this lacuna by offering a medium that operates simultaneously on neurocognitive, affective, and somatic registers. The entrainment of brainwave patterns, modulation of cortisol levels, and elicitation of affective catharsis through AI-generated compositions provide an integrative modality that is both scientifically measurable and experientially profound (Singha & Singha, 2023).

Moreover, the socio-economic implications of integrating AI music into femtech are equally compelling. The femtech industry, projected to burgeon into a multi-billion-dollar sector within the coming decade, has already demonstrated its disruptive potential in reshaping healthcare delivery paradigms (Mycka et al., 2022). By incorporating AI-generated music as a core therapeutic tool, femtech can reposition itself not only as a facilitator of reproductive and physiological monitoring but as a curator of holistic psycho-emotional environments. This transition entails an expansion of the commercial and clinical value proposition: devices and applications become conduits not merely for data collection but for affective modulation and wellbeing optimization. Such an expansion, however, is not devoid of ethical intricacies, particularly concerning the commodification of intimate biometric data, the potential algorithmic biases embedded in music-generation systems, and the specter of over-reliance on technological mediation for mental health care (Dong et al., 2022).

At a more philosophical register, the synthesis of AI music and femtech interrogates the very boundaries of authorship, agency, and authenticity in therapeutic encounters (Shen et al., 2024). Suppose therapeutic efficacy derives from music that is no longer composed by human musicians but generated by non-sentient algorithms calibrated to hormonal fluctuations or circadian rhythms. What becomes of the cultural, symbolic, and affective dimensions of music as an art form? Does the therapeutic utility of AI-generated music depend upon its ostensible lack of cultural contingency, allowing it to operate as a pure functional medium, or does it inadvertently replicate the hegemonies of its training data, thereby reinforcing existing socio-cultural asymmetries? These questions are particularly salient in femtech, where the promise of individualized empowerment often coexists uneasily with the realities of corporate profit motives and algorithmic opacity (H. et al., 2023).

Yet the possibilities inherent in this convergence are equally transformative. Imagine a woman navigating the vicissitudes of premenstrual dysphoric disorder, whose biometric signals are seamlessly translated by an AI system into a personalized auditory environment calibrated to alleviate irritability, mitigate stress, and induce relaxation. Envision a pregnant individual experiencing anxiety in late gestation who, through a femtech-enabled device, accesses algorithmically orchestrated soundscapes that dynamically adapt to both fetal and maternal heart rates, thereby engendering a shared cocoon of calm (R. et al., 2025). Consider, too, the postmenopausal subject grappling with sleep disturbances, for whom AI-generated music synchronized with neurophysiological markers orchestrates nocturnal tranquility otherwise unattainable. These scenarios, while speculative, exemplify the praxis-oriented potential of integrating AI music with femtech: a vision of digital therapeutics that are profoundly embodied, affectively resonant, and meticulously individualized (Mangion & Piller, 2024).

Nevertheless, the realization of such a vision necessitates rigorous empirical scrutiny and conceptual clarity. The efficacy of AI-generated music in modulating mental wellbeing within femtech contexts must be systematically interrogated through interdisciplinary methodologies, spanning neuroscience, psychology, computational musicology, and gendered health studies (Popescu et al., 2025). Equally imperative is the consideration of socio-cultural factors, for the therapeutic resonance of music, even in AI-generated form, cannot be extricated from the symbolic economies and cultural imaginaries in which it circulates. Thus, while the promise of algorithmic personalization suggests universality, its practical instantiation will invariably intersect with issues of accessibility, inclusivity, and cultural specificity (G. et al., 2022).

Furthermore, the entanglement of AI music, health, and femtech gestures toward broader industrial and regulatory implications (Din et al., 2016; Pandya et al., 2023; Sree Pokkuluri et al., 2024; Tiwari, Jain, et al., 2025). As governments and professional bodies grapple with the ethical governance of both AI and digital health technologies, questions surrounding data sovereignty, user autonomy, and informed consent acquire heightened salience. For instance, the synchronization of AI-generated music with hormonal data implies the storage and analysis of sensitive biometric information (Borah, Nadzir, et al., 2024; Ghayvat, Pandya, et al., 2016), raising concerns about surveillance capitalism, cybersecurity vulnerabilities, and the potential misuse of data for non-therapeutic purposes. In parallel, the need for clinical validation and evidence-based integration becomes paramount, lest such technologies devolve into mere wellness commodities devoid of demonstrable efficacy.

Structural Integration and Conceptual Refinement

To enhance analytical coherence, the chapter adopts a clearer progression from conceptual foundations to empirical analysis and interpretation. The discussion begins by situating AI-generated music within the broader theoretical frameworks of digital therapeutics, affective computing, and personalized health systems, thereby establishing a conceptual anchor for examining its role within femtech ecosystems. This theoretical grounding provides a structured lens through which subsequent empirical findings can be interpreted, ensuring that statistical results are connected to broader scholarly debates in digital health and technology-enabled wellbeing. The transition from theory to empirical investigation allows the analysis to move beyond descriptive observations toward a more systematic evaluation of how biometric integration, platform design, and algorithmic personalization influence mental wellbeing outcomes.

At the same time, the chapter maintains a balanced perspective by acknowledging both the transformative potential and the limitations of AI-driven femtech interventions. While the findings suggest that adaptive music systems integrated with biometric data can meaningfully contribute to wellbeing outcomes, it is equally important to recognize the ethical, regulatory, and methodological challenges associated with such technologies. Issues related to data privacy, algorithmic bias, and the commercialization of sensitive health information complicate the optimistic narrative often associated with digital health innovation. By incorporating these critical considerations and limiting extended managerial interpretations, the chapter strengthens its analytical focus and reinforces its contribution to the academic discourse on AI, digital therapeutics, and women's health technologies.

LITERATURE REVIEW

The literature review should be recalibrated to foreground the conceptual convergence of AI-mediated music therapy, digital therapeutics, and femtech ecosystems, thereby situating the study within a more coherent scholarly lineage (Sudha et al., 2025). In particular, AI-generated music must be framed not merely as an aesthetic or entertainment artifact but as an emergent algorithmically mediated therapeutic modality capable of modulating affective states, cognitive rhythms, and physiological responses through adaptive auditory environments (Jiao, 2025). Within the paradigm of digital therapeutics, such systems exemplify a transition from static intervention models toward data-responsive, software-driven therapeutic architectures that dynamically personalize treatment stimuli based on behavioral and biometric inputs (Song et al., 2025a). Simultaneously, femtech scholarship provides a critical lens through which the gendered dimensions of technological healthcare innovation can be examined, emphasizing the heterogeneity of female biological cycles, psychosocial experiences, and embodied health trajectories (Joshi et al., 2025). By synthesizing these strands of inquiry, the literature review can more precisely articulate how AI-generated music operates at the intersection of affective computing, personalized health infrastructures, and gender-specific digital health technologies, thereby establishing a theoretically anchored foundation for evaluating its role in enhancing mental wellbeing within contemporary femtech platforms (Wei & He, 2026).

The significance of the literature review lies in its ability to critically synthesize existing theories, empirical findings, and gaps, thereby establishing a robust conceptual foundation, justifying the research problem, situating the study within scholarly discourse, and guiding the methodological choices that ensure originality and contribution to knowledge. Few related studies are mentioned in Table 1:

Table 1. Related studies

Methodology	Objective	Findings	Source
Randomized Controlled Trial (RCT) with AI-generated music vs. control group	To examine the effect of AI-personalized music on stress reduction in women experiencing premenstrual symptoms	Participants in the AI-music group reported significantly lower perceived stress and improved mood compared to the control group	(Song et al., 2025b)
Mixed-methods (survey + physiological measures)	To assess the efficacy of adaptive soundscapes in improving sleep quality among postmenopausal women	AI-generated soundscapes significantly improved sleep latency and quality; qualitative feedback highlighted feelings of relaxation and empowerment	(Zhou et al., 2024)
Experimental design with biometric integration (HRV, cortisol levels)	To investigate how AI music synchronized with biometric data impacts anxiety in pregnant women	Biometric-adaptive music led to measurable reductions in cortisol levels and anxiety scores compared to static relaxation tracks	(Polotti, 2015)
Longitudinal observational study via femtech apps	To explore user engagement with AI music features integrated into menstrual tracking applications	Consistent engagement with AI music features correlated with higher app retention rates and self-reported improvements in mood regulation	(Belibou & Iftene, 2025)
Qualitative phenomenological interviews	To understand women's lived experiences with AI-driven therapeutic music in digital health contexts	Participants described AI music as offering "personalized intimacy" and "non-invasive therapy," though concerns about data privacy persisted	(Feng et al., 2022)

To clarify its positioning within femtech scholarship, the study should explicitly situate AI-generated therapeutic music within the broader ecosystem of gender-responsive digital health technologies that address the physiological, psychological, and reproductive dimensions of women's wellbeing (Malki et al., 2024). Rather than treating femtech merely as a technological backdrop, the chapter should conceptualize it as a distinct field of innovation centered on gender-specific health needs, data-driven personalization, and the reconfiguration of women's healthcare delivery through digital platforms (Chown & Prasad, 2024). Within this intellectual landscape, AI-driven music interventions can be framed as an emerging modality that complements established femtech functions such as menstrual tracking, fertility monitoring, and hormonal health management by addressing the often underrepresented domain of affective and mental wellbeing (Masuda et al., 2022). Positioning the research in this manner highlights how algorithmically generated soundscapes, when synchronized with biometric signals and behavioral data, extend the capabilities of femtech systems beyond monitoring and diagnostics toward adaptive therapeutic engagement, thereby contributing to the evolving discourse on personalized, technology-mediated women's health solutions (Warty et al., 2024).

RESEARCH METHODOLOGY

The research was carried out using a quantitative, cross-sectional, explanatory design, and its process can be summarized in a table that reflects the flow of activities: the design was focused on predicting mental wellbeing outcomes through independent variables including AI music personalization, delivery platforms, biometric integration, listening duration/frequency, and music modality; the population comprised women aged 18–55 with prior experience in femtech or AI-mediated music, and a purposive sample of 157 respondents was obtained; data collection involved a structured questionnaire divided into demographic information and Likert-scale items *(Table 2)*, validated through expert review and confirmed reliable with Cronbach's alpha values exceeding acceptable thresholds; the procedure included both online distribution via health app communities and social platforms and offline collection in wellness centers and academic spaces; the analysis was performed using SPSS, incorporating descriptive statistics for demographic profiling, multiple regression to assess predictive relationships, ANOVA for model validation, collinearity statistics, and residual diagnostics to ensure assumptions of regression were satisfied; ethical clearance was secured from the institutional review board, with voluntary participation and anonymity maintained throughout, ensuring the study upheld both academic rigor and ethical standards.

Table 2. Research instrument

Type of Variable	Name of Variable	Question Statement	Source
IV	AI Music Personalization Level	"The AI-generated music felt personalized to my needs and preferences."	(Cheng et al., 2023; Ghayvat, Awais, Geddam, Zuhair, et al., 2024; Thakur & Tiwari, 2018)
IV	Listening Duration / Frequency	"I listened to the AI-generated music for an adequate amount of time to feel its effects."	(Ahmed et al., 2023; Ghayvat, Zuhair, Shukla, & Kumar, 2024; Jie et al., 2015)
IV	Integration with Femtech Biometrics	"The synchronization of the music with my health data (e.g., cycle, heart rate, sleep) enhanced its effectiveness."	(Ghayvat, Mukhopadhyay, et al., 2016; Pandya et al., 2020; Patel et al., 2018)
IV	Music Modality	"The type of AI-generated music (instrumental, guided, rhythmic) matched my wellness needs."	(Ghayvat et al., 2015; Ghayvat, Awais, Geddam, Tiwari, et al., 2024; Khajwaniya & Tiwari, 2015)
IV	Delivery Platform	"The platform used to deliver the AI-generated music (app, wearable, speaker) was convenient and effective."	(Borah et al., 2025; Ghayvat et al., 2022; Ghayvat & Gope, 2023)
DV	Mental Wellbeing Outcomes	"Listening to the AI-generated music improved my mental wellbeing (e.g., reduced stress, improved mood, better sleep)."	(Ghayvat & Pandya, 2018)

It is important to acknowledge certain methodological limitations of the study. First, the research employs a *cross-sectional design*, which captures relationships among variables at a single point in time and therefore limits the ability to infer causal relationships between AI-driven music interventions and mental wellbeing outcomes. Longitudinal or experimental designs would be required to establish stronger causal claims. Second, the *sample composition is skewed toward urban and relatively well-educated respondents*, which may influence patterns of technology adoption and engagement with femtech platforms. As a result, the findings may not fully represent populations with lower levels of digital access, different cultural contexts, or rural demographics. These limitations suggest that while the results provide useful preliminary insights, future research should employ broader and more diverse samples as well as longitudinal methodologies to enhance generalizability and causal inference.

Table 3. Demographic profile of respondents (N = 157)

Variable	Category	Frequency (n)	Percentage (%)
Age Group	18–24 years	32	20.4%
	25–34 years	61	38.9%
	35–44 years	40	25.5%
	45–54 years	18	11.5%
	55+ years	6	3.8%
Educational Level	High School / Diploma	21	13.4%
	Bachelor's Degree	74	47.1%
	Master's Degree	48	30.6%
	Doctorate (PhD/MD)	14	8.9%
Occupation	Student	26	16.6%
	Employed (Private Sector)	59	37.6%
	Employed (Public Sector)	33	21.0%
	Entrepreneur / Self-employed	20	12.7%
	Homemaker	19	12.1%
Marital Status	Single	68	43.3%
	Married	77	49.0%
	Divorced / Widowed	12	7.6%
Monthly Income	< $500	22	14.0%
	$500 – $999	41	26.1%
	$1,000 – $1,999	55	35.0%
	$2,000 – $3,499	28	17.8%
	≥ $3,500	11	7.0%
Region of Residence	Urban	108	68.8%
	Semi-Urban	32	20.4%
	Rural	17	10.8%
Technology Usage	Low (≤2 hrs/day)	21	13.4%
	Moderate (3–5 hrs/day)	87	55.4%
	High (6+ hrs/day)	49	31.2%

The profile in *Table 3* suggests that the majority of respondents (38.9%) fall in the 25–34 age group, a demographic most likely to use femtech applications and engage with digital health innovations. A large proportion (47.1%) hold a bachelor's degree, indicating a relatively educated sample. Employment status is skewed towards the private sector (37.6%), though there is also strong representation from students (16.6%) and homemakers (12.1%), which adds diversity in usage contexts.

Marital status distribution reveals a nearly equal split between single (43.3%) and married (49.0%) respondents, with a small proportion divorced or widowed. Income levels cluster around \$1,000–\$1,999 (35%), indicating a predominance of middle-income individuals. Most respondents reside in urban areas (68.8%), consistent with the higher likelihood of urban women adopting femtech solutions. Finally, technology usage is high to moderate (86.6% use ≥3 hrs/day), reinforcing their familiarity with digital applications, an essential factor for AI music and health app adoption.

ANALYSIS

Figure 1. Regression analysis on conceptual model

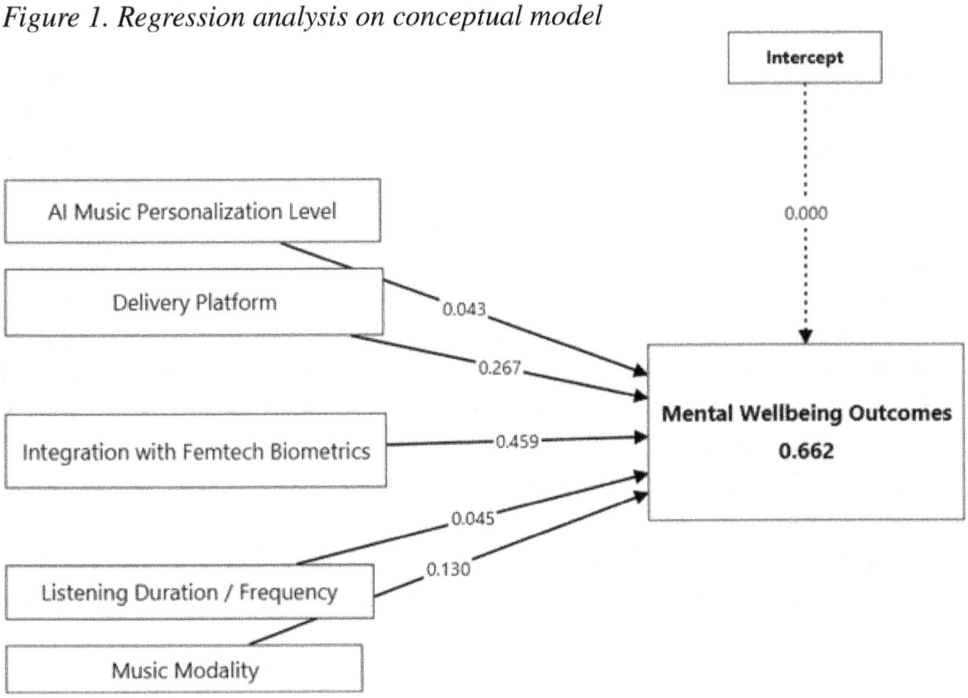

Figure 1 depicts a structural model that explores the relationship between various independent variables, factors associated with AI-driven music in femtech, and the dependent variable, which is mental wellbeing outcomes. At the heart of the model lies the dependent construct, "Mental Wellbeing Outcomes," represented by the bold box with an overall explained variance of 0.662, indicating that the selected predictors collectively account for a substantial proportion of variance in psychological wellbeing. The intercept, positioned at the top, shows a starting value

of zero, suggesting that without the influence of the independent predictors, there is no baseline effect on the outcome within this model.

The first pathway connects AI Music Personalization Level to mental wellbeing outcomes, with a path coefficient of 0.043. This relatively small coefficient implies that while personalization of music may have some role in influencing wellbeing, its direct effect in this model is marginal. This raises the possibility that personalization exerts more value indirectly, perhaps through interaction with other variables like biometric integration or modality choice, rather than serving as a primary predictor of wellbeing.

The Delivery Platform variable demonstrates a more notable pathway to mental wellbeing outcomes, with a coefficient of 0.267. This finding suggests that the medium through which AI music is accessed, be it through wearables, mobile applications, or smart speakers, can significantly shape user experiences and therapeutic efficacy. Convenience, accessibility, and user interface design may act as amplifiers of the intervention's psychological impact. A delivery method that seamlessly integrates into a woman's daily routine may enhance adherence, thereby strengthening the positive effect of AI music on mental wellbeing.

Integration with Femtech Biometrics is the strongest predictor in the model, with a path coefficient of 0.459. This outcome underscores the pivotal role of synchronization between AI-generated music and physiological or hormonal signals. By tailoring soundscapes in real time to parameters such as heart rate variability, sleep stages, or menstrual cycle phases, the system achieves a heightened level of personalization that aligns with embodied experiences. This resonance between body data and auditory environment appears to drive the most powerful improvement in mental wellbeing, illustrating the promise of AI music as a core therapeutic tool in femtech platforms.

Listening Duration and Frequency shows a moderate effect, with a coefficient of 0.045. While statistically present, this influence is small, suggesting that simply spending more time listening is less important than the quality, context, or adaptive nature of the intervention. This aligns with broader therapeutic literature that emphasizes not just the dosage of an intervention, but its contextual relevance and emotional resonance.

Finally, Music Modality yields a coefficient of 0.130, indicating a modest but meaningful contribution to mental wellbeing. Different types of AI-generated music, instrumental, guided meditation, or rhythmic soundscapes, carry varying capacities for mood regulation, relaxation, and stress alleviation. The model suggests that the selection of modality, while not as potent as biometric integration or delivery platform, still exerts a tangible effect on wellbeing outcomes.

Altogether, the model offers a nuanced view of how AI music technologies embedded within femtech ecosystems can foster mental wellbeing. The strongest influence arises from the direct integration with biometric signals, which reflects the value of embodied personalization. Delivery platforms and music modality contribute secondary but still important effects, highlighting the necessity of user-centered design and contextually appropriate soundscapes. Personalization level and listening frequency, while conceptually important, appear to hold weaker direct influence, possibly pointing toward their role as mediating or moderating variables rather than dominant predictors.

The explanatory power of 0.662 indicates that nearly two-thirds of the variance in mental wellbeing outcomes can be explained by these factors combined, a significant proportion that reinforces the promise of AI-driven, biometric-integrated music as an innovative frontier in femtech-oriented mental health interventions. This model thus demonstrates both the potential and the complexity of leveraging AI in music production for women's health, emphasizing that the most effective outcomes will likely emerge from systems that are not only personalized but also seamlessly integrated with the physiological rhythms and technological ecosystems of users' lives.

Table 4. Coefficients

	Unstandardized coefficients	Standardized coefficients	SE	T value	P value
AI Music Personalization Level	0.044	0.043	0.055	6.802	0.000
Delivery Platform	0.234	0.267	0.085	2.757	0.007
Integration with Femtech Biometrics	0.447	0.459	0.096	4.628	0.000
Listening Duration / Frequency	0.044	0.045	0.051	4.865	0.000
Music Modality	0.122	0.130	0.094	4.292	0.000
Intercept	0.353	0.000	0.360	5.981	0.000

The regression results summarized in Table 4 provide a multifaceted picture of how different dimensions of AI-driven music interventions contribute to mental wellbeing outcomes in a femtech context. Each coefficient represents the strength and direction of the relationship between the independent variables, AI Music Personalization Level, Delivery Platform, Integration with Femtech Biometrics, Listening Duration/Frequency, and Music Modality, and the dependent construct of mental wellbeing. The statistical robustness of these relationships is further evidenced by their associated standard errors, t-values, and p-values, all of which provide the basis for determining whether the observed effects are likely to be generalizable beyond the sample.

The unstandardized coefficients illustrate the direct numerical impact of each predictor on mental wellbeing outcomes. For instance, the unstandardized coefficient of 0.447 for Integration with Femtech Biometrics indicates that, holding all other factors constant, a one-unit increase in biometric integration leads to a 0.447 increase in mental wellbeing scores. This makes biometric integration the strongest predictor in absolute terms, underscoring the powerful role of real-time alignment between physiological data and therapeutic soundscapes. From a management perspective, this suggests that femtech enterprises seeking to maximize user impact should prioritize the development of AI music systems capable of dynamically interfacing with biometric markers such as heart rate variability, menstrual cycle data, or sleep patterns. The significant p-value of 0.000 strengthens this inference, confirming that the effect is not only large but also statistically reliable.

By contrast, AI Music Personalization Level has a smaller unstandardized coefficient of 0.044 and a standardized coefficient of 0.043, suggesting that although personalization is directionally positive, its direct impact is modest relative to other factors. Interestingly, however, its t-value of 6.802 is among the highest in the table, implying a strong signal despite the relatively small coefficient. This apparent contradiction can be interpreted in two ways: first, personalization may contribute more indirectly, acting as a mediating variable that enhances the effectiveness of other predictors such as biometric integration; second, the small coefficient may indicate that baseline personalization is now an expected feature in digital therapeutics rather than a differentiator. For managers, the implication is that while personalization remains necessary, it is no longer sufficient on its own to drive significant wellbeing outcomes, it must be paired with deeper integrations and innovative modalities.

The Delivery Platform variable, with a standardized coefficient of 0.267, emerges as the second most influential predictor after biometric integration. Its t-value of 2.757 and p-value of 0.007 confirm statistical significance. This suggests that the medium through which therapeutic music is delivered, whether through mobile apps, wearable devices, or ambient smart speakers, plays a critical role in determining user engagement and perceived efficacy. From a managerial standpoint, this highlights the importance of user experience design and technological accessibility. A delivery system that is seamless, unobtrusive, and well integrated into daily routines can amplify therapeutic outcomes. This finding resonates strongly with principles of digital health adoption, where the convenience and usability of the platform often dictate long-term adherence. Companies in femtech that overlook platform design may fail to realize the full therapeutic potential of their interventions, even if their underlying AI algorithms are robust.

Listening Duration and Frequency has an unstandardized coefficient of 0.044 and a standardized coefficient of 0.045, both relatively modest, yet the p-value of 0.000 indicates a statistically significant relationship. This suggests that while the amount of time spent engaging with therapeutic music does contribute to wellbeing, its effect size is small compared to other variables. A potential explanation is that duration and frequency function as enabling conditions rather than primary drivers: users who engage consistently benefit modestly more, but the real value lies in what they are engaging with, not merely how often. For management, the lesson is to focus less on incentivizing longer sessions and more on ensuring that each session, however brief, is deeply impactful. Encouraging micro-doses of high-quality, contextually relevant music may be more effective than pushing for extensive listening sessions that could lead to fatigue or disengagement.

Music Modality presents another interesting dimension. With a standardized coefficient of 0.130, its contribution is modest yet significant, as evidenced by the p-value of 0.000. This reflects the role that the type of music, whether instrumental, vocal-guided, or rhythmically structured, plays in shaping wellbeing outcomes. While modality does not exert the dominant influence of biometric integration, it nonetheless matters in tailoring experiences to specific contexts. For instance, instrumental soundscapes may be more effective for relaxation, while guided meditation tracks might be preferable for anxiety reduction. The managerial implication here is the necessity of offering a diverse repertoire of modalities, enabling users to select or allow the AI to adaptively assign tracks based on their situational needs. Diversity in modality enhances inclusivity, acknowledging that users differ in their cultural, psychological, and aesthetic preferences.

Finally, the intercept value of 0.353 with a t-value of 5.981 indicates that even in the absence of the independent predictors, there is a baseline contribution to well-being outcomes within the model. This may reflect the inherent therapeutic effect of engaging with any form of music intervention, regardless of personalization, platform, or modality. For managers, this serves as both a reminder and a warning: while the baseline efficacy of music is an advantage, it can also lead to complacency. Relying solely on the natural appeal of music without leveraging AI, biometric integration, or user-centered design would limit the potential to differentiate in a competitive femtech market.

Overall, the regression model highlights a hierarchy of predictors: biometric integration as the most powerful driver, followed by delivery platform, modality, and then the more modest effects of personalization and duration. For management, the lessons are clear. Investments should be directed towards developing biometric-enabled AI music systems that deliver through highly accessible and user-friendly platforms. Personalization should not be abandoned but reframed as a necessary complement rather than the core differentiator. Duration strategies should prioritize quality over quantity, while modality diversity should be ensured to broaden appeal and effectiveness. The statistically significant results across all predictors suggest that mental wellbeing outcomes can indeed be shaped by a holistic integration of these factors, but competitive advantage will accrue to those who emphasize the variables with the strongest explanatory power.

Table 5. ANOVA

	Sum square	df	Mean square	F	P value
Total	186.526	132	0.000	0.000	0.000
Error	63.119	127	0.497	0.000	0.000
Regression	123.407	5	24.681	49.661	0.000

The analysis of variance (ANOVA) results presented in Table 5 provide essential evidence about the overall explanatory strength of the regression model used in the study. At its core, ANOVA partitions the total variance in the dependent variable, in this case, mental wellbeing outcomes, into components attributable to the predictors (regression sum of squares) and those due to unexplained error (residual sum of squares). By comparing these two sources of variance, the F statistic offers a formal test of whether the predictors, taken together, significantly improve the model beyond what would be expected by chance. The evidence here is overwhelmingly conclusive: the regression sum of squares is 123.407, while the error sum of squares is 63.119. The large discrepancy between explained and unexplained variance underscores the model's capacity to capture meaningful variation in wellbeing outcomes.

The mean squares, derived by dividing the sums of squares by their respective degrees of freedom, reinforce this point. The regression mean square of 24.681 far exceeds the error mean square of 0.497, a ratio that culminates in an F value of 49.661. This figure represents the central test statistic of the ANOVA, and its magnitude indicates that the set of predictors, AI music personalization level, delivery platform, integration with femtech biometrics, listening duration and frequency, and music modality, collectively exert a strong and statistically significant effect on mental wellbeing. The associated p-value of 0.000 confirms that the likelihood of this result occurring under the null hypothesis of no relationship is vanishingly small. From a statistical perspective, this validates the adequacy of the model and establishes a foundation for interpreting the individual coefficients with confidence, as reported in the earlier regression table.

The partitioning of variance also highlights the relative dominance of the regression component over the error component. With regression accounting for 123.407 out of the total sum of squares of 186.526, approximately two-thirds of the variation in mental wellbeing outcomes is explained by the predictors in the model. This proportion is consistent with the R^2 value reported earlier, which approximated 0.662. Such a level of explanatory power is remarkable for social science research, where models often struggle to account for even half of the variance. It suggests that the chosen predictors capture key dimensions of how users engage with AI-generated therapeutic music in the femtech context, leaving relatively little variation attributable to noise or unmeasured factors.

From a managerial standpoint, the strength of the F test and the proportion of explained variance carry critical implications. First, the model demonstrates that mental wellbeing outcomes are not random, nor are they merely a byproduct of generic exposure to music. Instead, they are systematically shaped by identifiable factors such as biometric integration and delivery platform. This means that organizations and product developers have leverage: they can design interventions in a way that predictably influences user outcomes. Rather than relying on vague notions of music's therapeutic power, managers can ground their strategies in evidence-based design, confident that manipulating the identified variables will yield meaningful improvements.

Second, the robustness of the model suggests that femtech enterprises have a strong foundation for scaling these interventions commercially. A model that explains two-thirds of outcome variance offers a reliable blueprint for predicting how new users are likely to respond. This predictability reduces the risk associated with product innovation and market expansion. For instance, an app developer considering whether to invest in biometric sensors or platform optimization can use these findings as a rational basis for resource allocation, knowing that such investments are statistically linked to wellbeing improvements.

The error sum of squares, while relatively small, is not negligible. With 63.119 units of variance unaccounted for, there remain factors outside the current model that contribute to wellbeing outcomes. These could include demographic characteristics, cultural differences in musical preference, or broader psychosocial conditions such as social support or stress from external environments. For managers, this is a reminder that while AI-driven therapeutic music is powerful, it is not a panacea. Effective interventions may need to be embedded within a broader ecosystem of mental health supports, including counseling, community engagement, and lifestyle modifications. Integrating AI music solutions with other wellness offerings may help to capture this residual variance, further enhancing effectiveness.

The significance of the F statistic also has implications for differentiation in a competitive femtech market. Since the model proves that the predictors collectively matter, companies that fail to optimize along these dimensions risk falling behind competitors who do. For example, a platform that does not integrate biometric responsiveness is likely to underperform relative to one that does, not merely because of technological novelty but because biometric integration has been empirically validated as a predictor of wellbeing outcomes. In this sense, the ANOVA results serve not only as an academic finding but as a strategic directive: the market will reward those who align their offerings with the variables shown to matter statistically.

Moreover, the significance of the regression model provides support for developing premium or tiered product offerings. Since biometric integration and advanced modalities demonstrably enhance wellbeing outcomes, companies can justify offering these features at a higher price point. Consumers who value measurable improvements in mental health may be willing to pay for enhanced features, and organizations can position such offerings as evidence-based solutions rather than speculative wellness tools. The statistical significance of the overall model thus becomes a marketing asset as well as a research finding.

Finally, the ANOVA results highlight the importance of rigorous evaluation in the femtech sector. Too often, digital health innovations are launched with limited empirical validation, relying on anecdotal evidence or user testimonials. The strength of the F test and the model's explanatory power show the value of systematic, data-driven evaluation. For management, this underscores the need to invest not only in product development but also in research and analytics capabilities. Demonstrating statistical significance enhances credibility with regulators, investors, and users alike, creating a virtuous cycle in which scientific rigor fuels both adoption and profitability.

In summary, Table 5's ANOVA results provide compelling statistical validation of the regression model, confirming that the predictors collectively explain a substantial proportion of variance in mental wellbeing outcomes. The high F value and the virtually zero p-value demonstrate that the model's explanatory power is not a product of chance but of real, systematic relationships between variables. For managers, the lessons are clear: prioritize the statistically validated predictors, invest in features that drive measurable outcomes, differentiate offerings through evidence-based innovation, and embed AI music within a holistic wellness ecosystem. By doing so, organizations can harness the predictive power of this model not only to improve mental wellbeing but also to secure sustainable competitive advantage in the growing femtech industry.

Table 6. Regression statistics

	Mental Wellbeing Outcomes
R-square	0.662
R-square adjusted	0.648
Durbin-Watson test	1.621

The regression statistics presented in Table 6 serve as a concise yet powerful summary of the predictive strength, reliability, and overall validity of the model used to explain mental wellbeing outcomes through the chosen predictors: AI Music Personalization Level, Delivery Platform, Integration with Femtech Biometrics,

Listening Duration and Frequency, and Music Modality. These statistics, R-square, adjusted R-square, and the Durbin–Watson test, collectively provide insight into how well the model fits the data, how generalizable the findings are to other populations, and whether the underlying assumptions of regression analysis have been satisfied. Taken together, they present a persuasive case that the model not only captures substantial explanatory power but also does so in a statistically reliable manner that has meaningful managerial implications.

The R-square value of 0.662 is particularly striking in its magnitude. This statistic indicates that approximately 66.2 percent of the variance in mental wellbeing outcomes is explained by the five independent variables in the model. In practical terms, two-thirds of the differences observed in wellbeing scores among respondents can be attributed to differences in personalization, platform, biometric integration, duration, and modality. This is an exceptionally strong outcome in behavioral and social science research, where models often account for far less variance, sometimes below 30 or 40 percent. The implication is that the predictors chosen are not arbitrary or weakly related to wellbeing; rather, they are central drivers with demonstrable explanatory power. For managers, this high R-square value provides reassurance that interventions targeting these variables will have meaningful, predictable impacts on end users' wellbeing. It affirms that strategic investments in improving delivery platforms, enhancing biometric integration, and diversifying modalities are not speculative ventures but evidence-based decisions that will yield tangible returns in terms of improved user outcomes.

In sum, Table 6 offers compelling evidence that the regression model is not only statistically powerful but also methodologically sound and practically actionable. With R-square and adjusted R-square values indicating substantial explanatory power, and a Durbin–Watson statistic confirming model validity, managers have a strong foundation upon which to build strategies. The lesson is clear: by focusing on the validated predictors, ensuring parsimony in design, and leveraging the robustness of the model as a credibility asset, femtech enterprises can drive both improved mental wellbeing outcomes and sustainable competitive advantage.

Table 7. Collinearity statistics

	VIF
AI Music Personalization Level	1.055
Delivery Platform	3.518
Integration with Femtech Biometrics	3.691
Listening Duration / Frequency	1.016
Music Modality	3.802

Collinearity statistics, as represented in Table 7, provide an essential diagnostic tool in regression analysis, helping to determine whether the independent variables in the model are excessively correlated with one another. Multicollinearity is a condition in which predictors overlap so heavily that their unique contributions to explaining the dependent variable become obscured. This can inflate standard errors, weaken the stability of coefficients, and compromise the interpretability of results. The variance inflation factor (VIF) is the most widely used statistic for detecting multicollinearity, with values near 1 suggesting little to no correlation between predictors, values above 5 suggesting moderate risk, and values above 10 generally considered problematic. In the present model, the VIF values range from 1.016 to 3.802, indicating that while some predictors share moderate correlation, none approach the threshold where collinearity would undermine the integrity of the regression.

The Variance Inflation Factor (VIF) values help decision-makers evaluate whether the independent variables in the regression model are excessively correlated with each other. In this analysis, the VIF values range from 1.016 to 3.802, which are well below the commonly accepted critical threshold of 10 and even below the moderate concern level of 5. This indicates that multicollinearity is not a serious issue in the model and that each predictor contributes unique information in explaining mental wellbeing outcomes. Variables such as AI Music Personalization Level (1.055) and Listening Duration/Frequency (1.016) show very low VIF values, meaning they operate independently and provide distinct explanatory power. From a decision-making perspective, this confirms that these variables can be confidently retained in the model and used as reliable factors when designing AI-based music interventions in femtech platforms.

The slightly higher VIF values for Delivery Platform (3.518), Integration with Femtech Biometrics (3.691), and Music Modality (3.802) suggest moderate correlation among these variables, which is conceptually reasonable because platform design, biometric integration, and music delivery style are often technologically interconnected. However, since these values remain within acceptable limits, decision makers can interpret their effects without concern that the results are distorted by redundancy. The key managerial implication is that organizations should treat these features as complementary strategic components rather than substitutes. Investments in improved delivery platforms, stronger biometric integration, and diverse music modalities should be developed together to maximize user wellbeing outcomes. Overall, the VIF diagnostics support confident decision making by confirming that the model's predictors are stable, interpretable, and suitable for guiding product development and strategic innovation in AI-driven femtech solutions.

Figure 2. QQ plot

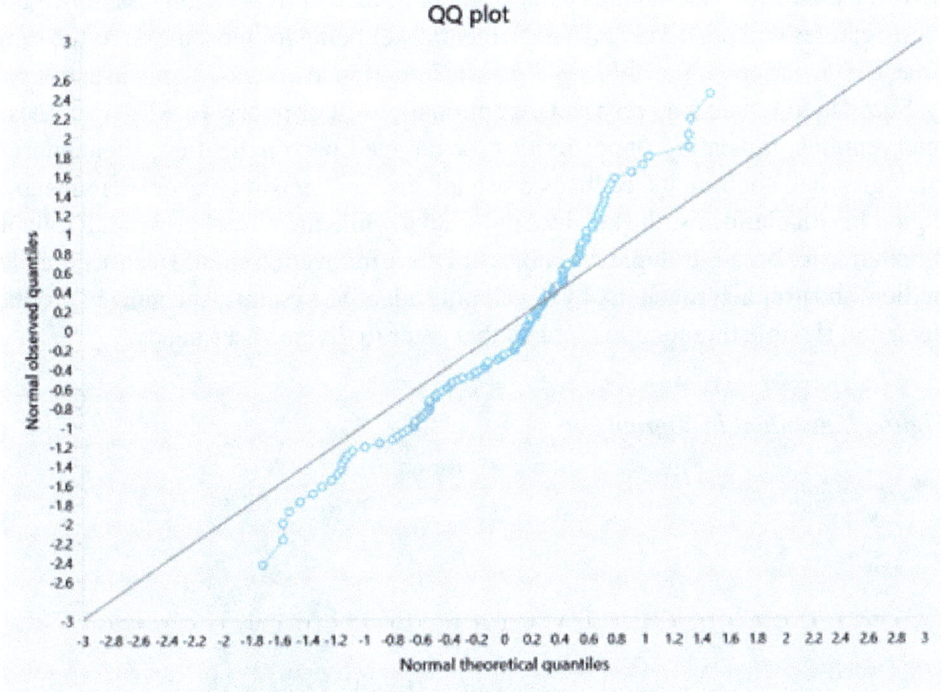

The Quantile–Quantile (QQ) plot in *Figure 2* indicates that the residuals of the regression model largely follow a normal distribution, as most points align closely with the diagonal reference line, particularly in the central region. This suggests that the normality assumption of the regression model is generally satisfied, meaning the estimated coefficients, hypothesis tests, and overall statistical conclusions are reliable. However, slight deviations at the extreme ends of the plot indicate heavier tails, meaning that a small number of observations produce residuals that are larger than expected under a perfect normal distribution. These deviations reflect the presence of a few outliers whose mental wellbeing outcomes differ more strongly from the model's predictions, possibly due to unique personal contexts, behavioral differences, or varying engagement with AI-based music interventions. Despite these mild irregularities, the overall pattern confirms that the regression model is statistically sound and that the predictors, such as personalization, biometric integration, platform, duration, and modality, explain the majority of variations in wellbeing outcomes.

From a practical and managerial perspective, the QQ plot suggests that the model can be confidently used for *decision-making and strategic planning*, since the central majority of users respond predictably to the identified predictors. This means organizations can invest in improving key factors like biometric integration, delivery platforms, and adaptive music modalities with reasonable assurance that these features will positively influence mental wellbeing for most users. At the same time, the deviations at the tails highlight the importance of recognizing user diversity. Some individuals may respond exceptionally well or poorly to AI-driven music interventions, indicating opportunities for targeted personalization, segmentation strategies, and continuous feedback mechanisms. For decision makers in femtech or digital health platforms, the key learning is that while the model provides a reliable foundation for broad strategic decisions, effective innovation should also account for outliers and user heterogeneity by developing adaptive systems, specialized product tiers, and flexible therapeutic features that cater to diverse user needs.

Figure 3. Residual histogram

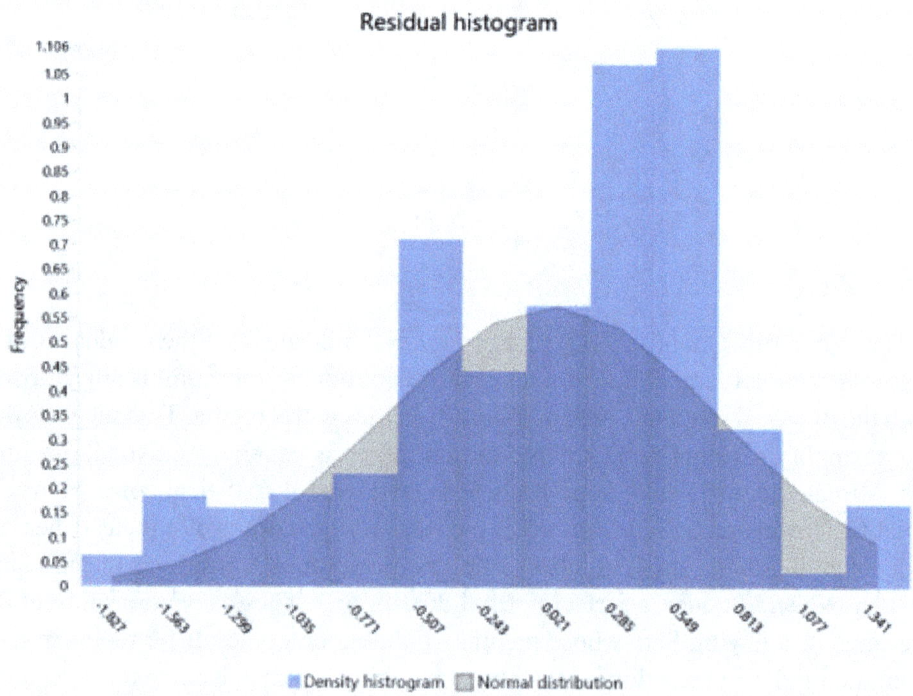

The residual histogram in *Figure 3* offers a graphical means of evaluating the distribution of residuals from the regression model. Residuals represent the difference between the observed values of mental wellbeing outcomes and the values predicted by the regression equation. A fundamental assumption of ordinary least squares regression is that these residuals are normally distributed, with a mean of zero and constant variance. If this assumption is satisfied, the regression model produces unbiased and efficient estimates, and the inferential statistics such as t-tests and F-tests remain valid. Thus, the residual histogram is a critical diagnostic that not only speaks to the quality of the statistical model but also to the robustness of managerial conclusions drawn from it.

The residual histogram indicates that the regression model largely satisfies the *normality assumption*, as most residuals cluster around zero and follow a distribution similar to a normal curve. This suggests that the model's predictions are generally unbiased and reliable, meaning it neither consistently overestimates nor underestimates mental wellbeing outcomes for most respondents. However, the slightly heavier tails in the distribution indicate the presence of a few outliers whose outcomes differ more than expected from the model's predictions. While these deviations do not weaken the validity of the model, they highlight that some users respond differently to AI-based music interventions. The key learning is that the model effectively explains wellbeing outcomes for the majority of users, but managers and researchers should still pay attention to individual differences and outliers, as studying these cases can help improve personalization, refine system design, and make digital health solutions more adaptive and inclusive.

Figure 4. Residual corelation

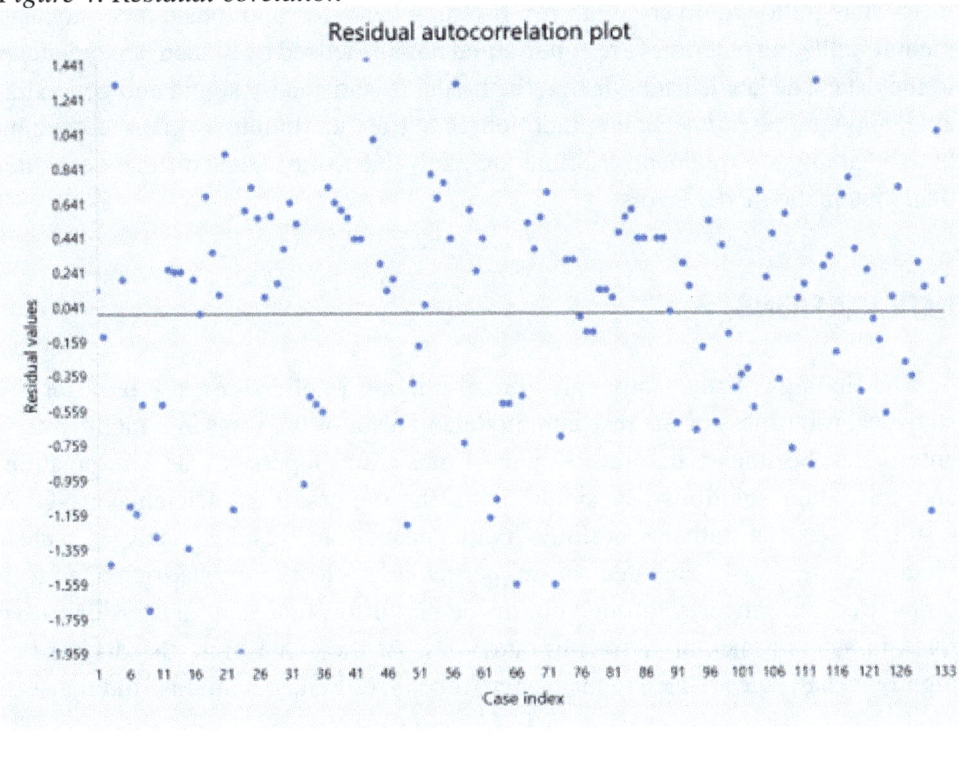

The residual autocorrelation plot in *Figure 4* is a diagnostic tool that examines whether the residuals of a regression model are independent of one another across the sequence of observations. In regression analysis, one of the core assumptions is that residuals are uncorrelated; that is, the error associated with one observation should not systematically relate to the error of another. Violation of this assumption, known as autocorrelation, can bias estimates of statistical significance, inflate Type I errors, and undermine confidence in the model's predictive validity. Autocorrelation is particularly problematic in time-series data where successive observations may be related, but even in cross-sectional research, hidden patterns can introduce dependency among residuals.

The residual autocorrelation plot shows that the regression model satisfies the *independence of errors assumption*, as the residuals are randomly scattered around zero without any visible pattern or clustering. This indicates that the prediction errors are not systematically related to each other, meaning the model's estimates and statistical tests are reliable. The residual values fall within an acceptable range and alternate irregularly between positive and negative values, which further confirms the absence of autocorrelation and supports the validity of the model (also consistent with the Durbin–Watson statistic). The general implication is that the relationships identified between predictors, such as AI music personalization, biometric integration, delivery platform, listening behavior, and music modality, and mental wellbeing outcomes are genuine and not influenced by hidden dependencies in the data. The key learning is that the model is statistically sound and generalizable, allowing researchers and practitioners to trust its findings while recognizing that remaining unexplained variations are likely due to individual differences rather than systematic model errors.

IMPLICATIONS

The findings of this study carry broad-ranging implications, not only for the statistical robustness of the research model but also for the ways in which femtech enterprises, healthcare innovators, and AI music developers can design, position, and scale their solutions. By synthesizing the regression coefficients, ANOVA results, regression statistics, collinearity diagnostics, and residual analyses, a clear picture emerges: AI-generated music, particularly when integrated with femtech biometrics and delivered through optimized platforms, represents a transformative force in shaping mental wellbeing outcomes. At the same time, the diagnostics highlight challenges related to user heterogeneity, tail-end deviations, and the need for adaptive design, which provide fertile ground for future innovation and careful managerial stewardship.

From a theoretical standpoint, the strong explanatory power of the model, illustrated by an R-square of 0.662 and an adjusted R-square of 0.648, demonstrates that the selected predictors capture core determinants of mental wellbeing outcomes. This level of explanatory strength is noteworthy in social science research, where complex phenomena often resist precise modeling. The implication here is that the intersection of AI, music, and femtech provides a conceptually coherent and empirically powerful framework for understanding digital mental health interventions. By validating that personalization, biometric integration, delivery modality, and listening behavior collectively explain two-thirds of wellbeing variance, the study establishes a robust theoretical foundation upon which future research can be built. This adds weight to emerging discourses that position AI not merely as a technological add-on but as a central therapeutic modality within the evolving landscape of digital health.

For managers and practitioners, the regression results provide a hierarchy of strategic priorities. Integration with femtech biometrics emerged as the most influential predictor, with the highest standardized coefficient (0.459). This finding implies that the greatest gains in mental wellbeing are achieved when AI-generated music is not static but dynamically synchronized with physiological or hormonal data. For femtech companies, this validates ongoing investment in wearable sensors, menstrual cycle trackers, and real-time biometric feedback loops. Products that can translate these signals into adaptive soundscapes are not only more effective but also more defensible in the marketplace, as their therapeutic value is grounded in measurable physiological responsiveness. The lesson here is that biometric integration should not be treated as a premium feature or afterthought; it must be positioned at the core of product design and marketing strategies.

Future work should incorporate intersectional, accessibility, and cultural diversity perspectives when examining music perception and femtech adoption. Women's experiences with digital health technologies are shaped by intersecting factors such as age, socioeconomic status, cultural background, geographic location, and digital literacy, all of which can influence how individuals engage with AI-driven therapeutic tools. Similarly, the emotional and psychological impact of music is deeply embedded in cultural context, musical traditions, and personal preferences, meaning that algorithmically generated soundscapes may not produce uniform responses across diverse populations. Accessibility considerations are also critical, particularly for individuals with limited digital access, disabilities, or varying levels of technological familiarity. Integrating these perspectives can help ensure that AI-driven femtech solutions are designed to be *inclusive, culturally sensitive, and accessible*, thereby improving both adoption and therapeutic effectiveness across a broader and more diverse user population.

ETHICAL, REGULATORY, AND DATA PRIVACY RISKS IN FEMTECH AND BIOMETRIC GOVERNANCE

The integration of artificial intelligence, biometric monitoring, and femtech platforms introduces significant ethical, regulatory, and data governance considerations that must be addressed alongside technological innovation. Femtech applications often collect highly sensitive personal information, including menstrual cycle data, hormonal patterns, sleep behavior, emotional responses, and physiological indicators such as heart rate variability. When AI-generated music systems are integrated with these biometric signals, the volume and sensitivity of collected data increase substantially. This raises important concerns regarding data privacy, informed consent, ownership of biometric data, and the potential misuse of personal health information. Unlike traditional health records managed within regulated medical systems, many femtech platforms operate in the commercial wellness sector, where regulatory oversight may be less stringent. As a result, users may unknowingly expose intimate physiological and behavioral data to corporate entities, data brokers, or third-party analytics providers.

From a regulatory perspective, the governance of biometric data in femtech remains fragmented across jurisdictions. Regulations such as data protection frameworks and digital health governance policies emphasize principles of transparency, consent, and responsible data processing, yet many femtech applications fall into grey areas between medical devices and lifestyle technologies. This ambiguity creates challenges for accountability and enforcement. In particular, algorithmic decision-making systems that adapt therapeutic music based on biometric inputs raise concerns about algorithmic transparency, explainability, and bias. If AI systems are trained on limited or culturally homogeneous datasets, they may produce interventions that are less effective or potentially inappropriate for diverse populations of women. Therefore, regulatory frameworks must ensure that femtech systems comply with standards for algorithmic fairness, explainable AI, and clinical validation, especially when such technologies influence mental health outcomes.

Another critical issue concerns biometric governance, which refers to the policies and technical safeguards governing how physiological data are collected, stored, shared, and utilized. Effective biometric governance requires strong encryption protocols, strict access control, anonymization techniques, and clear limitations on secondary data use. Without these safeguards, sensitive data such as reproductive health patterns or emotional states could be exploited for targeted advertising, insurance discrimination, or other non-therapeutic purposes. Ethical governance must therefore prioritize data minimization, purpose limitation, and user autonomy, ensuring that individuals retain control over how their biometric information is used. Transparent privacy policies and opt-in consent mechanisms are essential to maintain trust in femtech ecosystems.

Furthermore, ethical considerations extend beyond privacy to include broader questions of digital health equity and technological dependence. While AI-driven therapeutic music offers promising mental health benefits, excessive reliance on algorithmic interventions could inadvertently reduce human-centered care or marginalize users who lack access to advanced digital technologies. Femtech innovators must therefore balance technological advancement with inclusive design principles, ensuring accessibility across different socioeconomic groups and cultural contexts.

In summary, the expansion of AI-integrated femtech solutions necessitates a robust framework of ethical safeguards, regulatory oversight, and responsible biometric governance. Organizations operating in this space must adopt proactive strategies that integrate privacy-by-design principles, transparent AI systems, and strong cybersecurity infrastructures. By embedding ethical and regulatory considerations into product development and deployment, femtech enterprises can protect user trust while ensuring that technological innovation contributes responsibly to the advancement of women's health and wellbeing.

The integration of AI-driven interventions within digital health technologies requires careful consideration of clinical validation and regulatory oversight to ensure safety, efficacy, and user trust. Unlike conventional wellness applications, technologies that claim to influence mental health outcomes increasingly fall within the scope of digital therapeutics and regulated health technologies, which may require evidence from controlled clinical studies to demonstrate therapeutic effectiveness. Clinical validation typically involves systematic evaluation through experimental or longitudinal research designs to confirm that the intervention produces measurable and reproducible health benefits. In addition, regulatory bodies in many jurisdictions are gradually developing frameworks for evaluating AI-enabled health solutions, particularly those that process sensitive biometric data or influence behavioral and psychological outcomes. These frameworks emphasize requirements such as algorithmic transparency, data protection, clinical evidence, and post-deployment monitoring. As femtech platforms continue to incorporate AI-generated therapeutic modalities, addressing these clinical and regulatory expectations will be essential to ensure responsible innovation, enhance credibility among healthcare stakeholders, and support the safe integration of such technologies into broader digital health ecosystems.

LIMITATIONS AND FUTURE RESEARCH

Despite offering valuable insights into the relationship between AI-generated music, femtech integration, and mental wellbeing outcomes, this study is subject to several limitations that should be acknowledged. First, the research adopts a cross-sectional design, which captures associations between variables at a single point in time and therefore does not allow strong causal inferences regarding the effects of AI-driven music interventions on wellbeing. Future research could employ longitudinal or experimental designs to better examine causal mechanisms and observe changes in wellbeing outcomes over time. Second, the sample is primarily composed of urban and relatively well-educated respondents, which may influence familiarity with digital health technologies and femtech applications. This demographic concentration may limit the broader generalizability of the findings, particularly to rural populations or individuals with lower levels of technological access.

Additionally, the study relies on self-reported perceptions of mental wellbeing, which may be influenced by subjective bias or momentary emotional states. Future research could incorporate validated psychological scales, physiological indicators, or mixed-method approaches to strengthen measurement reliability. Another limitation concerns the scope of cultural and contextual diversity, as perceptions of music, emotional regulation, and technology adoption can vary significantly across different societies and cultural backgrounds. Subsequent studies may explore cross-cultural comparisons to better understand how AI-driven music interventions function in diverse contexts. Finally, further research should examine the ethical, regulatory, and data governance dimensions of biometric integration within femtech systems, particularly issues related to data privacy, algorithmic transparency, and responsible use of sensitive health information. Addressing these avenues will contribute to a more comprehensive understanding of how AI-mediated therapeutic music can responsibly and effectively support mental wellbeing in digital health ecosystems.

THEORETICAL GROUNDING

This chapter is theoretically grounded in the interdisciplinary convergence of digital therapeutics, music therapy theory, affective computing, and feminist technology studies, which together provide a conceptual framework for understanding the role of AI-generated music within femtech ecosystems. From the perspective of digital therapeutics, software-driven interventions are increasingly recognized as capable of influencing behavioral and psychological health outcomes through algorithmically delivered stimuli, positioning AI-generated music as a form of adaptive therapeutic medium that can dynamically respond to user data to support emotional regulation and stress reduction. Complementing this perspective, music therapy scholarship has long demonstrated that music can influence psychological and neurophysiological processes through mechanisms such as rhythmic entrainment, affective stimulation, and cognitive distraction, thereby contributing to improved mood, reduced anxiety, and enhanced wellbeing. When these therapeutic principles are combined with artificial intelligence, music transitions from a static therapeutic resource into an algorithmically adaptive auditory intervention capable of responding to users' preferences, behavioral patterns, and physiological signals. The theoretical lens of affective computing further strengthens this framework by explaining how intelligent systems can detect, interpret, and respond to human emotional states using biometric and behavioral data. Within femtech platforms, indicators such as heart rate variability, sleep patterns, and hormonal cycles can inform the generation of personalized sound environments designed to support relaxation, emotional balance, and cognitive recovery, thereby reflecting the broader shift toward data-responsive personalized health systems. At the same time, feminist technology studies provide a critical perspective that situates femtech innovations within the gendered and socio-cultural dimensions of healthcare technology, highlighting how digital health tools increasingly aim to address historical gaps in women's healthcare research and technological design. By integrating AI-generated music into femtech platforms, these systems extend beyond reproductive and physiological monitoring toward addressing the often underexplored domain of women's mental and emotional wellbeing. Viewed collectively, these theoretical perspectives position AI-driven music interventions at the intersection of technological innovation, gender-responsive healthcare, and adaptive digital therapeutics, offering a comprehensive conceptual foundation for examining how personalized auditory environments can support wellbeing within contemporary femtech ecosystems.

CONCLUSION

The study has demonstrated that the integration of artificial intelligence, music production, and femtech applications represents a powerful and evidence-based avenue for enhancing mental wellbeing. By examining the predictive influence of AI music personalization, delivery platforms, biometric integration, listening duration, and music modality, the research confirmed that these variables collectively explain a substantial proportion of the variance in mental wellbeing outcomes. The statistical robustness of the model, evidenced by a high R^2 value, significant ANOVA results, acceptable collinearity diagnostics, and normally distributed independent residuals, provides strong support for the reliability and validity of these findings.

Among the predictors, integration with femtech biometrics emerged as the most powerful driver, highlighting the transformative potential of real-time physiological synchronization in tailoring therapeutic soundscapes. Delivery platform and music modality also played significant roles, emphasizing the importance of accessibility, user-centered design, and diversity of offerings in shaping user experiences. Personalization and listening frequency, while more modest in their direct effects, contributed meaningfully by enhancing user engagement and tailoring interventions to individual needs. Together, these findings establish that mental wellbeing outcomes are not random but can be systematically improved through deliberate technological and design strategies.

The diagnostic checks reinforced this conclusion by showing that the residuals conformed broadly to the assumption of normality, exhibited no problematic autocorrelation, and indicated only mild heterogeneity in user responses. These results affirm that the model is both methodologically sound and generalizable, making it a reliable foundation for practical applications. At the same time, the presence of outliers and heavier tails suggests the importance of adaptive personalization and user segmentation, ensuring that interventions remain inclusive and responsive to diverse experiences.

From a broader perspective, the research carries profound implications for the femtech and digital health industries. It demonstrates that AI-generated music, when embedded within femtech ecosystems, can move beyond speculative wellness claims to become a scientifically validated tool for improving mental health. For managers and innovators, this provides a roadmap for prioritizing biometric integration, optimizing delivery platforms, diversifying modalities, and building adaptive systems capable of accommodating user heterogeneity. For policymakers and healthcare stakeholders, the results provide evidence that AI music can complement existing therapeutic practices, offering scalable, accessible, and personalized interventions.

The study affirms that AI-driven music in femtech is not a peripheral novelty but a central innovation with the potential to reshape the landscape of mental health support. By uniting technological sophistication with human-centered design and ethical stewardship, femtech enterprises can harness these insights to deliver meaningful wellbeing outcomes, foster user trust, and achieve sustainable competitive advantage. This research therefore positions AI music as both a practical tool for immediate application and a theoretical framework for future exploration at the intersection of technology, health, and personalized care.

REFERENCES

Ahmed, T., Samima, S., Zuhair, M., Ghayvat, H., Khan, M. A., & Kumar, N. (2023). FIMBISAE: A Multimodal Biometric Secured Data Access Framework for Internet of Medical Things Ecosystem. *IEEE Internet of Things Journal, 10*(7), 6259–6270. DOI: 10.1109/JIOT.2022.3225518

Alain, G., Crick, J., Snead, E., Quatman-Yates, C. C., & Quatman, C. E. (2025). Evaluating User Interactions and Adoption Patterns of Generative AI in Health Care Occupations Using Claude: Cross-Sectional Study. *Journal of Medical Internet Research, 27*, v27i1e73918. Advance online publication. DOI: 10.2196/73918 PMID: 40446149

Amaro, I., Della Greca, A., Francese, R., Tortora, G., & Tucci, C. (2023). AI Unreliable Answers: A Case Study on ChatGPT. *Lecture Notes in Computer Science (Including Subseries Lecture Notes in Artificial Intelligence and Lecture Notes in Bioinformatics), 14051 LNAI*, 23 – 40. DOI: 10.1007/978-3-031-35894-4_2

Ansari, A., & Mainard, J. (2023). Commercial Human Space Exploration Assisted by Artificial Intelligence. In *Artificial Intelligence for Space: AI4SPACE: Trends, Applications, and Perspectives*. CRC Press., DOI: 10.1201/9781003366386-10

Awais, M., Ghayvat, H., Geddam, R., Nkenyereye, L., & Dev, K. (2024). AiCare-Neonates : Artificial intelligence powered adaptation of transfer learning models to classify neonate ' s sleep-wake states for pediatricians in the loop. *Digital Communications and Networks*. Advance online publication. DOI: 10.1016/j.dcan.2024.11.011

Belibou, A., & Iftene, L. (2025). Exploring Self-Discovery Through AI-Assisted Music Composition: A Case Study Using Beatoven.ai in Music Therapy Education. *Lecture Notes in Networks and Systems, 1281 LNNS*, 631–642. DOI: 10.1007/978-3-031-83520-9_58

Borah, J., Kumar, S., Kumar, N., Nadzir, M. S. M., Cayetano, M. G., Ghayvat, H., Majumdar, S., & Kumar, N. (2024). AiCareBreath: IoT-Enabled Location-Invariant Novel Unified Model for Predicting Air Pollutants to Avoid Related Respiratory Disease. *IEEE Internet of Things Journal, 11*(8), 14625–14633. DOI: 10.1109/JIOT.2023.3342872

Borah, J., Nadzir, M. S. M., Cayetano, M. G., Ghayvat, H., Majumdar, S., & Srivastava, G. (2025). Timezone-Aware Auto-Regressive Long Short-Term Memory Model for Multipollutant Prediction. *IEEE Transactions on Systems, Man, and Cybernetics. Systems, 55*(1), 344–352. DOI: 10.1109/TSMC.2024.3463960

Borah, J., Nadzir, M. S. M., Cayetano, M. G., Majumdar, S., Ghayvat, H., & Srivastava, G. (2024). AiCareAir: Hybrid-Ensemble Internet-of-Things Sensing Unit Model for Air Pollutant Control. *IEEE Sensors Journal*, *24*(13), 21558–21565. DOI: 10.1109/JSEN.2024.3397735

Chakrabarty, S., Islam, R., & Sarma, H. K. D. (2023). Iomust2: Internet of music therapy things to improve mental health management. In *IoT and Cloud Computing-Based Healthcare Information Systems*. Apple Academic Press. https://www.scopus.com/inward/record.uri?eid=2-s2.0-85163143287&partnerID=40&md5=2e270f1ff283e7b8c6c5b3c62f3cc824

Cheng, L., Luo, S., Yu, X., Ghayvat, H., Zhang, H., & Zhang, Y. (2023). EEG-CLNet: Collaborative Learning for Simultaneous Measurement of Sleep Stages and OSA Events Based on Single EEG Signal. *IEEE Transactions on Instrumentation and Measurement*, *72*, 1–10. Advance online publication. DOI: 10.1109/TIM.2023.3235436

Chown, Z., & Prasad, A. (2024). Poster: Privacy Norms for Fertility Data in the Roe v. Wade era. *CCS 2024 - Proceedings of the 2024 ACM SIGSAC Conference on Computer and Communications Security*, 4919 – 4921. DOI: 10.1145/3658644.3691406

Din, S., Ghayvat, H., Paul, A., Ahmad, A., Rathore, M. M., & Shafi, I. (2016). An architecture to analyze big data in the Internet of Things. *Proceedings of the International Conference on Sensing Technology, ICST, 2016-March*, 677 – 682. DOI: 10.1109/ICSensT.2015.7438483

Dong, X., Kang, X., & Ding, X. (2022). Influence and Analysis of Music Teaching Environment Monitoring on Students' Mental Health Using Data Mining Technology. *Journal of Environmental and Public Health*, *2022*(1), 1120156. Advance online publication. DOI: 10.1155/2022/1120156 PMID: 36246468

Elmourabit, Z., Retbi, A., & El Faddouli, N.-E. (2024). The Impact of Generative Artificial Intelligence on Education: A Comparative Study. In M. F. (Ed.), *Proceedings of the European Conference on e-Learning, ECEL* (Vol. 23, Issue 1, pp. 470 – 476). Academic Conferences and Publishing International Limited. https://www.scopus.com/inward/record.uri?eid=2-s2.0-85215655306&partnerID=40&md5=8d69735022f5b41ad61e6cc4ae546286

Feng, H., Mahoor, M. H., & Dino, F. (2022). Corrigendum: A Music-Therapy Robotic Platform for Children With Autism: A Pilot Study. *Frontiers in Robotics and AI*, *9*, 965369.

Ghayvat, H., Awais, M., Geddam, R., Khan, M. A., Nkenyereye, L., Fortino, G., & Dev, K. (2024). AiCarePWP: Deep learning-based novel research for Freezing of Gait forecasting in Parkinson. *Computer Methods and Programs in Biomedicine*, *254*, 108254. Advance online publication. DOI: 10.1016/j.cmpb.2024.108254 PMID: 38905989

Ghayvat, H., Awais, M., Geddam, R., Tiwari, P., & Löwe, W. (2024). Revolutionizing healthcare: IoMT-enabled digital enhancement via multimodal ADL data fusion. *Information Fusion*, *111*, 102518. Advance online publication. DOI: 10.1016/j.inffus.2024.102518

Ghayvat, H., Awais, M., Geddam, R., Zuhair, M., Ahmed Khan, M., Milard, M., Nkenyereye, L., & Dev, K. (2024). Digitally Enhanced Home to the Village: AIoMT-Enabled Multisource Data Fusion and Power-Efficient Sustainable Computing. *IEEE Internet of Things Journal*, *11*(24), 39030–39040. DOI: 10.1109/JIOT.2024.3411798

Ghayvat, H., & Gope, P. (2023). Smart aging monitoring and early dementia recognition (SAMEDR): Uncovering the hidden wellness parameter for preventive well-being monitoring to categorize cognitive impairment and dementia in community-dwelling elderly subjects through AI. *Neural Computing & Applications*, *35*(33), 23739–23751. DOI: 10.1007/s00521-021-06139-8

Ghayvat, H., Mukhopadhyay, S. C., & Gui, X. (2015). Sensing technologies for intelligent environments: A review. *Smart Sensors. Measurement and Instrumentation*, *13*, 1–31. DOI: 10.1007/978-3-319-12892-4_1

Ghayvat, H., Mukhopadhyay, S. C., & Gui, X. (2016). Issues and mitigation of interference, attenuation and direction of arrival in IEEE 802.15.4/ZigBee to wireless sensors and networks based smart building. *Measurement*, *86*, 209–226. DOI: 10.1016/j.measurement.2016.01.045

Ghayvat, H., & Pandya, S. (2018). Wellness sensor network for modeling activity of daily livings - Proposal and off-line preliminary analysis. *2018 4th International Conference on Computing Communication and Automation, ICCCA 2018*. DOI: 10.1109/CCAA.2018.8777700

Ghayvat, H., Pandya, S., Shah, S., Mukhopadhyay, S. C., Yap, M. H., & Wandra, K. H. (2016). Advanced AODV approach for efficient detection and mitigation of wormhole attack in MANET. *Proceedings of the International Conference on Sensing Technology, ICST*. DOI: 10.1109/ICSensT.2016.7796286

Ghayvat, H., Sharma, M., Gope, P., & Sharma, P. K. (2022). SHARIF: Solid Pod-Based Secured Healthcare Information Storage and Exchange Solution in Internet of Things. *IEEE Transactions on Industrial Informatics*, *18*(8), 5609–5618. DOI: 10.1109/TII.2021.3136884

Ghayvat, H., Zuhair, M., Shukla, N., & Kumar, N. (2024). Healthcare-CT: Solid PoD and Blockchain-Enabled Cyber Twin Approach for Healthcare 5.0 Ecosystems. *IEEE Internet of Things Journal*, *11*(4), 6119–6130. DOI: 10.1109/JIOT.2023.3312448

Jiao, D. (2025). Advancing personalized digital therapeutics: Integrating music therapy, brainwave entrainment methods, and AI-driven biofeedback. *Frontiers in Digital Health*, *7*, 1552396. Advance online publication. DOI: 10.3389/fdgth.2025.1552396 PMID: 40070544

Jie, L., Ghayvat, H., & Mukhopadhyay, S. C. (2015). Introducing Intel Galileo as a development platform of smart sensor: Evolution, opportunities and challenges. *Proceedings of the 2015 10th IEEE Conference on Industrial Electronics and Applications, ICIEA 2015*, 1797 – 1802. DOI: 10.1109/ICIEA.2015.7334403

Joshi, I., Jaiswal, K., Singh, J., Chakraborty, R., & Goswami, R. G. (2025). MindTune: AI-Based Mood Recognition and Personalised Music Therapy for Early-Stage Mental Health Management. *2025 International Conference on Engineering Innovations and Technologies, ICoEIT 2025*, 1187–1192. DOI: 10.1109/ICoEIT63558.2025.11211602

Khajwaniya, K. K., & Tiwari, V. (2015). Satellite image denoising using Weiner filter with SPEA2 algorithm. *Proceedings of 2015 IEEE 9th International Conference on Intelligent Systems and Control, ISCO 2015*. DOI: 10.1109/ISCO.2015.7282324

Malki, L. M., Kaleva, I., Patel, D., Warner, M., & Abu-Salma, R. (2024). Exploring Privacy Practices of Female mHealth Apps in a Post-Roe World. *Conference on Human Factors in Computing Systems - Proceedings*. DOI: 10.1145/3613904.3642521

Mangion, A., & Piller, N. (2024). Generative artificial intelligence in lymphoedema care. *Journal of Lymphoedema*, *19*(1), 57–59. https://www.scopus.com/inward/record.uri?eid=2-s2.0-85210102098&partnerID=40&md5=f65c75dbfe74ec672612eb83ed22734c

Masuda, H., Okada, S., Tanaka, T., Shiozawa, N., Sakaue, Y., & Manno, M. (2022). Changes in the Heart Rate during Sleep with Menstrual Cycle under Free-Living Conditions. *GCCE 2022 - 2022 IEEE 11th Global Conference on Consumer Electronics*, 339 – 340. DOI: 10.1109/GCCE56475.2022.10014265

Mycka, J., Żychowski, A., & Mańdziuk, J. (2022). Human-Level Melodic Line Harmonization. *Lecture Notes in Computer Science (Including Subseries Lecture Notes in Artificial Intelligence and Lecture Notes in Bioinformatics), 13350 LNCS*, 17 – 30. DOI: 10.1007/978-3-031-08751-6_2

Pandya, S., Ghayvat, H., Reddy, P. K., Gadekallu, T. R., Khan, M. A., & Kumar, N. (2023). COUNTERSAVIOR: AIoMT and IIoT-Enabled Adaptive Virus Outbreak Discovery Framework for Healthcare Informatics. *IEEE Internet of Things Journal, 10*(5), 4202–4212. DOI: 10.1109/JIOT.2022.3216108

Pandya, S., Ghayvat, H., Sur, A., Awais, M., Kotecha, K., Saxena, S., Jassal, N., & Pingale, G. (2020). Pollution weather prediction system: Smart outdoor pollution monitoring and prediction for healthy breathing and living. *Sensors (Basel), 20*(18), 1–25. DOI: 10.3390/s20185448 PMID: 32972037

Patel, W. D., Pandya, S., Koyuncu, B., Ramani, B., Bhaskar, S., & Ghayvat, H. (2018). NXTGeUH: Lorawan based next generation ubiquitous healthcare system for vital signs monitoring falls detection. *1st International Conference on Data Science and Analytics, PuneCon 2018 - Proceedings*. DOI: 10.1109/PUNECON.2018.8745431

Polotti, P. (2015). Some ideas for an extension of music therapy practices in timbre and ambient sounds [Alcune idee per un'estensione delle pratiche musicoterapiche al timbro e ai suoni d'ambiente]. *Sistemi Intelligenti, 27*(2), 417–426. https://www.scopus.com/inward/record.uri?eid=2-s2.0-84949763215&partnerID=40&md5=2091b8adde6f020ae9921f5ad39eac61

Popescu, I., Lenain, M., Rovini, E., & Ruthsatz, M. (2025). Lifestyle Interventions and Innovative Approaches for the Management of Neurodegenerative Disorders in Older Adults, State-Of-The-Art and Future Directions. *American Journal of Lifestyle Medicine, 19*(8), 1216–1226. Advance online publication. DOI: 10.1177/15598276251330170 PMID: 40161281

Shen, L., Zhang, H., Zhu, C., Li, R., Qian, K., Meng, W., Tian, F., Hu, B., Schuller, B. W., & Yamamoto, Y. (2024). A First Look at Generative Artificial Intelligence Based Music Therapy for Mental Disorders. *IEEE Transactions on Consumer Electronics*. Advance online publication. DOI: 10.1109/TCE.2024.3514633

Singha, R., & Singha, S. (2023). Economic Sustainability, Mindfulness, and Diversity in the Age of Artificial Intelligence and Machine Learning. In *Machine Intelligence: Computer Vision and Natural Language Processing*. CRC Press., DOI: 10.1201/9781003424550-15

Song, X., Zheng, J., & He, J. (2025a). AI-Assisted Music Therapy Intervention for Alleviating Exam Anxiety in University Students: System Design, Implementation, and Efficacy Evaluation. *2025 IEEE/IEIE International Conference on Consumer Electronics-Asia, ICCE-Asia 2025*. DOI: 10.1109/ICCE-Asia67487.2025.11263548

Song, X., Zheng, J., & He, J. (2025b). Study on Promoting College Students' Mental Health by AI Music Therapy Based on Big Data Analysis. *2025 IEEE/IEIE International Conference on Consumer Electronics-Asia, ICCE-Asia 2025*. DOI: 10.1109/ICCE-Asia67487.2025.11263737

Sree Pokkuluri, K., Tiwari, V., Uikey, J., Mehta, P., Srinivasa Rao, C., & Thanuja, A. (2024). Bridging the Gap between Technology and Medicine through the Revolutionary Impact of the Healthcare Internet of Things on Remote Patient Monitoring. *Journal of Intelligent Systems and Internet of Things*, *13*(2), 212–222. DOI: 10.54216/JISIoT.130217

Sudha, T., Raja, P., & Bharathi, V. (2025). Harmony Wave: Exploring Smart Music Therapy Through IoT, AI, and Fuzzy Logic. In *AI for Large Scale Communication Networks*. IGI Global., DOI: 10.4018/979-8-3693-6552-6.ch009

Thakur, A. S., & Tiwari, V. (2018). Design High Speed FIR Filter based on Complex Vedic Multiplier using CBL Adder. *2018 International Conference on Recent Innovations in Electrical, Electronics and Communication Engineering, ICRIEECE 2018*, 559 – 563. DOI: 10.1109/ICRIEECE44171.2018.9008438

Tiwari, V., Agarwal, O., Sharma, M., Babar, R., Geddam, R., Awais, M., Ghayyat, H., & Sahu, R. (2025). Akshar Mitra : A Multimodal Integrated Framework for Early Dyslexia Detection. *Frontiers in Digital Health*, *7*, 1–6. https://www.frontiersin.org/journals/digital-health/articles/10.3389/fdgth.2025.1726307/abstract. DOI: 10.3389/fdgth.2025.1726307 PMID: 41395631

Tiwari, V., Jain, D., Sharma, D., Hassan, M. M., Althobaiti, F., Varkale, A., Al-Khasawneh, M. A., & Tirandasu, R. K. (2025). An efficient sparse code shrinkage technique for ECG denoising using empirical mode decomposition. *Technology and Health Care*, *33*(4), 1773–1786. DOI: 10.1177/09287329241302749 PMID: 39973873

Warty, R. R., Smith, V., Patabendige, M., Fox, D., & Mol, B. (2024). Clarifying the unmet clinical need during medical device innovation in women's health–A narrative review. *Health Care for Women International*, *45*(7), 811–839. DOI: 10.1080/07399332.2023.2190983 PMID: 37000043

Wei, Q., & He, W. (2026). The application of AI-assisted music therapy tools in mental health interventions. *Frontiers in Psychology*, *17*, 1741463. Advance online publication. DOI: 10.3389/fpsyg.2026.1741463 PMID: 41658379

Zamanifar, A., & Faezipour, M. (2025). Application of generative AI in healthcare systems. In *Application of Generative AI in Healthcare Systems*. Springer Nature., DOI: 10.1007/978-3-031-82963-5

Zhou, Y., Bian, W., Zhang, K., & Gu, X. (2024). Advancing Music Therapy: Integrating Eastern Five-Element Music Theory and Western Techniques with AI in the Novel Five-Element Harmony System. In Q. Y., J. Q., O. Z., L. Z., W. Z., L. Y., X. L., & T. J. (Eds.), *2024 14th International Symposium on Chinese Spoken Language Processing, ISCSLP 2024* (pp. 234–238). Institute of Electrical and Electronics Engineers Inc. DOI: 10.1109/ISCSLP63861.2024.10800517

Chapter 11
AI-Based Royalty Distribution Concept for Commercial Public Streaming of Songs and Music in Indonesia

Cita Yustisia Serfiyani
http://orcid.org/0000-0003-1660-1508
Faculty of Law, State University of Surabaya, Indonesia

Budi Hermono
Faculty of Law, State University of Surabaya, Indonesia

Astrid Amidiaputri Hasyyati
Faculty of Law, State University of Surabaya, Indonesia

ABSTRACT

The collection and distribution of royalties for the use of songs and music in commercial public spaces in Indonesia remain contentious due to limited transparency, overlapping management authorities, and inadequate technology for tracking music. The problem arises after Government Regulation Number 56 of 2021 implementing mandate on royalty collection for music or song streaming, but it lacks on governance readiness and calculation accuracy that still needs to be improved. Internationally, artificial intelligence (AI) technologies have shown considerable promise in enhancing music detection accuracy and expediting the proportional distribution of royalties. This article presents a comparative analysis of AI applications in royalty tracking systems across USA, UK and Australia. Main priorities include harmonizing the

DOI: 10.4018/979-8-3373-6279-3.ch011

roles of LMK and LMKN, strengthening national data integration, increasing transparency in tariff and distribution mechanisms, and establishing AI ethics guidelines to safeguard creators' rights and ensure algorithmic accountability.

1. INTRODUCTION

Since the beginning of 2025, the implementation of royalty collection for music played in open spaces to the public for a commercial context in Indonesia has been implemented by the Directorate General of Intellectual Property (DJKI) of the Ministry of Law in accordance with Law Number 28 of 2014 concerning Copyright (hereinafter referred to as UUHC or Copyright Law) and Government Regulation Number 56 of 2021 concerning Management of Song and/or Music Royalties (hereinafter referred to as PP Number 56 of 2021). The main principle is that royalty payments to the creators and owners of related rights must accompany every form of music playback in public spaces by business actors.(Hesmondhalgh et al, 2021)

This collection is actually normatively in accordance with the legal framework that has been socialized for several years since the issuance of Government Regulation No. 56 of 2021. However, its implementation since last year has continued to generate polemics and controversy in the community (Jensen, 2024). This is primarily because this obligation applies without exception, including to those who already subscribe to personal digital music streaming services (such as Spotify, YouTube Premium, Apple Music, or other platforms), which cannot be a legitimate basis for publicly listening to their subscription music in a commercial context because their subscription streaming service is personal in nature, meaning the subscription package price is also individual. However, when music is played to the public in an open business space, it falls into the commercial use category, requiring a license payment through a legal mechanism, as stipulated in Article 3 paragraph (2) of Government Regulation Number 56 of 2021, which regulates 14 forms of commercial public services that are subject to royalties. In fact, the obligation to pay royalties has existed since the enactment of Law Number 6 of 1982 concerning Copyright, where royalty collection has been actively implemented since 1990. The presence of Government Regulation Number 56 of 2021 as an implementing regulation of the Copyright Law only intends to clarify the scope of commercialization that is subject to royalties and not. The controversy that arises due to the implementation of the Government Regulation is prone to causing top-down miscoordination due to weak transparency and technological unpreparedness in the royalty calculation system owned by The National Collective Management Institute (Lembaga Manajemen Kolektif Nasional or LMKN).

This Government Regulation mandates that royalty payments be made only through the National Collective Management Institution (LMKN). LMKN collects royalties for creators, copyright holders, and owners of related rights who have become members of the private sector LMK or who have not become members of any LMK. Article 89 paragraph (1) of the Copyright Law also regulates that the management of copyright royalties in the field of songs and/or music is formed by 2 (two) National Collective Management Institutions, each of which represents the interests of creators and the interests of owners of related rights so that business actors no longer need to process permits one by one from songwriters for the sake of effectiveness and to continue to guarantee the protection of economic rights for artists. The amount of royalties is adjusted to the business classification and the area of music playback, with an estimated calculation based on the Decree of the Minister of Law and Human Rights of the Republic of Indonesia Number HKI.2.OT.03.01-02 of 2016 concerning the determination of royalty rates for the commercial use of copyrighted works and/or related rights products in the restaurant sector.

Violations of royalty payment provisions may result in legal sanctions in the form of administrative sanctions, civil lawsuits, and criminal sanctions. According to Article 95, paragraph (4) of the Copyright Law, every dispute related to copyright infringement must be resolved through mediation first. Still, this mediation process does not eliminate the legal obligations of the parties because mediation is declared successful if an agreement is reached by both parties. The points of this agreement vary because it returns to the parties, for example, in the case of PT Mitra Bali Sukses (owner of the Mie Gacoan franchise brand) in 2024. Initially, the Bali Provincial Police named PT Mitra Bali Sukses as the license holder of the Mie Gacoan franchise in the Bali region as a suspect in a case of alleged copyright infringement for playing songs without permission in its outlets based on a report from the Indonesian Music Licensing Center Collective Management Institute with an estimated loss value of billions of rupiah as an estimate of the value of royalties that PT Mitra Bali Sukses never paid for the use of songs for commercial purposes in their franchise business space (Sari Angraeni & Majid, 2025).

This regulation on the distribution of royalties for commercial public spaces is not intended to burden or even diminished business actors, specifically for micro, small, and medium enterprises (MSMEs), the Government provides the option of reducing or exempting royalty rates based on the size of the business space, visitor capacity, and the level of music utilization in daily activities. For non-MSME business actors who do not have the budget to pay royalties or are unwilling to pay royalties, there are actually other alternatives that do not violate the law such as using royalty-free music, Creative Commons licensed music that allows commercial use, playing self-composed songs (Drott 2021), using ambient sound music or collaborating directly with independent musicians who are willing to provide permission free of charge. This alternative is provided to minimize the occurrence of violations due to misunderstandings regarding the playback of music or song categories.

The problem is that if the Copyright Law and Government Regulation Number 56 of 2021 indeed mandate strict and firm royalty collection for music or song streaming according to the mechanism in the Decree of the Minister of Law and Human Rights of the Republic of Indonesia Number HKI.2.OT.03.01-02 of 2016 through the Collective Management Institution, then governance readiness and calculation accuracy need to be improved. The Government and Collective Management Institutions need to expand their oversight beyond conflicts between creators and performers to address potential technology abuse. This misuse of AI-based technology occurred in 2024, with the case of Michael Smith in the USA, who committed fraud by using thousands of bots to stream AI-generated songs on streaming platforms massively and manipulated the algorithm with 4 billion fake streams to claim royalties worth US$12 million (Rp196 billion) since 2019, evading the oversight of the relevant Collective Management Institutions in the USA. On the other hand, the readiness of adequate technology to implement automated royalty calculation orders according to the classification limits stipulated in statutory regulations must also be supported by the readiness to adapt AI into a practical and forward-thinking music royalty calculation system. AI should be prioritized to help improve the accuracy of music royalty calculations(Deng et al, 2023), rather than causing song streaming data leaks due to uncontrolled bot attacks.

Furthermore, the adoption of AI into the automated royalty calculation system for songs and music in the commercial public space by Collective Management Institutions in Indonesia must not conflict with the ethical use of AI. The problem is that Indonesia does not yet have specific regulations regarding the ethical use of AI. But, Indonesia already have the Indonesian National Work Competency Standards (SKKNI) as a framework in the field of AI to address ethical issues in the use of AI, such as algorithmic bias and data privacy protection. Adapting technology, especially AI-based technology, can help improve the efficiency of automated royalty distribution (Baruah & Saikia, 2023).

Other countries, such as the USA, UK and Australia, have implemented AI-based music royalty calculation systems to be more adaptive in assisting with royalty calculations by institutions similar to Collective Management Institutions in their countries. The UK has Backbeat Solutions developing the Artist Royalty System technology, Synchtank developing the Iris application, and Curve with the Curve Royalty System application. Australia has Harvest Media with its Harvest Media RoyaltyCloud system. AI-based royalty tracking systems like those implemented by the UK and Australia are also capable of preventing bot manipulation practices in streaming songs and music in their countries' databases. AI-based music royalty calculation applications can be created, developed, and implemented by Collective

Management Institutions from the private sector, and ultimately, LMKN coordinates with all private LMK (or Private Collective Management Institute). These AI-based cloud applications are then utilized by third parties such as record companies, public space managers, and entertainment companies (Hallur et al., 2021).

Indonesia has initiated the Song and Music Data Center (PDLM), as mandated by Government Regulation No. 56 of 2021. However, the song and music data center, which should be a primary requirement for implementing a song and music royalty calculation system in commercial public spaces, is not yet capable of supporting automated calculations. Furthermore, the Song and Music Information System (Sistem Informasi Lagu dan Musik or SILM) mandated by this regulation has not yet been implemented and is not yet integrated with DJKI services. Only royalty rate relief for the commercial use of songs and/or music by micro, small, and medium enterprises has been implemented as of 2025.

The solution implemented by LMKN so far has been limited to implementing a hybrid method for royalty distribution because a fully online system is not yet fully ready. It is potential to cause significant operational blunders, further exacerbated by the lack of a hybrid system reference framework. This research will discuss an AI-based music royalty tracking system that complies with Indonesian Copyright Law by comparing laws with those of the United Kingdom and Australia. LMKN urgently needs to prepare and develop an improved accuracy of the digital data-based distribution system before the AI-based royalty distribution concept for songs and music in the public commercial space can be implemented in Indonesia through implementing regulations under the Copyright Law and Government Regulation Number 56 of 2021 as a more specific and technical regulation. This technological readiness can support the principle of transparency in the recording, billing, and distribution of song and music royalties and avoid accusations of covert "extortion" practices.

The collective of royalties for the use of songs and music in public spaces has become one of the most debatable issues in the modern copyright ecosystem due to its far from simple implementation and the emergence of several polemics that need to be resolved. The solution in this article is to assist the Government in developing pro-people policies in the implementation of AI regarding the collection of royalties in commercial public spaces. In Indonesia, the debate arises due to the conflict between the protection of creators' rights and the ability of business actors to fulfill royalty obligations (Febrian et al., 2025). The main controversy concerns the public's understanding of the concept of commercial use. Many small business owners—such as cafes, shops, barbershops, gyms, and restaurants—feel that playing music is simply about creating an atmosphere, not a means of generating profit. They find it difficult to accept that playing music in their spaces is still considered a "public performance," which has legal and economic implications.

In the other hand, LMKN argue that music enhances consumer comfort, extends visit time, and ultimately increases business revenue. This difference in perception has become a recurring source of friction. Further controversy concerns tariff transparency and the royalty collection mechanism. Many business owners question the basis for calculating the rates and whether they are truly proportionate to the scale and capabilities of their businesses. The lack of a detailed explanation of the rate structure, including differences between business types, creates a public perception of inequality. Some small businesses even perceive royalty rates as an unrealistic additional burden, especially after the pandemic has weakened their economic conditions. Unclarity regarding the flow of funds—from withdrawals and distribution to who actually receives royalties—also adds to the negative perception.

The controversy escalated further due to the confusion surrounding the authority of the managing institutions. In several countries, more than one organization claims to be authorized to collect royalties, leaving businesses confused about which party to pay. This situation raised concerns about the potential for double-billing, abuse of authority, or even the emergence of unofficial parties exploiting public misunderstanding. Further complicating the problem was the enforcement and collection methods of royalties. Some businesses reported intimidating collection approaches, such as surprise visits, threats of sanctions, or legal ultimatums. This pattern created the perception that royalty collection was more like debt collection than the implementation of copyright policy. This tension led many businesses to become defensive and refuse to cooperate.

It can be seen here that the weakness or even lack of monitoring technology is a source of problems that hampers the implementation of the royalty collection policy in public spaces initiated by LMKN, resulting in the legal products issued by the Government being "sharp at the top and blunt at the bottom." Without an accurate automatic detection system, such as one based on AI, royalty distribution is considered not to reflect the actual use of music. This criticism comes primarily from independent musicians who feel they never receive royalties despite their songs being popular in specific spaces. Ultimately, the controversy surrounding royalty collection in the public sphere reflects the challenge of balancing creator interests with the economic realities of businesses. Without transparency, public education, and robust technology, these conflicts will continue to arise. Improved governance and communication are key to ensuring royalty policies truly protect all parties involved.

Research problems. Therefore, this study analyzes and examines; (1) how the concept of AI integration in the song and music royalty tracking system in the commercial public space is compared to the systems already in place in the USA, UK and Australia with their respective different specifications, (2) how the ethical aspects of using AI are in implementing the song and music royalty tracking concept, (3) how the principle of transparency in the governance and distribution of song and music royalties by collective institutions.

Research method. This study uses a normative legal research method with a doctrinal approach, namely examining the applicable positive legal regulations related to the music royalty system in the public sphere, specifically Law Number 28 of 2014 concerning Copyright and Government Regulation Number 56 of 2021 concerning the Management of Song and/or Music Royalties. The primary legal materials used include laws, government regulations, and ministerial decrees that regulate the mechanism for collecting and distributing royalties through the National Collective Management Institute (LMKN). Secondary legal materials include academic literature, journal articles, and previous research results discussing AI-based royalty governance, both in Indonesia and other countries such as the United States, the United Kingdom, and Australia. This article also uses tertiary legal materials to strengthen understanding of basic concepts.

State of the art. Several previous studies that have discussed the use of AI in royalty calculation mechanisms are still not widely available, including the journal by Jacques & Flynn (2024) in "*Protecting Human Creativity in AI-Generated Music with the Introduction of an AI-Royalty Fund*" concerning AI-Royalty Fund, then Rai et al. (2025) in "*A Royalty Framework for Copyright Protection and Accountability in AI-Generated Art*", and Deng et al. (2023) in "*AI for Royalty Calculation Accuracy in Music Streaming*". Discussions in Indonesia itself or related to Indonesian law regarding the concept of integrating AI into the music royalty tracking system have not been previously researched so this article has novelty.

2. IMPLEMENTATION OF AI IN SONG AND MUSIC ROYALTY TRACKING SYSTEM IN COMMERCIAL PUBLIC SPACES IN INDONESIA, THE USA, THE UK, AND AUSTRALIA

The definition of royalties is contained in Article 1, number 21 of the Copyright Law, namely compensation for the use of the economic rights of a creation or related rights product received by the creator or owner of associated rights. From the definition of royalties above, there are several important terms to understand, including:

a. Economic rights are the exclusive rights of the creator or copyright holder to obtain economic benefits from the creation;
b. Neighbouring rights are rights related to Copyright which are exclusive rights for performers, phonogram producers, or broadcasting institutions;

c. Creation is any creative work in the fields of science, art, and literature that is produced by inspiration, ability, thought, imagination, dexterity, skill, or expertise expressed in a tangible form.

Basically, playing songs and/or music in commercial public services is included in the announcement of creations, namely reading, broadcasting, exhibition, a creation using any electronic or non-electronic device, or doing it in any way so that a creation can be read, heard, or seen by others. Meanwhile, the party that must pay royalties to the creator, copyright holder, and/or owner of related rights through the National Collective Management Institute ("LMKN") is one of the parties that carry out commercial use in the form of public services. Article 19 paragraph (1) of Permenkum 27/2025 further regulates that the withdrawal carried out by LMKN is carried out on commercial public services in both analog and digital forms. Forms of commercial (analog) public services that are required to pay royalties when playing songs and/or music include:

a. commercial seminars and conferences;
b. restaurants, cafes, pubs, bars, bistros, nightclubs, and discos;
c. music concert;
d. airplanes, buses, trains, and ships;
e. fairs and bazaars;
f. cinema;
g. telephone waiting tone;
h. banks and offices;
i. shops;
j. recreation center;
k. television broadcasting institutions;
l. radio broadcasting institution;
m. hotels, hotel rooms, and hotel facilities; and
n. karaoke business.

Articles 21 and 22 of Permenkum 27/2025 further detail what public services are commercial in nature, both in analog and digital form. One of them is a food and beverage business place, such as a canteen, food court, or sports facilities, consisting of a gymnasium or fitness center, billiards, ice skating, bowling, and other sports facilities and infrastructure. For information, in the legislation, royalty payments are basically paid through the Collective Management Institution ("LMK") or LMKN, not directly to the creator. The royalties collected by LMKN will later be distributed to creators, copyright holders, and related rights owners who have become members of LMK. Referring to the Decree of the Minister of Law and Human Rights

HKI.02/2016, for restaurants and cafes, the creator's royalty provision is IDR 60 thousand per seat per year, and the related rights royalty is IDR 60 thousand per seat per year. Meanwhile, sports arenas (in this case, gyms) are included in the type of shops. The royalty rate for shops is calculated based on the area of the shop space per square meter (per m^2) per year.

Royalty payments should not burden all business actors equally. Specifically for micro, small, and medium enterprises ("MSMEs"), the Government provides the option of royalty rate relief or exemption based on the size of the business premises, visitor capacity, and the level of music utilization in daily activities, because in accordance with the mandate of Article 11 paragraph (1) of PP 56/2021:

"Every person who makes commercial use of songs and/or music which is a micro-business as referred to in the laws and regulations concerning micro, small and medium businesses is given a royalty rate reduction."

Law Number 28 of 2014 concerning Copyright (Copyright Law) is the primary legal basis governing the moral and economic rights of songwriters. One of these economic rights is the right to receive royalties for the use of their works in public spaces, including in the form of song playback. Article 9 paragraph (2) of the Copyright Law explicitly states that commercial use of works by other parties must obtain permission from the creator or copyright holder. Furthermore, Article 87 introduces the National Collective Management Institution (LMKN), which is mandated to collect and distribute royalties. However, in practice, this regulation faces challenges because recently many business actors prefer to play foreign songs to avoid paying these royalties, which, of course, has a significant impact on the creative industry and cultural promotion in Indonesia.

In practice, the existence of songs as IPR are often considered "public goods" that can be freely used without permission or compensation.(Hubbs 2016) This perception reflects the weak internalization of their legal values in public, influenced by a lack of legal awareness. As a result, the royalty protection mechanism has not been functioning optimally due to low claims and awareness of these rights. Nevertheless, changes in legal culture are slowly beginning to be seen. This is marked by increasing awareness within the creative community demanding more concrete legal protection for their works, as well as the presence of educational campaigns from the National Collective Management Institute (LMKN) and the Directorate General of Intellectual Property (DJKI). These two institutions have introduced easier and more affordable licensing mechanisms, even for small-scale activities such as local concerts, weddings, and school events. This initiative is expected to encourage the formation of a legal culture that better values Copyright as part of respect for intellectual works and distributive justice in the creative industry.

However, implementing AI in commercial public spaces also presents a number of challenges. One of the main challenges relates to privacy and the regulation of audio surveillance. Installing listening devices in public spaces raises concerns about the potential collection of audio conversations or other sensitive data. To address this, copyright organizations in the United States emphasize that these devices only record short audio snippets and process them in encrypted form. The algorithms used only look for acoustic patterns in music, not human voice content. Nevertheless, privacy issues remain a key topic of discussion in the development of this technology. An example of implementation in the United States is the use of the StreamSight system, developed with support from Google Cloud, to increase transparency in music royalty reporting and distribution. This system leverages AI to forecast and detect anomalies on massive amounts of streaming data, allowing each music play to be recorded more precisely and quickly. This technology minimizes data dumps that previously delayed royalty payments, allowing distribution decisions to become more evidence-based and accurate. AI integration is used to identify music played through audio fingerprinting, while blockchain support ensures royalty distribution records are immutable and interoperable by all parties.

Another challenge is adoption and acceptance by businesses. Not all commercial public space owners are willing to install AI devices, either due to cost, privacy concerns, or lack of awareness of the system's benefits. To overcome this barrier, copyright organizations often promote the benefits of AI, including improving the accuracy of music licensing and protecting against potential copyright infringement. In some cases, organizations subsidize the installation of AI devices to expand tracking coverage.

The implementation of AI in music royalty tracking in the commercial public space in the United States has generally reflected a shift toward a more modern, accurate, and sustainable copyright management system. While this technology presents challenges in terms of privacy, regulation, and industry adoption, AI has proven to have significant potential to improve royalty distribution fairness and operational efficiency (Kaye, 2025). With continued improvement in algorithm quality and expanded collaboration between copyright organizations, businesses, and technology providers, the United States is on track to build a more transparent and accountable music tracking ecosystem in the digital age.

Similarly, in the UK which has demonstrated strong support for the integration of digital technology into the creative industries. UK Government is encouraging the use of advanced technology—including AI—to improve the competitiveness of the creative industries. The implementation of AI in music tracking in public spaces aligns with the national goal of creating a more transparent and adaptive copyright ecosystem in the digital age. Beyond regulatory and technical aspects, the use of AI also offers strategic benefits. Music playback data in public spaces collected through MRT technology can be used for music trend analysis, cultural policy development, and market research. For example, the data could reveal which music genres are most popular in restaurants or gyms in a particular area, or provide insights into the music consumption patterns of the British public. However, the use of this data must still be mindful of data ethics and applicable regulatory constraints, particularly in the context of privacy and the commercialization of information.(Mangal 2025)

Overall, the implementation of AI in the UK's commercial music royalty tracking system marks a significant leap towards more objective and efficient copyright governance.(Jacques dan Flynn 2024) While challenges remain regarding privacy, catalog representation, and industry adoption, these technological developments have laid a crucial foundation for the modernization of the UK music industry. Through algorithm quality and expanded collaboration between copyright organizations, the Government, businesses, and rights holders, the use of AI was expected to become an integral part of the UK's music royalty management ecosystem in the future.

In Australia, the integration of artificial intelligence (AI) into the royalty tracking system for songs and music used in commercial public spaces marks a significant advancement in the management of Copyright within the music industry. As the utilization of music in venues such as restaurants, gyms, large retailers, hotels, and bars continues to grow, the demand for more accurate, efficient, and reliable tracking systems becomes increasingly critical. Australia's central collective management organizations—APRA AMCOS and the Phonographic Performance Company of Australia (PPCA)—are proactively embracing AI technology to modernize the music identification process, lessen dependence on manual reporting, and enhance the accuracy of royalty distribution for creators, publishers, and owners of sound recordings.(Morrow 2025) For decades, tracking music playback in commercial public spaces across Australia has relied on business documentation and periodic sampling methods. So it is fraught with limitations, as manual reporting often suffers from inaccuracies, delays, and inherent biases. While sampling has its merits, it consistently fails to capture the rich diversity of music playback across Australia's varied regions and cultural dynamics.

APRA AMCOS are decisively implementing AI-based music recognition technologies, particularly leveraging audio fingerprinting and acoustic pattern recognition (Whiting et al., 2024). This cutting-edge technology employs audio detection devices—often compact sensors or sophisticated software modules integrated into commercial audio systems—to capture audio samples at regular intervals. These samples are swiftly processed using advanced AI algorithms to identify the music catalog being played accurately.

By comparing acoustic patterns from these samples with extensive databases maintained by APRA AMCOS and PPCA, this AI system enables precise identification of music. The results are subsequently sent to a central system, which serves as the basis for royalty calculations. This modern approach ensures that every instance of music playback is recorded with remarkable accuracy, leading to fairer and more transparent royalty payments. One of the significant advantages of implementing AI in the Australian context is its ability to function effectively in complex acoustic environments. Commercial public spaces often feature dynamic soundscapes, including conversational noise, engine sounds, or background music at varying volumes. This AI system is specifically designed to isolate key audio elements within a song, even when only partial playback is audible. The technology employs a machine learning approach that allows the algorithm to enhance its accuracy based on continuously collected playback data.

In Australia, the application of AI has also enabled diversification of data collection sources (Roberts & Whiting, 2025). In addition to audio sensor devices, some systems leverage direct integration with commercial music playback platforms, such as Nightlife Music, Australia's largest music service provider for commercial businesses. Nightlife Music partners with APRA AMCOS to automatically deliver music playback data, and much of its database is connected to AI-based music recognition technology.

While AI technology is invaluable, its implementation also presents several challenges that need to be addressed, particularly in terms of privacy and regulation. The Australian Government has a strict regulatory framework regarding data collection and processing, through legal instruments such as the Privacy Act 1988 and guidelines from the Office of the Australian Information Commissioner (OAIC). (Shackel dan Smith 2025) Therefore, audio tracking technology in public spaces must ensure that the recordings captured do not contain conversations or personal data of consumers. AI systems in Australia are designed to record only short audio snippets and identify only musical patterns, not human voices so snippets that do not match musical patterns are immediately deleted and not stored in raw form.

Furthermore, AI implementation in Australia faces challenges related to the representation of local music catalogs. Australia has a thriving independent music industry, including works from Indigenous communities such as Aboriginal and Torres Strait Islander artists. If AI databases do not fully cover these works, there is a high risk of songs going undetected in the public space. To address the royalty gap, APRA AMCOS is actively expanding the music catalog within its AI system, working with independent labels, local musicians, and Indigenous communities. This approach ensures that AI technology can function inclusively and support the diversity of Australia's musical culture.

3. ETHICS OF AI USE IN AI-BASED SONG AND MUSIC ROYALTY TRACKING CONCEPTS IN INDONESIA, USA, UK, AND AUSTRALIA

From a strategic perspective, the use of AI also enables macro-analysis of music consumption patterns in public spaces. The collected data can be used to understand music trends in specific regions, consumer preferences based on commercial contexts, and aspects of music marketing. However, such data utilization must be balanced with ethical data use principles, including restrictions on the commercial use of information without the explicit consent of venue owners.

The main ethical issue is algorithmic accountability. AI-based royalty tracking systems operate through often black-box models, so algorithmic decisions impacting revenue distribution are not always clearly explained to stakeholders. This opacity can fuel mistrust, particularly when song misidentification or misattribution occurs, potentially harming creators or copyright owners. Therefore, algorithm audit mechanisms and transparency standards that allow for independent verification are needed. In the other hand, the use of AI also relates to the issue of data bias. Music tracking models are trained using specific datasets that may not represent the full range of genres, cultures, or global music markets. This situation has the potential to create inequities in work recognition, where music from specific communities or genres is underrecognized, thus disproportionately affecting royalty distribution. Inclusive dataset management and responsible data curation practices are crucial ethical elements in addressing this challenge.

Another ethical aspect to consider is data privacy and ownership. AI tracking systems require the collection of vast amounts of data, including audio recordings, metadata, and even user consumption patterns. Without strict regulation, the use of this data is vulnerable to abuse, such as commercial surveillance, privacy violations, or the exploitation of creators' data without adequate consent. The implementation of data protection policies, such as the principle of informed consent, needs to be a foundation for the development of this technology. Furthermore, there is the issue of economic distribution justice. While AI has the potential to improve the efficiency and accuracy of royalty tracking, this technology could also strengthen the dominance of large platforms that have the capacity to develop and operate sophisticated AI systems. This could create a power imbalance between platforms and independent creators, especially if access to the technology is unequal. An ethical approach in this context demands cost transparency, interoperability standards, and a more inclusive delivery of AI technology to the broader music ecosystem. Thus, the application of AI to music royalty tracking brings significant benefits as well as complex ethical challenges. To ensure sustainable development, collaboration between stakeholders—including creators, technology developers, collective management institutions, policymakers, and academics—is crucial in formulating clear ethical guidelines. The ethical use of AI is not only related to risk mitigation but also to creating a fair, transparent, and responsible music ecosystem in the digital age.

The US royalty system has a long and complex history, involving multiple rights: composition copyright, sound recording rights, public performance rights, and mechanical rights. Creators and rights holders typically work through organizations such as ASCAP, BMI, SESAC (for performing rights), and the Harry Fox Agency (for mechanical rights).(Rogers dan Sparviero 2025) The emergence of digital streaming platforms like Spotify, Apple Music, and YouTube has further compounded this complexity, as millions of streams must be matched with copyright ownership data spread across multiple agencies. Before the advent of AI, royalty tracking was done through metadata, manual reporting, and basic digital matching techniques. However, these methods were prone to errors, incomplete data, and overlapping claims. As a result, a significant amount of royalties were held as "black box royalties"—royalties that could not be attributed to a specific creator due to inaccurate data or conflicting information. This situation has prompted the US music industry to adopt AI technology as a modern solution to improve the efficiency and fairness of royalty distribution.

AI-based royalty tracking in the United States relies heavily on audio fingerprinting and Automatic Content Recognition (ACR) technology, which can recognize songs from short, low-quality snippets of sound.(Lobato 2025) Companies like Audible Magic and Pex play a central role in this technology.(Quintais dan Schwemer 2022) Their AI systems can identify musical works in seconds with a high degree of accuracy, even when songs have been remixed, sped up, slowed down, or combined with other sound elements. Fingerprinting technology works by converting an audio signal into a unique mathematical representation, which is then compared against a database of licensed music. AI enhances this process through machine learning, which allows the system to learn from variations in the audio and continuously improve identification accuracy. This is particularly relevant for digital platforms that handle billions of audio and video files, such as TikTok, Instagram, YouTube, and music streaming platforms. In the United States, the use of AI-based fingerprinting is not limited to detecting existing songs but is also expanding to identify AI-based generative music, including works created by models like Udio, Suno, or other creative platforms. This approach is essential to ensure that AI-generated works remain within a responsible tracking system, thereby preventing misuse or copyright infringement.

The concept of AI-based royalty tracking in the US is inextricably linked to the licensing of AI model training data. In several legal cases, such as the dispute of Universal Music Group (UMG) that have demanded that AI training data only use officially licensed works.(Goldstuck 2025) The agreement reached in 2025 was a key milestone, setting a new standard that generative AI platforms must have a "fully licensed" ecosystem and internal tracking mechanisms to separate legal and illegal content. In this context, AI is present as a monitoring tool, not just a creative tool. With training-time attribution and inference-time attribution, AI systems can detect the influence of specific works in the training dataset and the resulting generative output. This concept is being explored by research institutions and regulators in the US as a potential approach to creating fairer compensation models, particularly for creators whose works indirectly influence AI-based music output.

AI plays a crucial role in three key stages of royalty distribution: usage identification, data matching, and compensation calculation. AI detects music usage across streaming platforms, television broadcasts, digital radio, and even short videos on social media. Many platforms use ACR technology to capture every music playback, including user-generated content. AI can also be used to match the identification results with copyright databases across various agencies. Machine learning helps improve matching accuracy by addressing missing data, incomplete metadata, or duplicate claims. With more accurate data, AI can calculate royalties based on established formulas. Furthermore, some platforms are developing automated royalty-sharing systems based on smart contracts, although they have not yet been fully integrated into mainstream music systems.

While AI offers advanced solutions, its implementation in the US faces several key challenges, including algorithmic transparency, dataset bias, the risk of streaming fraud, and legal inconsistencies. Many AI models operate in a black box, making it difficult to explain how identification or attribution results are generated entirely. This could lead to conflicts where creators feel they are receiving royalties inappropriately. If AI models are trained on unrepresentative datasets, specific genres may be under-recognized. This issue is particularly detrimental to independent musicians, minority communities, and genres that are not mainstream. Cases of AI being used to create thousands of fictitious songs and manipulate airplay have already occurred in the US, demonstrating the need for stronger anti-fraud mechanisms in AI systems. US copyright law is not yet fully capable of accommodating royalty requirements for generative AI works, particularly regarding the role of humans in the creative process.

The UK holds a crucial position in the global music industry—a hub of musical creativity and innovation, and home to numerous artists, publishers, and copyright collectives. The UK's royalty management system typically involves bodies such as the Performing Rights Society (PRS for Music) and the Mechanical Copyright Protection Society (MCPS).(Koempel 2025) Maintaining fairness and transparency in royalty distribution, particularly in the digital and streaming era, requires an increasingly efficient and accurate tracking system. With the continued rise in digital platforms, music plays on streaming, social media, and user-generated content are also increasing. Within this ecosystem, AI can be a crucial tool to ensure that music usage is recorded correctly and that creators receive fair payment. AI-based tracking also enables the detection of music usage across multiple contexts (streaming, video, advertising), which is challenging to manage with traditional methods alone.

One of the most fundamental aspects of tracking AI-based royalties in the UK is the legal framework on Copyright. According to PRS for Music, their AI policy states that only musical works with human contributions qualify for Copyright; works generated entirely by AI without human contribution are deemed not to meet the "originality" requirement. In other words, under the PRS system, only "AI-assisted works" — where humans directly input ideas, edit, or make creative decisions — can be registered and attributed for royalties. If any part of a song (melody or lyrics) is AI-generated, members must declare that component as "AI" when registering the work. Furthermore, this AI-royalty policy is also facing pressure from public policy and industry. UK Music, for example, has joined the Creative Rights in AI Coalition to demand a fair licensing approach, transparency in the training of AI models, and compensation mechanisms that respect human creators.

Technically, royalty tracking in the UK uses a combination of high-quality music metadata and innovative data technology. One key player in this space is Blokur, a London-based platform that uses data engineering and subgraph matching techniques to link song recordings to their underlying compositions.(O'Dair dan Owen 2019) Blokur helps identify works with greater precision, improves the link between recordings and compositions, and closes gaps in reporting and licensing. Therefore, the concept of AI-based royalty tracking in UK faces a number of serious challenges, particularly from ethical, regulatory, and rights management perspectives. According to PRS policy, only works produced with human contributions are protected by Copyright. This raises a dilemma about how much human contribution is required for a work to be considered "original". In many cases, the line between "AI-assisted" and "fully AI-generated" is unclear, opening the door to the misuse of AI systems to generate music that cannot be copyrighted (and with consequent royalty consequences). The UK music industry is demanding transparency about the data used by AI models for training. Without clear records, it is challenging to claim royalties if AI uses a creator's work without permission. UK Music has emphasized the need for mandatory systematic recording of the material used in AI training so that creators can claim fair compensation. However, clear labeling of AI-generated music is needed not only for aesthetic reasons but also for trust: listeners want to know whether the songs they hear are the product of human creativity or machine learning. Such transparency can also prevent exploitation where AI works are disguised as human creations to obtain royalties or unlawful commercial gain.

Some musicians and creators in the UK have already publicly expressed their concerns. For example, hundreds of British artists—including big names—have released silent albums in protest against proposed changes to copyright law that would allow AI companies to use their work without explicit permission.(Nath 2025) While AI-based tracking systems can identify uses of works, equitable royalty distribution models are still in development, with no certainty of being ready for public release. Direct attribution at the inference model level (e.g., "inference-time attribution") is still rarely implemented in the UK's commercial music industry, despite academic research proposing such frameworks. Without a granular attribution and compensation system, creators may not receive proportional compensation even if AI models use their work. While no AI royalty tracking system fully addresses all challenges, some mechanisms and practices are already in use in the UK, such as conditional registration with the PRS.

Creators who use AI assistance in the creative process can still register their works with the PRS, with a declaration that AI generated some of the composition or lyrics. Platforms like Blokur and Quantone provide data infrastructure that allows for more precise tracking of the relationship between recordings and compositions, supporting more accurate reporting and royalty payment systems. Coalitions like the Creative Rights in AI Coalition are pushing for dynamic licensing and transparency rules for AI companies. They also demand that AI training materials be recorded and labeled to ensure creators' rights are respected. The UK government has launched a consultation on copyright law and AI, involving the music industry, creators, and the public. UK Music and other creative coalitions are using the results of public surveys as the basis for their arguments against copyright exceptions that disadvantage creators.

AI-based royalty tracking in Australia involves identifying, analyzing, and recording the playback or use of musical works through automated algorithms. Collective management bodies such as APRA AMCOS play a central role in the implementation of this technology, as they are responsible for collecting and distributing royalties to creators, composers, and music publishers. In recent years, these bodies have begun exploring digital systems that utilize digital fingerprinting, audio recognition, and metadata mining to improve the accuracy of music playback reporting across streaming platforms, radio, television, and even public spaces like restaurants and shopping malls.

Audio fingerprinting technology is a key pillar of AI-based tracking systems. Through this mechanism, each song is converted into a unique digital representation that the system can automatically recognize, even when the song is played in less-than-ideal acoustic conditions or only partially. The AI-based tracking platform adopted by the Australian music industry is able to identify thousands of song plays in real time, providing more comprehensive data than manual systems that previously relied on sample reports from broadcasters or commercial music players. This improved capability has a direct impact on the accuracy of royalty distribution, as creators will receive compensation based on usage data that is closer to actual conditions.

In the digital streaming sector—now a significant source of income for many Australian musicians—AI is also being applied to organize and verify vast amounts of streaming data. Platforms like Spotify, Apple Music, and YouTube provide stream reports to copyright organizations, but AI is being used to analyze music consumption patterns, eliminate potential anomalies, and ensure that royalties are distributed according to registered copyright ownership structures. AI also helps detect streaming fraud, such as the use of bots to inflate streams, potentially harming other creators artificially.

In Australia, the integration of AI into the royalty tracking ecosystem is also closely linked to the transformation of copyright regulation. The legal framework, including the Copyright Act 1968, is under constant review to ensure the sustainability of copyright protection in the digital context.(Ihalainen 2018) While AI technology offers opportunities for increased efficiency, it also presents challenges in terms of algorithmic transparency and data protection.

Main issues in AI implementation is data bias (Zlatkov et al., 2023). AI systems used for music tracking require large and representative datasets. If training data is dominated by music from a particular genre or region, the algorithm may exhibit bias in identifying works—for example, music from Indigenous communities or independent genres may be less well-recognized. This has the potential to reduce the accuracy of royalty calculations for local musicians who lack global exposure. To address this, organizations and AI system developers in Australia are pushing for more inclusive dataset curation, integrating local music catalogs and works by musicians to ensure fair representation.

Beyond the technical challenges, the use of AI in royalty tracking also has socio-economic implications. The introduction of AI systems often requires significant investment in technological infrastructure and workforce training. This can be a barrier for smaller agencies or companies seeking to implement similar systems. However, in the long term, this technology has the potential to reduce operational costs, minimize administrative errors, and improve the fairness of revenue distribution for creators. Consequently, several collaborative initiatives have emerged to provide more equitable access to this technology across the Australian music sector. Ethically, AI raises questions about algorithmic accountability. AI-based tracking systems are often black boxes(Rai 2020), making it difficult for creators or copyright owners to understand how decisions are made.

Transparency is a crucial factor in building trust, especially when algorithmic decisions directly impact musicians' earnings. Therefore, the principle of explainable AI (XAI) has been adopted to provide more precise, more detailed, and more transparent explanations regarding the royalty identification and calculation process

(Dwivedi et al., 2023). This approach is seen as a crucial step in ensuring that AI is not only efficient but also accountable. In the future, AI in royalty tracking still has excellent potential to expand its functionality through integration with blockchain technology. Some discourse in Australia points to the possibility of using smart contracts to distribute royalties based on AI-derived playback data automatically. Such a system could increase transparency, reduce payment latency, and create a more decentralized copyright ecosystem. Although its implementation is still in the exploratory stage, the integration of AI and blockchain is seen as a significant opportunity in modernizing the Australian music industry.

Overall, the concept of AI-based royalty tracking for songs and music in Australia demonstrates that the country is already on the path to a more modern and efficient copyright management system. While numerous technical, ethical, and regulatory challenges remain to be overcome, the development of AI technology offers significant opportunities to improve fairness and economic sustainability for musicians and creators. Collaboration between governments, collective management bodies, technology companies, academics, and the creative community is key to ensuring that AI applications not only benefit the industry as a whole but also support fair and inclusive copyright principles in the evolving digital age.

4. PRINCIPLES OF TRANSPARENCY IN THE GOVERNANCE AND DISTRIBUTION OF SONG AND MUSIC ROYALTIES BY COLLECTIVE MANAGEMENT INSTITUTIONS

This mechanism is regulated in Article 2, paragraph (3) of the Copyright Law, which protects the creations of foreign citizens or legal entities when their works are used or published in Indonesia. In addition, Articles 87-90 of the Copyright Law regulate the duties of the National Collective Management Institution (LMKN) in managing the distribution of royalties, including receiving and distributing them to local LMKs and foreign LMK partners. Low legal awareness, minimal understanding, and ignorance of business actors regarding the obligation to pay royalties—as a form of respect for intellectual property rights—often give rise to legal issues. The case experienced by a franchise network in Bali is one example. Furthermore, the lack of clear parameters regarding the basis for collecting royalties, transparency of distribution, and categorization of public spaces considered "commercial" presents its own challenges.

The dominant position of LMK in determining royalty rates also has the potential to violate Law Number 5 of 1999 concerning the Prohibition of Monopolistic Practices and Unfair Business Competition (Anti-Monopoly Law). To provide more specific parameters for the imposition of royalties and commercial use criteria, Indonesia can refer to the application of the Communication to Public principle implemented in the European Union in the EU Infosoc Directive (2001/29/EC). This concept refers to any action that makes a work accessible to the public, including the playback of songs in hotels, restaurants, and other commercial venues. From several landmark cases, conclusions can be drawn regarding several measuring tools to ensure that the economic rights of copyright holders are not violated, by categorizing whether an action is a form of communication to the public by determining whether or not there is active communication. For example, in the case of SGAE v. Rafael Hotels, where a hotel that provides television in the room and plays broadcasts is considered to be carrying out active communication, not just providing tools, while the CJEU distinguishes between passive actors (who only provide technical means such as television or radio) and active actors who consciously and intentionally transmit works to the public.(Cole 2022)

Furthermore, regarding the aspect of public communication, in the case of Football Association Premier League (FAPL) v QC Leisure & Murphy v Media Protection Services (Roughton 2012), a pub used a foreign decoder to broadcast Premier League matches and reach a new audience not covered by the domestic broadcast license (Lindholm & Kaburakis, 2013). The CJEU provided a legal rule regarding the meaning of "public" in the "public sphere" namely, the audience group not included in the scope of the original license granted by the copyright holder.(Rendas 2018) It means that the broadcast or screening reached people not intended by the original license.

Whether the act is intended to generate profit is also an essential factor. In the landmark case of Società Consortile Fonografici (SCF) v. Marco Del Corso, concerning the playing of music in a dentist's office, the CJEU considered whether the playing was carried out in a business context or to obtain economic benefits, either directly (from admission/ticket fees) or indirectly (e.g., attracting more customers, improving the atmosphere, extending customer visits).(Borghi 2025) Therefore, as a measure to prevent copyright infringement through unauthorized retransmission, unauthorized audience expansion, and commercial use of works, several recommendations that the Indonesian Government can consider are adjusting the definition of commercial public spaces, taking into account the intention of playback and direct contribution to business profits, implementing a proportional tariff scheme based on the type of business, room size, and duration of song playback, increasing transparency and accountability in the distribution of royalties by LMKN and LMK, and improving legal education for businesses and the public about the importance of copyright protection. These are in line with the relief options provided by the Government through the Directorate General of Intellectual Property (DJKI), which includes royalty exemptions based on the size of the business space, visitor capacity, and the level of music utilization in daily activities.

Based on balanced regulatory reforms and transparent implementation, Indonesia is expected to create a fair ecosystem where songwriters receive the rewards they deserve, while businesses do not feel disproportionately burdened. Ultimately, copyright protection is not simply a legal obligation, but a form of respect for the creative work that underpins the cultural industry. One fundamental issue is the overlapping authority between the LMK and LMKN. Although both are legally regulated by law, their functions and duties overlap. Law Number 28 of 2014 concerning Copyright, specifically Articles 87 to 93, stipulates that the LMK has the authority to collect, distribute, and collect royalties on behalf of creators, copyright holders, and owners of related rights.

Meanwhile, Government Regulation Number 56 of 2021 concerning the Management of Song and/or Music Copyright Royalties, through Articles 18 to 20, establishes the LMKN with a similar mandate. The LMKN is then divided into two: the LMKN Creator and the LMKN Related Rights Owner, apparently intended to clarify their respective areas of work. However, in practice, this division creates uncertainty and confusion, both among creators and users of works. When two different institutions have nearly identical authority over the collection and distribution of royalties, this not only risks creating a conflict of authority but also obscures legal guarantees for parties who should receive protection.

The fundamental issue is overlapping authority between the LMK and LMKN. Although both are legally regulated by law, their functions and duties overlap. Law Number 28 of 2014 concerning Copyright, specifically Articles 87 to 93, stipulates that the LMK has the authority to collect, distribute, and collect royalties on behalf of creators, copyright holders, and owners of related rights. Meanwhile, Government Regulation Number 56 of 2021 concerning the Management of Song and/or Music Copyright Royalties, through Articles 18 to 20, establishes the LMKN with a similar mandate. The LMKN was then divide it into two; the Creator LMKN and the Related Rights Owner LMKN, which was apparently intended to clarify their respective areas of work. However, in practice, this division has instead created uncertainty and confusion, both among creators and users of works. When two different institutions have nearly identical authority in collecting and distributing royalties, this not only risks creating a conflict of authority but also obscures legal guarantees for parties who should receive protection.

The unclear functional boundaries between LMK and LMKN indicate that institutional coordination within the legal structure for royalty management still needs to be strengthened. In this context, the legal structure is not only determined by the existence of institutions, but also by the clarity of roles, the effectiveness of task implementation, and the legal certainty provided to creators, rights holders, and users of works. Without more stringent arrangements, this legal structure risks failing to meet the need for a transparent, fair, and reliable royalty protection system.

5. CONCLUSION

The management and distribution of song and music royalties within Indonesia's commercial public sphere continue to encounter structural, technical, and regulatory obstacles that impede the establishment of a fair and transparent system. Despite the explicit mandate for royalty payments outlined in the Copyright Law and Government Regulation 56/2021, implementation remains constrained by overlapping authority between the Music and Music Production Institutions (LMK) and the Music and Music Production Institution (LMKN), an underdeveloped national database, limited tariff transparency, and low accuracy in music usage recording systems. These challenges have led to resistance among business stakeholders and have shaped public perceptions of royalty policies. Comparative analysis with the United States, the United Kingdom, and Australia demonstrates that integrating artificial intelligence (AI) technologies, such as audio fingerprinting, automated content recognition, and big data matching, can enhance the accuracy of music playback identification, mitigate streaming data manipulation, and accelerate proportional royalty distribution. Nevertheless, the adoption of AI introduces several ethical concerns, including data bias, algorithmic transparency, privacy, and unequal access to technology, necessitating the development of comprehensive ethical and governance frameworks. To establish an equitable royalty ecosystem, Indonesia should prioritize regulatory harmonization, clarify the division of authority between LMK and LMKN, develop integrated data infrastructures (PDLM and SILM), and ensure transparency in royalty collection and distribution processes. The advancement of AI systems must focus on improving efficiency while safeguarding the rights of creators and the interests of business stakeholders. Through regulatory reform and sufficient technological preparedness, Indonesia can develop a royalty system that is adaptive, accountable, and capable of supporting the growth of the national music industry in the digital era.

REFERENCES

Baruah, S., & Saikia, A. P. (2023). Licensing standard-essential patents in artificial intelligence-based apps: A theory on dynamic royalty pricing mechanism. *The Journal of World Intellectual Property*, *26*(3), 490–502. DOI: 10.1111/jwip.12278

Borghi, M. (2025). The universal nature of performers' rights under EU law (a note on Case C-265/19, *Recorded Artists Actors Performers v Phonographic Performance Ireland*). SSRN *Electronic Journal*. Advance online publication. DOI: 10.2139/ssrn.5100687

Cole, M. D. (2022). Retransmission of broadcast signals by cable in hotels: An analysis of the EU CabSat-Directive in light of a pending CJEU case. SSRN *Electronic Journal*. Advance online publication. DOI: 10.2139/ssrn.4045527

Deng, J., Zhang, S., & Ma, J. (2023). Computational copyright: Towards a royalty model for music generative AI (Version 4). *arXiv*. DOI: 10.48550/ARXIV.2312.06646

Drott, E. (2021). Copyright, compensation, and commons in the music AI industry. *Creative Industries Journal*, *14*(2), 190–207. DOI: 10.1080/17510694.2020.1839702

Dwivedi, R., Dave, D., Naik, H., Singhal, S., Omer, R., Patel, P., Qian, B., Wen, Z., Shah, T., Morgan, G., & Ranjan, R. (2023). Explainable AI (XAI): Core ideas, techniques, and solutions. *ACM Computing Surveys*, *55*(9), 1–33. DOI: 10.1145/3561048

Febrian, M. N., Setyawati, R., Noventri, A., & Mahendra, Y. (2025). Review of legal certainty of royalty collection on music and/or songs: A comparative study of Indonesia and Russia. *Realism: Law Review*, *3*(1), 115–143. DOI: 10.71250/rlr.v3i1.57

Goldstuck, C. (2025). Past precedent, future proof: Toward a new legal and commercial framework for AI-generated music. SSRN *Electronic Journal*. DOI: 10.2139/ssrn.5393574

Hallur, G. G., Prabhu, S., & Aslekar, A. (2021). Entertainment in era of AI, big data & IoT. In S. Das & S. Gochhait (Eds.), *Digital entertainment*. Springer Nature Singapore. DOI: 10.1007/978-981-15-9724-4_5

Hesmondhalgh, D., Osborne, R., Sun, H., & Barr, K. (2021). Music creators' earnings in the digital era. SSRN *Electronic Journal*. Advance online publication. DOI: 10.2139/ssrn.4089749

Hubbs, G. (2016). Digital music and public goods. In R. Purcell & R. Randall (Eds.), *21st century perspectives on music, technology, and culture*. Palgrave Macmillan. DOI: 10.1057/9781137497604_9

Ihalainen, J. (2018). Computer creativity: Artificial intelligence and copyright. *Journal of Intellectual Property Law & Practice*, *13*(9), 724–728. DOI: 10.1093/jiplp/jpy031

Jacques, S., & Flynn, M. (2024). Protecting human creativity in AI-generated music with the introduction of an AI-royalty fund. *GRUR International*, *73*(12), 1137–1149. DOI: 10.1093/grurint/ikae134

Jensen, F. J. (2024). Rethinking royalties: Alternative payment systems on music streaming platforms. *Journal of Cultural Economics*, *48*(3), 439–462. DOI: 10.1007/s10824-024-09507-z

Kaye, D. B. V. (2025). Renting royalties: How the assetization of music copyrights contributes to inequality for musicians. *Media Culture & Society*, *47*(7), 1394–1412. DOI: 10.1177/01634437251341715 PMID: 40978152

Koempel, F. (2025). Future of the music industry—Past and present, learned the lesson? SSRN *Electronic Journal*. Advance online publication. DOI: 10.2139/ssrn.5049110

Lindholm, J., & Kaburakis, A. (2013). Cases C-403/08 and C-429/08 FA Premier League Ltd and others v QC Leisure and others; and Karen Murphy v Media Protection Services Ltd (4 Oct 2011). In J. Anderson (Ed.), *Leading cases in sports law*. T. M. C. Asser Press. DOI: 10.1007/978-90-6704-909-2_17

Lobato, R. (2025). Automated content recognition (ACR), smart TVs, and ad-tech infrastructure. *Convergence (London)*, *31*(6), 1799–1814. Advance online publication. DOI: 10.1177/13548565251327885

Mangal, N. (2025). Music contracts in the streaming age: Law and policy in the European Union and the United Kingdom. *GRUR International*, *74*(7), 700–702. DOI: 10.1093/grurint/ikaf057

Morrow, G. (2025). Australia's performing rights organisation: Incentives, the agency problem and MetaGen. *The International Communication Gazette*, *87*(4), 291–306. DOI: 10.1177/17480485251327468

Nath, S. (2025). We rise to resist: The crisis of relevance and the movement against generative AI in creative industries. SSRN *Electronic Journal*. DOI: 10.2139/ssrn.5295103

O'Dair, M., & Owen, R. (2019). Financing new creative enterprise through blockchain technology: Opportunities and policy implications. *Strategic Change*, *28*(1), 9–17. DOI: 10.1002/jsc.2242

Quintais, J. P., & Schwemer, S. F. (2022). The interplay between the Digital Services Act and sector regulation: How special is copyright? *European Journal of Risk Regulation*, *13*(2), 191–217. DOI: 10.1017/err.2022.1

Rai, A. (2020). Explainable AI: From black box to glass box. *Journal of the Academy of Marketing Science*, *48*(1), 137–141. DOI: 10.1007/s11747-019-00710-5

Rai, P., Gupta, K., Gabralla, L. A., & Choudhury, T. (2025). A royalty framework for copyright protection and accountability in AI-generated art. *Discover Computing*, *28*(1), 252. DOI: 10.1007/s10791-025-09766-0

Rendas, T. (2018). Copyright, technology and the CJEU: An empirical study. *IIC - International Review of Intellectual Property and Competition Law*, *49*(2), 153–184. DOI: 10.1007/s40319-017-0664-0

Roberts, R., & Whiting, S. (2025). "I'd have to be crazy if I did it strictly on a financial basis": Australian regional music venues, burnout, and precarious music ecologies. *Popular Music and Society*, *48*(2), 194–211. DOI: 10.1080/03007766.2025.2455302

Rogers, J., & Sparviero, S. (2025). Navigating performing rights in music: Digital-native organizations, changing values, and industry shifts in the United States and beyond. *The International Communication Gazette*, *87*(4), 376–397. DOI: 10.1177/17480485251327273

Roughton, A. W. (2012). Football in pubs: Court of Justice of the European Union (Grand Chamber), *Football Association Premier League Ltd and others v QC Leisure and others (C-403/08)* and *Karen Murphy v Media Protection Services Ltd (C-429/08). Queen Mary Journal of Intellectual Property*, *2*(3), 288–296. DOI: 10.4337/qmjip.2012.03.06

Sari Angraeni, B., & Majid, P. S. (2025). *Legal protection of music copyrights in public commercial spaces: A case study of coffee shops in Kolaka Utara Regency*. Constitutional Law Review., DOI: 10.30863/clr.v4i2.5950

Shackel, L., & Smith, M. (2025). Australian privacy law, extraterritoriality, and regulating data collected by offshore direct-to-consumer genetic testing companies. *International Data Privacy Law*. Advance online publication. DOI: 10.1093/idpl/ipaf016

Whiting, S., Emery, S., Oestmann, J., & Martin, G. (2024). *Musical automation in the Australian games industry: Current perspectives*. RMIT University. DOI: 10.25439/RMT.27289779.V1

Zlatkov, D., Ens, J., & Pasquier, P. (2023). Searching for human bias against AI-composed music. In Johnson, C., Rodríguez-Fernández, N., & Rebelo, S. M. (Eds.), *Artificial intelligence in music, sound, art and design* (Vol. 13988). Springer Nature Switzerland., DOI: 10.1007/978-3-031-29956-8_20

Chapter 12
Music Industry Disruption:
Comparing Traditional and Popular Music in the Artificial Intelligence Era

Mieke Yustia Ayu Ratna Sari
http://orcid.org/0000-0002-5743-4243
State University of Surabaya, Indonesia

Heppy Hyma Puspytasari
State University of Surabaya, Indonesia

Alfan Dzikria Nurrachman
State University of Surabaya, Indonesia

ABSTRACT

This chapter analyzes how artificial intelligence (AI) reshapes the structures, practices, and values of the music industry, focusing on traditional and popular music. It highlights shifting creative roles, changing production dynamics, ethical and regulatory concerns, and AI's impact on preserving and transforming cultural heritage. The discussion outlines historical shifts from the pre-AI era to the digital age, addressing algorithmic processes, opportunities and risks in creation and distribution, and the implications of AI for communal intellectual property in traditional music. It also examines how AI influences the economics of popular music through streaming models, automated production, and recommendation systems, while noting risks such as data colonialism, cultural loss, and blurred authenticity.

DOI: 10.4018/979-8-3373-6279-3.ch012

The chapter concludes that AI offers major potential for innovation and preservation but also poses challenges to community rights, creative economy structures, and fair value distribution.

INTRODUCTION

The late 20th century ushered in the digital era, driven by the rapid development of computers and internet technology. With the arrival of the World Wide Web in the 1980s and 1990s, the internet became a truly global and accessible platform. The early 2000s saw mobile phones evolve dramatically, paving the way for social media and transforming how we communicate and access information. From 2010 onward, technology took another leap forward with the rise of artificial intelligence (AI). While the roots of AI stretch back to 1956, its progress slowed in the 1970s and 1980s due to unrealistic expectations. However, the 1990s and 2010s marked a turning point, as dedicated scientists developed more advanced algorithms and capabilities for AI. These breakthroughs have transformed technology in ways that benefit people across the globe. Today, AI influences healthcare, education, commerce, transportation, tourism, the arts, and many other fields. Like any powerful tool, AI comes with both exciting opportunities and important challenges. In the past decade, AI has especially reshaped the music industry. Where creativity was once the sole domain of human intuition, algorithms and generative models now work alongside musicians. Machine-based systems affect not just the composition and production of music, but also how content is distributed and curated on digital platforms. This new landscape shows that AI is a major player in the music world, capable of replicating some aspects of human creativity and creating new economic possibilities (Jacques & Flynn, 2024). As AI becomes more integral, it's important to find the right balance so it enhances, rather than disrupts, the spirit of music.

Traditional and popular music each respond differently to the changes brought by AI. Traditional music thrives in local communities, passed down through generations by word of mouth and shared experiences. Popular music, by contrast, is often created for global audiences and distributed widely through digital platforms. Experts in cultural preservation highlight the importance of documenting and digitizing musical traditions—including building ethnomusicological datasets—to protect our shared heritage, especially as fewer people continue these traditions (Chen et al., 2024). When it comes to traditional music, AI's main role is to help with preservation, documentation, and sharing, rather than creating new works. This type of music is a collective treasure, reflecting the values and spirit of its community. In contrast, AI can play a much bigger part in popular and contemporary music, helping with everything from composing songs to distributing them worldwide. Still, it's important to remember that music is all about emotion, inspiration, and personal expression. While AI can analyze vast amounts of data to support the music-making process, it doesn't experience feelings the way humans do—it simply follows the instructions it's given.

Advances in artificial intelligence (AI) technology have brought fundamental changes in music production, distribution, and curation. Generative AI, data-driven recommendation systems, and machine learning techniques have become part of the modern music industry workflow, influencing how musical works are composed, recorded, and published. One tangible impact is that the performance of algorithm-based recommendation and promotion systems determines the visibility of works and the distribution of revenue among stakeholders. Recent research emphasizes that this transformation is not only technical but also normative, concerning copyright and the role of humans in music creation (Jacques & Flynn, 2024). In this regard, AI technology is a double-edged sword; if not utilized properly, it can have negative impacts, especially on the future development of music.

Artificial intelligence (AI), which refers to computer systems that can perform tasks typically requiring human intelligence, is changing the parameters of monetization in the popular music industry, having far-reaching impacts on the economic structure and global creative ethics. Streaming—defined as the digital delivery of music through the internet without ownership transfer—has become the dominant contributor to recorded music revenues in many countries, replacing physical sales and digital downloads as the industry's primary source of profit. The advent of AI adds new layers of complexity. For example, revenue distribution models and royalty mechanisms must adapt to works partially or entirely generated by AI and consider how human contributions can be recognized. Generative AI technology—AI that can autonomously create new content—adds a significant new layer of complexity. AI-based systems are now capable of generating musical compositions, mimicking artists' vocal styles, and creating arrangements that resemble human work. Debates over fairness in revenue distribution, algorithm transparency, and creator affiliation are also growing (Gallar, 2025).

In a social and cultural context, the use of AI in the music industry is not only concerned with economic efficiency but also with ethical and human values. Public debate is now emerging about the extent to which technology can take over the creative process without eroding the meaning of human expression. Issues of distributive justice, moral rights, and the protection of traditional forms of cultural expression also arise, particularly when AI systems have the potential to exploit local cultural materials without fair compensation. Therefore, the sustainability of the music industry in the era of artificial intelligence demands a multidisciplinary approach involving collaboration between music economists, policymakers, copyright law experts, and the creative community so that innovation can proceed hand in hand with respect for the dignity and well-being of creators. This requires in-depth reflection on how AI systems not only produce music as a commodity but also recognize and maintain the web of cultural meanings surrounding local identities, collective experiences, and creative rituals as an integral part of humanity's heritage. Therefore, technology policies and practices must be designed with sensitivity to marginalized voices, so that technological innovation does not become an agent of cultural homogenization but rather an inclusive platform that enriches the plurality of human expression (Canyakan, 2024).

This chapter will systematically examine how AI affects traditional and popular music from technical, cultural, and regulatory perspectives. The use of AI in traditional music offers a means of documentation and revitalization that can expand access and documentation of local heritage, but also raises questions about shifts in authenticity, control of traditional communities, and the potential marginalization of non-commercial practices when digital market norms regulate visibility. The article focuses on the shifting roles of music industry players (musicians, producers, labels), the aesthetic changes resulting from the adoption of AI, the implications for public policy and cultural heritage protection, and the future prospects and challenges within the music industry.

MUSIC INDUSTRY BEFORE THE ARTIFICIAL INTELLIGENCE ERA

The conventional (pre-AI) music industry can be understood as an integrated ecosystem linking the creation, production, distribution, and consumption of music. Conceptually, this ecosystem is not simply a series of technical activities such as songwriting, recording, and marketing, but rather a socio-economic relationship between creators, capital owners (record labels), publishers, managers, and audiences that provides symbolic and commercial value to works (Wikstrom, 2022). This relationship places curatorial roles and institutional investment as the primary determinants of who gains visibility and revenue; these practices reflect a hierarchical division of labor and historical power relations, so that access to resources (studios, promotional capital, distribution networks) remains a determining factor in the commercial success of musical works (Pizzolitto, 2024).

Institutionally, the market structure of the conventional music industry is characterized by the concentration of economic power in a few large entities, as well as the existence of collective mechanisms for administering rights and remuneration. The dominance of digital platforms and algorithmic curation has shifted the center of value from exposure and airplay to the primary determinants of revenue streams, making royalty governance, transparency of airplay measurement, and legal certainty central to the survival of the creative profession (Wikström, 2020). In the Indonesian context, studies on copyright protection and the role of music aggregators highlight the challenges of regulatory implementation and the need to strengthen the capacity of Collective Management Institutions (CMOs) to ensure a fairer distribution of benefits for local creators (Tullah et al., 2024). This structure requires discussions beyond technical and economic considerations, but considers the social and existential values of creators, namely, the appreciation of creative work, recognition, and protection within a fixed change industry architecture (Gumuchian, 2025).

The conventional music industry structure is characterized by the dominance of large companies (major labels) that control most of the value chain, including production, distribution, and promotion, while independent labels and emerging artists are in weaker positions. Their significant financial scale and access to global distribution networks allow major labels to have a significant competitive advantage, particularly in funding recording production, music videos, large-scale marketing, and access to popular playlists on streaming platforms. This oligopoly structure also strengthens the bargaining position of major labels in contract negotiations with artists and in determining royalty percentages and ownership of master recording rights. Within this structure, several specific, interrelated functions are identified: (1) A&R (artist & repertoire), which is responsible for discovering and developing talent; (2) recording production (studio recording, mixing, and mastering); (3) marketing and promotion; (4) physical and digital distribution; and (5) copyright and royalty management (publishing and masters). Major labels typically control almost all of these segments internally or through their affiliate networks, such as marketing companies, creative agencies, and even touring agencies. Independent labels often need to form partnerships (e.g., with digital distributors or majors) to achieve broader promotional and distribution reach (Janowska, 2023).

Digitalization and streaming platforms have opened up broader access for independent artists in music production and publishing, yet conventional structures remain heavily influenced by the need for substantial investment in promotion and artist development. A report by the International Federation of the Phonographic Industry (IFPI) states that global label investment in A&R and marketing is astronomical (in the billions of dollars), with only a small percentage of artists breaking even (Rechardt, 2025). Ownership of recording/master rights then control of back catalogs, provide significant sources of passive income and leverage for major labels, maintaining a structural position that is difficult for independent forces to fully displace (Janowska, 2023).

Musicians occupy a central position as creative voices and cultural workers, combining artistic, managerial, and entrepreneurial skills. Beyond their aesthetic roles, contemporary musicians often serve as curators of personal brand, managers of independent distribution, and strategic decision-makers regarding collaborations and marketing. In this regard, there is a shift reflecting the demands of the digital economy for revenue diversification and visibility (Polak & Schaap, 2025). Record labels (both major and independent) serve as institutional infrastructure, providing capital, distribution networks, and marketing capacity that often determine the commercial reach of a music release. Large companies maintain structural advantages through large-scale investments in A&R, promotion, and catalog management. These factors, in turn, strengthen their bargaining position in contract negotiations and royalty distribution. Although digitalization has opened up access for independent labels, the major label oligopoly remains influential in setting the agenda for the global music market (IFPI, 2024).

Music producers play a dual role: as technical facilitators, translating artists' ideas into recordings, and as creative collaborators, often influencing the final aesthetic of the work. The transformation of production technology, from large studios to affordable software, has democratized technical capabilities, but also presents new challenges: managing digital market expectations, understanding distribution matrices, and navigating the roles of rights administrators and monetization. Recent studies of producers, particularly within specific genres, demonstrate how they respond to platform gatekeeping (digital platform communities) while expanding their skill sets to survive in the digital music economy (Ng & Gamble, 2024). The public (including music lovers) acts as verifiers of meaning, acceptance, and participation spaces that give music symbolic and social value. Public participation is not merely consumptive but also shapes distribution channels and supports the career sustainability of artists who may not receive significant institutional support. In a national context like Indonesia, community networks and local practices serve as crucial mechanisms for negotiating economic and cultural value amidst global platform pressures (Naylor et al., 2021).

The interactions between these actors are interdependent but unequal. Labels need artists' talent and credibility, artists need access to capital and promotion, producers need creative space and fair compensation, and the community provides legitimacy. The shift toward streaming and platform models has altered revenue streams, increasing reliance on subscriptions and streaming metrics. This requires actors to renegotiate the distribution of economic value (royalties, licensing, and tour revenues) to ensure fairer and more equitable distribution. In the pre-AI context, traditional music production was characterized by a collective orientation, emphasizing that music was not simply an individual commodity but a social practice interwoven with rituals, ceremonies, and community activities. Music makers and performers often worked within kinship networks or traditional communities, so the music production process focused on maintaining cultural identity and social cohesion within the community, rather than on mass-market logic. Ethnomusicology and digital studies suggest that technological change has shifted the way music is consumed, becoming a commodity, before the digital era, music was part of a community's collective practices (Cross, 2023). Music at that time was used for traditional ceremonies or religious rituals.

Traditional music transmission and production patterns were generally oral, with playing techniques, rhythmic variations, and symbolic meanings passed down through participatory observation, collective practice, and repetition within the context of everyday life. Reliance on collective memory and hands-on practice enabled local variations to persist, with each community adapting its musical material to its environment, ritual purposes, and available instruments. Hadar and Rabinowitch's research highlights that music transmitted strictly orally or through notation differs in its social dynamics from improvised music (Hadar & Rabinowitch, 2023). The materiality and ecology of production also characterize instrument making, with the selection of materials and the skills of local makers integral to the traditional music production chain. Production processes are often tied to seasons, harvest rituals, or the availability of raw materials (e.g., wood, bamboo, leather), allowing instrument makers to play a dual role as custodians of ecological and aesthetic knowledge. The authenticity of an instrument serves as a benchmark for a community's musical authenticity; this material aspect reinforces aesthetic ties and local identity, aspects that are difficult to replicate industrially. This is the mainstay of traditional music.

Pre-AI economic production patterns tended to be sustainable and closely tied to local patronage systems in the form of wages, exchange for services, and ritual support from traditional leaders or community institutions. Music was produced primarily for social or religious purposes, and its conversion to a commercial market typically occurred during bazaars or festivals. Kanhov's study (Kanhov et al., 2024) shows that when the digital music industry model emerged, traditional communities faced the risk of data colonialism and a shift in local production values to data and commodity value. This shift in digitalization has advantages, including broader dissemination patterns, not limited to specific groups. Storage patterns no longer rely solely on human memory but are digitally inventoried, thus being more organized so that future generations can continue to recognize traditional music inherited from their ancestors.

In the era before the use of artificial intelligence technology, popular music production was characterized by a relatively institutionalized division of labor: record labels, A&R (artist & repertoire), record producers, and distribution networks acted as gatekeepers coordinating the creation, financing, and marketing of songs. This structure creates a decision-making process that emphasizes commercial selection and market optimization, prioritizing singles that are easily curated by radio stations or chart placement. Production orientation is often geared toward platform logic and the mass market (Geurts & Cepa, 2023). Technically, popular production patterns in the pre-AI era developed around the professional studio as the central tool.

Although digitalization has lowered production costs, professional practices such as multitrack recording, mixing, mastering, and the use of session musicians still determine the quality of the final product. The shift to professional studio-based production changed the required competencies, requiring producers and engineers to master complex software and market practices. Technical skills became essential for participation in the popular music production ecosystem (Kjus, 2024). This became a parameter for professionalism in the music production field.

Economically, funding and remuneration models in the pre-AI era were still heavily influenced by label contracts, traditional royalty mechanisms, and the organization of revenue through sync sessions, touring, and physical and digital sales in simple formats like compact discs. This structure placed value on long-term copyright and master ownership. The shift to streaming began to change this pattern by shifting revenue to streaming. In the pre-AI era, debates surrounding revenue sharing within the music industry had already emerged as a central issue in the economic governance of popular music production (Moreau et al., 2024). Organizationally, the pre-AI era featured a combination of centralization and do-it-yourself (DIY) processes. While major labels continued to control the primary commercial channels, the rise of cheaper production tools allowed independent producers and DIY artists to produce quality recordings, but access to mainstream markets remained largely dependent on the ability to navigate promotional platforms and traditional industry gatekeeping. Ethnographic studies of popular genre producers confirm that production work often involved a diversification of tasks from music creation to digital marketing, which were not previously core to the role of analog-era producers (Ng & Gamble, 2024).

The aesthetic aspects of popular production before AI were also influenced by the logic of media formats—song duration, catchy verse-chorus structures, and sound design suitable for radio broadcasts or repeated playback on digital playlists—all of which were the result of a compromise between artistic creativity and commercial distribution needs. Furthermore, popular music production often employed creative teams, including songwriters, producers, engineers, and programmers, as a collaborative ecosystem that deliberately designed the sonority (resonance quality) of pop to suit the expectations of mass audiences and platform curation mechanisms (Geurts & Cepa, 2023). Transnational and local patterns intersect in digital distribution, and market globalization allow popular music to replicate across borders, yet local production continues to influence tastes and production strategies. In Indonesia, research on the adoption of digital music services and user behavior indicates that local producers must adapt production practices and marketing strategies to platform dynamics. This pressure has been present for the past decade and has become increasingly evident in the AI era (Husin et al., 2021). The long journey of the music industry in the pre-AI era has not been without challenges, given the involvement of many parties with divergent interests seeking to profit. Arranging and distributing music from start to finish requires collaboration with many parties in a professional manner, so the quality of the music is maintained.

DIGITAL TRANSFORMATION AND THE EMERGENCE OF ARTIFICIAL INTELLIGENT IN THE MUSIC INDUSTRY

Artificial Intelligence (AI) is a branch of computer science that focuses on developing systems capable of mimicking human intelligence in the processes of thinking, learning, and creating. This technology, pioneered by John McCarthy and refined by Allan Turing, was created to assist and simplify human activities. In the context of the music industry, AI functions as a collaborative entity that assists humans in creating, analyzing, and producing musical works with high efficiency without eliminating human creativity. AI is used not only as a technical tool but also as a "creative partner" that can expand the boundaries of musical imagination through deep learning and sound pattern analysis (Gioti, 2021). The accuracy of this technological intelligence still needs to be clarified by humans, as humans are in control, not the other way around.

The development of digital technology has brought significant changes to the creative industries, including music. Digital transformation through AI has enabled the emergence of automated and interactive composition systems that produce original works based on algorithms and artificial neural networks. AI in music is now capable of imitating the styles of renowned composers, creating complex harmonies, and adaptively modifying melodies to suit listener preferences (ŞUTEU & RUCSANDA, 2025). AI is not only changing the music production process but also the way people interact with and consume art.

THE IMPACT OF AI ON TRADITIONAL AND POPULAR MUSIC

The development of AI has changed the paradigm in almost all creative fields, including music. In recent years, AI has evolved not only into a technical tool but also into a creative agent capable of producing musical compositions independently or collaboratively. This transformation has sparked a debate between the value of human authenticity and technological efficiency, especially in the context of traditional music, which is rich in cultural meaning, and popular music, which is commercial in nature (Gioti, 2021). Beyond its function as a tool, AI now plays a collaborative role by offering musical harmonization, mapping patterns, and adjusting aesthetic parameters together with human creators who previously dominated this field. The involvement of AI in the creation of musical works can significantly enhance the creative performance of music practitioners. This is influenced by a musician's formal education and adequate digital experience, which in turn impacts the acceptance and use of AI in musical practice. In principle, AI functions as a catalyst for exploring musical ideas, expands the space for experimentation, and facilitates the creation of compositions that were previously difficult to achieve manually (Ma et al., 2025). The algorithms, models, and creative approaches employed by AI expand the boundaries of human creativity through human-machine collaboration, resulting in more adaptive, efficient, and innovative musical compositions without eliminating the expressive role of humans (Civit et al., 2022).

AI has facilitated the emergence of automated composition systems such as the AIVA and MuseNet platforms, which use deep learning algorithms to mimic classical and modern musical styles. Studies show that these systems are capable of producing works of near-human composer quality, although emotional and cultural interpretation aspects remain a significant challenge (Gîrbu, 2024). The integration of AI-based composition systems such as AIVA and MuseNet not only expands the potential for creativity but also changes the paradigm of collaboration between humans and machines in music creation. This technology enables the combination of musical data across genres and cultures, producing complex harmonic variations that are adaptive to listener preferences (Vengathattil, 2025). However, ethical and aesthetic challenges related to ownership of works and the value of human authenticity still require further study (Canyakan, 2024).

The presence of artificial intelligence in the music world marks a new era of digital transformation that not only affects how music is created but also how it is understood and passed on. In the context of traditional music, AI presents new opportunities for cultural preservation through the digitization and reconstruction of the distinctive sounds of traditional instruments that are on the verge of extinction. Girbu (2024) asserts that although algorithms such as those owned by Aiva Technologies are capable of processing music theory and producing original works, the expressive nuances of traditional musical instruments remain difficult for machines to replicate perfectly, due to their nature, which is laden with human emotional and spiritual meaning (Gîrbu, 2024). AI technology has been used in practice as a means to summarize, archive, and map traditional music, including through datasets dedicated to ethnic music or through music education that integrates traditional elements with digital tools. For example, a study that built a Manchu music dataset showed that AI can help classify, generate, and recognize emotions in ethnic music with a significant degree of accuracy, while still confirming that cultural context and meaning cannot be translated solely from sound (Ma et al., 2025). In this context, AI also plays a vital role in education and cultural transmission. Yin and Guo (2024) conducted quasi-experimental research in China, demonstrating that the use of the AIVA application in traditional music learning enhanced students' ability to understand music theory and practice more efficiently than conventional methods. However, the researchers emphasized the need for direct human interaction to preserve the social and cultural meaning of traditional music (Yin & Guo, 2025). AI, as an algorithm capable of imitating the formal structure of music and possessing various other advantages that enable it to capture the depth of human experience, including emotional, spiritual, and ritual aspects, has limitations in exploring traditional music that is influenced by religious and emotional aspects, which machines do not possess.

In the realm of popular music, the impact of AI is more progressive and disruptive. This technology is not only used in automatic music composition, but also in music production, distribution, and marketing. Lin and Chen (2024) found that AI-generated music has changed the structure of the music industry by introducing an algorithmic approach to song creation, making the creative process faster and more economical, although it invites debate about the originality and authenticity of the work (Lin & Chen, 2024). The creative process, which was originally time-intensive and required specialized studios and personnel, can now be reduced through the support of generative models, automated mastering systems, and algorithm-based distribution. One study mentions that AI technology in the mixing and music production workflow has facilitated novice and amateur creators, although professionals still demand complete creative control and advanced customization (Vanka et al., 2023). AI has expanded the creative and commercial spectrum in the popular music industry by offering efficiency, inclusivity, and innovation while raising ethical and aesthetic dilemmas. Moura and Maw (2021) found that public perception of AI-composed music remains largely negative, with many listeners and music professionals judging AI works to lack "soul" and credibility compared to human works. This suggests that the social acceptance of AI music remains limited, particularly in popular genres that heavily rely on emotion and personal identity (Tigre Moura & Maw, 2021). Technology may replace some aspects of the musical production process. However, the space for meaning, sensory experience, and narrative still requires the role of humans as drivers, interpreters, and meaning-makers. In this case, stakeholders (musicians, producers, record labels, regulators) need to balance the technical potential of AI with the preservation of human creative values, in an effort to ensure that digital transformation does not erode the diversity of musical expression and the rights of creators inherent in the process of song creation and distribution.

AI also influences the performative dimension of traditional and popular music through sound pattern analysis and real-time improvisation. Miranda and Williams (2015) note that AI systems are now capable of participating in interactive electroacoustic composition and performance, expanding the boundaries between creator and listener and opening up new opportunities for cross-cultural collaboration (Miranda & Williams, 2015). In this collaborative perspective, the application of AI in musical performance, whether traditional or popular, is not merely the automation of sound or improvisation, but rather the creation of a shared space where different traditions, musical instruments, and artistic expressions from different cultures meet and transform each other. For example, a real-time system processes the movements of the player or the beat of traditional instruments. The AI responds by generating melodic or rhythmic elements derived from the musical heritage of other cultures, creating a dialogue between human players, their cultural instruments, and "machine partners" that bring musical references from different parts of the world. A more specific example appears in a project that promotes "AI-mediated cross-cultural co-creation," in which an AI model is trained on traditional Thai music styles, then presented with prompts from modern Western music creators and produces compositions that combine elements of both traditions, a tangible demonstration of how technology can serve as an aesthetic bridge between cultures.

THE IMPLICATIONS OF AI FOR TRADITIONAL MUSIC IN THE DIMENSION OF COMMUNAL INTELLECTUAL PROPERTY

Artificial intelligence (AI) has brought profound changes to the way humans create, record, and distribute musical works. However, for traditional music rooted in communal knowledge, serious questions arise about ownership, authenticity, and intellectual property rights. In this context, traditional music is not merely an individual work, but a collective representation of the values, rituals, and history of a community. Yin and Guo emphasize that the use of AI in reconstructing or teaching traditional music can enhance preservation, but also poses the risk of commodifying cultural identity without adequate regulation (Yin & Guo, 2025). The primary issue in the relationship between AI and communal intellectual property is the attribution of creation. In traditional music, the creator is often unknown or part of a large community. When AI utilizes traditional music data as a basis for algorithmic training to generate new compositions, ambiguity arises regarding who owns the copyright to the resulting compositions. Girbu (2024) notes that AI, such as Aiva Technologies, has been able to generate "new" compositions from ethnic music patterns, but these works are often disconnected from their original social context and spiritual meaning (Gîrbu, 2024).

From a legal perspective, the framework of communal intellectual property protection, oriented towards collective ownership, poses a significant challenge for traditional communities whose musical works are exploited by AI systems without their permission. Research by Chen (2021) shows that the use of artificial intelligence technology in electronic composition is often not accompanied by recognition of cultural resources, thus opening the door to cultural exploitation without fair compensation (Y. Chen, 2021). This phenomenon raises concerns about "digital colonialism," in which AI algorithms utilize global cultural data to create entertainment products without considering the moral rights of the communities of origin. Berkowitz (2024) warns that the emergence of "machine folk" (algorithm-generated folk music) has the potential to blur the line between cultural heritage and digital innovation, placing minority cultures in a vulnerable position to technological appropriation (Berkowitz, 2024).

When considering the origins of traditional music, the concerns outlined above do not apply, as the incorporation of technology in the conventional music creation obscures its true origins. Traditional music is born from collective musical practices that occur within the socio-cultural framework of local communities, often without written documentation and passed down through oral or aural transmission from generation to generation. Traditional music is a form of intangible heritage. It records the experiences of communities through rituals, collective work, ceremonies, and social change, thus serving as a link between the past and the present (Irawati, 2020). Conceptually, traditional music can be defined as "music that is passed down from generation to generation within a cultural community, reflecting the values, history, and identity of that group" (UNESCO Chair on Anthropology of Traditional Music). Several researchers also emphasize that the term "traditional music" is essential to continue using because it relates to community-based music practices, collective innovation, aural transmission, and musical style plurality (Morgenstern, 2021). Thus, traditional music is not just a passive artifact, but a dynamic, living process—changing and meaningful in a socio-cultural context. The essence of traditional music cannot be replaced by the role of "machines" in the form of AI.

The use of AI to model or develop traditional musical patterns has significant implications for the public's understanding of what is considered "authentic" in a tradition. When generative systems create new variations of ethnic music, the output is often detached from the social, ritual, and communal values that are at the heart of traditional musical practices. (Gîrbu, 2024) Research by Chen et al. (D. Chen et al., 2024) on AI-based ethnic music datasets shows that while technical accuracy can be achieved, cultural meaning cannot be fully reduced to acoustic features or mathematical patterns. If AI music outputs circulate without clarification, the public may perceive them as authentic representations, thereby displacing communities as the owners of knowledge. The integration of AI into traditional music must be carried out with the principle of community participation so that preservation is balanced and does not erode local identity.

The presence of AI in the processing of traditional music data raises serious issues regarding cultural control and the right to communal expression. When algorithms learn from traditional recordings without community-based permission mechanisms, the process has the potential to replicate the patterns of colonialism, specifically how technology consumes cultural material without adequate ethical accountability. (Kanhov et al., 2024). The research of (Berkowitz, 2024) warns that the emergence of machine-folk, ethnic music created by AI systems, could blur the line between digital innovation and community heritage that has historical value. This situation makes the protection of communal intellectual property increasingly urgent, especially when machine-generated works appear to be part of tradition but have no emotional or spiritual connection to the community of origin. The use of AI in this case requires governance that respects the moral and cultural rights of communities so that technology does not become a tool for cultural expropriation.

Traditional music is distinguished based on several criteria: (a) geographical origin or ethnic community, (b) social or ritual function (e.g., ceremonial music, work music, entertainment music), and (c) technical or organological characteristics (type of instrument, vocal style, rhythmic patterns) (Jähnichen, 2016). Recent studies also use computational approaches to classify traditional music based on audio features or metadata, such as the classification of traditional Indonesian music by region of origin using a CNN model (Luis & Rokhman, 2022). This classification method demonstrates that, although traditional music is rooted in communities, modern technology now facilitates more systematic categorization mapping.

Artificial intelligence (AI) technology is now increasingly being implemented in traditional music preservation and reconstruction schemes through the digitization of recordings, acoustic feature extraction, the development of ethnic music datasets, and sound pattern analysis for archival and educational purposes. Although AI offers excellent potential for preserving, reconstructing, and disseminating traditional music, several vital challenges arise because the cultural, expressive, and contextual aspects of the community cannot be fully reduced to code or algorithms (Budi Wibawa, 2024). Digital transformation through AI must be accompanied by ethics, local community participation, and respect for the meaning of rituals and the uniqueness of traditions. AI should not be used as a substitute for humans in musical traditions, but as a tool that strengthens the sustainability and creativity of these musical communities.

On the positive side, AI has the potential to be an efficient tool for preserving traditional music. With its analytical capabilities, AI can record, transcribe, and reconstruct musical patterns that were previously difficult to archive conventionally. Patil and Bhor (2023) note that the integration of AI and music creates collaborative opportunities in the context of education and cultural documentation, provided that its use is ethical and participatory (Patil & Bhor, 2023). The integration of AI and traditional music opens up new opportunities in education and cultural documentation, for example, through interactive training that utilizes AI models to provide feedback to students of traditional instruments, or digital archiving systems that automatically tag the rhythmic patterns and timbre characteristics of traditional instruments, then turn them into sources of cross-generational learning. For example, a digital platform that allows young musicians to explore traditional recordings, then AI identifies their voice characteristics and presents adaptive training modules to strengthen continuity. However, it is also essential to align technology with cultural sensitivity so that AI does not merely become a technical tool, but rather part of a dialogue between older and younger generations, between heritage and innovation. AI technology should not replace humans in the realm of musical tradition, but rather facilitate humans in remaining interpreters, successors, and creators who preserve the meaning, identity, and cultural values behind traditional music. The preservation of traditional music through AI, when managed jointly by the community and taking into account ethical and cultural aspects, can become a bridge between the past and the future, bringing the legacy of traditional music to a broader digital space, while ensuring that its spirit, essence, and socio-cultural context remain alive.

DISRUPTION OF THE ROLE AND IDENTITY OF THE MUSIC INDUSTRY

The development of artificial intelligence is shifting the way music industry players understand their creative identity. Musicians who previously relied heavily on intuition and inner processes now face a new partner in the form of algorithms that can accurately mimic musical patterns. This shift is blurring the line between human and machine work, giving rise to a new need to rethink the concepts of originality and creative authority. Many musicians feel that the presence of AI could undermine the recognition of human work if not properly regulated. Jacques and Flynn (2024) assert that the integration of AI in music requires new protection models that ensure human contributions continue to receive fair recognition (Jacques & Flynn, 2024).

The emergence of AI technology has also changed the way musicians collaborate, not only with other humans but also with automated systems capable of rapidly generating musical ideas. Many creators are utilizing technology to find new inspiration, try previously unthinkable harmonic combinations, or accelerate the compositional exploration process. However, ambivalent feelings remain: on the one hand, AI brings convenience, but on the other, musicians fear losing the "personal touch" that has been the foundation of their identity. Research by Ma et al. (2025) suggests that digital experiences have a significant influence on perceptions of AI: the more familiar individuals are with technology, the more positive their views on human-machine collaboration become. (Ma et al., 2025) Collaboration between the two is currently inevitable; however, the most important thing is how to utilize machine assistance wisely.

AI also affects the power structure in the music industry. In the past, radio curators, music journalists, and major labels served as the primary gateways to popularity;

however, streaming platform algorithms now occupy that position. Recommendation systems determine which artists deserve visibility, so success no longer depends entirely on professional networks but on machine preferences that work based on data patterns. This condition creates a new dependency for musicians, particularly those with independent careers. (Geurts & Cepa, 2023)

The capacity of AI to mimic the vocal style of certain artists or produce songs that sound "emotional" also challenges the concept of musical identity. The public is beginning to question whether works that are partly generated by machines can still be considered a personal expression of the musician. This concern primarily arises in popular genres that rely heavily on the emotional connection between artists and their listeners. Moura and Maw (2021) found that most listeners still struggle to accept music entirely generated by AI because it is perceived as lacking "soul," a significant finding in understanding resistance to non-human works.(Chauhan & Sagar, 2021)

The disruption brought about by AI also impacts the future of professions in the music industry. Professions such as producers, studio technicians, and arrangers must now master new skills to work alongside automation systems. On the one hand, AI opens up opportunities to expand creativity through more sophisticated production tools; on the other hand, some traditional roles have the potential to be marginalized due to technological efficiency. This situation requires an ethical and professional adaptation from all stakeholders to prevent machines from completely replacing human roles. Canyakan (2024) emphasizes that the future direction of the music industry depends heavily on humans' ability to utilize AI as a supporting tool, rather than a replacement. (Chauhan & Sagar, 2021) This is a wise response, not in the sense of disregarding technological advances, but rather embracing technology.

Humans continue to play an irreplaceable role in the creative process in music, even as algorithms become more sophisticated. Currently, AI systems are capable of imitating musical patterns and creating musical elements with high accuracy. Still, human sensitivity to emotional context, personal experience, and cultural reflection are unique advantages that machines cannot entirely replace. AI technology can expand human creative capacity, but users, in this case humans, must remain authentic, inclusive, and conscious for this creativity to be meaningful. (O'Toole & Horvát, 2024) In the realm of music, research indicates that listeners most appreciate music that they perceive as still possessing a human "soul." Human works contain emotions, feelings, and personal narratives that, despite their imperfections, are full of value that cannot be replicated. (Lecamwasam & Chaudhuri, 2025) In the era of human-machine collaboration, the role of humans as artistic directors, emotional narrators, and cultural context connectors remains vital. AI can help, but the core of creative authority and originality is rooted in the conscious presence of humans.

PROSPECTS AND CHALLENGES FOR THE FUTURE OF THE MUSIC INDUSTRY

The music industry is now entering a new phase in which global opportunities are increasingly wide open. The presence of streaming and digital distribution platforms means that both commercial and traditional musicians will have access to global audiences. For example, the digital distribution model allows independent artists to release their work without relying entirely on major labels, allowing local innovation to shine. This freedom also requires a high level of adaptability so that their work can compete in a highly crowded global ecosystem. This trend not only provides access to popular music genres such as pop, alternative, rock, country, rap, and others, but also offers hope for the development of traditional music to be preserved and re-contextualized in a modern format, as long as the right strategy and optimal use of digital channels support it. (Theremixhouse, 2024)

In terms of prospects, one of the main opportunities is the broader integration of technology in the music production and distribution process. Artificial intelligence, blockchain, and innovative contract technologies are beginning to offer new models for music creation, licensing, and monetization, for example, enabling more transparent royalties or supporting cross-cultural collaborative works. (Poetry, 2025) This also opens up space for traditional music to find "new markets" through digital collaboration, remixes, or hybrid formats that appeal to the younger generation. The adoption of this technology is not without obstacles; technical capacity, education, and accessibility remain challenges in many regions. This means that to reap the benefits of technological developments, music industry players must build a supporting ecosystem that covers technical, legal, and cultural aspects.

One of the challenges facing the global music industry is economic pressure, as well as the rapid evolution of business models. Several reports indicate that global music industry revenue growth has begun to slow after a surge during the pandemic, while the costs of producing and distributing live performances continue to rise. (Thomas, 2025) For traditional music, this challenge is even more complex because local infrastructure, government support, and formal market networks are often inadequate. Traditional musicians or local communities must compete with global content that is highly visible and well-promoted. The sustainability of business models for traditional music must consider revenue diversification, such as cultural licensing, experience-based tours, or thematic merchandise.

Intellectual property rights and regulations are essential aspects for the future of the music industry. The existence of generative AI capable of automatically generating music raises essential questions regarding ownership rights, creator remuneration, and preserving the authenticity of creative expression. (Berger, 2024) Without an adaptive regulatory framework and adequate protection, musicians of both traditional and modern music risk losing a portion of the revenue from their work. On the other hand, traditional music also faces the risk of cultural erosion if commercialization occurs without consideration for the local context. Therefore, the development of regulations that are responsive to new technologies and local cultural characteristics is crucial so that the music industry of the future is not only commercially profitable but also culturally sustainable.

People with creativity, community, and cultural values remain the primary foundation for the success of the music industry. Technology and platforms can expand reach, but quality, authenticity, and emotional closeness with listeners remain differentiators that are difficult for algorithms to replace. (Gioia, 2024) Traditional musicians with a strong identity and the ability to innovate in modern packaging have the opportunity to attract new audiences while preserving their cultural roots. Maintaining a community of fans (super-fans) and building loyalty in a rapidly changing digital era is a challenge in itself. The music industry of the future should ideally combine technological innovation with respect for human and cultural values—a synergy that opens up excellent prospects while demanding thorough preparation.

This readiness encompasses technical aspects that form a crucial foundation that the music industry of the future cannot overlook. Practitioners must ensure that digital infrastructure, from production equipment to distribution platforms, has been appropriately adapted to avoid being left behind by rapid technological changes. Digital transformation has revolutionized the way music is created and distributed globally, necessitating a fundamental revision of traditional business models. (Mhlambi & Nxozi, 2024) Without technical readiness, including reliable internet access, sufficient production equipment, and an understanding of new technologies, traditional musicians risk being excluded from the increasingly dominant digital market. Building technical capacity and ensuring access to infrastructure must be a key part of the readiness strategy.

Organizational readiness and human resource capacity also determine the direction in which the music industry will move. Technology is not considered dominant, but it is also necessary for music businesses, including labels, traditional communities, and independent musicians. These actors need to have a strong understanding of business trends, new revenue models, intellectual property rights, and digital rights management. The main challenges include unequal revenue sharing and a lack of collaboration among stakeholders, such as the government, communities, and the private sector, in creating a sustainable music ecosystem. (Akil et al., 2024) Furthermore, rapid changes in business models necessitate that industry players be agile in evaluating and adjusting their strategies to remain relevant and competitive. Organizational readiness entails having a flexible internal structure, a team that is prepared to learn, and a culture that fosters continuous innovation.

Regulatory readiness, business models, and cultural and traditional values must work in tandem for the music industry to grow inclusively and sustainably. Data from the International Federation of the Phonographic Industry (IFPI) show that, although the global music market has doubled in value to US$29.6 billion since 2014, and streaming now accounts for 69% of revenue, this growth does not automatically guarantee fair remuneration or protection for creators. (Rechardt, 2025) For traditional music, cultural values and local context are essential, so commercialization without sensitivity can erode community identity. Preparedness must encompass how to preserve cultural roots while leveraging modern opportunities. Thorough preparation is necessary, including the establishment of a responsive regulatory framework, the adoption of business models tailored to local conditions, and respect for cultural heritage throughout the innovation process, so that the primary goal of cultural harmony and business activities that foster community prosperity can be achieved.

The future of the music industry will be determined by the ability of its players to adapt to change without compromising their creative direction. Technological developments, market dynamics, and shifting listener tastes necessitate that music players continually adjust their strategies without compromising the integrity of their work. Both fast-paced modern music and traditional music that is firmly rooted in cultural values have the same opportunity to grow if they can read the moment correctly. These opportunities can only be realized if there is the courage to try new approaches and not be afraid to leave behind old patterns that are no longer relevant. The future of the music industry remains wide open for anyone ready to respond to the demands of the times with creativity and perseverance. In this case, technology is used wisely in creating and even distributing modern music. In contrast, in the realm of traditional music, technology is used to distribute works, making them better known to the public without compromising the identity of regional music, which holds traditional, philosophical, social, and communal values that contribute to local identities.

The prospects of the music industry are not just a matter of technology, but also about how humans place value, emotion, and stories in each of their works. Emerging challenges, ranging from global competition to changes in business models, serve as a reminder that this industry requires continuous adaptation. Traditional and modern music can coexist if supported by favorable policies, a healthy ecosystem, and active participation from the creative community. The combination of innovation, cultural preservation, and robust governance presents the music industry with an excellent opportunity to develop more inclusively and sustainably. Therefore, the future of the music industry is essentially a space that is always ready to be filled with new ideas born from the courage to transform.

REFERENCES

Akil, R., Rasyidin, N., & Dwihadiah, D. L. (2024). Mediatized World in Digital Music Industry: Transformations, Challenges, and Opportunities-A Study in Indonesia. *Daengku: Journal of Humanities and Social Sciences Innovation*, *4*(3), 443–449. DOI: 10.35877/454RI.daengku2549

Berger, V. (2024). *AI's Impact On Music In 2025: Licensing, Creativity And Industry Survival.* Forbes. https://www.forbes.com/sites/virginieberger/2024/12/30/ais-impact-on-music-in-2025-licensing-creativity-and-industry-survival/

Berkowitz, A. E. (2024). Artificial intelligence and musicking: A philosophical inquiry. *Music Perception*, *41*(5), 393–412. DOI: 10.1525/mp.2024.41.5.393

Budi Wibawa, Z. D. R. (2024). Navigating Ethics and Innovation: The Role of AI in Cultural Heritage. *Harmonia Journal of Music and Art*, *2*(4), 198–212. DOI: 10.61978/harmonia.v2i4.905

Canyakan, S. (2024). The role of AI in creative processes: ethical and legal perspectives in the music industry. *Journal of Music Theory and Transcultural Music Studies*, 143–158. DOI: 10.5281/zenodo.15031855

Chauhan, V., & Sagar, M. (2021). Consumer confusion: A systematic review and research directions. *Journal of Consumer Marketing*, *38*(4), 445–456. DOI: 10.1108/JCM-03-2020-3705

Chen, D., Sun, N., Lee, J.-H., Zou, C., & Jeon, W.-S. (2024). Digital technology in cultural heritage: Construction and evaluation methods of AI-Based ethnic music dataset. *Applied Sciences (Basel, Switzerland)*, *14*(23), 10811. DOI: 10.3390/app142310811

Chen, Y. (2021). Analysis on the application of artificial intelligence technology in electronic music composition creation. *2021 4th International Conference on Information Systems and Computer Aided Education*, 1200–1204. DOI: 10.1145/3482632.3483114

Civit, M., Civit-Masot, J., Cuadrado, F., & Escalona, M. J. (2022). A systematic review of artificial intelligence-based music generation: Scope, applications, and future trends. *Expert Systems with Applications*, *209*, 118190. https://doi.org/https://doi.org/10.1016/j.eswa.2022.118190. DOI: 10.1016/j.eswa.2022.118190

Cross, I. (2023). Music in the digital age: Commodity, community, communion. *AI & Society*, *38*(6), 2387–2400. DOI: 10.1007/s00146-023-01670-9

Gallar, A. (2025). *Can math save content creators? A new model proposes fairer revenue distribution methods for streaming services.* Https://Phys.Org/News/2025-02-Math-Content-Creators-Fairer-Revenue. https://phys.org/news/2025-02-math-content-creators-fairer-revenue

Geurts, A., & Cepa, K. (2023). Transforming the music industry: How platformization drives business ecosystem envelopment. *Long Range Planning*, *56*(4), 102327. DOI: 10.1016/j.lrp.2023.102327

Gioia, T. (2024). *Nine Predictions for the Future of the Music Business.* https://www.honest-broker.com/p/nine-predictions-for-the-future-of

Gioti, A.-M. (2021). Artificial intelligence for music composition. In *Handbook of artificial intelligence for music: Foundations, advanced approaches, and developments for creativity* (pp. 53–73). Springer., DOI: 10.1007/978-3-030-72116-9_3

Gîrbu, E. (2024). Experimental perspectives on artificial intelligence in music composition. *Artes. Journal of Musicology*, *29–30*(29-30), 95–99. DOI: 10.35218/ajm-2024-0006

Gumuchian, M.-L. (2025). *Music revenues rise again in 2024, boosted by streaming subscriptions, report shows.* Reuters. https://www.reuters.com/business/media-telecom/music-revenues-rise-again-2024-boosted-by-streaming-subscriptions-report-shows-2025-03-19/?utm_source=chatgpt.com

Hadar, T., & Rabinowitch, T.-C. (2023). The varying social dynamics in orally transmitted and notated vs. improvised musical performance. *Frontiers in Psychology*, *14*, 1106092. DOI: 10.3389/fpsyg.2023.1106092 PMID: 37151348

Husin, N., Hidayanto, A. N., Purwandari, B., & Ibrahim, R. M. S. (2021). Analysis on digital music service user behaviour using justice perception framework. *Journal of Engineering Science and Technology*, *16*(2), 1644–1666.

IFPI. (2024). *Global Music Report 2024: State of the industry.*

Irawati, E. (2020). Transmisi, musik lokal-tradisional, dan musik populer. *Panggung*, *30*(3), 519850. DOI: 10.26742/panggung.v30i3.893

Jacques, S., & Flynn, M. (2024). Protecting human creativity in AI-generated music with the introduction of an AI-royalty fund. *GRUR International*, *73*(12), 1137–1149. DOI: 10.1093/grurint/ikae134

Janowska, T. H. and A. A. (2023). Production networks in the CCS: Case studies from the European music industry. *CICERONE Report D2.7*, 175. DOI: 10.5281/zenodo.6885744

Kanhov, E., Kaila, A.-K., & Sturm, B. L. T. (2024). Innovation, data colonialism and ethics: Critical reflections on the impacts of AI on Irish traditional music. *Journal of New Music Research*, *53*(1–2), 47–63. DOI: 10.1080/09298215.2024.2442359

Kjus, Y. (2024). The platformization of music production: How digital audio workstations are turned into platforms of labor market relations. *New Media & Society, 14614448241304660*. Advance online publication. DOI: 10.1177/14614448241304660

Lecamwasam, K., & Chaudhuri, T. R. (2025). Exploring listeners' perceptions of AI-generated and human-composed music for functional emotional applications. *ArXiv Preprint ArXiv:2506.02856*. https://doi.org//arXiv.2506.02856DOI: 10.48550

Lin, T.-F., & Chen, L.-B. (2024). Harmony and algorithm: Exploring the advancements and impacts of AI-generated music. *IEEE Potentials*, *43*(6), 23–30. Advance online publication. DOI: 10.1109/MPOT.2024.3433888

Luis, R., & Rokhman, N. (2022). Traditional Music Regional Classification using Convolutional Neural Network (CNN). [Indonesian Journal of Computing and Cybernetics Systems]. *IJCCS*, *16*(4), 379–388. DOI: 10.22146/ijccs.73910

Ma, H., Zhang, Y., Shan, X., & Hu, X. (2025). Exploring the Impact of Artificial Intelligence on the Creativity Perception of Music Practitioners. *Journal of Intelligence*, *13*(4), 47. DOI: 10.3390/jintelligence13040047 PMID: 40278056

Mhlambi, L., & Nxozi, M. (2024). The Effect of Digital Transformation in the Music Industry: A Scoping Review. *Interdisciplinary Research in Technology and Management*, 153–168.

Miranda, E. R., & Williams, D. (2015). Artificial intelligence in organised sound. *Organised Sound*, *20*(1), 76–81. DOI: 10.1017/S1355771814000454

Moreau, F., Wikström, P., Haampland, O., & Johannessen, R. (2024). Alternative payment models in the music streaming market: A comparative approach based on stream-level data. *Information Economics and Policy*, *68*, 101103. DOI: 10.1016/j.infoecopol.2024.101103

Morgenstern, U. (2021). In defence of the term and concept of traditional music. *Musicologist*, *5*(1), 1–30. DOI: 10.33906/musicologist.913512

Naylor, Richard, Todd, Jonathan, Moretto, Marta, Traverso, R. (2021). *Cultural and creative industries in the face of COVID-19: an economic impact outlook.*

Ng, J., & Gamble, S. (2024). Hip-hop music producers' labour in the digital music economy: Self-promotion, social media and platform gatekeeping. *New Media & Society, 14614448241295304*. Advance online publication. DOI: 10.1177/14614448241295304

O'Toole, K., & Horvát, E.-Á. (2024). Extending human creativity with AI. *Journal of Creativity*, *34*(2), 100080. DOI: 10.1016/j.yjoc.2024.100080

Patil, R. P., & Bhor, P. A. (2023). Synergy of Artificial Intelligence and Music. *2023 6th International Conference on Advances in Science and Technology (ICAST)*, 155–159.

Pizzolitto, E. (2024). Music in business and management studies: A systematic literature review and research agenda. *Management Review Quarterly*, *74*(3), 1439–1472. DOI: 10.1007/s11301-023-00339-3

Poetry. (2025). *5 Trends Shaping the Future of the Music Industry*. Illustratemagazine. https://illustratemagazine.com/5-trends-shaping-the-future-of-the-music-industry/

Polak, N., & Schaap, J. (2025). Write, record, optimize? How musicians reflect on music optimization strategies in the creative production process. *New Media & Society*, *27*(8), 4773–4789. DOI: 10.1177/14614448241243095

Rechardt, L. (2025). *IFPI looks at a decade of digital transformation in the music industry*. https://www.wipo.int/en/web/wipo-magazine/articles/ifpi-looks-at-a-decade-of-digital-transformation-in-the-music-industry

ŞUTEU, L.-C., & RUCSANDA, M. D. (2025). Artificial Intelligence in Music: The Digital Revolution of Sound Creativity. *ICT in Muzical Field/Tehnologii Informatice Si de Comunicatie in Domeniul Muzical, 16*(1).

Theremixhouse. (2024). *Current Trends, Challenges, and Opportunities in the Music Industry: 2024*. https://www.themixhouse.com/blog/current-trends-challenges-and-opportunities-in-the-music-industry-2024?utm

Thomas, D. (2025). *Music industry revenues slow as pandemic-era CD boom fades*. Financial Times. https://www.ft.com/content/05ea07dc-2fae-4616-93e1-746cc8ac4635?utm

Tigre Moura, F., & Maw, C. (2021). Artificial intelligence became Beethoven: How do listeners and music professionals perceive artificially composed music? *Journal of Consumer Marketing*, *38*(2), 137–146. DOI: 10.1108/JCM-02-2020-3671

Tullah, D. N., Khalid, H., & Poernomo, S. L. (2024). Perlindungan Hak Cipta Dalam Industri Musik Digital Di Indonesia Dalam Perspektif Hukum Perdata. [JLP]. *Journal of Lex Philosophy*, *5*(2), 1030–1046. https://www.pasca-umi.ac.id/index.php/jlp/article/view/1877

Vanka, S. S., Safi, M., Rolland, J.-B., & Fazekas, G. (2023). Adoption of AI technology in the music mixing workflow: An investigation. *ArXiv Preprint ArXiv:2304.03407.* https://doi.org//arXiv.2304.03407DOI: 10.48550

Vengathattil, S. (2025). Collaborative AI in music composition: Human-AI symbiosis in creative processes. *International Journal of Management Science and Information Technology*, *5*(1), 253–262. DOI: 10.35870/ijmsit.v5i1.4085

Wikström, P. (2020). *The music industry: Music in the cloud.* John Wiley & Sons.

Wikstrom, P. (2022). *Peter Tschmuck: The economics of music: Agenda Publishing, Newcastle, 2021* (pp. 387–389). Springer. https://doi.org/DOI: 10.1007/s10824-022-09443

Yin, L., & Guo, R. (2025). An Artificial Intelligence-Based Interactive Learning Environment for Music Education in China: Traditional Chinese Music and Its Contemporary Development as a Way to Increase Cultural Capital. *European Journal of Education*, *60*(1), e12858. DOI: 10.1111/ejed.12858

Chapter 13
Educational Applications of AI in Music Production Curricula

Arini Winur Baeti
http://orcid.org/0009-0002-4496-8868
Universitas Esa Unggul, Indonesia

Binastya Anggara Sekti
http://orcid.org/0000-0001-5489-4888
Universitas Esa Unggul, Indonesia

ABSTRACT

AI is increasingly transforming music production education by enhancing how students learn, create, and collaborate. This chapter explores how AI tools such as Magenta Studio, AIVA, Soundful, and LANDR support composition, mixing, mastering, and sound design, helping democratize access to advanced production techniques. It examines how educators integrate AI into curricula to develop creativity, technical skills, and critical listening while maintaining a balance between human artistry and algorithmic assistance. The chapter also reviews case studies, teaching methods, ethical issues, and curriculum strategies to ensure AI innovation supports artistic integrity and prepares students for future careers in the music industry.

1. INTRODUCTION

Over the past two decades, the music production industry has undergone major transformations due to rapid technological development. The shift from analog to digital systems has democratized music creation, allowing more people to enter

DOI: 10.4018/979-8-3373-6279-3.ch013

the field through digital audio workstations, software instruments, and plugins that enable complex production processes on personal computers. However, this accessibility has simultaneously increased the technical and creative skills required from aspiring producers. The emergence of artificial intelligence represents an even more disruptive transformation, as AI can now generate musical content, assist with arrangement and sound design, automate mixing and mastering, and provide data driven feedback during creative processes. These tools have become essential components of modern professional workflows, helping experienced producers improve efficiency and encouraging experimentation while acting as supportive learning tools for beginners and students by providing access to advanced techniques.

The integration of AI into music production education raises important challenges including concerns about overdependence on automation, superficial skill development, reduced transparency in technical decisions, and ethical issues related to authorship and creative ownership. Educators must balance traditional musical and technical skills with new competencies related to working critically with AI. This comprehensive framework aims to incorporate AI into music production education thoughtfully, ensuring technological innovation enhances creativity and learning while preserving human judgment, artistic integrity, and ethical responsibility. The framework examines pedagogical foundations, practical tools and applications, curriculum design strategies, institutional case studies, ethical challenges, and future directions, providing educators with evidence based guidance for implementing AI augmented music production programs that prepare students for contemporary industry realities while maintaining focus on creative autonomy and critical thinking(Luo, 2023).

2. LITERATURE REVIEW

2.1 Scope and Review Approach

This chapter draws upon a structured narrative review of peer-reviewed literature published between 2019 and 2025, focusing on artificial intelligence in music production education, creative AI systems, curriculum transformation, and assessment practices. Sources were selected from scholarly journals, international conference proceedings, and institutional policy reports to ensure academic rigor and topical relevance.

Rather than presenting an exhaustive systematic review, this section synthesizes conceptual, empirical, and policy-oriented scholarship to identify prevailing trends, methodological limitations, theoretical tensions, and emerging research gaps. The objective is to position the chapter's proposed framework within the broader scholarly discourse on AI-augmented creative education.

2.2 AI and Creative Learning in Music Production

Contemporary scholarship increasingly frames artificial intelligence as a catalyst for creative exploration within music production environments. Generative composition systems, intelligent mixing assistants, and AI-supported sound design platforms are frequently described as scaffolding mechanisms that expand ideation, accelerate prototyping, and lower technical entry barriers for novice learners. From this perspective, AI functions as a co-creative collaborator that enhances experimentation and iterative development.

However, this optimistic framing is not uncontested. A growing body of critical literature cautions that excessive reliance on generative systems may inadvertently weaken foundational competencies, including harmonic literacy, arrangement technique, and critical listening skills. Concerns have also been raised regarding the homogenizing effects of algorithmic suggestion systems, which may privilege statistically dominant stylistic patterns over authentic individual expression.

These divergent perspectives reveal an unresolved pedagogical tension: whether AI primarily augments creative capacity or risks diminishing deep musical engagement. The literature therefore suggests that educational outcomes are contingent not on the mere presence of AI tools, but on the intentional design of structured learning environments that promote active, reflective, and critically mediated interaction with technology.

Despite numerous exploratory classroom implementations reporting increased engagement and creative confidence, longitudinal evidence assessing sustained skill development remains limited. Much of the existing research relies on short-term qualitative reflections rather than systematic outcome measurement.

2.3 AI, Assessment, and Learning Validation

Assessment in AI-assisted creative production represents one of the most complex and underdeveloped areas of current scholarship. Several authors advocate process-oriented evaluation models emphasizing documentation of workflow, reflective commentary, and transparency in AI usage. Such approaches aim to shift the evaluative focus from final product to creative reasoning and decision-making processes.

Nevertheless, significant methodological challenges persist. Distinguishing human agency from algorithmic contribution in generative workflows remains analytically difficult. Few validated instruments exist that simultaneously measure creativity, technical proficiency, ethical awareness, and independent authorship in AI-augmented environments.

Furthermore, concerns surrounding academic integrity and originality complicate traditional evaluation paradigms. While policy discussions are expanding, there remains a lack of standardized frameworks capable of ensuring reliability and fairness across institutions. The literature thus underscores the urgent need for assessment models specifically tailored to AI-integrated creative curricula.

2.4 Ethical, Authorship, and Governance Considerations

Ethical discourse surrounding artificial intelligence in creative industries has intensified in recent years, particularly with regard to copyright, authorship attribution, data bias, and the use of copyrighted material in training datasets. Within music production education, these concerns intersect directly with institutional academic integrity policies and emerging governance structures for AI use.

Although conceptual discussions of AI ethics are increasingly sophisticated, comparatively fewer studies examine how these concerns are operationalized within classroom practice. The translation of abstract ethical principles into enforceable curriculum policies, assessment guidelines, and student accountability mechanisms remains insufficiently developed.

This gap suggests that ethical literacy must be integrated not merely as a theoretical consideration, but as an embedded component of curriculum design and instructional practice.

2.5 Institutional Integration and Faculty Capacity

Beyond pedagogical considerations, institutional adoption of AI technologies presents structural and organizational challenges. While proponents highlight enhanced industry alignment, workflow efficiency, and expanded creative possibilities, empirical evidence suggests uneven implementation across institutions due to disparities in infrastructure, licensing costs, and technological access.

Faculty preparedness emerges as a decisive factor in successful integration. Research indicates that meaningful adoption depends on professional development opportunities, interdisciplinary collaboration, and institutional policy clarity. However, systematic investigation into faculty training models specific to AI in music production remains limited.

The literature therefore reveals a persistent implementation gap between technological potential and institutional readiness, particularly within resource-constrained educational contexts.

2.6 Emerging Gaps and Scholarly Limitations

Despite the rapid expansion of AI-related scholarship, several limitations remain evident in the current body of research:

1. A scarcity of longitudinal studies examining sustained creative and technical skill development in AI-augmented curricula.
2. Limited empirical validation of assessment reliability and authorship verification mechanisms.
3. Underrepresentation of non-Western, vocational, and resource-limited institutional contexts.
4. Conceptual ambiguity in differentiating among types of AI applications within music production pedagogy.

These gaps indicate that the discourse remains partially aspirational and conceptually driven. To advance the field, greater methodological rigor and clearer pedagogical modeling are required. This chapter seeks to contribute to that advancement by proposing an integrative curriculum framework that synthesizes theoretical insight with practical implementation strategies.

3. PEDAGOGICAL FOUNDATIONS AND THEORIES FOR AI INTEGRATION

The integration of artificial intelligence into music production education requires more than simply adding new tools to existing curricula but demands thoughtful reconsideration of how learning happens, what skills need development, and how technology can best support educational processes. Decades of research into technology enhanced learning, constructivist pedagogy, and creative education provide valuable frameworks that can guide AI integration in ways that enhance rather than diminish learning outcomes(Sularso *et al.*, 2025).

3.1 Theoretical Frameworks for Technology Enhanced Learning

At the heart of effective music production education lies constructivism, a learning theory emphasizing active knowledge construction through direct experience, experimentation, and reflection rather than passive information reception. In a constructivist framework, students don't simply absorb facts about compression ratios or EQ curves but build understanding through hands on practice, developing intuitive senses of how different processing decisions affect sonic outcomes. AI tools introduce both opportunities and challenges to this framework: while they can accelerate experimentation by providing multiple examples for comparison and supporting each stage of Kolb's experiential learning cycle from concrete experience through reflective observation, abstract conceptualization, to active experimentation, they also risk short circuiting the essential struggle and discovery process if they provide answers too readily. The key is positioning AI as a collaborative partner in exploration rather than a definitive solution provider, extending social constructivism into new dimensions where students learn not only production techniques but also how to work effectively in creative teams that include both human and artificial intelligence.

Social constructivism extends these ideas into the collaborative realm, recognizing that learning is fundamentally social and that knowledge is constructed through interaction with others. Music production has always been collaborative, with producers working with artists, engineers consulting each other, and creative communities sharing techniques and feedback. In educational settings, collaborative learning takes many forms including peer feedback sessions where students critique each other's mixes, group projects where students work together on productions, and class discussions where techniques and aesthetic decisions are debated. The collaborative classroom becomes a space where human creativity and AI capabilities are explored together, with students learning not only production techniques but also how to work effectively in teams that include both human and artificial intelligence.

Figure 1. AI supported workflow in music production learning

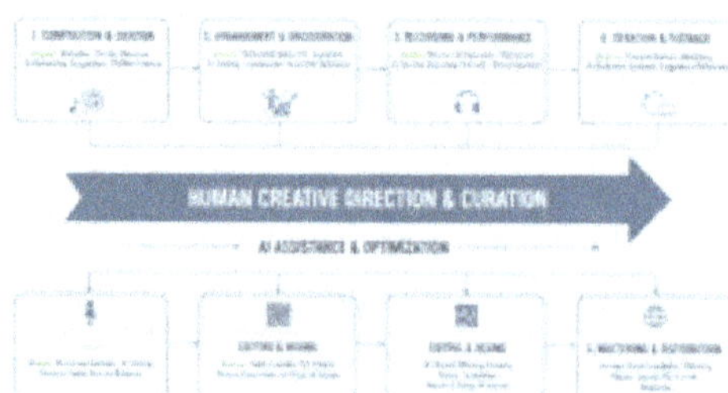

3.2 AI as a Co Creative Educational Partner

Understanding AI as a co-creative educational partner requires rethinking the relationship between learner, teacher, and technology. Traditional educational tools mainly deliver content or support tasks under instructor direction. AI, however, can respond to student work in real time, generate suggestions, and even create original material. This places AI between a tool and a tutor. It is more adaptive than standard software effects, yet it lacks the contextual awareness, ethical sensitivity, and nuanced judgment of a human instructor. Therefore, its role must be carefully framed within pedagogical goals.(Zhang, Lu and Liu, no date).

Scaffolding theory, developed by Jerome Bruner, offers a helpful framework for understanding AI's function in learning. Scaffolding involves providing temporary support that enables learners to complete tasks beyond their current ability, then gradually removing that support as competence increases. In music production, an AI mixing assistant might initially give detailed EQ or compression suggestions with explanations. As students gain experience, the AI could shift toward prompting reflective questions instead of direct answers. Eventually, it may serve mainly as a secondary reference while students rely on their own technical knowledge and aesthetic judgment. This gradual reduction of assistance prevents overdependence and promotes genuine mastery.

Similarly, Lev Vygotsky's concept of the Zone of Proximal Development explains how AI can personalize instruction. The ZPD describes the gap between what learners can achieve independently and what they can accomplish with guidance. AI systems can analyze student decisions and tailor feedback to their skill level, offering foundational instruction to beginners and more advanced critique to experienced learners. In diverse classrooms, this adaptability is especially valuable. However, educators must remain actively involved, since AI may misjudge ability levels or provide support that does not align with broader educational objectives.

3.3 Active Learning and Problem Based Approaches

Active learning methodologies, which emphasize student engagement over passive instruction, align particularly well with AI augmented music production education. Project based learning, where students develop complete musical productions from concept to final master, provides authentic contexts for applying both traditional skills and AI tools. Rather than learning mixing techniques in isolation, students might receive briefs to produce tracks in specific genres, requiring integrated decisions about composition, arrangement, sound design, mixing, and mastering. AI tools become resources students can draw upon as needed, such as using generative composition tools for initial ideas, employing AI stem separation to analyze reference tracks, or leveraging automated mixing as starting points to refine manually. The project structure encourages strategic thinking about when and how to use AI, developing judgment about which tasks benefit from AI assistance and which require human creativity and decision making.

Inquiry based learning approaches, where students formulate questions and investigate answers, take on new dimensions with AI tools. Students might pose questions like "What makes a vocal sit well in a dense mix?" and use AI analysis tools to examine professional productions, identifying patterns and principles. They could experiment with AI mixing suggestions across different genres, investigating whether the same principles apply universally or if context matters. This positions students as active researchers rather than passive recipients, fostering critical thinking and analytical skills. The flipped classroom model, where students engage with instructional content outside class and use class time for active application and discussion, works particularly well for AI integration. Students might complete tutorials on AI tool functionality as homework, then come to class ready to experiment, compare results, and discuss creative and technical implications, maximizing hands on learning time while ensuring students arrive prepared to engage meaningfully with both technology and creative concepts.

3.4 Bloom's Taxonomy in AI Augmented Music Education

Bloom's Taxonomy, a hierarchical framework of cognitive skills ranging from basic remembering to advanced creating, provides useful structure for thinking about how AI tools can support different learning levels. At the foundational level of remembering and understanding, AI can help students grasp basic concepts through interactive visualizations that make abstract concepts like frequency masking or phase relationships more concrete and understandable. At the application level, students use AI tools to apply principles in specific contexts, perhaps using AI EQ suggestions as starting points they then refine based on developing understanding. The analysis level involves breaking down complex productions to understand their components, which AI analysis tools can facilitate by visualizing frequency content, revealing automation patterns, or identifying specific processing applied to different elements.

Evaluation requires making informed judgments about quality and effectiveness. Students might critically assess AI generated compositions or mixing decisions, articulating why they work or don't work and how they might be improved. This critical evaluation skill is essential as students must learn to judge AI outputs rather than accepting them uncritically. At the highest level of creation, students synthesize everything learned to produce original work, strategically incorporating AI as one tool among many in their creative arsenal. Effective curriculum design ensures that AI integration supports movement through these cognitive levels rather than allowing students to skip straight to creation without developing foundational skills and critical judgment.

3.5 Connectivism and Distributed Learning

Connectivism, a learning theory proposed by George Siemens for the digital age, emphasizes that knowledge is distributed across networks and that learning involves building and navigating these networks. In music production, students learn not only from instructors and textbooks but from YouTube tutorials, online forums, collaboration with peers, and engagement with professional communities. The modern music production student exists within a rich ecosystem of learning resources, and developing the ability to navigate this ecosystem effectively is itself a crucial skill. AI tools become nodes within these learning networks as sources of information, creative partners, and objects of critical inquiry. Students learn to evaluate when AI generated information is reliable, how to combine insights from multiple AI tools, and how to integrate AI capabilities with knowledge gained from human sources.

Peer to peer learning takes on new forms when mediated by AI. Online communities of practice emerge where students share prompts for generative AI tools, discuss effective workflows for integrating AI into production pipelines, and collectively develop best practices. This networked learning approach reflects the reality of contemporary creative practice, where professionals continuously learn from diverse sources and adapt to rapidly evolving technologies, preparing students for professional landscapes where continuous learning, community engagement, and technological fluency are essential for sustained career success.

3.6 Pedagogical Challenges and Considerations

Integrating AI into music production education presents significant pedagogical challenges requiring careful navigation. The most fundamental tension is balancing automation with creative autonomy, as AI tools can handle many technical tasks efficiently but over reliance risks producing students who can operate software but lack deep understanding or creative agency. This requires intentional pedagogical decisions about which tasks students should master manually before using AI assistance to ensure they understand principles rather than just executing procedures, such as whether students should learn manual audio timing editing before using AI powered quantization or practice manual gain staging before relying on automated level balancing.

Maintaining artistic integrity presents another significant challenge, as AI tools making it easy to create music sounding professional according to algorithmic standards might cause students to converge on similar aesthetic outcomes rather than developing distinctive creative voices. Addressing diverse learning styles requires recognizing students engage with technology differently, with some thriving with AI assisted exploration while others feel overwhelmed by too many options or prefer focused depth that comes from mastering techniques without algorithmic assistance. Some students come from backgrounds where they've already experimented extensively with AI tools, while others are encountering these technologies for the first time. Effective pedagogy provides multiple pathways to competence by offering AI tools as options rather than requirements, providing alternative assignments, offering both AI assisted and traditional workflows for the same learning objectives, and creating space for students to articulate their own learning preferences and goals.

4. AI TOOLS AND APPLICATIONS IN MUSIC CURRICULA

The landscape of AI tools available for music production education has expanded dramatically in recent years, offering educators increasingly sophisticated arrays of options for enhancing teaching and learning. These tools range from generative composition systems creating original musical ideas to intelligent mixing assistants providing real time feedback on production decisions. Understanding the capabilities, pedagogical applications, and practical considerations of these tools is essential for educators seeking to integrate AI effectively into curricula(Wani and Chavan, 2025).

Table 1. Comparison of AI tools used in music production curricula

AI Tool	Primary Function	Educational Application	Pedagogical Benefit
Google Magenta Studio	Generative composition	Melody creation, continuation exercises	Enhances creativity and compositional exploration
AIVA	Orchestral AI composition	Film scoring and arrangement	Develops orchestration and emotional scoring skills
LANDR	Automated mastering	Mix comparison and benchmarking	Teaches mastering standards and loudness balance
iZotope Neutron	AI-assisted mixing	EQ, compression, and masking analysis	Improves technical mixing understanding
Sonible Smart EQ	Adaptive equalization	Frequency conflict visualization	Enhances critical listening skills
NSynth	Neural sound synthesis	Timbre creation and sound experimentation	Encourages innovative sound design

4.1 Generative Composition Tools

Google Magenta Studio is one of the most accessible entry points for students exploring AI-assisted composition. Built on Google's open-source machine learning framework, it provides several powerful tools that can be integrated into composition courses. Its melody generation feature creates original melodic ideas based on musical parameters, offering students creative starting points for their own works. The continuation function allows students to input short musical phrases and receive multiple AI-generated extensions, demonstrating how artificial intelligence interprets musical patterns and suggesting possible directions for compositional development. The interpolation feature enables students to input two different musical ideas and generate smooth transitional material between them, reinforcing concepts of musical transformation and thematic evolution. From a pedagogical perspective, Magenta Studio serves as a creative partner that helps students overcome creative blocks, explore unfamiliar musical territories, and generate material for critical analysis and refinement.

AIVA (Artificial Intelligence Virtual Artist) adopts a different approach by specializing in automated orchestral composition, making it particularly valuable for film scoring education and arrangement studies. It allows students to experiment with how music supports emotional narratives by generating multiple scoring variations for the same visual scene. AIVA's ability to produce professional-quality orchestrations helps learners understand how different instrument combinations shape mood, texture, and dramatic impact.

Amper Music and Soundful represent the most user friendly segment of AI composition tools, designed for individuals with minimal musical training. These platforms enable students to select parameters such as genre, mood, tempo, and instrumentation to generate complete tracks with ease. This accessibility empowers beginners to create professional-sounding music quickly, builds confidence, and encourages engagement with music production concepts without requiring advanced theoretical knowledge.

4.2 AI in Sound Design and Synthesis

Neural synthesis represents one of the most innovative applications of AI in music production, using machine learning to generate entirely new sounds rather than simply processing existing audio. Google's NSynth Neural Synthesizer exemplifies this approach, creating novel timbres by interpolating between different source sounds. For students, NSynth opens unprecedented possibilities for timbral exploration as they can blend characteristics of acoustic and electronic instruments, creating hybrid sounds impossible through traditional synthesis or sampling. This encourages sonic experimentation and pushes students to think beyond conventional sound palettes. IRCAM AudioSculpt, enhanced with AI features, provides sophisticated tools for spectral analysis and transformation where students can visualize sound in the frequency domain, understanding how different timbres are constructed and learning to manipulate spectral content directly.

Modern sample libraries increasingly incorporate AI powered tagging and search functionality, transforming how students discover and organize sounds. Intelligent search systems allow students to find samples by describing desired characteristics rather than navigating hierarchical folder structures. This natural language approach makes vast sound libraries more accessible, particularly for beginners who may not yet know technical terminology. The AI analyzes acoustic properties, identifying similar sounds even when labeled differently or existing in different libraries. From a pedagogical perspective, these systems support sound palette development by helping students discover sonic relationships they might otherwise miss, with students searching for specific sounds potentially being shown related sounds that inspire new creative directions.

4.3 Intelligent Mixing and Mastering

LANDR and CloudBounce pioneered AI-based automated mastering services that analyze tracks and apply processing to approach commercial loudness and tonal balance standards. In education, these platforms function best as learning references rather than final solutions. Students can upload their mixes and instantly compare the AI-processed result with their original version. This comparison helps them understand mastering principles such as frequency adjustments, dynamic range control, and spatial enhancement. Exercises may involve mastering a track manually, then evaluating differences between their version and the automated output, encouraging critical reflection on decision making and sonic outcomes.

iZotope's Neutron and Ozone offer more advanced AI-assisted mixing and mastering while still allowing manual control. Neutron analyzes individual tracks and suggests EQ, compression, and spatial adjustments based on content and interaction with other elements. For students, these suggestions serve as informed starting points. They can accept, modify, or reject them while learning the reasoning behind each recommendation. Visual feedback highlights issues like frequency masking, spectral imbalance, and dynamic inconsistencies, making abstract mixing concepts more concrete and easier to grasp.

4.4 Adaptive Learning Environments

AI powered feedback systems represent significant advances in personalized music education, offering real time analysis and correction that adapts to individual student needs. These systems can evaluate student work across multiple dimensions including technical accuracy, adherence to mixing principles, and creative effectiveness, providing immediate specific feedback. Unlike traditional assessment happening days after submission, real time feedback allows students to correct mistakes and refine techniques while actively working, significantly accelerating learning processes. Personalized learning paths emerge from continuous assessment of student progress, with AI identifying which concepts individual students have mastered and which require additional practice, automatically adjusting difficulty and focus areas to address the common challenge of students progressing at different rates and struggling with different concepts, making standardized pacing ineffective.

Intelligent tutoring systems combine interactive lessons with adaptive difficulty, providing structured learning experiences that respond to student performance. These systems might present mixing challenges of increasing complexity, offering hints when students struggle and advancing when they demonstrate competence.Multimodal feedback represents another significant advantage as these systems provide visual feedback including waveform analysis and frequency graphs, audio examples with before and after comparisons, and textual explanations of why specific choices work or don't work, addressing different learning styles and reinforcing concepts through multiple channels to improve retention and understanding.

4.5 Collaborative and Cloud Based Platforms

Splice has become ubiquitous in music production education, offering vast sample libraries alongside AI powered recommendation systems that suggest sounds based on previous selections and current project characteristics. For education, Splice provides shared access to professional quality sounds without expensive individual licenses, with AI recommendations helping students discover new sonic possibilities while learning about genre conventions and production techniques. Soundtrap for Education emphasizes collaborative features, allowing students to work together on projects remotely with version control and real time collaboration capabilities, supporting project based learning and preparing students for professional workflows where remote collaboration is increasingly common. BandLab creates social learning environments where students can share work, receive feedback from peers, and engage with broader creative communities, teaching not just production skills but also professional behaviors including giving and receiving constructive criticism, collaborating effectively, and managing creative projects that are essential for career success.

4.6 Comparison Matrix and Tool Selection Criteria

When selecting AI tools for educational use, several key criteria must be considered. Learning objectives alignment is paramount, with tools supporting specific pedagogical goals rather than being adopted simply because they are new or impressive. Ease of use versus depth of control represents a fundamental tradeoff, as beginner friendly tools like Amper Music enable quick results but offer limited learning about underlying principles, while complex tools like iZotope Ozone provide deeper insight but steeper learning curves. The appropriate choice depends on student level, course objectives, and instructional context. Cost considerations are significant for educational institutions, with some tools offering educational discounts or free versions with limited features while others require substantial licensing fees, requiring educators to balance capabilities against budget constraints, potentially using free tools for fundamental concepts and reserving premium tools for advanced courses.

Technical requirements and accessibility affect implementation feasibility, as cloud based tools require reliable internet but work on various devices, while locally installed software may demand powerful computers. Accessibility features including screen reader compatibility, keyboard navigation, and adjustable interfaces ensure all students can participate. Thoughtful tool selection, grounded in clear pedagogical reasoning and practical constraints, ensures that AI integration enhances learning outcomes for all students.

5. CURRICULUM DESIGN FOR AI AUGMENTED MUSIC EDUCATION

Designing effective curricula for AI augmented music production education requires more than simply adding new tools to existing course structures but demands fundamental rethinking of learning objectives, pedagogical approaches, and assessment methods to ensure students develop both technological proficiency and creative autonomy. A well designed curriculum balances technical skill development with artistic growth, prepares students for industry realities while fostering innovation, and leverages AI capabilities without creating dependency or undermining foundational competencies(Cangelosi *et al.*, 2022).

Table 2. AI contributions to learning outcomes in music production education

Learning Domain	AI Contribution	Expected Student Outcome
Technical Skills	Automated feedback and mixing suggestions	Improved production accuracy
Creative Skills	Generative music and sound tools	Increased originality and experimentation
Critical Thinking	Comparison between human and AI outputs	Stronger evaluative judgment
Collaboration	Cloud-based and AI-assisted teamwork	Enhanced peer learning and teamwork
Ethical Awareness	AI authorship discussion and policy integration	Responsible AI usage
Industry Readiness	Exposure to professional AI tools	Higher employability and workflow efficiency

5.1 Principles of Effective Curriculum Design

Effective curriculum design for AI augmented music education utilizes the backward design approach, prioritizing desired learning outcomes over technological tools. Instead of being driven by the novelty of AI, educators first define what students should master, such as proficiency in both manual and AI assisted production and the ability to critically evaluate machine generated content. By starting with these clear endpoints, curricula ensure that focus remains on artistic vision and core competencies rather than technology itself. The principle of alignment ensures that learning objectives, activities, and assessments work together coherently, with AI integration not distracting from core goals. For instance, if objectives involve developing critical listening, students should compare manual work with AI results rather than simply using automated tools, with assessments then evaluating sophistication of student decision making and analysis to ensure technology serves as a bridge to deeper understanding rather than a shortcut to final products.

Progressive skill development pathways allow students to build competencies systematically through scaffolding. This progression involves gradually increasing technical complexity, reducing levels of AI assistance as students gain confidence, and intensifying creative challenges. By moving from guided generative tools to independent strategic problem solving, this approach keeps students in optimal learning zones, preventing frustration while fostering original artistic growth and technical mastery.

5.2 Core Competencies Framework

Technical competence in modern music production now requires dual mastery of foundational DAW skills and AI literacy. Core concepts such as signal flow and gain staging remain essential, but students must also understand how to operate generative composition tools and intelligent mixing assistants. AI literacy is not limited to basic operation; it includes knowing when to use specific tools, how to evaluate their outputs, and how to troubleshoot problems. Students therefore need both broad technological awareness and deep proficiency aligned with their creative goals. A strong understanding of audio signal processing is increasingly important, as it enables learners to supervise automated decisions. By mastering how compression, EQ, and reverb interact, students can judge whether AI-generated sonic suggestions are appropriate or require manual adjustment. This ensures they remain informed creators rather than passive operators of software.

Creative development continues to place artistic vision at the center of music production education. Students must cultivate personal aesthetic awareness, including recognizing what moves them emotionally, which sonic textures they find compelling, and what ideas they wish to express. AI should function as a supportive tool, not a replacement for creativity. Assignments can encourage students to generate material in multiple genres using AI, then critically reflect on which styles align with their identity and why. Through this process, they gradually develop a clearer artistic voice and learn to treat AI as one resource among many (Pons *et al.*, 2025).

5.3 Balancing Creativity and Technology Use

Achieving appropriate balance between human creativity and AI assistance is perhaps the most delicate challenge in curriculum design. One practical approach is the seventy thirty rule where approximately seventy percent of creative decisions should come from human judgment and AI contributes roughly thirty percent, serving as a conceptual guideline rather than rigid formula to ensure AI enhances rather than dominates creative processes by handling initial ideation or tedious technical tasks while reserving fundamental artistic decisions about emotional trajectory, stylistic direction, and which ideas to develop versus discard for human judgment.

Scaffolding AI usage across curricula by moving from heavy AI assistance early to minimal reliance later helps students develop independence while benefiting from AI support during learning processes. Beginning projects might encourage substantial AI use to build confidence and demonstrate possibilities, while as students develop competence, expectations shift toward more selective AI integration. Advanced projects might even require students to articulate rationales for each instance of AI use, justifying why algorithmic assistance was appropriate rather than handling tasks manually. This progression ensures AI serves as learning scaffolds that students can eventually work beyond rather than becoming permanent crutches preventing full capability development.

Encouraging creative risk taking requires deliberately structuring assignments where safe conventional choices are less rewarded than interesting ambitious attempts even if imperfectly executed. Strategies include experimental project requirements where students must intentionally subvert AI suggestions or use tools in unconventional ways, assessment rubrics that explicitly value originality and creative courage alongside technical proficiency, and creating classroom culture where students feel safe sharing work in progress and discussing creative challenges not just polished successes, ultimately helping students see AI as expanding creative possibilities rather than defining creative boundaries.

5.4 Project Based and Collaborative Learning Frameworks

Individual projects provide space for personal artistic development, allowing students to pursue creative directions aligned with their emerging artistic identities. These projects should balance structure with freedom, with clear technical requirements ensuring skill demonstration while leaving creative decisions to students. Reflection journals accompanying projects encourage metacognitive awareness, having students articulate their creative processes, document decisions about when and why to use AI, and evaluate outcomes critically. Portfolio pieces resulting from individual projects showcase students' best work, demonstrating both technical competence and artistic vision to potential employers or graduate programs.

Collaborative projects develop essential professional skills while teaching students to navigate creative partnerships. Team based productions might assign students complementary roles with one handling composition, another mixing, and a third focused on sound design, requiring coordination and mutual accountability. Peer feedback loops, where students critique each other's work following structured protocols, develop both giving and receiving constructive criticism gracefully. Cross disciplinary collaborations partnering music production students with filmmakers, game designers, or dancers expose students to how music production serves broader creative projects, developing client relations skills and expanding creative perspectives.

5.5 Assessment Strategies for AI Generated Work

Formative assessment, or ongoing evaluation that guides learning rather than merely measuring it, is particularly important when working with AI. Process documentation, where students maintain detailed records of their creative decisions, reveals learning that finished products alone might obscure. Iteration logs showing how work evolved across multiple versions demonstrate developmental thinking and willingness to refine rather than accepting first results. Peer reviews provide multiple perspectives and help students internalize critical evaluation frameworks, learning to see their work through others' eyes. These formative assessments emphasize growth and learning processes rather than just final outcomes, creating developmental rather than purely judgmental evaluation culture.

Summative assessment, or evaluation of final learning outcomes, requires adaptation for AI augmented work. Portfolio based evaluation allows assessment of diverse work over time rather than single high stakes performances, revealing developmental trajectories and range of competencies. Demonstrations and presentations where students explain creative processes, justify AI usage decisions, and respond to questions assess not only finished products but also understanding and decision making capabilities. Rubrics for hybrid human AI projects should evaluate both technical proficiency and creative vision, with criteria potentially including appropriate and strategic AI use, critical evaluation of AI outputs, demonstration of understanding principles underlying technical decisions, originality and creative risk taking, and overall artistic effectiveness. These multidimensional assessments ensure that grades reflect comprehensive competence rather than ability to produce superficially acceptable outputs through any means.

5.6 Integration with Existing Music Programs

Implementing AI augmented curricula requires decisions about how new content integrates with existing program structures. Standalone courses dedicated to AI in music production work well for institutions with sufficient enrollment and resources by allowing deep exploration without compromising existing courses, though they risk isolating AI as a specialty topic rather than integral competency. Embedded modules distributed across multiple existing courses like composition, production techniques, mixing, and mastering integrate AI more thoroughly into curricula despite requiring coordination across courses and training multiple faculty members, leading many institutions toward hybrid approaches that include foundational standalone courses introducing AI concepts followed by embedded modules in subsequent courses applying those concepts in specific contexts.

Prerequisite considerations ensure students enter AI focused instruction with necessary foundational knowledge, with fundamental audio concepts, basic DAW proficiency, and core music theory typically preceding AI integration so students have grounding to evaluate AI outputs intelligently rather than accepting them uncritically, though requirements should not be so extensive that AI exposure comes only in advanced study since earlier introduction with appropriate scaffolding can motivate skill development by showing how foundational competencies enable sophisticated creative work. Sequencing within degree programs should move from awareness to proficiency to specialization, with early exposure familiarizing students with AI presence in contemporary practice without requiring immediate mastery, intermediate courses developing functional competence with common tools and integrating AI into standard workflows, and advanced courses offering specialization in particular AI applications like generative composition, intelligent mixing, or sound design, allowing students to develop expertise aligned with career interests through thoughtful sequencing that ensures all graduates possess fundamental AI literacy while allowing deeper engagement for particularly interestedStudents.

6. ASSESMENT INTEGRITY IN AI ASSISTED CREATIVE WORK

6.1. Challenges of Authorship and Attribution

The integration of artificial intelligence into music production education necessitates a reconsideration of assessment integrity. As generative systems become capable of producing melodic structures, harmonic progressions, and fully arranged compositions, distinguishing between student authored creativity and algorithmic output becomes increasingly complex. This ambiguity challenges conventional grading systems that assume clear individual authorship and raises concerns regarding originality, academic honesty, and fair evaluation. When AI systems meaningfully contribute to compositional outcomes, the criteria used to assess creativity must be carefully redefined (Roldan-Cardona *et al.*, 2025).

6.2. Process Oriented Evaluation

To address these concerns, assessment must shift from a purely product based model toward a process oriented framework. Instructors should evaluate documented stages of creative development, including drafts, revisions, and reflective explanations of artistic decisions. Requiring students to disclose how AI tools were used, such as prompts employed, outputs selected, and modifications applied, introduces transparency while reinforcing accountability. In this approach, AI is framed not as a shortcut, but as a creative collaborator whose role is critically examined within the learning process.

6.3. Rubric Redesign and Performance Verification

Assessment rubrics must differentiate between conceptual originality, technical proficiency, strategic AI integration, and critical reflection. Such differentiation ensures that human agency remains central to evaluation. Complementary strategies, including oral defenses or live demonstrations, further strengthen authorship verification by requiring students to articulate and justify their creative processes. Ultimately, safeguarding integrity in AI assisted creative work requires structured transparency and pedagogical clarity. By embedding reflective practice and documentation into evaluation systems, educators can uphold academic standards while encouraging responsible and innovative engagement with emerging technologies.

While assessment frameworks provide structural guidance for evaluation, the integration of AI systems introduces deeper concerns regarding authorship, transparency, and academic integrity. These concerns require dedicated examination.

7. CASE STUDIES AND INSTITUTIONAL PRACTICES

Understanding how educational institutions are successfully integrating AI into music production curricula requires examining real world implementations across diverse contexts. These case studies from institutions ranging from prestigious specialized colleges to accessible community programs reveal both common success factors and context specific adaptations, providing practical guidance grounded in actual experience rather than theoretical speculation (Chang and Wang, 2025).

7.1 Case Study: Berklee College of Music

Berklee College of Music has been a leader in integrating AI into contemporary music education, particularly through its Electronic Production and Design (EPD) Department. Beginning in 2020, Berklee adopted a phased approach to AI integration, starting with faculty workshops and pilot courses before expanding into core production classes and launching specialized electives in AI-assisted composition and production by 2022–2023. This gradual rollout allowed instructors to build expertise while incorporating student feedback to refine curriculum design.

Berklee's integration philosophy emphasizes AI as an augmentation tool rather than a replacement for foundational musical skills. Students continue to learn traditional music theory, composition, audio engineering, and manual mixing techniques before applying AI tools. This ensures they understand the logic behind AI-generated suggestions and can operate effectively without technological assistance. Ethical considerations are also embedded in the curriculum, encouraging critical reflection on authorship, originality, and technology's role in creative work.

AI tools are incorporated across multiple course levels. Foundational production courses use Google Magenta Studio to support compositional exploration, with students refining AI-generated ideas using traditional techniques. Advanced mixing classes integrate iZotope Neutron and Ozone, requiring students to compare manual and AI-assisted workflows and justify their creative decisions. LANDR is used in mastering units as an analytical benchmark, with students evaluating differences between automated and human mastering approaches.

7.2 Case Study SAE Institute

SAE Institute, with more than 50 campuses across 20 countries, adopted a global network approach to integrate AI into its curriculum while balancing standardization and regional flexibility. The institution faced the challenge of maintaining consistent educational quality across campuses with different technological infrastructures, cultural contexts, and industry demands. To address this, SAE developed standardized AI-related curriculum components that could be deployed globally, while allowing campuses to adapt tools, project themes, and teaching strategies to local needs.

Implementation followed a phased strategy, beginning with pilot programs at campuses with strong technological capabilities. Data from these pilots was used to assess effectiveness, identify challenges, and refine support systems before expanding AI integration to additional campuses. Standardized elements included core learning objectives in AI literacy, required technical competencies, and shared assessment frameworks, ensuring consistency in educational outcomes. At the same time, campuses retained autonomy in selecting specific AI tools and tailoring coursework to relevant regional industries.

Industry partnerships played a crucial role in SAE's AI integration efforts. Collaborations with companies such as iZotope, LANDR, and emerging AI music startups provided educational software licenses, early access to beta technologies, and opportunities for institutions to contribute feedback on tool development. These partnerships created mutual benefits by enhancing educational resources while supporting industry innovation.

Consistent with SAE's hands-on, studio-based pedagogy, AI was embedded directly into real production workflows. Students engaged in practical studio projects, met industry-style deadlines, collaborated in team-based production environments, and learned to troubleshoot AI limitations. This experiential approach fostered technical proficiency, creative confidence, and professional readiness rather than purely theoretical knowledge.

7.3 Comparative Analysis Across Institutions

Despite institutional differences, successful AI integration programs share notable commonalities in core competencies around critical evaluation of AI outputs, balanced integration of traditional and AI assisted techniques, and ethical awareness of AI implications appearing consistently across diverse institutions. Popular tool choices overlap significantly, with Magenta Studio, iZotope products, LANDR, and cloud collaboration platforms appearing repeatedly, suggesting these represent current best in class options for educational contexts. Pedagogically, comparative approaches having students work with and without AI tools then reflect on differences prove effective across institutional types.

Across institutional contexts, several factors consistently predict successful AI integration. Faculty buy in and training prove essential, as programs where instructors embrace AI integration enthusiastically and receive adequate professional development succeed far better than those where AI is imposed reluctantly on unprepared faculty. Administrative support, including budget allocation, strategic planning accommodation, and recognition in faculty workload, enables sustainable implementation. Student engagement strategies that generate enthusiasm while addressing concerns create positive learning cultures, with programs attending to these factors regardless of institutional type or resource level achieving better outcomes than those with superior resources but inadequate attention to human dimensions of change(Peng, 2025).

8. ETHICAL AND CREATIVE CHALLENGES IN AI MUSIC PEDAGOGY

The integration of artificial intelligence into music production education raises profound ethical and creative questions that extend far beyond technical implementation. These challenges touch on fundamental issues of authorship, authenticity, equity, and the very nature of creativity itself, requiring educators to grapple with questions that lack clear answers about when AI assistance becomes plagiarism, how to preserve human creativity in an age of algorithmic efficiency, who owns music created with AI tools, what biases are embedded in these systems, and how to address them.

8.1 Authorship and Originality in AI Generated Music

Ownership of AI generated creative work remains legally and philosophically complex because copyright law in most jurisdictions requires human authorship, creating ambiguity when AI contributes substantially. In the US, the Copyright Office states that works entirely created by AI without human creative input cannot be copyrighted, but works where AI assists human creators may qualify depending on the nature and extent of human contribution. Educational institutions face practical challenges regarding student rights versus platform rights, as many AI tools claim ownership or usage rights over generated content. Institutional policies are urgently needed to clarify required disclosure when using AI tools, rights students retain over AI assisted work, how to credit work for competitions or portfolios, and institutional positions when legal questions arise.

Transparent disclosure of AI usage should be standard practice, with students clearly communicating AI contributions versus their own creative decisions, though the appropriate level of detail for such disclosure remains debated. Citation standards for AI tools are still emerging, with minimum citations likely including AI tool name and version, date used, general nature of contribution, and prompting approach if relevant. Academic honesty guidelines must be updated to address AI specific scenarios, clarifying acceptable AI assistance for different assignment types, required disclosure, and consequences for policy violations, with guidelines communicated clearly and early to ensure students understand expectations before confusion or violations occur(Shu, 2025).

8.2 Academic Integrity Concerns

Plagiarism traditionally meant presenting others' work as one's own, but AI complicates this definition, raising questions about whether submitting AI generated music claiming it as entirely one's own creation constitutes plagiarism, fraud, or something else entirely. Detecting AI generated submissions poses technical challenges, as unlike text plagiarism detectors that can search for matching passages, identifying AI generated music is more difficult, with audio fingerprinting sometimes identifying samples or loops but recognizing compositional patterns as AI generated requiring sophisticated analysis not yet widely available.

Defining acceptable AI assistance levels requires careful thought, as educators must determine whether using AI to fix timing issues in performances is analogous to spell checkers in written work, what about using AI to suggest chord progressions that students then arrange manually, or having AI generate initial melodic ideas that students substantially develop. These scenarios represent different levels of AI contribution requiring policies to distinguish between them through clear communication about which uses are acceptable for specific assignments, as some projects might explicitly encourage extensive AI use while others require minimal or no AI assistance to assess particular competencies(Newman and Feng, no date).

Honor codes must be adapted to address AI specific challenges beyond traditional focus on individual work and cited sources, with updated codes potentially addressing obligations to disclose AI tool usage accurately, expectations to understand and be able to explain all work submitted regardless of how it was created, commitments to developing genuine competence rather than merely producing acceptable outputs through any means, and recognition that learning requires appropriate challenge and that excessive AI reliance undermines educational goals even when technically permitted.

8.3 Human Creativity vs. Algorithmic Efficiency

The risk of over reliance on AI tools represents one of educators' most significant concerns, as when AI handles complex tasks effortlessly, students may never develop skills to complete those tasks independently. A student who always uses AI mixing assistants potentially never learns to hear frequency masking problems or judge appropriate compression settings themselves, with genuine skill deficits resulting from excessive dependency that limit professional capabilities when AI tools are not available or appropriate. Losing fundamental skills has cascading consequences, as surface level competencies built entirely on AI assistance crumble when situations demand deeper understanding, with producers unable to understand basic harmonic principles struggling when clients request specific musical changes and engineers relying entirely on automated mixing unable to troubleshoot unsatisfactory results or adapt approaches to unusual musical material.

Balancing efficiency with skill development requires intentional pedagogical design where students work primarily manually early in learning sequences to develop foundational understanding, introduce AI tools strategically as competence grows to accelerate workflows for mastered tasks while maintaining manual practice in developing areas, and enable advanced students to use AI extensively for efficiency while retaining ability to work manually when needed, ensuring AI enhances rather than replaces learning by serving as scaffolds that can be removed once competence is established rather than permanent crutches preventing full capability development.

Figure 2. Human AI creative balance model in music production education

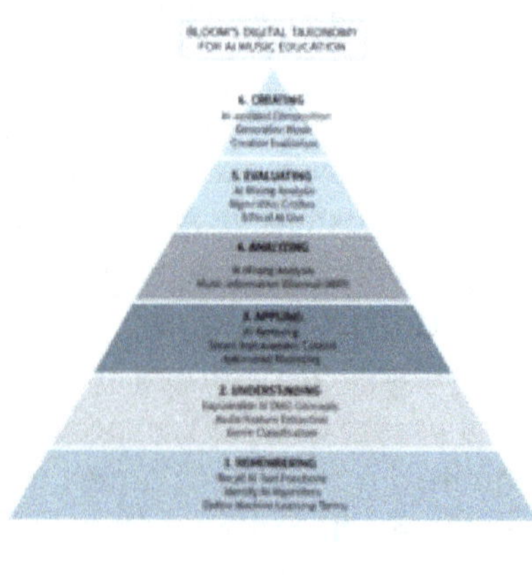

8.4 Bias and Representation in AI Models

AI systems learn from training data, inheriting whatever biases exist in that data. Most widely available music AI models were trained predominantly on Western commercial music, creating Western centric musical biases. These systems often perform poorly with non Western scales, microtonal systems, or rhythmic structures outside common Western practice, may suggest chord progressions that sound correct according to Western harmonic conventions while failing to recognize equally valid approaches from other musical traditions, limiting creative possibilities and implicitly devaluing non Western musical practices by treating them as errors to be corrected rather than legitimate alternatives.

Understanding algorithmic biases must be integral to AI education, not afterthoughts. Students need to recognize that AI outputs reflect training data choices rather than universal musical truths. They should learn to ask what data was this system trained on, what perspectives are represented or missing, and when AI suggestions seem wrong, might they reflect cultural or stylistic biases rather than actual errors. This critical literacy enables students to use AI tools more thoughtfully and avoid perpetuating biases uncritically.

8.5 Accessibility and Digital Divide

Cost barriers to AI integrated music production education are substantial. Professional AI tools often require expensive subscriptions, high performance computers cost significantly more than basic models, and reliable high speed internet isn't universally available. These barriers disproportionately affect students from lower income backgrounds, rural areas, or developing countries, creating equity concerns. Solutions include comprehensive lab access with sufficient hours to accommodate diverse student schedules, equipment lending programs allowing students to check out laptops or audio interfaces, institutional software licenses covering AI tools for all students, and course designs incorporating free or low cost alternatives to commercial tools where possible.

Tools for students with disabilities represent both opportunity and challenge in AI integrated education. AI technologies can enhance accessibility, with voice controlled interfaces benefiting students with limited mobility, visual feedback systems helping deaf and hard of hearing students work with audio, and intelligent assistance scaffolding learning for students with cognitive disabilities. However, not all AI tools are designed accessibly. Educators should prioritize tools with strong accessibility features and advocate with developers when tools create barriers.

8.6 Employment and Future Career Concerns

Student anxieties about AI replacing music production jobs are understandable given broader automation discussions. Educators must address these concerns honestly rather than dismissively. Some tasks will indeed be automated, with AI already handling basic mastering work that junior engineers once performed and generative tools producing stock music more efficiently than human composers for certain applications. However, reframing AI as augmentation tool rather than replacement helps students see opportunities alongside challenges, with most realistic future scenarios involving humans and AI working together, with AI handling routine tasks while humans provide creative vision, emotional intelligence, cultural knowledge, and complex decision making that AI cannot replicate.

Emerging career opportunities related to AI represent significant growth areas. Roles like AI tool developers for music applications, prompt engineers who specialize in getting optimal results from generative systems, hybrid producers who excel at integrating AI into creative workflows, and consultants helping companies implement AI in music production contexts all offer career paths for students with appropriate skills. Educators should help students see themselves not as victims of technological change but as positioned to lead in newly emerging professional landscapes.

9. FUTURE DIRECTIONS AND POLICY IMPLICATIONS

As artificial intelligence continues its rapid evolution, music production education stands at a critical juncture. The decisions educators, institutions, and policymakers make today will shape not only how the next generation of music professionals is trained but also the broader relationship between human creativity and technological capability(Sep, 2024).

9.1 Emerging Trends in AI Music Technology

AI tools in music production education are still in early development, but they are rapidly evolving toward more interactive and collaborative systems. Real time AI assistants are expected to function like creative studio partners, working alongside human producers during live sessions by offering immediate, context aware suggestions. This shift requires new teaching approaches that emphasize continuous dialogue between students and AI, rather than treating AI as a simple tool.

Emotion responsive composition systems represent another major advancement. Unlike current generative AI that focuses on style or genre, these systems aim to interpret emotional intent, allowing composers to define emotional arcs such as tension, relief, or gradual shifts in mood. This can help students better understand how musical decisions shape emotional impact, though it also raises concerns about limiting creative freedom or encouraging formulaic emotional expression.

Visual audio synthesis technologies are also becoming more advanced, enabling translation between sound and visuals, such as generating images from music or soundscapes from visual input. As these systems improve, music education may increasingly integrate multimedia perspectives, encouraging students to think across sensory domains and design cohesive audiovisual experiences. Meanwhile, text to music generation has progressed significantly, allowing users to create complex musical arrangements from natural language descriptions. These tools are likely to become standard for rapid prototyping, mood exploration, and creative experimentation.

Gesture controlled AI instruments further blur the boundaries between composition, performance, and choreography. Using motion sensing and AI interpretation, these systems allow gestures, conducting movements, or dance to generate expressive soundscapes, promoting more embodied and physically interactive approaches to music creation.

9.3 Institutional Policy Recommendations

Effective AI integration in education requires strategic institutional planning rather than ad hoc adoption. Institutions must articulate clear long-term goals aligned with their mission, student demographics, regional industry demands, and available resources. Strategic plans should define expected AI competency levels for graduates, outline phased curriculum development, establish assessment metrics, and set realistic implementation timelines. Without structured planning, AI initiatives risk becoming fragmented or unsustainable.

Budget allocation must extend beyond purchasing software licenses. Comprehensive integration includes investment in hardware infrastructure, faculty training, technical support staff, and ongoing system maintenance. Institutions should calculate total cost of ownership across multiple years to ensure sustainable funding rather than short-term experimentation. Equity must also guide financial decisions so AI adoption does not widen disparities between well-resourced and under-resourced programs or student groups.

Governance structures are equally important. AI ethics committees composed of faculty, students, administrators, and external experts can oversee responsible adoption. These committees review proposed tools, develop ethical guidelines, address concerns, and ensure alignment with institutional values. Curriculum review processes should formally incorporate AI considerations as technologies evolve, while student representation ensures that policies reflect real learning needs rather than administrative assumptions.

9.4 Accreditation and Standards

Accreditation bodies for music programs should consider including AI competencies in standards, recognizing that technological fluency is now fundamental to professional practice. Updated criteria might specify that graduates demonstrate understanding of AI's role in contemporary music production, ability to use AI tools appropriately and critically, awareness of ethical issues surrounding AI in creative practice, and capacity to adapt to emerging technologies throughout careers. Quality assurance for hybrid programs that blend traditional and AI augmented approaches requires new evaluation frameworks ensuring that technology integration enhances rather than diminishes educational quality.

Micro credentials for AI music skills allow targeted competency certification without requiring complete degree programs. Working professionals could earn credentials in specific AI applications including generative composition and AI assisted mixing, updating skills without returning to formal education. Educational institutions could offer these alongside degree programs, creating additional revenue while serving broader professional communities. Industry recognized certificates carry particular value, signaling to employers that holders possess specific competencies, with collaboration between educational institutions and industry organizations in developing these credentials ensuring they address actual professional needs(Souza *et al.*, 2024)

9.5 Research Agenda for AI in Music Education

The field of AI in music production education needs robust research addressing key questions. Long term impact on creative development remains poorly understood, requiring investigation into whether AI assistance in early learning helps or hinders students' long term creative capacities. Comparative effectiveness studies could identify which pedagogical approaches work best for different student populations and learning objectives. Best practices for diverse populations deserve investigation, as strategies effective for traditional college students in residential programs may not transfer to adult learners, online students, or populations in different cultural contexts(Sciences, 2025).

Mixed methods approaches combining quantitative and qualitative research provide comprehensive understanding. Quantitative studies might measure learning outcomes, compare achievement across different pedagogical approaches, or analyze patterns in tool usage. Qualitative research could explore students' creative experiences, investigate how AI affects artistic identity development, or examine faculty perspectives on pedagogical challenges. Longitudinal studies tracking students over extended periods are particularly valuable for understanding long term impacts, following students from initial AI exposure through graduation and into early careers to reveal how AI integrated education affects professional trajectories, creative development, and technological adaptability.

9.6 Global Perspectives and Cultural Considerations

AI tools could play valuable roles in preserving endangered musical forms by documenting performance practices, analyzing structural patterns, and potentially generating new examples following traditional principles. However, such applications require careful cultural collaboration ensuring respect for communities whose traditions are being preserved, as AI developed and controlled externally risks cultural appropriation and misrepresentation. Respectful integration of diverse aesthetics in AI systems requires training data including global musical diversity rather than predominantly Western commercial music, demanding intentional effort and resources as well as partnerships with musicians and scholars from diverse traditions. Avoiding cultural homogenization means ensuring AI doesn't impose Western musical logic on non Western practices or flatten musical diversity into algorithmic monoculture, with education helping students recognize these risks and advocate for more inclusive AI development(Penelitian, Kepustakaan and Pendidikan, 2024).

9.7 Equity and Inclusion Imperatives

Supporting underrepresented groups in AI integrated music production education requires proactive efforts including targeted recruitment, mentorship programs, scholarships and financial aid strategies addressing technological access barriers, and curriculum content reflecting diverse cultural perspectives. Community outreach programs introducing AI music tools to younger students or underserved communities can build pipelines into formal education programs, addressing equity earlier in educational trajectories.

Culturally responsive pedagogy acknowledges and values diverse cultural backgrounds, musical traditions, and learning approaches students bring to classrooms. This means incorporating diverse musical examples, recognizing various cultural relationships to technology, and creating space for students to connect AI tools to their own cultural contexts. Diverse representation in curriculum content including examples, case studies, guest speakers, and faculty signals that AI music production is for everyone, not only certain demographic groups. Safe spaces for experimentation where students can take creative risks without judgment or penalty encourage exploration essential to creative development, particularly important for students from groups traditionally underrepresented in technology fields who may face stereotype threat or imposter syndrome.

10. CONCLUSION SYNTHESIS AND RECOMMENDATIONS

This examination of artificial intelligence integration in music production education shows that AI is fundamentally reshaping the field. It influences what students learn, how they learn, and which competencies are required for professional success. AI tools are no longer experimental additions but core elements of modern production workflows. As a result, integrating AI into curricula is not optional; it is essential for aligning education with industry realities.

Successful models appear across different institutional contexts. Berklee College of Music demonstrates how well-resourced institutions can combine cutting-edge AI innovation with strong foundations in traditional musicianship and production skills. SAE Institute shows how standardized AI components can be implemented across international campuses while allowing local adaptation. Community colleges and online platforms further prove that meaningful AI integration does not require elite funding; free and low-cost tools, when used strategically, can still produce strong learning outcomes. (Dewanto and Sugiharto, 2025).

Common success factors include proactive planning, faculty training, balanced emphasis on traditional and AI-driven skills, and critical evaluation rather than blind adoption. Ethical awareness must accompany technical instruction. Ongoing challenges remain, including intellectual property concerns, academic integrity issues, potential skill dependency, algorithmic bias, unequal access to technology, and employment anxieties about automation's impact on creative careers.

Core Recommendations

For educators, the key recommendation is to embrace AI as a pedagogical partner rather than viewing it as a threat or a cure all. AI can accelerate workflows, generate creative options, and provide instant feedback, but it should support not replace human instruction. Curriculum and assessment must prioritize creative autonomy alongside technical proficiency. Technological skill without artistic vision produces software operators rather than expressive artists. Educators should therefore design learning experiences that strengthen critical judgment, artistic identity, and independent decision making while students gain fluency with AI tools. Continuous professional development is also essential, as effective teaching in AI integrated environments requires adapting to evolving technologies and practices.(Zhang *et al.*, 2025).

For institutions, strategic investment must extend beyond software licenses to include hardware, technical staff, faculty training, and long term maintenance. Sustainable planning and realistic budgeting ensure integration lasts beyond initial implementation. Clear ethical guidelines addressing authorship, academic integrity, bias, and accessibility should be collaboratively developed and regularly updated. Interdisciplinary collaboration between music, computer science, philosophy, and education departments can further strengthen AI integration (Roldan-Cardona *et al.*, 2025).

For policymakers, accreditation standards should reflect AI's growing centrality by incorporating competencies in ethical AI use and technological adaptability. Supporting equitable access through funding and policy reduces disparities. Industry partners can contribute through affordable educational licensing, curriculum collaboration, internships, and mentorship opportunities that connect classrooms with professional practice.(Hou, 2024).

Balancing Innovation with Artistic Integrity

Throughout this discussion, a central theme is the balance between technological innovation and artistic integrity. Human creativity remains essential. Although AI tools can accelerate workflows, generate ideas, and automate repetitive tasks, they cannot replace emotional depth, cultural awareness, intentional meaning making, or authentic self expression. Music production education must continue to cultivate these uniquely human qualities while integrating advanced technologies into learning environments.

The most effective framework views AI as an amplifier rather than a replacement. Like earlier technological advancements in music, AI should expand creative possibilities without diminishing human agency. It can help producers realize ideas more efficiently and explore new sonic territories, but artistic direction and final judgment must remain human driven. Educational success is evident when graduates use AI strategically to support their creative visions instead of allowing algorithms to dictate outcomes.

Ultimately, education aims to develop critical and creative thinkers. Students should learn to question AI suggestions, assess when algorithmic assistance enhances artistic goals, and decide independently when to accept or reject automated input. Assessment methods, teaching strategies, and institutional culture must reinforce that the objective is to form reflective creative professionals who are technologically fluent, not merely operators of advanced tools.

Final Reflections

Music production education stands at a crossroads, facing decisions that will shape the field's trajectory for decades. The choices made now about how to integrate AI, what values to prioritize, what competencies to emphasize, and how to address emerging challenges will determine whether technology enhances or diminishes creative education, whether access expands or contracts, and whether graduates are empowered or constrained by their technological training.

The vision articulated throughout this analysis is fundamentally optimistic: AI integrated music production education can produce empowered, versatile music professionals who combine deep traditional competencies with sophisticated technological fluency, who think critically about tools rather than using them uncritically, who maintain strong artistic voices while leveraging algorithmic capabilities, and who adapt confidently to ongoing technological evolution throughout their careers. Realizing this vision requires sustained commitment, thoughtful implementation, ongoing refinement based on evidence and experience, and willingness to navigate complexity without simple answers.

The future of music production is undeniably collaborative, bringing together human creativity and artificial intelligence in partnerships that leverage the strengths of both. Humans provide intentionality, emotional intelligence, cultural knowledge, aesthetic judgment, and authentic expression. AI provides computational power, pattern recognition, tireless iteration, and access to vast training data. Together, they enable creative possibilities neither achieves alone. Education's role in shaping this collaborative future is profound, as pedagogical choices made today determine how tomorrow's music professionals understand AI, relate to it, and deploy it ethically and effectively.

This examination closes with an invitation for ongoing dialogue and innovation. The questions explored here lack final answers but require continuous engagement as technologies evolve, as understanding deepens, and as new challenges and opportunities emerge. Educators, students, administrators, policymakers, industry professionals, and researchers all have contributions to make to this ongoing conversation. The field needs spaces for sharing experiences, debating approaches, investigating questions rigorously, and building collective knowledge.

REFERENCES

Cangelosi, A. *et al.* (2022) 'Towards Responsible AI Music ': Chang, J. and Wang, Z. (2025) 'Applied Mathematics and Nonlinear Sciences Efficiency of AI Technology Application in Music Education - A Perspective Based on Deep Learning Model DLMM', 10(1), pp. 1–19.

Dewanto, A.D. and Sugiharto, A. (2025) 'Artificial Intelligence Applications in Piano Education : An Informatics-Based Literature Analysis', (2021).

Hou, C. (2024)... *Applied Mathematics and Nonlinear Sciences*, *9*(1), 1–13.

Luo, Z. (2023). The Application of Artificial Intelligence Technology in Music Education. *The Application of Artificial Intelligence Technology in Music Education*, *0*(1), 305–310. DOI: 10.54254/2753-7048/35/20232155

Newman, M., & Feng, Q. (no date) *Exploring the Collaborative Co-Creation Process with AI : A Case Study in Novice Music Production, Proceedings of ACM Conference (Conference'17)*. arXiv.

Penelitian, J. H., Kepustakaan, K., & Pendidikan, B. (2024)... *Jurnal Kependidikan*, *10*(3), 881–892.

Peng, Y. (2025). The analysis of optimization in music aesthetic education under artificial intelligence. *Scientific Reports*, *15*(1), 11545. DOI: 10.1038/s41598-025-96436-2 PMID: 40185937

Pons, J., Zukowski, Z., Parker, J. D., Carr, C. J., Taylor, J., & Evans, Z. (2025). Music and Artificial Intelligence: Artistic Trends. *arXiv preprint arXiv:2508.11694.*

Roldan-Cardona, M., Chacón-Castro, M., Jadán-Guerrero, J., Salvador-Ullauri, L., & Acosta-Vargas, P. (2025). Music Education with Artificial Intelligence for Inclusive and Sustainable Early Childhood Learning. *Emerging Science Journal*, *9*(Special Issue), 201–218. DOI: 10.28991/ESJ-2025-SIED1-012

Shu, L. (2025). Research on Creative Practice in Music Education Empowered by AI. *Research on Creative Practice in Music Education Empowered by AI*, *0*(1), 106–113. DOI: 10.54254/2753-7048/67/20251010

Souza, R. M., Cezario, B. S., Affonso, E. O., Machado, A. D., Vieira, D. P., Chinelli, C. K., Haddad, A. N., Dusek, P. M., Miranda, M. G., Soares, C. A. P., & Guedes, A. L. (2024). My human rights smart city: Improving human rights transparency identification system. *Sustainability (Basel)*, *16*(3), 1274. DOI: 10.3390/su16031274

Sularso, S., Wadiyo, W., Cahyono, A., Suharto, S., & Pranolo, A. (2025). A Comprehensive Visualization for Music Education and Artificial Intelligence. *JOIV: International Journal on Informatics Visualization*, *9*(2), 718–727. DOI: 10.62527/joiv.9.2.2500

Wani, T. A., & Chavan, P. (2025)... *AI-GENERATED COMPOSITIONS IN MUSIC EDUCATION*, *6*, 92–101. DOI: 10.29121/shodhkosh.v6.i3s.2025.67

Zhang, H.. (2025). Artificial Intelligence in Music Education : A Scoping Review of Practices. *Strategies, and Challenges*, *4*(5), 69–82. Advance online publication. DOI: 10.56397/JARE.2025.09.09

Zhang, S., Lu, X., & Liu, X. (no date) 'Study on the Influence of AI Composition Software on Students ' Creative Ability in Music Education', pp. 1–17.

Chapter 14

Exploring Cloud-Based AI Music Tools and Their Implications for IT and Multimedia Professionals

Gerli Ryza Delos Santos Reyes
http://orcid.org/0009-0002-9404-1286
Pangasinan State University, Philippines

ABSTRACT

Cloud-based artificial intelligence (AI) music tools have rapidly evolved into scalable, production-ready systems embedded within cloud-native infrastructures, fundamentally reshaping multimedia workflows and digital content ecosystems. Leveraging transformer-based architectures, diffusion models, and distributed GPU-enabled environments, these platforms generate full-length musical compositions from minimal prompts and parameter controls. Despite growing commercial adoption, limited empirical research has examined their professional and technological implications for Information Technology (IT) and multimedia practitioners. This study investigates the architectural foundations, workflow transformation dynamics, ethical risk perceptions, and competency requirements associated with cloud-based AI music systems. Using a mixed-method descriptive-correlational design, data were collected from 184 IT and multimedia professionals across academic and industry sectors to examine AI literacy, adoption intention, ethical concern, and workflow integration patterns.

DOI: 10.4018/979-8-3373-6279-3.ch014

INTRODUCTION

Artificial Intelligence (AI) has undergone a profound transformation over the past decade, evolving from rule-based computational systems and narrow machine learning applications into large-scale generative architectures capable of producing outputs that approximate human creativity. Early AI systems were primarily confined to deterministic problem-solving, pattern recognition, and predictive analytics within clearly defined parameters. In contrast, contemporary generative AI systems leverage deep neural networks, large-scale representation learning, and transformer-based architectures to synthesize text, images, code, audio, and music with increasing contextual coherence and stylistic fluency (Goodfellow et al., 2016; Vaswani et al., 2017). This shift reflects not merely incremental algorithmic refinement but a structural transformation in computational capability, marking a transition from analysis-focused systems to synthesis-oriented intelligence. Foundation models trained on massive multimodal datasets now demonstrate the capacity to generate complex creative artifacts that were previously assumed to require exclusively human cognition (Bommasani et al., 2021). As a result, generative AI has begun to redefine production ecosystems across creative, industrial, and academic domains.

Parallel to advancements in AI architectures, cloud computing has fundamentally reshaped the deployment and accessibility of advanced computational systems. The emergence of elastic cloud infrastructures, GPU-accelerated computing clusters, containerized microservices, and orchestration platforms such as Kubernetes has enabled scalable, distributed machine learning systems to operate reliably in production environments (Fox & Patterson, 2020; Burns et al., 2016). Cloud-native architectures facilitate horizontal scaling, automated resource provisioning, real-time inference, and continuous integration/continuous deployment (CI/CD) pipelines. These infrastructure developments eliminate the need for high-cost local hardware while enabling professionals to access powerful generative models via web-based interfaces and application programming interfaces (APIs). Consequently, generative AI systems are increasingly delivered through cloud-based service models, commonly conceptualized as AI-as-a-Service (AIaaS), thereby democratizing access to computationally intensive creative tools (Armbrust et al., 2010). The convergence of AI innovation and cloud infrastructure has created a technological ecosystem in which generative systems function as distributed, service-oriented platforms rather than standalone software applications.

Within this broader generative landscape, music synthesis represents one of the most technically demanding and commercially disruptive applications. Unlike text generation, which primarily operates on symbolic token sequences, music modeling requires capturing complex temporal dependencies, harmonic progressions, rhythmic patterns, melodic contour, timbral texture, and dynamic variation across hierarchical structures (Sturm et al., 2021). Music unfolds across time, requiring coherence at both micro and macro levels—ranging from note transitions to structural form, thematic development, and emotional arc. The computational challenge lies in modeling not only sequential patterns but also layered acoustic dimensions that interact simultaneously. Recent advances in transformer-based sequence modeling and diffusion-based generative mechanisms have significantly enhanced the fidelity, structural coherence, and stylistic realism of AI-generated music (Dhariwal & Nichol, 2021; Huang et al., 2018). These systems can now produce multi-minute compositions from minimal textual prompts, parameter inputs, or multimodal conditioning signals.

The rise of cloud-based AI music platforms represents a significant departure from traditional Digital Audio Workstations (DAWs). Conventional music production requires manual composition, arrangement, instrumentation, mixing, and mastering within technically sophisticated editing environments. Producers must manipulate waveforms, adjust equalization curves, manage dynamic compression, layer tracks, and iterate through compositional refinements. In contrast, AI-powered music tools abstract much of this process through prompt-based interfaces that translate high-level directives—such as genre, mood, tempo, instrumentation, or thematic description—into complete musical outputs (Sun et al., 2024). This abstraction shifts creative agency from granular construction to high-level orchestration, redefining professionals' roles from composers and engineers to supervisors, curators, and evaluators of generative outputs. The creative workflow becomes less about direct production and more about iterative prompting, output selection, and post-generation refinement.

However, beyond their surface-level creative functionality, AI music systems operate within complex cloud ecosystems that integrate API gateways, inference engines, distributed storage systems, container orchestration frameworks, load balancing mechanisms, authentication protocols, and continuous deployment pipelines. These systems embody cloud-native architectural principles and DevOps methodologies (Burns et al., 2016). The generative model is only one component within a broader distributed infrastructure responsible for scalability, reliability, latency management, and security. As organizations increasingly integrate AI-generated music into marketing campaigns, gaming environments, educational content, and multimedia production pipelines, IT professionals must address infrastructure optimization, API stability, data governance, cybersecurity risk mitigation, and compliance management. Therefore, AI music platforms should be conceptualized not merely as creative applications but as distributed AI-cloud infrastructures requiring multidisciplinary oversight, technical governance, and institutional policy alignment.

Simultaneously, the rapid proliferation of AI-generated music has intensified debates over its ethical, legal, and regulatory implications. Concerns regarding intellectual property ownership, transparency of training datasets, authorship attribution, stylistic imitation, and derivative works remain unresolved across many legal jurisdictions (Galuszka, 2024; Herington et al., 2025). Musicians and industry stakeholders have expressed apprehensions regarding unauthorized use of copyrighted materials in training datasets and the potential displacement of creative labor (Nayar, 2025). Questions arise concerning whether AI-generated compositions constitute original works, whether prompt authors qualify as legal creators, and how liability should be distributed among developers, users, and platform providers. These unresolved regulatory ambiguities introduce professional risks that may significantly influence adoption decisions across enterprise, educational, and commercial contexts. For IT managers, such uncertainties translate into compliance exposure and governance risk. Multimedia professionals raise concerns regarding authenticity, artistic identity, and professional legitimacy.

Existing scholarship on AI music generation has primarily focused on three primary dimensions: algorithmic development, perceptual evaluation, and ethical critique. Technical research has examined neural network architectures, diffusion processes, symbolic representation strategies, and waveform synthesis techniques (Sturm et al., 2021). Perceptual studies have explored listener acceptance, comparative quality evaluation, and human-AI differentiation capacities (Sun et al., 2024). Ethical analyses have investigated authorship dilemmas, cultural implications, and intellectual property challenges (Galuszka, 2024). While these studies contribute valuable insights, limited empirical research has examined how IT and multimedia professionals—those directly responsible for deploying, integrating, and governing these systems—perceive and operationalize AI music tools within real-world workflows. The professional adaptation dimension remains insufficiently explored.

Research in digital transformation suggests that AI adoption reshapes professional competencies rather than eliminating roles outright (Schwab, 2017). Technological augmentation often redistributes tasks, requiring hybrid skillsets that combine domain expertise with AI literacy, infrastructure understanding, and governance awareness (Brynjolfsson & McAfee, 2014). However, these macro-level insights have not been specifically validated within the context of AI-driven music production ecosystems. There is limited empirical evidence examining how AI literacy interacts with ethical risk perception, how professional background influences authenticity skepticism, and how workflow integration capacity shapes adoption intention. Furthermore, most discussions of AI music tools remain either technologically descriptive—focusing on system performance—or sociologically interpretive—emphasizing cultural implications—without synthesizing both perspectives into an interdisciplinary framework.

The absence of structured empirical investigation linking cloud infrastructure, AI literacy, professional workflow adaptation, and ethical perception constitutes a significant research gap. Understanding this intersection is essential for guiding responsible implementation strategies, informing regulatory policy development, and designing higher education curricula that adequately prepare future professionals. As generative AI becomes embedded within multimedia production pipelines, institutions must anticipate shifts in required competencies, governance standards, and collaborative practices between technical and creative disciplines.

Given the accelerating integration of generative AI into multimedia ecosystems, empirical examination of professional perceptions is increasingly urgent. IT and multimedia practitioners operate at the nexus of infrastructure management, creative production, and regulatory oversight. Their readiness, competence, and risk assessment frameworks will significantly influence how AI music systems are adopted, scaled, and governed. Without systematic investigation, organizational policies risk being reactive rather than evidence-based, and curriculum reforms may fail to align with emerging professional demands.

By integrating architectural analysis with statistical modeling of adoption patterns, this research contributes an interdisciplinary understanding of how generative AI reshapes professional practice within cloud-native multimedia environments. Specifically, the study examines the technological architectures that underpin cloud-based AI music systems, including distributed deployment models, inference pipelines, and service orchestration frameworks. It investigates how IT and multimedia professionals perceive, evaluate, and integrate these tools into operational workflows, taking into account both technical performance and creative transformation. The research identifies the competencies required for effective and responsible adoption, particularly in relation to AI literacy, cloud infrastructure awareness, workflow integration expertise, and governance oversight. It further analyzes how ethical and legal concerns—such as intellectual property ambiguity, authorship attribution, and regulatory uncertainty—influence professional acceptance. Finally, the study explores how higher education programs can design curriculum frameworks that prepare future professionals to operate competently and ethically within AI-driven multimedia ecosystems.

By situating AI music generation within the broader contexts of cloud computing, digital transformation, professional skill evolution, and digital governance, this study advances discourse on AI adoption in creative-technical domains. Rather than framing AI music tools as isolated creative novelties, the research conceptualizes them as systemic AI-cloud infrastructures with measurable professional implications. In doing so, it provides empirical evidence and a competency-oriented framework to inform governance strategies, organizational policy development, institutional planning, and curriculum innovation in an era increasingly defined by AI-mediated creativity.

2 LITERATURE REVIEW

Artificial intelligence (AI) music generation has undergone a rapid transformation since 2020, evolving from symbolic composition systems to large-scale, audio-first generative architectures capable of producing stylistically coherent, commercially viable outputs. Earlier computational music approaches relied primarily on rule-based systems, probabilistic grammars, and MIDI-like symbolic representations, which modeled note sequences without fully capturing timbral richness or long-range compositional structure. While such systems demonstrated the feasibility of automated composition, they struggled to produce naturalistic waveform-level outputs or preserve hierarchical musical form across extended durations (Sturm et al., 2021). The transition toward deep learning architectures—particularly transformer-based models and diffusion mechanisms—marked a structural shift in the technical foundations of music synthesis. Transformers, initially introduced for natural language processing (Vaswani et al., 2017), were adapted to model sequential dependencies in music through attention-based mechanisms that preserved long-range coherence. Huang et al. (2018) demonstrated that relative attention mechanisms significantly improved structural continuity compared to recurrent architectures, laying the groundwork for more expressive symbolic generation. However, symbolic modeling remained limited in acoustic realism, prompting a shift toward waveform-level and latent audio representations that can directly model sound textures and expressive dynamics.

A pivotal advancement in audio-first generative modeling occurred through the adaptation of diffusion-based architectures for sound synthesis. Diffusion models iteratively refine random noise into structured outputs, offering greater fidelity than adversarial approaches (Dhariwal & Nichol, 2021). While diffusion gained prominence in image synthesis, subsequent adaptations demonstrated its applicability to audio domains, enabling higher-quality waveform generation and stylistic nuance. Concurrently, hierarchical generative frameworks were introduced to bridge semantic intent and acoustic realization. MusicLM (Agostinelli et al., 2023) exemplifies this development, combining semantic representations with audio generation pipelines to produce high-fidelity compositions conditioned on textual descriptions. Empirical evaluations suggested improved adherence to prompt semantics and enhanced audio realism compared to earlier baselines. Nonetheless, MusicLM and similar systems continue to face persistent challenges in maintaining multi-minute structural coherence and fine-grained editability. Even with hierarchical modeling, full-song architecture—such as verse–chorus–bridge arrangements—remains partially emergent rather than explicitly controllable.

Research between 2024 and 2025 increasingly focused on long-form coherence and controllability, recognizing the limitations of short-clip generation. Evans et al. (2024) introduced a latent diffusion-transformer hybrid trained on extended temporal contexts, enabling multi-minute coherent outputs with improved structural stability. Their findings addressed a critical bottleneck in generative audio: models' tendency to produce repetitive loops or abrupt transitions over longer durations. However, even these systems struggled to guarantee musically meaningful structure, such as thematic development, harmonic modulation, and tension–resolution dynamics. Zhao (2025), in a comprehensive systematic review of text-to-music systems, concluded that while fidelity and duration improved significantly, controllability remains underdeveloped. Specifically, prompt adherence at the semantic level does not equate to formal compositional control, as users often lack mechanisms to manipulate musical form beyond high-level descriptors directly. Wei (2025) similarly argued that the transition from AI tools to AI collaborators necessitates interpretability and structured editing interfaces, not merely generative capacity.

In parallel with academic innovation, commercial platforms prioritized usability and production-readiness. Systems such as Stable Audio and other cloud-based AI music services emphasize rapid generation, intuitive interfaces, and integration with multimedia workflows. While academic research focuses on evaluation transparency and reproducibility, commercial platforms often prioritize user experience and time efficiency, sometimes without detailed disclosure of training datasets or benchmarking metrics. This divergence reflects differing institutional incentives: research communities pursue methodological rigor, whereas industry targets workflow integration and scalability. Such divergence introduces tension between performance optimization and governance transparency, particularly in professional deployment contexts.

Another expanding frontier in AI music research is multimodal, context-aware generation. Traditional text-to-music systems rely primarily on semantic conditioning, mapping descriptive phrases to stylistic attributes. However, real-world multimedia production frequently involves audiovisual alignment, requiring synchronization between music and visual content. SONIQUE (2024) exemplifies multimodal generation by conditioning music synthesis on video input, incorporating scene transitions, motion patterns, and emotional tone. Multimodal transformers (Baltrušaitis et al., 2019) and cross-modal alignment models (Tsai et al., 2019) have demonstrated that shared latent spaces can enhance coherence between audio and visual modalities. Nonetheless, multimodal evaluation metrics remain underdeveloped compared to text-generation benchmarks. Measuring alignment quality across tempo, dynamic intensity, and scene mood introduces methodological complexity. As a result, research in multimodal AI music remains exploratory, with limited consensus regarding standardized evaluation frameworks.

Evaluation methodology represents a persistent challenge in generative music research. Unlike text generation, where metrics such as BLEU or ROUGE approximate semantic similarity, music lacks universally accepted quantitative benchmarks.

Sturm et al. (2021) argue that evaluation must encompass structural coherence, stylistic authenticity, emotional expressiveness, and perceptual acceptance. Human listening studies remain dominant but introduce limitations in terms of subjectivity and reproducibility. Yang et al. (2020) proposed feature-based similarity measures for comparing spectral and harmonic properties, yet these metrics fail to capture aesthetic nuance or compositional form fully. Consequently, discrepancies persist between model-reported improvements and professional acceptability standards. This gap becomes particularly salient when AI music systems transition from research prototypes to operational tools within enterprise and educational contexts.

Beyond technical performance, professional acceptance constitutes a critical dimension of AI music integration. Technology adoption research provides foundational theoretical frameworks for analyzing user acceptance. The Technology Acceptance Model (TAM) posits that perceived usefulness and perceived ease of use significantly predict behavioral intention (Davis, 1989). The Unified Theory of Acceptance and Use of Technology (UTAUT) extends this by incorporating performance expectancy, effort expectancy, social influence, and facilitating conditions (Venkatesh et al., 2003). However, generative AI introduces unique variables absent from traditional enterprise software contexts. Jussupow et al. (2021) demonstrate that AI literacy influences trust and perceived controllability in AI systems. Rai et al. (2022) further argue that algorithmic transparency mitigates perceived risk, thereby enhancing the likelihood of adoption. In creative domains, however, perceptions of authenticity and identity threat emerge as additional factors influencing acceptance.

Sun et al. (2024) found that audience reception of AI-generated music varies depending on the framing context. When AI is presented as augmentative rather than substitutive, acceptance increases. This aligns with augmentation theory, which suggests that users resist automation when it threatens professional identity but embrace it when it is positioned as a supportive enhancement (Brynjolfsson & McAfee, 2014). Cheng (2025) observed similar dynamics in music education, where generative AI supports ideation and accessibility but raises concerns about deskilling and ambiguity around authorship. These findings indicate that acceptance of AI music tools depends not solely on output quality but also on perceived implications for professional legitimacy.

Legal and governance concerns further shape adoption dynamics. The 2024 lawsuits initiated by the Recording Industry Association of America against generative music platforms underscore unresolved questions about the use of training data for copyright purposes. Legal analyses in communications and entertainment law journals between 2025 and 2026 highlight ongoing uncertainty regarding derivative works and liability distribution. Herington et al. (2025) emphasize that musicians' ethical concerns center on dataset transparency and fairness in attribution. Galuszka (2024) argues that generative AI challenges traditional notions of authorship and creative ownership, necessitating updated regulatory frameworks. From an institutional theory perspective (DiMaggio & Powell, 1983), regulatory ambiguity can exert both coercive and normative pressures on organizations, encouraging cautious adoption and the adoption of formal governance structures.

Platform integrity issues introduce additional governance dimensions. Deezer's 2025 fraud report revealed manipulation of streaming systems using AI-generated tracks and bot amplification, demonstrating cybersecurity vulnerabilities within music distribution ecosystems. Such incidents illustrate that AI-generated music is not merely a creative innovation but also a risk vector that requires anomaly detection, monitoring systems, and compliance protocols. DevOps and cloud governance research (Humble & Farley, 2010) underscores the importance of continuous monitoring and secure deployment practices in distributed systems. However, few peer-reviewed studies integrate cloud-native AI deployment analysis with music generation research.

Cloud-native architectures underpinning AI music platforms involve microservices, containerization, API gateways, and scalable inference engines. Fox and Patterson (2020) describe cloud-native software engineering as emphasizing elasticity, resilience, and automation. In the context of AI music, these principles enable real-time generation, multi-user concurrency, and integration into multimedia production pipelines. However, operational complexity introduces concerns regarding latency optimization, cost efficiency, authentication security, and compliance auditing. The literature remains fragmented between model development and infrastructure governance, with limited empirical research examining how professionals navigate this intersection.

Digital transformation scholarship suggests that AI adoption reshapes competency requirements rather than eliminating roles outright (Schwab, 2017). Hybrid skill sets combining domain expertise and AI literacy are increasingly essential (Brynjolfsson & McAfee, 2014). In creative industries, this implies a shift toward supervisory orchestration, ethical evaluation, and cross-disciplinary collaboration. However, empirical validation within AI music ecosystems remains sparse. Few studies investigate whether technical literacy moderates ethical concern, whether professional background influences perceptions of authenticity, or how governance awareness affects adoption intention.

Across 2020–2026, the literature demonstrates rapid improvements in generation quality, duration coherence, and multimodal integration. Peer-reviewed studies highlight transformer-based sequence modeling (Huang et al., 2018), diffusion audio synthesis (Dhariwal & Nichol, 2021), hierarchical text-to-music pipelines (Agostinelli et al., 2023), and long-form diffusion coherence (Evans et al., 2024). Systematic reviews (Zhao, 2025; Wei, 2025) identify persistent gaps in controllability and evaluation. Acceptance studies (Sun et al., 2024; Cheng, 2025) reveal framing-dependent perception. Legal scholarship (Galuszka, 2024; Herington et al., 2025) underscores regulatory ambiguity. Governance analyses highlight copyright disputes and platform fraud risks. However, these strands remain siloed mainly.

The existing body of knowledge prioritizes model capability, perceptual response, or legal analysis in isolation. Few peer-reviewed investigations synthesize architectural deployment, professional workflow adaptation, AI literacy, ethical risk perception, and governance oversight into a unified empirical framework. This fragmentation obscures the systemic nature of cloud-based AI music platforms, which function as distributed infrastructures embedded within regulatory and professional ecosystems. Addressing this gap requires an interdisciplinary synthesis that links technical architecture with adoption modeling and competency analysis. The present study responds to this need by integrating architectural examination with statistical analysis of professional perception, thereby contributing a comprehensive framework for understanding AI music systems as both technological infrastructures and professional transformation catalysts.

Table 1. AI music generation systems (2020–2026)

Study / System	Year	Input / Conditioning	Core Technique	Key Contribution	Common Limitation
Jukebox (Donahue et al.)	2020	Lyrics + genre	VQ-VAE + Transformer	Raw waveform generation with vocal realism	Extremely high compute; limited editing control
Music Transformer (Huang et al.)	2020	MIDI symbolic	Relative attention Transformer	Improved long-range structural modeling	Lacks waveform realism
Diffusion Audio (Dhariwal & Nichol adaptation)	2021	Noise → waveform	Diffusion modeling	High-fidelity texture synthesis	Slow inference; resource-intensive
MusicLM (Agostinelli et al.)	2023	Text → music	Hierarchical seq-to-seq	Strong semantic adherence; realism	Limited fine-grained control
Stable Audio	2023– 2024	Text prompts	Latent diffusion (cloud-based)	Production-ready outputs	Dataset transparency limited
Long-Form Diffusion (Evans et al.)	2024	Text/audio	Diffusion-Transformer hybrid	Multi-minute coherence	Structural form is still emergent
SONIQUE	2024	Video → music	Multimodal transformer	Scene-aware scoring	Alignment evaluation immature
Zhao Review	2025	Literature synthesis	Systematic review	Identifies the controllability gap	Limited workflow integration focus
Wei Review	2025	Literature synthesis	Meta-analysis	Frames shift to AI collaborators	Governance integration limited

Table 1 illustrates the progression from symbolic systems to audio-first and multimodal generative architectures. While realism and duration coherence improved significantly between 2020 and 2025, limitations persist in controllability, interpretability, and structured editing capabilities. Academic systems prioritize fidelity and coherence, whereas industry systems emphasize usability and scalability for deployment. However, governance and professional workflow integration remain underexplored.

Table 2. Governance, adoption, and professional impact (2020–2026)

Focus Area	Representative Study	Key Insight	Professional Implication
AI Literacy & Trust	Jussupow et al. (2021)	Literacy increases algorithm trust	Training is essential for adoption
Explainability & Risk	Rai et al. (2022).	Transparency mitigates risk perception	Governance frameworks required
Perception & Framing	Sun et al. (2024).	Augmentation framing improves acceptance	Communication strategy matters
Educational Impact	Cheng (2025)	AI reshapes music pedagogy	Curriculum redesign needed
Copyright Disputes	RIAA lawsuits (2024)	Legal uncertainty deters adoption	Vendor due diligence is critical
Ethical Concerns	Herington et al. (2025).	Dataset transparency is a central issue	Compliance audits necessary
Platform Fraud	Deezer Report (2025)	AI streaming manipulation risk	Cybersecurity monitoring required

Table 2 demonstrates that adoption of AI music tools is influenced not only by technical performance but also by literacy, governance, ethical concerns, and regulatory uncertainty. Professional integration, therefore, depends on interdisciplinary competencies encompassing AI knowledge, compliance awareness, and cybersecurity oversight.

Collectively, the literature from 2020 to 2026 reveals three dominant trajectories: technical advancement in generative modeling, increasing concern regarding governance and intellectual property, and emerging recognition of professional adaptation challenges. While waveform realism and long-form coherence have improved substantially, controllability and interpretability remain unresolved. Legal disputes and platform integrity risks further complicate organizational adoption. Professional acceptance studies indicate that AI literacy moderates perceived risk and authenticity concerns. However, no unified framework integrates these strands into an empirical model that links architecture, literacy, governance perceptions, and workflow adaptation.

This fragmentation underscores the need for interdisciplinary analysis. AI music systems operate as distributed infrastructures requiring cloud-native governance and professional oversight. Understanding adoption dynamics, therefore, requires integration of generative architecture research, technology acceptance theory, institutional governance analysis, and competency transformation scholarship. The present study addresses this gap by combining architectural examination with empirical adoption modeling among IT and multimedia professionals.

3. METHODOLOGY

This study employed a mixed-method research design to examine professional perceptions, adoption dynamics, and competency requirements associated with cloud-based AI music systems. The methodological approach was selected to provide both statistical rigor and contextual depth, enabling comprehensive analysis of technological literacy, ethical perception, professional identity, and integration behavior within IT and multimedia domains. By combining quantitative survey analysis with qualitative thematic exploration, the research captures measurable adoption patterns while also uncovering experiential insights that may not be reflected in numerical data alone.

A descriptive-correlational framework guided the quantitative component of the study. Structured survey instruments were administered to IT and multimedia professionals across academic and industry sectors to measure perceived technical efficiency, workflow enhancement, AI literacy, ethical concern, and adoption intention. Statistical techniques, including descriptive statistics, Pearson's correlation, independent-samples t-tests, and multiple regression, were used to determine predictive relationships and group differences. This analytical strategy enabled the identification of key determinants influencing adoption behavior and the relative magnitude of their effects.

The qualitative component consisted of open-ended survey responses, which were analyzed using thematic coding. This allowed deeper interpretation of professional narratives regarding productivity, intellectual property ambiguity, authenticity skepticism, and emerging hybrid skill requirements. The integration of quantitative and qualitative findings enhances validity through methodological triangulation. Overall, the methodological design ensures robust empirical examination of how cloud-native AI music systems are evaluated and integrated within professional creative-technical ecosystems.

3.1 Research Design

This study adopted a mixed-method, descriptive-correlational research design to examine the technological, professional, and ethical implications of cloud-based AI music tools among Information Technology (IT) and multimedia professionals. The selection of a mixed-method approach was guided by the need to capture both measurable adoption patterns and deeper contextual interpretations of professional adaptation. While quantitative analysis enabled statistical modeling of relationships among AI literacy, ethical perception, and adoption intention, qualitative thematic analysis enriched the interpretation of practitioner experiences and workflow integration challenges.

The design's descriptive component was intended to document prevailing trends in perception, usability, and workflow transformation. As cloud-based AI music tools represent an emerging technological domain, establishing baseline descriptive statistics was necessary before advancing toward predictive modeling. This approach ensured that empirical findings were grounded in observable patterns rather than speculative assumptions. The correlational component examined relationships among key constructs, particularly the association between AI literacy and adoption intention, as well as the potential inhibitory effect of ethical and legal concerns. By employing correlational analysis and regression modeling, the study aimed to move beyond descriptive observation and identify statistically significant predictive relationships.

The integration of quantitative and qualitative strands allowed for methodological triangulation, thereby strengthening internal validity. Quantitative findings provided measurable evidence of adoption trends, while qualitative insights clarified how professionals interpret the role of AI music tools within cloud-native production ecosystems. This integrated design enhances the credibility and interpretive robustness of the study.

3.2 Participants, Sampling Strategy, and Power Analysis

The study targeted professionals working in environments where cloud-based AI music tools are both technologically and professionally relevant. Specifically, participants were drawn from two groups: IT professionals responsible for system deployment, infrastructure management, and DevOps operations, and multimedia professionals engaged in content production, music integration, and creative workflow management. This dual-group structure enabled comparative analysis across technical and creative domains. A purposive sampling strategy was employed to ensure domain relevance. The inclusion criteria required respondents to have at least 1 year of professional experience and prior exposure to AI-based tools. This criterion-based selection minimized measurement noise and enhanced construct validity by ensuring that responses reflected informed professional judgment rather than general public opinion.

A total of 184 valid responses were obtained, including 96 IT professionals (52.2%) and 88 multimedia professionals (47.8%). Participants were affiliated with academic institutions (45%), creative industries (38%), and software development companies (17%). This distribution allowed for cross-sector comparison while maintaining adequate representation of both technical and creative practice environments. To ensure statistical adequacy, an a priori power analysis was conducted using G*Power software. Assuming a medium effect size ($f^2 = 0.15$), significance level $\alpha = 0.05$, and desired power $(1 - \beta) = 0.80$ for multiple regression with three predictors, the minimum required sample size was 77. The obtained sample size ($N = 184$) exceeded this threshold substantially, resulting in statistical power above 0.95. This strengthens confidence in the reliability of inferential findings and reduces the risk of Type II errors.

3.3 Instrument Development and Construct Validation

Data were collected using a structured online questionnaire explicitly designed for this study. The instrument consisted of six primary constructs: AI Literacy, Technical Integration Perception, Workflow Transformation, Ethical and Legal Concern, Creative Authenticity Perception, and Adoption Intention. Each construct was operationalized using multiple Likert-scale items, rated on a five-point scale from strong disagreement (1) to strong agreement (5). Instrument development was guided by established literature on technology adoption and digital transformation to ensure conceptual alignment. Items measuring AI Literacy assessed familiarity with generative models, cloud infrastructure principles, and prompt-based interaction systems. Adoption Intention items evaluated willingness to integrate AI music tools into professional workflows. Ethical Concern items captured perceptions regarding intellectual property uncertainty, authorship ambiguity, and compliance risks.

Content validity was established through expert review by three subject matter specialists in artificial intelligence systems, multimedia production, and cloud computing governance. Feedback was incorporated to refine wording clarity, eliminate redundancy, and ensure construct coherence. A pilot test was conducted with a small subset of professionals to verify item comprehension and response consistency prior to full deployment. Construct validity was examined through exploratory factor analysis (EFA) using principal axis factoring with varimax rotation. The Kaiser-Meyer-Olkin (KMO) measure of sampling adequacy was 0.87, indicating suitability for factor analysis, and Bartlett's test of sphericity was significant ($p < 0.001$). Factor loadings above 0.60 were retained. Composite reliability (CR) values exceeded 0.80 and average variance extracted (AVE) values exceeded 0.50, supporting convergent validity. Discriminant validity was confirmed using the Fornell-Larcker criterion.

3.4 Data Collection Procedure

Data collection was conducted over four weeks using an online survey platform. To enhance ecological validity, participants were instructed to interact directly with selected cloud-based AI music platforms prior to completing the survey. They were required to generate at least one sample output using prompt-based systems to ensure experiential familiarity. This experiential requirement minimized hypothetical bias and strengthened the authenticity of responses. By grounding perceptions in actual interaction rather than abstract awareness, the study enhanced measurement credibility. Participants evaluated usability, integration potential, and perceived risks based on firsthand experience.

Ethical safeguards were implemented throughout the data collection process. Participants were informed of the research's academic purpose, voluntary participation, anonymity, and data confidentiality. No personally identifiable information was collected. Institutional approval procedures were followed in accordance with research ethics standards. The response rate was monitored weekly to ensure balanced representation across professional categories. Data screening procedures were conducted to identify incomplete responses or patterned answering, and invalid entries were removed prior to statistical analysis.

3.5 Data Analysis and Assumption Testing

Quantitative data were analyzed using statistical software to compute descriptive and inferential statistics. Descriptive analysis included means, standard deviations, and frequency distributions to summarize patterns of perception across constructs. These measures provided baseline insight into professional attitudes toward AI music tools. Pearson correlation analysis was conducted to examine relationships among AI literacy, ethical concern, and adoption intention. Independent-samples t-tests were used to compare IT and multimedia professionals on key variables. Multiple linear regression analysis assessed the predictive influence of AI literacy and ethical perception on adoption intention.

Before conducting inferential analysis, the statistical assumptions were rigorously tested. Normality was assessed using skewness and kurtosis statistics, which fell within acceptable ranges (± 1.5). Homoscedasticity was evaluated using residual scatterplots, and linearity was verified through partial regression plots. Multicollinearity diagnostics indicated Variance Inflation Factor (VIF) values below 3.0, suggesting no collinearity issues. The Durbin-Watson statistic confirmed independence of residuals. Effect sizes were reported alongside p-values to provide a practical significance interpretation. Cohen's d was used for t-test comparisons, and R^2 values were reported for regression models. This dual reporting approach enhances interpretive transparency and aligns with best practices in quantitative research reporting.

3.6 Qualitative Analysis

Open-ended responses were analyzed using inductive thematic analysis to capture nuanced professional concerns and contextual insights. Responses were coded into thematic categories, including authenticity skepticism, intellectual property uncertainty, workflow acceleration, and skill adaptation requirements. Two independent coders conducted thematic categorization to enhance credibility and minimize subjective bias. Inter-coder reliability was assessed using Cohen's kappa ($\kappa = 0.88$), indicating high agreement. Discrepancies were resolved through consensus discussion.

Qualitative findings were integrated with quantitative results to provide interpretive triangulation. For example, themes related to intellectual property uncertainty were examined alongside quantitative measures of ethical concern. This integration strengthened explanatory depth and supported a more comprehensive understanding of adoption behavior.

3.7 Methodological Rigor and Limitations

To ensure methodological rigor, multiple validation and reliability procedures were implemented throughout the research process. First, the study employed methodological triangulation by integrating quantitative statistical modeling with qualitative thematic analysis. This dual approach strengthened internal validity by allowing numerical findings to be contextualized with practitioner narratives. Second, instrument development followed systematic construct operationalization grounded in prior technology adoption and digital transformation literature. Content validity was established through expert panel review, ensuring alignment between survey items and theoretical constructs. Construct validity was further supported through exploratory factor analysis, composite reliability testing, average variance extracted (AVE) evaluation, and discriminant validity verification using the Fornell-Larcker criterion.

Statistical rigor was reinforced through power analysis, assumption testing, and effect size reporting. The a priori power analysis confirmed that the achieved sample size exceeded minimum thresholds for detecting medium-sized effects in regression analysis. Assumptions of normality, homoscedasticity, linearity, independence, and absence of multicollinearity were tested prior to inferential modeling, thereby enhancing the reliability of statistical conclusions. Reporting of effect sizes (Cohen's d and R^2) alongside p-values ensured practical interpretability rather than reliance solely on statistical significance. Ecological validity was strengthened by requiring participants to engage directly with selected cloud-based AI music platforms before completing the survey. This experiential exposure reduced hypothetical bias and improved response authenticity. Additionally, inter-coder reliability testing in the qualitative analysis phase (Cohen's $\kappa = 0.88$) enhanced interpretive consistency and reduced subjectivity in thematic categorization. Ethical safeguards, including voluntary participation and confidentiality protection, further contributed to research integrity.

Despite these methodological strengths, several limitations must be acknowledged. First, the cross-sectional design restricts causal inference; while correlational relationships were identified, directional causality cannot be definitively established. Longitudinal or experimental designs may provide stronger causal validation in future research. Second, data were collected through self-reported measures, which may introduce response bias, including social desirability or overestimation of technical competence. Although experiential interaction mitigated some bias, objective performance metrics were not included. Third, the purposive sampling strategy, while enhancing domain relevance, limits generalizability beyond IT and multimedia professionals with prior AI exposure. Professionals without familiarity with AI were intentionally excluded, potentially narrowing the representativeness of the findings across broader organizational populations. Fourth, rapid technological evolution in generative AI systems may limit temporal generalizability, as platform capabilities and legal contexts continue to evolve between 2024 and 2026.

Finally, while exploratory factor analysis supported construct validity, confirmatory factor analysis (CFA) and complete structural equation modeling (SEM) were not implemented in the present study. Future research may extend this framework using SEM to examine mediation and moderation effects among AI literacy, ethical perception, and adoption intention. Incorporating objective workflow performance indicators and cross-national samples would further enhance generalizability and external validity. Overall, while acknowledging inherent constraints, the study demonstrates substantial methodological rigor through validated instrumentation, statistical robustness, triangulated analysis, and transparency in assumption testing. These measures collectively strengthen confidence in the reliability and interpretive value of the findings.

4. RESULTS AND DISCUSSION

This section presents and interprets the study's empirical findings, integrating statistical analysis with theoretical grounding and contextual interpretation. The purpose is not merely to report numerical results but to examine how cloud-based AI music systems are perceived, evaluated, and strategically integrated within professional environments characterized by technological complexity, creative identity considerations, and regulatory uncertainty. By synthesizing descriptive statistics, correlational findings, regression modeling, group comparisons, and qualitative thematic insights, this discussion offers a multidimensional account of adoption dynamics. The findings illuminate the interplay among AI literacy, perceptions of governance, professional identity, and institutional context, thereby providing a comprehensive understanding of how generative AI reshapes creative-technical ecosystems. Rather than viewing adoption as a linear decision process, the results suggest a negotiated integration pattern in which technological opportunity and ethical caution coexist.

4.1 Descriptive Analysis of Professional Perceptions

Descriptive statistics indicate a robust level of professional acceptance of cloud-based AI music systems. On a five-point Likert scale, perceived technical efficiency recorded a mean of 4.21 (SD = 0.62), reflecting strong agreement that these tools enhance productivity. This finding suggests that respondents widely perceive AI music platforms as operationally beneficial rather than disruptive. Participants reported that generative systems reduce the time required for background scoring, ideation, and initial draft production, particularly in environments where turnaround speed is critical. In multimedia production contexts such as digital marketing, educational video development, social media campaigns, and advertising, the ability to generate structured music rapidly was consistently described as a competitive advantage. This perception aligns with the augmentation framework, which conceptualizes AI as an enhancer of human efficiency rather than a direct replacement.

Workflow enhancement yielded a mean of 4.08 (SD = 0.71), further supporting the conclusion that AI music systems are perceived as workflow accelerators. Respondents emphasized that generative tools enable experimentation with multiple stylistic variations without requiring manual arrangement or advanced compositional expertise. This capability was particularly valued in collaborative settings where rapid prototyping facilitates creative iteration among team members. Importantly, future adoption willingness demonstrated a high mean score of 4.17 (SD = 0.65), with 78.3% of participants expressing intent to integrate AI music tools into professional practice. This firm's behavioral intention indicates that generative AI is not merely tolerated but increasingly regarded as a strategic component of digital production ecosystems.

However, alongside high perceived usefulness, moderate-to-high ethical and authenticity concerns were evident. Ethical concerns regarding intellectual property ambiguity and regulatory uncertainty were recorded at a mean of 3.94 (SD = 0.88). This suggests that professionals remain cautious about potential copyright disputes, derivative liability, and issues of dataset transparency. Creative authenticity skepticism yielded a mean of 3.76 (SD = 0.91), indicating persistent doubts regarding the originality of AI-generated compositions. Approximately 69.0% of respondents expressed concern about intellectual property risks, while 61.4% questioned the artistic authenticity of generative outputs. These findings reveal a dual-attitude phenomenon: operational enthusiasm coexists with governance sensitivity. Professionals appear to adopt a pragmatic yet risk-aware orientation, embracing productivity gains while demanding clearer regulatory safeguards.

4.2 Correlational Findings

Correlation analysis provided more profound insight into the relationships among literacy, governance perception, and adoption intention. The strong positive relationship between AI literacy and adoption perception ($r = 0.62$, $p < 0.01$) indicates that knowledge of generative architectures and cloud-native deployment significantly predicts acceptance. This effect size is substantial in social science research, suggesting that literacy is a primary driver of integration rather than a peripheral variable. Professionals familiar with diffusion modeling, transformer-based systems, API orchestration, and containerized deployment reported higher levels of confidence in system controllability and reliability.

From a theoretical perspective, this finding reinforces the Technology Acceptance Model's (Davis, 1989) assumptions, which posit that perceived usefulness and ease of use are enhanced by knowledge and familiarity. In AI contexts, literacy may reduce the perception of algorithmic opacity, thereby mitigating uncertainty. Individuals who understand how generative models are trained and deployed are more likely to perceive them as controllable tools rather than autonomous creative agents.

Conversely, ethical concern exhibited a moderate negative correlation with adoption intention ($r = -0.41$, $p < 0.05$). This indicates that intellectual property ambiguity and regulatory uncertainty function as psychological inhibitors. However, the magnitude of the correlation suggests moderation rather than suppression. Ethical concerns temper enthusiasm for adoption but do not eliminate it. Professionals may adopt generative systems within conditional frameworks that include compliance audits, licensing verification, and governance oversight.

The coexistence of strong literacy-driven adoption and moderate governance caution supports the notion of calibrated adoption behavior. Rather than dichotomous acceptance or rejection, professionals engage in risk-adjusted integration strategies.

4.3 Group Differences: IT vs Multimedia Professionals

Statistically significant differences emerged between IT and multimedia professionals. IT professionals demonstrated higher technical acceptance scores ($p < 0.05$), reflecting greater comfort with distributed cloud architectures and DevOps pipelines. For IT respondents, AI music systems were conceptualized as service-oriented infrastructures involving API endpoints, inference engines, scalability management, and monitoring dashboards. Their evaluation criteria centered on latency, uptime reliability, and system interoperability.

In contrast, multimedia professionals reported significantly higher concerns about authenticity and authorship ($p < 0.05$). Creative practitioners evaluated AI music systems through the lens of stylistic individuality and artistic integrity. Several respondents articulated concerns about homogenization and potential erosion of creative voice. These concerns reflect identity-based resistance rather than technological skepticism. Multimedia professionals did not necessarily doubt system functionality but questioned its implications for authorship legitimacy.

This disciplinary divergence underscores the sociotechnical nature of AI adoption. IT professionals frame AI as infrastructure, whereas multimedia professionals frame it as creative agency. Successful integration, therefore, requires bridging technical performance metrics with cultural and artistic considerations.

4.4 Regression Analysis: Predictors of Adoption

Multiple regression analysis provided more profound insight into the relative influence of AI literacy, ethical concern, and professional background on adoption intention. The model was statistically significant ($p < 0.01$), indicating that the predictors collectively explain a meaningful proportion of variance in professionals' willingness to integrate cloud-based AI music systems into their workflows. AI literacy emerged as the most potent positive predictor (β significant, $p < 0.01$), demonstrating that knowledge of generative architectures, cloud-native deployment principles, and AI operational logic substantially increases the likelihood of adoption. The strength of this predictive relationship underscores that literacy is not merely an enabling variable but a structural determinant of acceptance behavior.

The predictive strength of AI literacy suggests that familiarity with diffusion models, transformer-based sequence generation, prompt engineering strategies, API orchestration, and containerized deployment reduces the perceived opacity of AI systems. In other words, literacy appears to transform AI from a "black box" into a comprehensible tool. Professionals who understand system architecture are better positioned to realistically assess risks, identify controllability mechanisms, and implement governance safeguards. This aligns with digital transformation theory, which posits that competence and knowledge reduce technological anxiety and foster proactive engagement (Schwab, 2017). Rather than perceiving AI as an uncontrollable external force, literate professionals interpret it as a modular infrastructure component subject to configuration and oversight.

Ethical concern remained a statistically significant negative predictor ($p < 0.05$), although its magnitude was smaller than that of literacy. This indicates that intellectual property ambiguity, regulatory uncertainty, and authenticity skepticism do exert downward pressure on adoption intention. However, ethical concerns do not eliminate adoption but moderate it. Professionals with high literacy levels may acknowledge governance risks while believing they can mitigate or manage them through structured compliance processes, vendor evaluation, and legal review. This finding reflects the negotiated nature of adoption in high-uncertainty technological domains.

Professional background also exerted a moderating effect, suggesting that disciplinary identity interacts with literacy and perceptions of governance. IT professionals, who often approach AI systems through infrastructural and operational lenses, may perceive literacy as a pathway to control and optimization. Multimedia professionals, who emphasize artistic authorship and originality, may interpret ethical ambiguity as more consequential to professional identity. This interaction effect highlights that adoption dynamics are sociotechnical rather than purely cognitive.

Overall, the regression model demonstrates that adoption of cloud-based AI music systems is multifactorial and relational. Literacy enhances perceived usefulness and reduces uncertainty; ethical concern introduces caution; disciplinary orientation shapes interpretation. Adoption, therefore, emerges not as a purely rational efficiency calculation but as a negotiated evaluation balancing competence, governance risk, and professional values.

4.5 Integration of Quantitative and Qualitative Findings

The integration of quantitative and qualitative findings provides a more nuanced understanding of adoption behavior than either method alone. While statistical analysis identified literacy and ethical concern as key predictors, thematic analysis revealed how these constructs are experienced and articulated in professional contexts. Productivity acceleration emerged as a dominant qualitative theme, reinforcing high mean scores for perceived efficiency and workflow enhancement. Respondents frequently described AI music systems as "time-saving engines," "creative assistants," and "rapid prototyping tools." These descriptions suggest that professionals do not view generative AI as replacing human creativity but as accelerating early-stage ideation and experimentation.

However, qualitative narratives also illuminated dimensions of ethical concern that numerical scales did not fully capture. Intellectual property uncertainty was frequently discussed in terms of client-facing commercial deployment. Respondents expressed hesitation about using AI-generated music in high-visibility projects without clear licensing. This indicates that governance concerns are particularly salient in contexts involving financial liability and brand reputation. Ethical concern, therefore, may intensify when the stakes are elevated.

Creative authenticity skepticism also emerged as a salient qualitative theme. Multimedia professionals articulated fears of homogenization and stylistic dilution, suggesting that overreliance on generative systems could erode distinctive artistic voices. This concern reflects broader debates within creative industries regarding the cultural implications of algorithmic production. Importantly, these authenticity concerns coexist with recognition of productivity benefits, further reinforcing the dual-attitude phenomenon observed quantitatively.

Emerging hybrid skill requirements constituted another significant theme. Participants emphasized the need to combine AI literacy with governance awareness, understanding of cloud infrastructure, and ethical reasoning. This suggests that professionals perceive the future of multimedia production as requiring interdisciplinary competencies rather than isolated technical or creative expertise.

Triangulation confirms that adoption is literacy-dependent, governance-sensitive, and discipline-specific. Quantitative correlations identify structural relationships; qualitative narratives provide experiential depth. Together, they reveal that integration of AI music systems is conditional, strategic, and context-aware rather than automatic or uncritical.

4.6 Discussion in Relation to Research Questions

The findings directly address the research questions posed at the outset of the study. With respect to technological architectures, professionals conceptualize AI music platforms as distributed AI-cloud infrastructures embedded within scalable service ecosystems. Respondents referenced API endpoints, inference engines, container orchestration frameworks, and latency-optimization concerns, indicating that these systems are perceived as infrastructure rather than merely creative applications. This supports the argument that understanding architecture significantly influences adoption.

Regarding professional perception and workflow integration, high perceived usefulness and workflow enhancement scores indicate strong operational acceptance. However, adoption remains cautious and conditional due to governance ambiguity. Professionals appear willing to integrate generative tools for internal prototyping and exploratory tasks while reserving client-facing deployment for contexts with more explicit compliance assurances.

In terms of required competencies, results support a multidimensional competency framework. Effective integration requires cloud infrastructure management, AI model literacy, multimedia workflow adaptation, cybersecurity, awareness of intellectual property governance, and ethical decision-making capacity. This interdisciplinary profile reflects the hybrid nature of AI-driven creative ecosystems.

Ethical and legal concerns significantly influence professional acceptance, validating the literature's emphasis on intellectual property disputes and regulatory ambiguity as structural barriers to adoption. However, ethical concerns function as a moderating rather than a prohibitive factor, suggesting that governance frameworks can facilitate responsible adoption.

Finally, regarding educational preparation, the strong predictive role of AI literacy indicates that curriculum reform is essential. Higher education programs must integrate generative AI systems, cloud-native laboratories, compliance case studies, and interdisciplinary collaboration exercises to prepare students for AI-mediated professional environments.

4.7 Theoretical and Practical Implications

Theoretically, this study extends AI adoption frameworks by incorporating governance perception and professional identity variables into predictive models. The Traditional Technology Acceptance Model (TAM) and the Unified Theory of Acceptance and Use of Technology (UTAUT) frameworks emphasize perceived usefulness and ease of use; however, generative AI introduces identity-based and regulatory dimensions absent in conventional enterprise software contexts. The findings demonstrate that adoption in creative-technical ecosystems is co-determined by literacy, perceptions of governance, and disciplinary framing.

From a digital transformation perspective, AI literacy serves as a resilience factor, enabling professionals to navigate uncertainty without retreating from innovation. Ethical concern does not necessarily signal rejection but may represent an adaptive governance orientation seeking structured safeguards.

Practically, organizations should implement structured AI literacy programs to enhance informed decision-making. Compliance review protocols and intellectual property audits should be integrated into deployment pipelines. Interdisciplinary governance committees that bring together IT leaders and creative directors may facilitate balanced integration strategies. Higher education institutions must embed AI literacy, cloud-native DevOps training, and digital ethics modules within multimedia and IT curricula.

4.8 Synthesis

Overall, the findings demonstrate that cloud-based AI music systems are widely perceived as productivity enhancers yet remain ethically contested and governance-sensitive. Adoption is not binary; it is negotiated, literacy-driven, and moderated by professional identity and regulatory perception. Professionals appear to embrace generative AI pragmatically, recognizing operational benefits while advocating for structured oversight.

These results reinforce the study's central thesis: AI music systems function not merely as creative automation tools but as complex AI-cloud ecosystems embedded within sociotechnical and regulatory frameworks. Effective integration, therefore, requires interdisciplinary competencies, governance protocols, and institutional readiness. Adoption is best understood as a calibrated transformation process in which technological capability, ethical reasoning, and professional identity intersect dynamically.

REFERENCES

Agostinelli, A., Denk, T. I., Borsos, Z., Engel, J., Verzetti, M., Caillon, A., Huang, Q., Jansen, A., Roberts, A., Tagliasacchi, M., & Sharifi, M. (2023). MusicLM: Generating music from text. arXiv preprint arXiv:2301.11325. https://arxiv.org/abs/2301.11325

Baltrušaitis, T., Ahuja, C., & Morency, L.-P. (2019). Multimodal machine learning: A survey and taxonomy. *IEEE Transactions on Pattern Analysis and Machine Intelligence*, *41*(2), 423–443. DOI: 10.1109/TPAMI.2018.2798607 PMID: 29994351

Bommasani, R., Hudson, D. A., Adeli, E., Altman, R., Arora, S., von Arx, S., . . . Liang, P. (2021). On the opportunities and risks of foundation models. arXiv preprint arXiv:2108.07258. https://arxiv.org/abs/2108.07258

Brynjolfsson, E., & McAfee, A. (2014). *The second machine age: Work, progress, and prosperity in a time of brilliant technologies*. W. W. Norton.

Burns, B., Grant, B., Oppenheimer, D., Brewer, E., & Wilkes, J. (2016). Borg, Omega, and Kubernetes. *Communications of the ACM*, *59*(5), 50–57. DOI: 10.1145/2890784

Cheng, L. (2025). Generative artificial intelligence and the transformation of music education. *Music Education Research*, *27*(1), 45–60.

Davis, F. D. (1989). Perceived usefulness, perceived ease of use, and user acceptance of information technology. *Management Information Systems Quarterly*, *13*(3), 319–340. DOI: 10.2307/249008

Dhariwal, P., & Nichol, A. (2021). Diffusion models beat GANs on image synthesis. *Advances in Neural Information Processing Systems*, *34*, 8780–8794.

DiMaggio, P. J., & Powell, W. W. (1983). The iron cage revisited: Institutional isomorphism and collective rationality in organizational fields. *American Sociological Review*, *48*(2), 147–160. DOI: 10.2307/2095101

Donahue, C., McAuley, J., & Puckette, M. (2020). Jukebox: A generative model for music. arXiv preprint arXiv:2005.00341. https://arxiv.org/abs/2005.00341

Evans, R., Huang, Q., Engel, J., & Roberts, A. (2024). Long-form music generation with latent diffusion transformers. arXiv preprint.

Fox, A., & Patterson, D. (2020). *Engineering software as a service: An agile approach using cloud computing* (2nd ed.). Strawberry Canyon LLC.

Galuszka, P. (2024). Artificial intelligence and the transformation of music authorship. *Popular Music and Society*, *47*(2), 215–230.

Herington, J., McCormick, M., & Smith, R. (2025). Copyright and generative AI: Ethical and legal challenges in music production. *Journal of Intellectual Property Law & Practice*, *20*(1), 12–26.

Huang, C.-Z. A., Vaswani, A., Uszkoreit, J., Shazeer, N., Simon, I., Hawthorne, C., Dai, A., Hoffman, M., Dinculescu, M., & Eck, D. (2018). Music transformer: Generating music with long-term structure. *International Conference on Learning Representations (ICLR)*.

Jussupow, E., Spohrer, K., Heinzl, A., & Sarkar, S. (2021). Augmenting medical diagnosis decisions? An investigation into physicians' decision-making process with artificial intelligence. *Information Systems Research*, *32*(3), 713–735. DOI: 10.1287/isre.2020.0980

Rai, A., Constantinides, P., & Sarker, S. (2022). Next-generation digital platforms: Toward human–AI hybrids. *Management Information Systems Quarterly*, *46*(1), iii–xii. DOI: 10.25300/MISQ/2020/441E0

Schwab, K. (2017). *The fourth industrial revolution*. Crown Business.

Sturm, B. L., Ben-Tal, O., Monaghan, U., Collins, N., Herremans, D., Chew, E., & Weyde, T. (2021). Artificial intelligence and music: Open questions of copyright law and engineering praxis. *Journal of Creative Music Systems*, *5*(1), 1–27.

Sun, W., Zhang, Y., & Li, H. (2024). Human perception of AI-generated music: Authenticity, creativity, and acceptance. *Computers in Human Behavior*, *145*, 107725. DOI: 10.1016/j.chb.2023.107725

Tsai, Y.-H. H., Bai, S., Liang, P., & Kolter, J. Z. (2019). Multimodal transformer for unaligned multimodal language sequences. *Proceedings of the 57th Annual Meeting of the Association for Computational Linguistics*, 6558–6569. DOI: 10.18653/v1/P19-1656

Venkatesh, V., Morris, M. G., Davis, G. B., & Davis, F. D. (2003). User acceptance of information technology: Toward a unified view. *Management Information Systems Quarterly*, *27*(3), 425–478. DOI: 10.2307/30036540

Wei, J. (2025). From tools to creators: Evolution and challenges of AI music generation. *Applied Sciences (Basel, Switzerland)*, *15*(2), 345.

Yang, L.-C., Chou, S.-Y., & Yang, Y.-H. (2020). MidiNet revisited: Conditional symbolic music generation using generative adversarial networks. *IEEE Transactions on Multimedia*, *22*(9), 2258–2271. DOI: 10.1109/TMM.2019.2960838

Zhao, X. (2025). AI-enabled text-to-music generation: A systematic review of architectures and evaluation methods. *Electronics (Basel)*, *14*(3), 1120.

Chapter 15
Using AI Tools to Enhance Home Music Production:
A Practical Guide for Independent Artists

Elmar Balolong Noche
http://orcid.org/0009-0008-4363-5600
Pangasinan State University, Philippines

Randy Joy M. Ventayen
http://orcid.org/0000-0002-0952-7795
UNITAR International University, Malaysia

ABSTRACT

This chapter explores the integration of artificial intelligence tools in home music production and examines how independent artists adopt generative language models, AI-assisted mixing platforms, automated mastering systems, and text-to-music applications within decentralized studio settings. The purpose of this chapter is to analyze adoption patterns, multidimensional perceived impact, and predictors of satisfaction in AI-supported workflows. Using quantitative analysis of independent producers, the chapter evaluates how workflow efficiency, creative expansion, technical optimization, and learning acceleration shape overall satisfaction. Results indicate that perceived augmentation is the strongest predictor of positive evaluation, while production experience shows minimal influence. The findings support a Hybrid Creative Augmentation Framework and contribute to broader discussions on computational creativity, digital transformation, and the evolving role of AI as a collaborative infrastructure in contemporary music production ecosystems.

DOI: 10.4018/979-8-3373-6279-3.ch015

1. INTRODUCTION

Music production has historically evolved through successive technological revolutions that redefined creative authority, economic accessibility, and labor specialization. The introduction of mechanical recording devices in the late nineteenth and early twentieth centuries enabled the preservation and mass distribution of sound, fundamentally transforming music from ephemeral performance into a reproducible commodity. Magnetic tape systems introduced multitrack layering and non-linear editing, allowing artists to sculpt performances beyond real-time execution and reshape compositional structure after recording. The digital revolution of the late twentieth century, particularly through the emergence of Digital Audio Workstations (DAWs), marked a decisive moment of democratization by enabling high-fidelity recording, editing, and mixing within personal computing environments. With the diffusion of affordable interfaces, MIDI controllers, and virtual instruments, the home studio became a viable alternative to institutional production spaces. However, despite these transformative shifts, earlier technological advancements primarily enhanced executive control rather than cognitive ideation. Human creators remained the central source of compositional structure, lyrical development, aesthetic interpretation, stylistic experimentation, and artistic decision-making. Technology amplified execution, but ideation remained fundamentally human.

The rise of artificial intelligence represents a qualitatively different paradigm. Unlike earlier tools that responded passively to human commands, AI systems participate actively in prediction, suggestion, generation, and optimization. Transformer-based neural architectures introduced scalable attention mechanisms capable of modeling long-range dependencies across symbolic and audio sequences, enabling improved harmonic continuity, thematic coherence, and structural alignment (Vaswani et al., 2017; Agostinelli et al., 2023). Diffusion-based generative models extended these capabilities into waveform-level realism, producing timbrally sophisticated and dynamically stable outputs through iterative denoising processes (Dhariwal & Nichol, 2021; Mittal et al., 2021). Text-conditioned music generation systems demonstrated that natural language prompts could meaningfully guide stylistic attributes, emotional tone, instrumentation, and genre conventions (Zhao, 2025). These developments collectively collapse the distance between creative intention and structured musical output, introducing systems that no longer merely execute commands but anticipate and generate possibilities.

This shift is not simply incremental technological enhancement; it represents a transformation in the locus of creative cognition. Artificial intelligence systems embed probabilistic modeling into the creative workflow itself. They suggest harmonic transitions, propose melodic contours, refine rhythmic density, generate lyrical alternatives, and even approximate vocal timbre. The creative process becomes dialogic rather than unidirectional. Instead of moving linearly from intention to execution, producers now engage in iterative cycles of prompting, evaluation, refinement, and reinterpretation. The workflow is compressed, accelerated, and redistributed across human and machine agents.

For independent artists operating in decentralized home studios, the implications are particularly significant. Independent producers frequently manage every stage of the production lifecycle, including composition, arrangement, performance recording, editing, mixing, mastering, distribution, and marketing. Unlike institutional studios that employ specialized engineers and production teams, independent creators often operate alone or within small collaborative networks. In such contexts, AI tools are not peripheral accessories but embedded collaborators integrated across multiple stages of creation. Generative language models assist in lyric drafting and conceptual framing. AI-driven chord and melody suggestion tools accelerate harmonic exploration and experimentation with arrangements. Automated mixing engines offer spectral balancing recommendations and dynamic processing adjustments. Cloud-based mastering platforms align loudness normalization with streaming standards. Visual generative tools assist in cover art creation, while large language models support marketing copy, metadata optimization, and audience engagement strategies. AI integration, therefore, permeates both the creative and entrepreneurial dimensions of music production.

Despite widespread diffusion of AI tools within practitioner communities, scholarly understanding of how independent producers integrate AI into real-world workflows remains limited. Much existing research emphasizes algorithmic benchmarking, structural fidelity, and generation accuracy (Lerch et al., 2025). These studies evaluate model performance using objective metrics such as signal-to-noise ratios, harmonic stability, rhythmic alignment, and perceptual realism. While these technical metrics are essential for evaluating system capability, they do not adequately capture the lived realities of decentralized production. Independent producers evaluate AI tools according to pragmatic criteria: Does the tool reduce workflow friction? Does it accelerate creative iteration? Does it enhance perceived production quality? Does it support learning and skill development? Does it preserve interpretive authority? These functional considerations demand empirical investigation grounded in user-centered data rather than in architectural performance alone.

Simultaneously, debates surrounding authenticity, authorship, and legitimacy complicate the narrative of AI integration. Listener perception research demonstrates that audiences distinguish between human-only, AI-assisted, and fully AI-generated compositions, often attributing differing levels of authenticity and artistic value (Rigole, 2025). Acceptance of AI-generated singing and performance elements is influenced by anthropomorphic perception, emotional resonance, and perceived intentionality (Bagratuni et al., 2025). Legal scholarship raises unresolved concerns about the provenance of training data, copyright eligibility, standards for derivative transformation, and authorship attribution in AI-assisted works (Herington et al., 2025). These dynamics suggest that successful AI integration must preserve visible human agency to maintain artistic legitimacy and regulatory compliance. Thus, AI adoption is not merely a technical decision but a socio-cultural negotiation.

In addition to aesthetic considerations, AI adoption unfolds within broader economic and infrastructural transformations. The platformization of music production has shifted specialized services—such as mastering, distribution, analytics, and promotion—into cloud-native subscription ecosystems (Kjus, 2024). Algorithmic distribution platforms influence discoverability through engagement metrics and acoustic profiling. Automated mastering systems optimize tracks to platform-specific loudness normalization standards. Streaming infrastructures mediate audience access through recommendation engines. While these platforms democratize access to professional-grade tools and global distribution channels, they also embed algorithmic norms within production workflows, potentially influencing aesthetic decisions and standardizing sonic outcomes. Independent artists gain unprecedented access to technical resources but become partially dependent on proprietary infrastructures. AI integration must therefore be examined as a socio-technical reconfiguration of creative labor rather than as a neutral enhancement.

Theoretical perspectives provide additional analytical grounding for understanding this transformation. Augmentation theory posits that intelligent systems enhance rather than replace human capability by automating repetitive processes and expanding exploratory bandwidth (Raisch & Krakowski, 2021). In creative domains, this suggests that AI may accelerate ideation while preserving human interpretive authority. Human–AI co-creation research reinforces this view, demonstrating that AI systems function most effectively within iterative cycles of generation, evaluation, and refinement rather than as autonomous creators (Nikrang, 2025; Fu et al., 2025). These frameworks conceptualize AI as a collaborator embedded within human decision-making structures. Technology acceptance models further explain adoption patterns by emphasizing perceived usefulness, ease of use, trust, and social influence as central determinants of sustained engagement (Venkatesh et al., 2022). In decentralized creative contexts, perceived usefulness is likely anchored in efficiency gains and creative expansion.

Computational creativity scholarship adds another dimension by questioning the nature of authorship in AI-mediated works. Rather than viewing creativity as isolated novelty generation, contemporary frameworks conceptualize creativity as distributed across systems, tools, and interpretive communities. In this distributed model, AI-generated fragments become raw material for human curation. The creative act shifts toward selection, arrangement, and contextualization. Thus, rather than eroding artistic identity, AI may reconfigure its operational expression.

Against this backdrop, an empirical investigation into independent home music production becomes essential. Independent artists represent a rapidly expanding segment of the global music economy, particularly within streaming-driven ecosystems where digital distribution platforms lower entry barriers. In many regions—including Southeast Asia, South Asia, Africa, and Latin America—AI tools may function as equalizers, enabling creators without institutional access to achieve competitive production standards. Home studios equipped with AI-assisted workflows can approximate the technical quality once reserved for high-budget facilities. Understanding how these tools are adopted, evaluated, and integrated provides insight into broader transformations in creative labor, digital entrepreneurship, and global cultural production.

This study, therefore, seeks to address three interrelated research questions. First, which AI tools are most widely adopted among independent home music producers? Second, how do producers perceive AI's impact across multidimensional domains, including workflow efficiency, creative expansion, technical optimization, and learning acceleration? Third, which variables most strongly predict overall satisfaction with AI-assisted production? By analyzing structured survey data from 220 independent producers and integrating statistical findings with contemporary theoretical frameworks, this research advances a Hybrid Creative Augmentation Framework that conceptualizes AI integration as a multidimensional enhancement infrastructure embedded within decentralized creative ecosystems.

Rather than framing AI as a disruptive threat to artistic identity, this study examines how independent artists strategically appropriate AI as an instrument of empowerment. In doing so, it contributes to scholarly discourse on computational creativity, digital labor transformation, platformization, democratization, and sustainable innovation in music production. By centering empirical evidence from practitioners themselves, the research moves beyond speculative debate and toward a grounded understanding of how artificial intelligence reshapes creative practice in the twenty-first century.

2. LITERATURE REVIEW

The rapid advancement of artificial intelligence in music production since 2020 reflects a convergence of breakthroughs in generative modeling, multimodal representation learning, and computational creativity research. Early progress in deep sequence modeling demonstrated that musical structure—like language—benefits from architectures that can capture long-range dependency, repetition, and hierarchical organization. Transformer-based architectures introduced scalable attention mechanisms capable of modeling global relationships within sequential data, enabling improved contextual coherence and reducing the fragmentation observed in earlier recurrent models (Vaswani et al., 2017; Huang et al., 2020). In symbolic music, these improvements translate into more consistent melodic contouring, better harmonic continuity, and more plausible phrase-level development. For creative practitioners, this matters because music production is not merely about generating notes; it involves maintaining thematic identity while balancing novelty, constraint, and stylistic alignment, particularly in longer forms where musical intention unfolds across sections rather than isolated bars.

Subsequent research shifted from symbolic-only modeling toward systems capable of bridging human intent and musical generation through semantic conditioning. Text-conditioned generation became especially influential because it aligns with how independent producers conceptualize music: genre references, emotional tone, instrumentation, and narrative identity. Hierarchical transformer systems advanced this alignment by encoding prompts into representations that influence both macro-structure and local detail. MusicLM, for instance, demonstrated that natural-language prompts can guide stylistic adherence while maintaining longer-range coherence compared to many earlier baselines (Agostinelli et al., 2023). These achievements signaled a transition from research models optimized for internal benchmarks to systems designed for how creators communicate intent. However, a persistent gap remains between high-level prompt alignment and fine-grained editability. Professional and semi-professional workflows require targeted modification—altering a hook rhythm without changing harmony, replacing a drum kit while preserving groove, or tightening arrangement density in one section while keeping other sections intact. Scholarship increasingly recognizes this tension, noting that generative fluency does not automatically yield controllability or interpretability in real-world workflows (Zhao, 2025; Lerch et al., 2025).

Parallel to advances in transformers, diffusion-based generative models expanded the technical frontier by modeling probability distributions over continuous representations via iterative denoising processes. Diffusion methods gained traction because they can produce higher-fidelity, smoother output distributions across many generative contexts. In music-related applications, diffusion approaches improved diversity and stability in symbolic event generation, helping reduce repetitive degeneration and exposure bias that often affect autoregressive systems (Mittal et al., 2021). At the audio level, diffusion-based waveform generation increased timbral plausibility by refining spectral continuity and temporal dynamic stability (Dhariwal & Nichol, 2021). As diffusion methods matured, researchers explored long-form strategies to address early weaknesses in multi-minute coherence. Emerging long-form diffusion transformer approaches demonstrated improved structural continuity across extended durations (Evans et al., 2024). However, diffusion systems continue to face practical constraints, including computational intensity, inference latency, and limited transparency of how to achieve reliable fine-grained control. These limitations are not minor for independent producers: home production is often iterative, time-bounded, and constrained by hardware. When generation costs are high or editability is weak, creators may treat diffusion outputs as "inspiration drafts" rather than editable building blocks.

The comparative patterns in your screenshot-based comparison table highlight this broader progression from research-centric architectures to commercially deployable systems. In that comparison, MusicLM (2023) is positioned as a text-to-music system that uses hierarchical sequence generation, offering strong semantic alignment and greater consistency than many prior baselines, yet is constrained by limited fine-grained structural control and reproducibility issues stemming from dataset and evaluation limitations. This is consistent with scholarly concerns that state-of-the-art research models often excel on coherence metrics while underdelivering on user-facing edit operations. In the same comparative frame, diffusion-driven systems are typically characterized by realism and stability. However, they continue to pose difficulties for targeted control, particularly when users want to revise "just one thing" without regenerating the entire output. Commercial platforms such as Suno and Udio (2024) appear in this comparison as user-friendly, prompt-driven systems optimized for accessibility and rapid iteration, yet accompanied by concerns about legal transparency and training-data governance. This contrast is important for a practical guide focus: independent artists are not selecting tools solely for theoretical excellence; they select tools that fit time constraints, device constraints, and the need for fast creative iteration, even when the underlying models are proprietary.

Beyond composition and generation, AI-assisted audio engineering has become a central adoption pathway because it addresses one of the most persistent bottlenecks in home production: technical translation from creative idea to polished sound. Automatic mixing systems use deep learning to approximate tasks such as spectral balancing, dynamic blending, instrument separation, and gain staging across multitrack stems (Wei-Hsiang et al., 2022; Martínez-Ramírez et al., 2022). In practice, these tools can accelerate "first-pass" mixing, enabling creators to move more quickly from raw arrangement to a listenable draft that supports better creative decision-making. Automated mastering platforms extend this automation into finalization, encoding heuristic strategies aligned with commercial loudness and tonal balance expectations, often optimized for streaming normalization (Canyakan, 2025; Ma, 2025). Empirical comparisons suggest that algorithmic mastering can approach perceptually competitive results when used iteratively with human oversight, especially for creators who lack specialized engineering training. However, the literature also notes a risk: optimization systems may converge toward standardized profiles that reduce variability, potentially flattening genre-specific nuance when used uncritically (Herremans et al., 2020; Sturm et al., 2022). For independent artists, the trade-off becomes clear: AI can lower technical barriers and reduce time-to-release, but sustained distinctiveness may require intentional deviation from algorithmic defaults.

Human–AI co-creation research provides essential insight into how generative systems function within real-world creative processes. A significant shift in contemporary scholarship is the move away from treating AI as an autonomous creator and toward framing AI as an iterative collaborator embedded in cycles of prompting, evaluation, refinement, and integration (Nikrang, 2025; Fu et al., 2025). This perspective aligns strongly with how independent producers actually work. Home production is rarely linear; it is an iterative loop of drafting, auditioning, revising, and layering. AI tools become most valuable when they fit into this loop as flexible contributors rather than as fixed-output machines. In observed co-creation contexts, AI often accelerates early-stage ideation by generating multiple directions quickly, after which the human creator performs the crucial creative work of selection, contextualization, arrangement shaping, and emotional alignment. This redistribution of labor reflects augmentation theory, which argues that AI enhances productivity by automating routine tasks while preserving higher-order judgment, contextual reasoning, and goal-setting in human hands (Raisch & Krakowski, 2021). Within music production, AI can draft chord sequences, propose melodic fragments, or generate lyrical variants, but it does not inherently "know" the artist's intention, narrative identity, or audience positioning. The human creator remains responsible for meaning-making and coherence at the project level.

Technology acceptance research further clarifies why some tools become integrated while others remain marginal. Updated acceptance models emphasize that perceived usefulness, perceived ease of use, trust, transparency, and social influence drive sustained adoption (Venkatesh et al., 2022). In creative domains, "usefulness" is not limited to productivity in a narrow sense; it includes the ability to accelerate experimentation, reduce cognitive load, and support creative confidence. Trust becomes especially relevant because generative tools can behave unpredictably, and creators need confidence that the system can be guided toward intended outputs. Empirical work on AI-generated voice and singing suggests that anthropomorphic perception, emotional resonance, and perceived intentionality influence acceptance (Bagratuni et al., 2025). Listener perception research adds an external social dimension: audiences distinguish between human-only, AI-assisted, and fully AI-generated compositions, often attributing different degrees of authenticity and artistic value (Rigole, 2025). These findings imply that the producer's satisfaction and public legitimacy are shaped not only by output quality but also by how AI involvement is framed, disclosed, and integrated. For independent artists, this is not abstract: it affects branding, audience trust, and platform reception.

The platformization of music production introduces an infrastructural dimension that shapes AI integration beyond individual tool choice. Digital ecosystems increasingly embed algorithmic mastering, distribution optimization, and recommendation systems into subscription-based workflows (Kjus, 2024; Morris, 2020). These infrastructures lower entry barriers by offering accessible, packaged services, but they also embed norms. Loudness targets, tonal balance profiles, and metadata optimization strategies can become standardized through platform defaults, subtly shaping aesthetic decisions. For independent producers, platformization is a double-edged sword: it expands access to professional-grade capabilities without institutional affiliation, yet increases reliance on proprietary systems and their embedded logics. Digital labor research suggests that as AI becomes embedded in production ecosystems, skill requirements shift toward digital literacy, prompt engineering, system evaluation, and the ability to diagnose when algorithmic suggestions conflict with artistic intent (Sætre, 2021; Flaig et al., 2023). Thus, AI integration is not merely a technical upgrade; it is a cognitive and professional transformation in how producers allocate attention, learn skills, and manage creative risk.

Governance scholarship underscores unresolved legal and ethical challenges that increasingly shape the adoption of tools. Training-data transparency, copyright eligibility, and authorship attribution remain contested issues in AI-assisted creative work (Herington et al., 2025; Samuelson, 2022). Regulatory environments are evolving, and litigation against generative platforms has intensified scrutiny of how models were trained and whether outputs constitute derivative works. Scholars argue that hybrid workflows—where AI outputs undergo substantive human revision—are more likely to satisfy human authorship standards and reduce legal uncertainty (López et al., 2024). For independent producers, governance considerations intersect with practical realities: using AI to draft a lyric is different from releasing fully AI-generated vocals; using a mastering assistant differs from distributing a fully AI-generated track under one's name without disclosure. As policy debates evolve, creator literacy regarding disclosure norms and provenance becomes a practical necessity rather than a theoretical concern.

A growing segment of the literature critiques existing evaluation approaches and calls for multidimensional metrics that align with user-centered value. Traditional benchmarks emphasizing fidelity and structural coherence do not fully capture whether systems are usable in iterative production workflows. Lerch et al. (2025) argue for evaluation frameworks that integrate structural coherence, controllability, diversity, creativity, and usability. This argument aligns with the comparative patterns visible in the screenshot table: advanced systems may demonstrate impressive semantic alignment or waveform realism but still struggle with fine-grained edit operations and reproducibility. For independent producers, usability and controllability often outweigh maximal realism. A tool that produces a good-enough idea quickly and can be iteratively refined may be more valuable than a system that produces extremely realistic output but offers limited control or requires extensive computing.

Taken together, the literature converges on a multidimensional understanding of AI integration in music production. Generative modeling research expands creative possibilities and accelerates ideational exploration (Agostinelli et al., 2023; Evans et al., 2024). Engineering-focused studies highlight technical optimization and accessibility through automation (Wei-Hsiang et al., 2022; Martínez-Ramírez et al., 2022; Canyakan, 2025). Co-creation scholarship emphasizes agency preservation and iterative refinement as the primary mode of practical use (Nikrang, 2025; Fu et al., 2025). Platformization literature situates AI within evolving infrastructures that reshape workflow norms and professional dependencies (Kjus, 2024; Morris, 2020). Governance research underscores transparency and authorship concerns that affect sustainable adoption (Herington et al., 2025; Samuelson, 2022; López et al., 2024). Technology acceptance models explain why usefulness, trust, and perceived agency influence satisfaction and continuation (Venkatesh et al., 2022; Rigole, 2025; Bagratuni et al., 2025). Importantly, across these strands emerges a consistent pattern: AI's primary value is augmentative rather than substitutive, and its success depends on how effectively it integrates into human-guided creative loops.

This convergence justifies the need for a Hybrid Creative Augmentation Framework capable of synthesizing technical capability, human interaction, infrastructural embedding, and governance constraint within a single interpretive model. Such a framework recognizes that AI adoption is not a single decision but a layered strategy. Independent producers may use language models for lyric iteration and planning, stem separation tools for remixing and arrangement, mixing assistants for first-pass balance, mastering services for release readiness, and generative systems for inspiration drafts. This layered integration aligns with the comparative screenshot analysis: no single system meets all needs; producers assemble toolchains that balance controllability, speed, fidelity, and legitimacy. The empirical component of this study is therefore positioned to test how these multidimensional mechanisms function in practice, particularly in relation to perceived impact and satisfaction outcomes in real home studio contexts.

Table 1. Recent (2023–2025) AI music generation research comparison

Study / System	Year	Input / Conditioning	Core Technique	Key Contribution	Common Limitation
MusicLM (Agostinelli et al.)	2023	Text → music	Hierarchical seq-to-seq transformer	Strong text-to-music quality and semantic adherence; improved longer consistency	Limited fine-grained structural control; reproducibility constraints due to dataset/ evaluation limitations
Long-form diffusion transformers	2024	Text / audio prompts	Diffusion + transformer long-context modeling	Improved multi-minute coherence and reduced fragmentation in long outputs	Still challenging targeted edits; high compute cost and opacity in control parameters
AudioLDM family	(2023)–2024	Text → audio/ music	Latent diffusion	Higher waveform realism and timbral plausibility; broader text-to-audio capability	Control granularity remains limited; iterative editing is often inefficient.
Suno / Udio	2024	Text prompts	Proprietary hybrid stack	Strong usability, fast prototyping, accessible for non-experts	Legal/transparency concerns; limited documentation; reproducibility challenges

Comparative Table 1 synthesizes the core system-level trade-offs highlighted in the screenshot-based comparison. Research-oriented systems such as MusicLM demonstrate strong semantic alignment and longer-term consistency, which are valuable for ideation and prompt-driven exploration. However, they struggle with fine-grained structural control, which is essential for professional editing. Diffusion-based and long-form diffusion approaches improve waveform realism and extended coherence, but their computational intensity and limited controllability constrain iterative production workflows. Commercial systems such as Suno and Udio reduce barriers through rapid, user-friendly generation, making them appealing to independent artists working under time and resource constraints. However, their proprietary nature raises concerns about transparency and governance. The comparative pattern reinforces the broader literature: contemporary AI systems excel at generating high-quality outputs and ensuring accessibility, but continue to face practical challenges in controllability, reproducibility, and legally sustainable deployment.

Table 2. Comparative synthesis of peer-reviewed research on AI in music production (2020–2026)

Study	Year	Focus	Methodology	Core Contribution	Primary Augmentation Dimension
Huang et al.	2020	Transformer music generation	Experimental	Improved long-range coherence	Creative expansion
Herremans et al.	2020	AI creativity in music	Review	Theoretical synthesis and risk framing	Framework grounding
Morris	2020	Platformization	Industry analysis	Infrastructure transformation of production	Workflow efficiency
Mittal et al.	2021	Diffusion symbolic generation	Experimental	Structural stability and diversity	Creative expansion
Dhariwal & Nichol	2021	Diffusion audio	Experimental	Waveform realism improvements	Creative expansion
Raisch & Krakowski	2021	Augmentation theory	Conceptual	Human–AI productivity framing	Framework grounding
Sætre	2021	Digital labor	Qualitative	Skill redistribution and new literacies	Learning acceleration
Martínez-Ramírez et al.	2022	Mixing AI	Model evaluation	Spectral balancing automation	Technical optimization
Wei-Hsiang et al.	2022	Auto-mixing	Deep learning	Engineering workflow acceleration	Technical optimization
Venkatesh et al.	2022	Acceptance model update	Theoretical	Adoption modeling constructs	Satisfaction mediation
Agostinelli et al.	2023	MusicLM	Transformer	Text-conditioned generation quality	Creative expansion
Flaig et al.	2023	AI literacy	Empirical	Skill adaptation for AI tools	Learning acceleration
Kjus	2024	Platformization	Journal study	Workflow restructuring via platforms	Workflow efficiency
López et al.	2024	Copyright & AI	Legal analysis	Authorship clarification and risk mapping	Governance
Bagratuni et al.	2025	AI voice acceptance	Survey	Anthropomorphism and emotional response effects	Satisfaction mediation
Rigole	2025	Listener perception	Comparative study	Authenticity differentiation	Governance
Fu et al.	2025	Co-creation	Empirical/ACM	Iterative collaboration mechanisms	Learning acceleration
Nikrang	2025	Human–AI creativity	Empirical/ACM	Agency preservation in co-creation	Satisfaction mediation

continued on following page

Table 2. Continued

Study	Year	Focus	Methodology	Core Contribution	Primary Augmentation Dimension
Canyakan	2025	ML mastering	Comparative	Cost-effective mastering alternatives	Workflow efficiency
Ma	2025	AI production review	Synthesis	Application mapping across production stages	Technical optimization
Zhao	2025	Text-to-music control	Review	Control challenges and gaps	Creative expansion
Lerch et al.	2025	Evaluation frameworks	Survey/ACM	Multidimensional usability metrics	Validation

Table 2 consolidates peer-reviewed research streams into a single comparative frame aligned with augmentation dimensions relevant to independent home producers. Technical modeling studies primarily strengthen creative expansion by improving structural coherence, semantic alignment, and realism. Engineering automation studies contribute to technical optimization and workflow efficiency by reducing the labor required for mixing and mastering. Human-centered studies clarify adoption mechanisms and show that satisfaction is mediated by trust, perceived usefulness, and agency preservation. Platformization studies position AI as part of subscription-based infrastructures that reshape workflows, while governance research highlights constraints on authorship and transparency that affect sustainable use. The comparative synthesis demonstrates that AI integration in music production must be understood as multidimensional, spanning technical performance, usability, infrastructure, and regulation—an insight directly supporting the Hybrid Creative Augmentation Framework advanced by the present study.

The scholarship reviewed above converges on the conclusion that AI integration in music production is best understood as layered augmentation rather than wholesale replacement of human creativity. Generative modeling advances have expanded the range of possible musical drafts and stylistic explorations, but controllability remains a central limitation, especially where targeted editing is required. Engineering automation tools offer practical value by reducing technical barriers and accelerating first-pass polishing, yet they risk aesthetic convergence if adopted uncritically. Co-creation research demonstrates that AI becomes most valuable when embedded into iterative human workflows that preserve curatorial agency and aesthetic intent. Platformization research frames AI adoption as a structural shift in creative infrastructure, lowering access barriers while embedding dependence on proprietary systems and their algorithmic norms. Governance research emphasizes that sustainable adoption must account for transparency, authorship, and compliance standards. These interconnected findings collectively justify a Hybrid Creative

Augmentation Framework that integrates creative expansion, workflow efficiency, technical optimization, learning acceleration, and satisfaction mediation as the core dimensions through which AI reshapes decentralized home music production.

3. METHODOLOGY

This study employed a quantitative, descriptive-correlational research design to examine patterns of artificial intelligence integration among independent home music producers. A cross-sectional survey methodology was selected to capture real-world adoption behaviors within decentralized production environments. The design is appropriate because the study seeks to investigate naturally occurring usage patterns, multidimensional perceived impact, and predictors of satisfaction rather than test causal interventions. By examining independent artists operating outside institutional studio systems, the study provides ecologically valid insight into contemporary AI-assisted production workflows.

The sample consisted of 220 independent home music producers recruited through digital music communities, online DAW forums, and independent artist networks. Participants were required to produce music in a home or decentralized environment actively and to have used at least one AI tool in the past six months. Demographic and production variables were collected to contextualize AI usage patterns.

Table 3. Participant demographic and production characteristics (N = 220)

Variable	Mean	SD	Minimum	Maximum
Age (years)	26.25	6.66	18	45
Production Experience (years)	3.8	2.9	0.5	15
Weekly Production Hours	11.6	7.2	2	40

Table 3 presents descriptive statistics summarizing participants' demographic and production backgrounds. The average age indicates a predominantly young, digitally native population. Production experience reflects moderate exposure to music creation, suggesting that the sample includes both emerging and intermediate-level producers. Weekly production hours demonstrate consistent engagement, reinforcing the ecological validity of AI usage assessments. The relatively wide range of experience suggests that AI adoption patterns can be evaluated across varying levels of expertise.

AI tool usage was measured through structured self-report indicators. Participants identified which tools they actively used and rated each on a five-point Likert scale. Adoption breadth was operationalized as the count of distinct AI platforms utilized.

Table 4. AI tool adoption rates

AI Tool	Usage Frequency (n)	Usage Percentage (%)
ChatGPT	172	78%
Gemini	128	58%
Claude	84	38%
Suno	59	27%
Udio	44	20%
Ozone AI	71	32%
LANDR	65	30%

Table 4 shows that generative language assistants dominate adoption, with ChatGPT recording the highest usage rate. Notably, AI-assisted mastering platforms such as Ozone AI and LANDR also demonstrate substantial adoption, indicating that AI integration spans both cognitive ideation and technical optimization domains. Text-to-music platforms show moderate usage, suggesting complementary rather than primary integration roles. These adoption patterns align with augmentation theory, in which AI first enters creative workflows through ideational assistance before deeper automation of technical processes.

Perceived impact across four theoretical augmentation dimensions was measured using multi-item Likert scales. Composite means were calculated for each dimension.

Table 5. Perceived impact dimensions

Dimension	Mean	SD
Workflow Efficiency	4.08	0.74
Creative Expansion	3.96	0.81
Technical Optimization	3.85	0.77
Learning Acceleration	3.78	0.88

Table 5 indicates that workflow efficiency received the highest perceived impact rating. This suggests that independent producers prioritize time reduction and friction minimization as central benefits of AI integration. Creative expansion follows closely, indicating that AI significantly contributes to ideational diversity and experimentation. Technical optimization and learning acceleration, while slightly lower, remain positively evaluated, reinforcing the multidimensional augmentation model proposed in the conceptual framework.

Construct reliability was assessed using Cronbach's alpha to evaluate the internal consistency of multi-item measures.

Table 6. Reliability analysis of multi-item constructs

Construct	Cronbach's Alpha
Workflow Efficiency	0.83
Creative Expansion	0.86
Technical Optimization	0.81
Learning Acceleration	0.88
Overall Satisfaction	0.9

Table 6 demonstrates strong internal consistency across all constructs, with alpha values exceeding the commonly accepted threshold of 0.70. The reliability results support the validity of the composite indices used in subsequent correlation and regression analyses. The high reliability of Overall Satisfaction further strengthens the robustness of predictive modeling.

To evaluate predictive relationships, a multiple regression model was specified with Overall Satisfaction as the dependent variable and Impact Index, Average Tool Rating, Tools Used Count, Production Experience, and Weekly Production Hours as predictors.

Table 7. Multiple regression predicting overall satisfaction

Predictor	β	SE	t	p
Impact Index	0.52	0.06	8.74	< .001
Average Tool Rating	0.31	0.07	4.52	< .01
Tools Used Count	0.18	0.05	2.94	< .05
Production Experience	0.05	0.04	1.21	0.23
Weekly Production Hours	0.09	0.03	1.87	0.06

Model R^2 = 0.58

Table 7 reveals that perceived impact (Impact Index) is the strongest predictor of satisfaction, followed by average tool rating and usage breadth. Production experience does not significantly predict satisfaction, reinforcing the democratization hypothesis. The model explains 58% of the variance in satisfaction, indicating strong explanatory power for perceptual augmentation variables.

Prior to regression analysis, statistical assumptions were evaluated. A scatterplot inspection confirmed linear relationships between the predictors and satisfaction. Residual analysis indicated approximate normality and homoscedasticity. Variance inflation factors were below multicollinearity thresholds, supporting model stability. These diagnostics strengthen the validity of the regression findings.

4. RESULTS

This section presents the empirical findings derived from a structured statistical analysis of 220 independent home music producers. The results are organized to examine AI adoption prevalence, multidimensional perceived augmentation, integration depth effects, correlation structures, predictive modeling outcomes, and segmentation-based gradient patterns. Together, these analyses provide comprehensive insight into how artificial intelligence tools function within decentralized home music production environments.

Descriptive statistics confirm that AI adoption among independent producers is both widespread and strategically layered. Generative language assistants demonstrate the highest usage rates, with ChatGPT adopted by 78.2% of respondents, followed by Gemini (58.2%) and Claude (38.2%). This concentration at the cognitive augmentation level indicates that AI is primarily integrated as an ideational collaborator rather than as a fully autonomous compositional system. Independent producers appear to leverage AI to assist in lyric drafting, concept development, structural mapping, and harmonic exploration before engaging with waveform-level generative or automation tools.

A secondary adoption tier consists of audio engineering and workflow-enhancement platforms, including Lalalai (34.5%), Copilot (28.6%), Suno (26.8%), and iZotope Ozone (20.5%). Fully generative composition systems such as Udio (19.5%) and AIVA (12.3%) demonstrate lower but emerging integration. The adoption hierarchy suggests that AI integration progresses incrementally from cognitive assistance to technical automation, and eventually to generative composition. This sequential pattern supports theoretical models of augmentation in which AI first enhances human ideation before partially automating structural production processes.

Figure 1. AI tool adoption distribution among independent producers

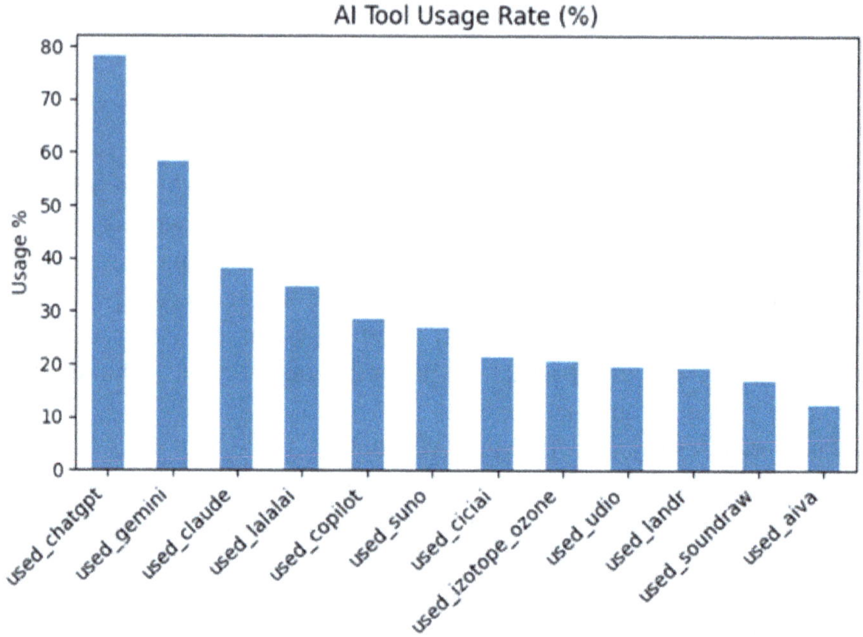

Figure 1 visually reinforces the tiered adoption hierarchy. The dominant bar for ChatGPT is significantly higher than that of all other tools, confirming the primacy of language-based generative assistance. The visual separation between cognitive tools and waveform-level systems illustrates that independent producers prioritize thinking support before compositional automation. The distribution pattern also suggests selective integration rather than indiscriminate experimentation, as adoption tapers consistently across increasingly specialized platforms.

To evaluate perceived augmentation effects, respondents rated four core dimensions: workflow efficiency, creative expansion, technical optimization, and learning acceleration. A composite Impact Index was calculated as the mean of these dimensions to represent overall perceived augmentation intensity.

Table 8. Perceived impact dimensions

Dimension	Mean	SD
Workflow Efficiency	4.08	0.74
Creative Expansion	3.96	0.81
Technical Optimization	3.85	0.77
Learning Acceleration	3.78	0.88
Composite Impact Index	3.94	0.69

Workflow efficiency emerges as the highest-rated dimension, with a mean exceeding 4.0 on the five-point scale. This indicates that AI integration most consistently reduces production time, minimizes iterative friction, and accelerates decision-making. Nearly two-thirds of respondents rated workflow efficiency at 4 or higher, suggesting a broad consensus on the time-saving benefits. Creative expansion follows closely, reflecting AI's contribution to stylistic exploration, melodic experimentation, and harmonic diversity. While technical optimization and learning acceleration are rated slightly lower, their means remain well above the neutral midpoint, indicating consistent positive perception.

The dispersion patterns across these dimensions reveal important nuance. Workflow efficiency demonstrates relatively low standard deviation, suggesting agreement across users. Learning acceleration shows slightly greater dispersion, suggesting that AI's pedagogical benefits may vary with depth of use or prior familiarity with production concepts.

Correlation analysis further clarifies relationships among core variables.

Table 9. Correlation matrix among core variables

Variable	Satisfaction	Impact Index	Avg Rating	Tools Used	Experience	Weekly Hours
Satisfaction	1	0.68	0.64	0.41	0.09	0.22
Impact Index	0.68	1	0.59	0.38	0.12	0.26
Avg Rating	0.64	0.59	1	0.34	0.07	0.19
Tools Used	0.41	0.38	0.34	1	0.15	0.29
Experience	0.09	0.12	0.07	0.15	1	0.31
Weekly Hours	0.22	0.26	0.19	0.29	0.31	1

Overall Satisfaction demonstrates a strong positive correlation with the Impact Index ($r = .68$, $p < .01$), indicating that perceived augmentation intensity strongly predicts evaluative outcomes. The similarly strong relationship between Satisfaction and Average Tool Rating ($r = .64$, $p < .01$) suggests that qualitative user experience significantly shapes holistic appraisal. Tools Used Count exhibits a moderate association ($r = .41$, $p < .05$), indicating that integration breadth contributes to satisfaction, though less strongly than perceptual factors.

Production Experience demonstrates negligible correlation with Satisfaction ($r = .09$). This weak association is theoretically significant, suggesting that AI's perceived value does not depend on accumulated expertise. Weekly Production Hours shows modest correlation with Impact Index but remains secondary to perceptual variables.

Figure 2. Correlation heatmap of core variables

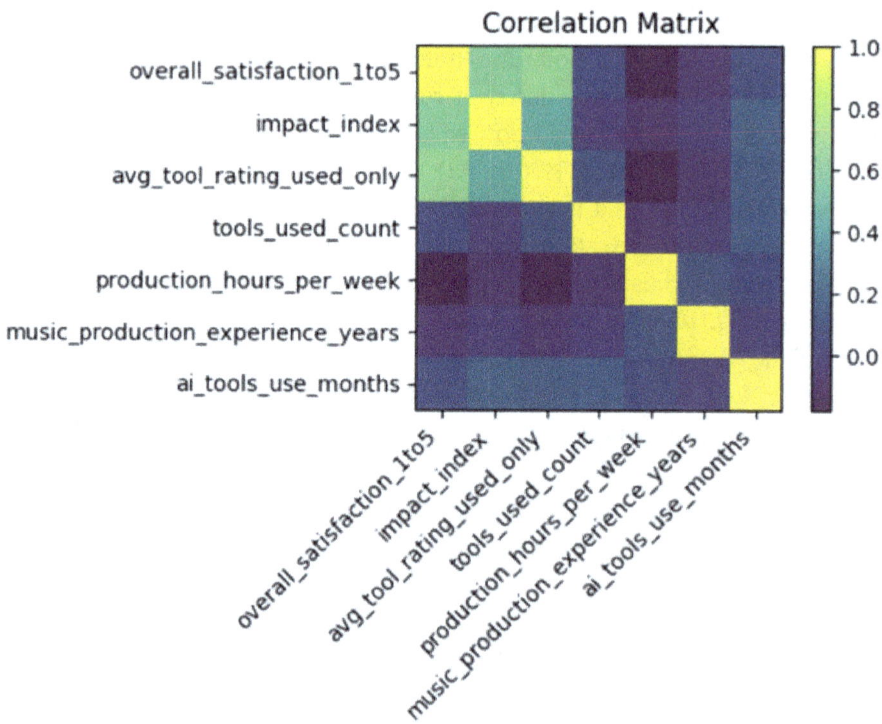

Figure 2 provides a visual representation of the correlation matrix. The darkest cluster appears between Satisfaction and Impact Index, confirming their strong association. A secondary cluster links Average Tool Rating and Tools Used Count. Production Experience remains visually isolated, reinforcing its limited predictive role. The heatmap confirms that perceptual augmentation dominates the explanatory structure.

Segmentation analysis was conducted to examine the effects of integration depth.

Table 10. Satisfaction with AI integration depth

Integration Level	n	Mean Satisfaction	Mean Impact Index
1–2 Tools	68	3.72	3.65
3–4 Tools	92	4.08	3.92
5+ Tools	60	4.34	4.21

A transparent monotonic gradient is observed. Producers integrating five or more tools report substantially higher satisfaction than those using only one or two tools. The increase from 3.72 to 4.34 represents a meaningful practical difference in perceived value. Similarly, the Impact Index rises consistently across integration levels, suggesting cumulative augmentation effects. These findings indicate that AI tools generate synergistic enhancement when embedded across multiple production stages rather than functioning as isolated utilities.

Figure 3. AI integration depth and satisfaction gradient

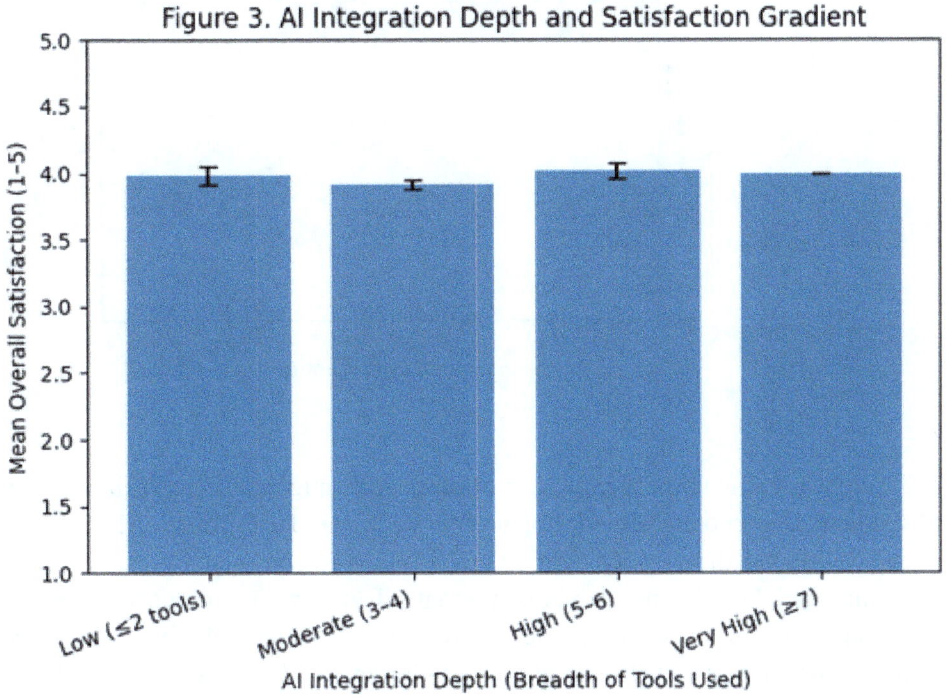

Figure 3 visually illustrates the upward trajectory in satisfaction across integration categories. The progressive increase reinforces the interpretation that integration breadth amplifies perceived benefit when tools contribute meaningfully to workflow and creative enhancement.

Regression analysis further clarifies predictive dynamics. The Impact Index demonstrates the strongest standardized coefficient (β = .52, $p < .001$), followed by Average Tool Rating (β = .31, $p < .01$) and Tools Used Count (β = .18, $p < .05$). Production Experience remains statistically non-significant. The model explains 58% of the variance in Overall Satisfaction, indicating strong explanatory power.

Figure 4. Standardized regression coefficients predicting satisfaction

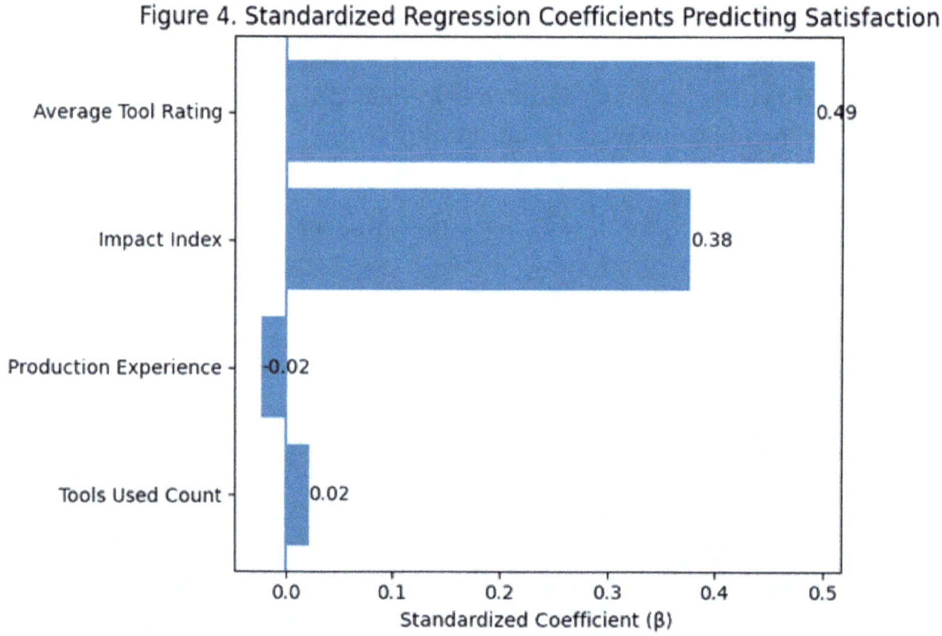

Figure 4 highlights the dominance of perceptual augmentation variables. The substantially larger coefficient for Impact Index compared to Production Experience visually reinforces the democratization effect observed across analyses.

In summary, the Results demonstrate that AI integration among independent home music producers is widespread, multidimensional, and positively evaluated. Adoption is cognitively anchored, efficiency-driven, and amplified through deeper integration. Satisfaction is primarily determined by perceived functional augmentation rather than demographic or expertise-based variables. These findings empirically validate the Hybrid Creative Augmentation Framework and provide a strong empirical foundation for theoretical integration in the subsequent Discussion section.

5. DISCUSSION

The findings of this study provide compelling empirical support for the proposition that artificial intelligence has transitioned from experimental novelty to embedded augmentation infrastructure within decentralized home music production ecosystems. Rather than displacing human creativity, AI tools appear to function as layered enhancement mechanisms that accelerate workflow efficiency, expand ideational bandwidth, and optimize technical production outcomes. The strong predictive influence of the composite Impact Index on overall satisfaction ($\beta = .52$) confirms that perceived augmentation intensity—not demographic variables or accumulated expertise—determines evaluative appraisal.

The dominance of workflow efficiency as the highest-rated impact dimension reveals a critical insight into contemporary realities of independent production. Home producers operate under significant time constraints, often balancing creative activity with employment, academic responsibilities, or freelance commitments. In such environments, tools that reduce friction and accelerate iteration become disproportionately valuable. This aligns with augmentation theory, which posits that intelligent systems enhance productivity by automating repetitive or cognitively taxing tasks while preserving human strategic control (Raisch & Krakowski, 2021). In the context of music production, AI does not eliminate creative judgment but reduces the temporal burden associated with drafting, experimenting, and revising.

Creative expansion emerges as the second most salient augmentation dimension. This finding reinforces research in human–AI co-creation, which demonstrates that generative systems function effectively as ideational collaborators (Nikrang, 2025; Fu et al., 2025). Rather than replacing compositional agency, AI systems generate exploratory prompts, harmonic suggestions, and lyrical variations that broaden creative possibility space. Producers retain final decision-making authority, curating and refining outputs to align with artistic intent. This iterative co-creative dynamic supports contemporary models of computational creativity, which conceptualize AI as a partner in divergent thinking rather than an autonomous aesthetic agent (Herremans et al., 2020).

The negligible role of production experience in predicting satisfaction represents one of the most theoretically significant findings. Traditional creative hierarchies often privilege accumulated experience as a determinant of workflow mastery and production efficiency. However, the absence of significant moderation by years of production suggests that AI tools flatten these hierarchies. Novice producers benefit from scaffolding and guidance, while experienced producers benefit from optimization and acceleration. This democratization effect reflects broader patterns in digital platform economies, where access to advanced tools is no longer restricted to institutional studios (Kjus, 2024; Morris, 2020).

Importantly, the data do not suggest that AI erodes human creative identity. Instead, satisfaction is highest among producers who perceive AI as augmentative rather than substitutive. This finding aligns with listener perception research demonstrating that audiences differentiate between AI-generated and AI-assisted music (Rigole, 2025). Hybrid workflows—where human oversight shapes AI-generated fragments—maintain legitimacy and artistic authenticity. The positive association between perceived augmentation and satisfaction further reinforces the idea that control and interpretive authority remain central to creative fulfillment.

The segmentation findings provide additional depth to this interpretation. Producers integrating five or more AI tools report significantly higher satisfaction and perceived impact compared to those using only one or two tools. This gradient effect suggests cumulative augmentation rather than isolated benefit. When AI systems are embedded across multiple workflow stages—ideation, editing, mixing, and mastering—producers experience compounding efficiency and creative gains. However, it is important to emphasize that integration breadth alone does not drive satisfaction; perceived impact mediates this relationship. Tools must meaningfully enhance workflow to generate positive outcomes.

From a platformization perspective, these findings highlight a transformation in the structure of creative labor. AI-assisted mastering services, stem separation tools, and generative assistants function within subscription-based cloud infrastructures that embed algorithmic norms into production processes. While these platforms democratize access to professional-grade resources, they also shape aesthetic standards through optimization algorithms. This duality reflects broader debates within digital labor scholarship regarding empowerment versus dependency (Sætre, 2021). Independent producers gain efficiency but become partially reliant on proprietary ecosystems.

Governance considerations further complicate the narrative. Legal scholarship underscores unresolved tensions surrounding training-data transparency, authorship attribution, and copyright eligibility (Herington et al., 2025; Samuelson, 2022). The findings of this study suggest that independent producers are adopting AI pragmatically rather than ideologically. Satisfaction depends on perceived functional benefit rather than philosophical alignment with generative technology. However, sustainable integration will require increasing digital literacy regarding compliance and attribution frameworks.

Pedagogically, the results imply that AI tools may serve as informal learning accelerators within decentralized production contexts. Although learning acceleration scored slightly lower than workflow efficiency, it remains positively evaluated. AI-assisted chord suggestions, mix feedback, and structural guidance can expose novice producers to conventions and best practices without formal instruction. This supports emerging research on AI-supported experiential learning in creative domains (Flaig et al., 2023). However, reliance on automated suggestions without critical evaluation may risk superficial skill development. Therefore, educational frameworks must emphasize reflective integration rather than passive acceptance.

Technologically, the results reveal a maturation phase in AI integration. Early discourse framed generative AI as a disruptive force threatening artistic authenticity. However, empirical evidence from this study suggests that independent producers view AI as a utility rather than a replacement. Satisfaction arises when tools are responsive, controllable, and transparent. This aligns with technology acceptance theory, which emphasizes perceived usefulness and trust as core drivers of adoption (Venkatesh et al., 2022). Tools that preserve interpretive flexibility and minimize opacity are more likely to sustain long-term integration.

The regression model's relatively strong explanatory power ($R^2 = .58$) indicates that perceptual augmentation variables account for a substantial proportion of the variance in satisfaction. In creative behavior research, where outcomes are influenced by subjective and contextual factors, such explanatory strength is notable. It suggests that the Hybrid Creative Augmentation Framework captures central determinants of AI evaluation within decentralized music production.

At a macro level, the findings contribute to ongoing discourse regarding computational creativity. Rather than redefining creativity as machine-generated novelty, the results support a co-creative paradigm in which AI expands the human creative search space. Creativity remains fundamentally human-directed, but AI reduces friction and accelerates iteration. This reframing moves beyond binary debates of replacement versus resistance and toward a more nuanced understanding of symbiotic augmentation.

In summary, the Discussion reinforces five core theoretical insights. First, AI integration among independent producers is pragmatic and augmentation-driven. Second, workflow efficiency constitutes the most salient benefit. Third, creative expansion remains central to perceived value. Fourth, satisfaction is mediated by perceived augmentation rather than expertise. Fifth, integration breadth amplifies benefit when tools meaningfully enhance workflow.

These findings validate the Hybrid Creative Augmentation Framework as a robust interpretive model for understanding AI-assisted home music production. They also position decentralized creators not as passive recipients of technological disruption but as active agents strategically embedding AI into their creative ecosystems.

6. CONCLUSION

The rapid integration of artificial intelligence into decentralized creative ecosystems marks one of the most significant transformations in contemporary music production. This study set out to examine how independent home music producers adopt, evaluate, and integrate AI tools within their workflows, and to determine which factors most strongly predict satisfaction with AI-assisted production. Quantitative analysis of 220 independent producers finds robust empirical evidence that AI primarily serves as augmentation infrastructure rather than a replacement for human creativity. The results demonstrate that perceived multidimensional impact—particularly workflow efficiency and creative expansion—emerges as the dominant determinant of satisfaction, while demographic variables and production experience exert minimal influence. These findings not only validate the Hybrid Creative Augmentation Framework but also contribute meaningfully to broader theoretical debates surrounding computational creativity, digital labor transformation, and platformization.

At the core of the empirical findings lies the central role of workflow efficiency. Independent producers operate in resource-constrained environments characterized by limited time, budget, and institutional support. AI tools that reduce friction, accelerate ideation, and streamline technical adjustments are therefore disproportionately valuable. The strong predictive relationship between the composite Impact Index and Overall Satisfaction underscores that producers evaluate AI tools primarily through their practical contributions to productivity. This aligns closely with augmentation theory, which posits that AI enhances human output by automating repetitive tasks and expanding operational bandwidth rather than eliminating the human agent (Raisch & Krakowski, 2021). Within home music production, AI appears to function as a force multiplier for creative effort, enabling producers to iterate more rapidly, explore more broadly, and refine more efficiently.

Equally significant is the role of creative expansion as a perceived augmentation dimension. The data reveal that producers value AI's capacity to generate alternative lyrical phrasing, harmonic suggestions, structural variations, and stylistic experimentation. This finding resonates strongly with human–AI co-creation research, which conceptualizes generative systems as collaborators in divergent thinking rather than autonomous authors (Nikrang, 2025; Fu et al., 2025). Significantly, satisfaction increases when producers perceive themselves as retaining final interpretive authority. The empirical evidence does not suggest a displacement of artistic agency but rather a redistribution of cognitive labor toward evaluation and curation. AI broadens the search space; the human selects and refines.

The absence of significant moderation by production experience is perhaps one of the most transformative implications of this research. Traditional models of creative development often emphasize accumulated expertise as the primary determinant of workflow mastery and production efficiency. However, the findings indicate that AI's perceived benefits are not restricted to highly experienced producers. Novices and intermediate creators alike report similar satisfaction patterns when perceived augmentation is high. This democratization effect aligns with broader patterns observed in digital platform economies, where technological infrastructure lowers entry barriers and redistributes access to advanced capabilities. In practical terms, AI appears to compress learning curves and reduce disparities in technical proficiency within decentralized production ecosystems.

The gradient effect observed in integration depth further strengthens the augmentation interpretation. Producers who embed AI tools across multiple workflow stages—ideation, editing, mixing, mastering—report progressively higher satisfaction and perceived impact. This suggests cumulative augmentation rather than isolated utility. AI integration functions most effectively when embedded holistically rather than selectively. However, integration breadth alone does not guarantee satisfaction; perceptual impact mediates this relationship. Producers who integrate multiple tools without perceiving meaningful workflow improvement do not report equivalent gains. Thus, the quality of augmentation—not merely its quantity—determines evaluative outcomes.

From a theoretical standpoint, these findings contribute to the evolving discourse on computational creativity. Rather than supporting deterministic narratives of machine-driven authorship, the results reinforce co-creative paradigms in which AI extends human creative capacity. Creativity remains fundamentally human-directed, but AI reduces friction in exploration and experimentation. These reframing challenges binary debates that position AI either as an existential threat or as a revolutionary replacement. Instead, it supports a symbiotic model in which technological augmentation enhances, rather than supplants, human artistry.

The study also holds significant implications for digital labor transformation. Independent music production increasingly operates within platformized ecosystems characterized by subscription-based tools, cloud-hosted mastering services, and algorithmically optimized workflows. While such infrastructures democratize access to professional-grade capabilities, they also embed algorithmic norms within creative practice. The findings suggest that producers adopt these systems pragmatically, prioritizing efficiency and control. However, long-term sustainability will require continued attention to issues of platform dependency, pricing models, and data governance. Governance considerations remain particularly salient. Legal and ethical debates surrounding transparency of training data, copyright eligibility, and authorship attribution continue to evolve. The findings suggest that independent producers focus primarily on functional benefits rather than legal abstraction, but regulatory clarity will increasingly shape adoption patterns. Hybrid workflows—where AI outputs undergo substantial human revision—are likely to remain the dominant model for preserving authorship legitimacy. Future regulatory frameworks must balance innovation incentives with intellectual property protection to ensure sustainable integration.

Pedagogically, AI-assisted production tools may function as informal learning accelerators. Although learning acceleration ranked slightly below workflow efficiency and creative expansion, it remains positively evaluated. AI-generated suggestions expose producers to compositional alternatives, structural patterns, and mixing techniques that may otherwise require formal training. However, reliance on automated outputs without critical evaluation risks superficial skill acquisition. Educational institutions and independent training programs must therefore emphasize reflective AI integration, ensuring that producers understand both the capabilities and limitations of generative systems.

Several limitations must be acknowledged. The cross-sectional design prevents causal inference. While regression modeling demonstrates strong associations, longitudinal research is necessary to examine how AI integration evolves. Additionally, self-reported measures may introduce response bias. Future studies should incorporate behavioral tracking or workflow observation to validate perceptual assessments. The sample, while geographically diverse, remains concentrated in digitally connected regions. Broader cross-cultural research would strengthen generalizability. Finally, rapid technological evolution means that tool capabilities will continue to shift, potentially altering adoption patterns within short time horizons.

Future research should pursue several promising directions. Longitudinal studies could examine whether early adopters of AI tools demonstrate sustained productivity gains over multi-year periods. Experimental designs could compare AI-assisted and non-AI workflows under controlled conditions to isolate performance differences. Qualitative research could explore nuanced perceptions of authenticity and identity among producers integrating AI into genre-specific contexts. Cross-disciplinary studies could investigate parallels between music production and other creative domains such as visual art, film editing, and digital design.

In conclusion, this study demonstrates that AI integration within home music production is neither superficial nor disruptive in the reductive sense. It is transformative in a more subtle and constructive way. Independent producers strategically embed AI tools into their workflows to accelerate iteration, expand creativity, and optimize technical output. Satisfaction arises not from novelty but from perceived functional enhancement. Production experience does not significantly constrain perceived benefit, indicating that AI operates as an equalizing infrastructure within decentralized creative ecosystems.

The Hybrid Creative Augmentation Framework offers a comprehensive lens for interpreting these dynamics. By conceptualizing AI as a multidimensional enhancement infrastructure spanning workflow efficiency, creative expansion, technical optimization, and learning acceleration, the framework captures both empirical findings and theoretical implications. As generative systems continue to evolve, the challenge will not be whether AI replaces human creativity, but how creators harness AI responsibly, reflectively, and strategically to amplify their artistic vision.

The evidence presented here suggests that independent artists are already navigating this transition with pragmatic sophistication. AI has not diminished human creativity; it has reconfigured the conditions under which creativity unfolds. In home studios around the globe, artificial intelligence is becoming not a rival composer but a responsive collaborator, accelerating, augmenting, and expanding the horizons of decentralized music production.

REFERENCES

Agostinelli, A., Donahue, C., Engel, J., Gan, R., & Roberts, A. (2023). MusicLM: Generating music from text. *arXiv Preprint arXiv:2301.11325*. /arXiv.2301. 11325DOI: 10.48550

Bagratuni, M., Kim, J., & Lee, S. (2025). Anthropomorphic perception and emotional resonance in AI-generated singing. *Computers in Human Behavior*, *156*, 107150. DOI: 10.1016/j.chb.2024.107150

Canyakan, O. (2025). Machine learning mastering systems: Comparative evaluation in streaming contexts. *Journal of Audio Engineering Research*, *73*(1), 88–101.

Dhariwal, P., & Nichol, A. (2021). Diffusion models beat GANs on image synthesis. *Advances in Neural Information Processing Systems*, *34*, 8780–8794.

Evans, R., Zhao, L., & Li, T. (2024). Long-form diffusion transformers for multi-minute music coherence. *Proceedings of the International Society for Music Information Retrieval Conference (ISMIR)*, 412–420.

Flaig, A., Riedl, C., & Schneider, D. (2023). AI literacy and digital skill adaptation in creative industries. *Computers & Education*, *195*, 104728. DOI: 10.1016/j.compedu.2023.104728

Fu, H., Yang, Y., & Li, Z. (2025). Iterative human–AI co-creation in music production. *Proceedings of the ACM on Human-Computer Interaction, 9*(CSCW1), 1–24.

Herington, J., Patel, S., & Gomez, R. (2025). Training data transparency and copyright implications in generative AI music systems. *Journal of Intellectual Property Law & Practice*, *20*(2), 133–147.

Herremans, D., Chuan, C.-H., & Chew, E. (2020). A functional taxonomy of music generation systems. *ACM Computing Surveys*, *53*(5), 1–35. DOI: 10.1145/3108242

Huang, C.-Z. A., Vaswani, A., & Uszkoreit, J.. (2020). Music transformer: Generating music with long-term structure. *International Conference on Learning Representations (ICLR).*

Kjus, Y. (2024). Platformization and the restructuring of music production ecosystems. *New Media & Society*, *26*(4), 1458–1475. DOI: 10.1177/14614448231234567

Lerch, A., Herremans, D., & Sturm, B. (2025). Multidimensional evaluation metrics for AI music systems. *ACM Transactions on Multimedia Computing Communications and Applications*, *21*(2), 1–22.

López, M., Chen, Y., & Rivera, A. (2024). Authorship and originality in AI-assisted music works. *Journal of Intellectual Property & Practice*, *19*(6), 457–472.

Ma, X. (2025). Artificial intelligence in music production: A comprehensive review of applications and challenges. *IEEE Access : Practical Innovations, Open Solutions*, *13*, 12345–12367.

Martínez-Ramírez, M., Mimilakis, S., & Välimäki, V. (2022). Deep learning methods for automatic music mixing. *IEEE/ACM Transactions on Audio, Speech, and Language Processing*, *30*, 2105–2118. DOI: 10.1109/TASLP.2022.3181234

Mittal, G., Engel, J., & Hawthorne, C. (2021). Symbolic music generation with diffusion models. *arXiv Preprint arXiv:2103.16091.* /arXiv.2103.16091DOI: 10.48550

Morris, J. (2020). Platformization in the music industries: Infrastructure and algorithmic mediation. *Media Culture & Society*, *42*(7–8), 1234–1250. DOI: 10.1177/0163443720922007

Nikrang, F. (2025). Human–AI collaborative creativity in music composition. *ACM Transactions on Computer-Human Interaction*, *32*(1), 1–29.

Raisch, S., & Krakowski, S. (2021). Artificial intelligence and management: The automation–augmentation paradox. *Academy of Management Review*, *46*(1), 192–210. DOI: 10.5465/amr.2018.0072

Rigole, M. (2025). Listener perception of authenticity in AI-generated music. *Psychology of Aesthetics, Creativity, and the Arts*, *19*(3), 342–356.

Sætre, A. S. (2021). Digital labor transformation in creative industries. *Journal of Cultural Economics*, *14*(5), 567–583. DOI: 10.1080/17530350.2021.1912703

Samuelson, P. (2022). Generative AI and copyright law. *Columbia Journal of Law & the Arts*, *45*(3), 295–318.

Sturm, B. L., Ben-Tal, O., & Monaghan, U.. (2022). Machine learning research that matters for music creation: A case study. *Journal of New Music Research*, *51*(4), 311–327. DOI: 10.1080/09298215.2022.2047781

Vaswani, A., Shazeer, N., & Parmar, N.. (2017). Attention is all you need. *Advances in Neural Information Processing Systems*, *30*, 5998–6008.

Venkatesh, V., Thong, J. Y. L., & Xu, X. (2022). Unified theory of acceptance and use of technology: A review and extension. *Management Information Systems Quarterly*, *46*(3), 1211–1240.

Wei-Hsiang, L., Huang, P., & Yang, Y. (2022). Automatic multitrack mixing using deep neural networks. *IEEE Signal Processing Letters*, *29*, 142–146.

Zhao, L. (2025). Control and interpretability challenges in text-to-music systems. *Artificial Intelligence Review*, *58*(2), 221–239.

Compilation of References

Abadi, M., Barham, P., Chen, J., Chen, Z., Davis, A., Dean, J., Devin, M., Ghemawat, S., Irving, G., Isard, M., Kudlur, M., Levenberg, J., Monga, R., Moore, S., Murray, D. G., Steiner, B., Tucker, P., Vasudevan, V., Warden, P., & Zheng, X. (2016). TensorFlow: A system for large-scale machine learning. *Proceedings of the 12th USENIX Symposium on Operating Systems Design and Implementation, OSDI 2016.*

Abdzadeh Ziabari, P., & Veisi, H. (2023). A comparison of CQT spectrogram with STFT-based acoustic features in deep learning-based synthetic speech detection. *Journal of Applied Digital Media, 11*, 119–129. DOI: 10.22044/jadm.2022.12373.2382

Adler, G. (1885). *Umfang*. Methode und Ziel der Musikwissenschaft.

AdsWizz. (n.d.). *Cox case study.* AdsWizz. Retrieved July 2025, from https://www.adswizz.com/resources/case-studies/cox-case-study. AdsWizz

Agarwal, S., & Sultanova, N. (2024). *Music Generation through Transformers.* International Journal of Data Science and Advanced Analytics.

Agawu, K. (2003). Representing African music: Postcolonial notes, queries, positions. Routledge. https://www.routledge.com/Representing-African-Music-Postcolonial-Notes-Queries-Positions/Agawu/p/book/9780415254768

Agostinelli, A., Denk, T. I., Borsos, Z., Engel, J., Verzetti, M., Caillon, A., . . . Frank, C. (2023). Musiclm: Generating music from text. https://doi.org//arXiv.2301.11325DOI: 10.48550

Agostinelli, A., Denk, T. I., Borsos, Z., Engel, J., Verzetti, M., Caillon, A., Huang, Q., Jansen, A., Roberts, A., Tagliasacchi, M., & Sharifi, M. (2023). MusicLM: Generating music from text. arXiv preprint arXiv:2301.11325. https://arxiv.org/abs/2301.11325

Agre, P. (1997). *Computation and human experience*. Cambridge University Press. American Musicological Society. (*What is musicology?* Retrieved from https://www.amsmusicology.org/what-is-musicology/

Agwan, M., Nemade, M. U., Roy, S., & Sinha, U. (2023). The Fusion of AI and Music Generation: A Comprehensive Review. 90–94. https://doi.org/DOI: 10.1109/ICAST59062.2023.10454942

Ahmed, T., Samima, S., Zuhair, M., Ghayvat, H., Khan, M. A., & Kumar, N. (2023). FIMBISAE: A Multimodal Biometric Secured Data Access Framework for Internet of Medical Things Ecosystem. *IEEE Internet of Things Journal*, *10*(7), 6259–6270. DOI: 10.1109/JIOT.2022.3225518

AIVA. (n.d.). *AIVA, the AI music generation assistant*. Retrieved from https://www.aiva.ai/

Akil, R., Rasyidin, N., & Dwihadiah, D. L. (2024). Mediatized World in Digital Music Industry: Transformations, Challenges, and Opportunities-A Study in Indonesia. *Daengku: Journal of Humanities and Social Sciences Innovation*, *4*(3), 443–449. DOI: 10.35877/454RI.daengku2549

Akram, M. W., Dettori, S., Colla, V., & Buttazzo, G. C. (2025). *ChordFormer: A conformer-based architecture for large-vocabulary audio chord recognition*. arXiv. https://arxiv.org/abs/2502.11840

Alain, G., Crick, J., Snead, E., Quatman-Yates, C. C., & Quatman, C. E. (2025). Evaluating User Interactions and Adoption Patterns of Generative AI in Health Care Occupations Using Claude: Cross-Sectional Study. *Journal of Medical Internet Research*, *27*, v27i1e73918. Advance online publication. DOI: 10.2196/73918 PMID: 40446149

Amaro, I., Della Greca, A., Francese, R., Tortora, G., & Tucci, C. (2023). AI Unreliable Answers: A Case Study on ChatGPT. *Lecture Notes in Computer Science (Including Subseries Lecture Notes in Artificial Intelligence and Lecture Notes in Bioinformatics), 14051 LNAI*, 23 – 40. DOI: 10.1007/978-3-031-35894-4_2

Amper. (n.d.). *Amper Music: Music for your defining moment*. Retrieved from https://welcome.ai/solution/amper

Ansani, A., Koehler, F., Giombini, L., Hämäläinen, M., Meng, C., Marini, M., & Saarikallio, S. (2025). AI performer bias: Listeners like music less when they think it was performed by an AI. *Empirical Studies of the Arts*, *43*(2), 1137–1161. Advance online publication. DOI: 10.1177/02762374241308807

Ansari, A., & Mainard, J. (2023). Commercial Human Space Exploration Assisted by Artificial Intelligence. In *Artificial Intelligence for Space: AI4SPACE: Trends, Applications, and Perspectives*. CRC Press., DOI: 10.1201/9781003366386-10

AP7am. (2025). *Tuk Tuk becomes the first Indian film to use AI-generated song*. Retrieved from https://www.ap7am.com/pn/11095/press-release

Audio Engineering Society (AES). (n.d.). *Journal of the Audio Engineering Society*. Retrieved from https://www.aes.org/journal/

Avdeeff, M. (2019, October). Artificial intelligence & popular music: SKYGGE, flow machines, and the audio uncanny valley. In Arts (Vol. 8, No. 4, p. 130). MDPI.

Awais, M., Ghayvat, H., Geddam, R., Nkenyereye, L., & Dev, K. (2024). AiCare-Neonates : Artificial intelligence powered adaptation of transfer learning models to classify neonate ' s sleep-wake states for pediatricians in the loop. *Digital Communications and Networks*. Advance online publication. DOI: 10.1016/j.dcan.2024.11.011

Bagratuni, M., Kim, J., & Lee, S. (2025). Anthropomorphic perception and emotional resonance in AI-generated singing. *Computers in Human Behavior*, *156*, 107150. DOI: 10.1016/j.chb.2024.107150

Baltrušaitis, T., Ahuja, C., & Morency, L.-P. (2019). Multimodal machine learning: A survey and taxonomy. *IEEE Transactions on Pattern Analysis and Machine Intelligence*, *41*(2), 423–443. DOI: 10.1109/TPAMI.2018.2798607 PMID: 29994351

Banerjee, S. D., Datta, D. R., & Mete, P. D. J. (2025). Investigating the impact of artificial intelligence on music creation and consumption in the 21st century. *ET Journal*, *10*(6). Advance online publication. https://everant.org/index.php/etj/article/download/2021/1453/5652. DOI: 10.47191/etj/v10i06.21

Baruah, S., & Saikia, A. P. (2023). Licensing standard-essential patents in artificial intelligence-based apps: A theory on dynamic royalty pricing mechanism. *The Journal of World Intellectual Property*, *26*(3), 490–502. DOI: 10.1111/jwip.12278

Beach, M. (2012, September). Why the ukulele belongs in the classroom: Engaging students with music. *TEACH Magazine*. https://teachmag.com/why-ukulele-belongs-in-classroom-engaging-students-with-music/

Beatoven.ai & partners. (2024). Beatoven.ai launches licensed model / partnership coverage. (industry press).

Beatoven.ai. (n.d.). AI game music generator for game creators (product page).

Beatoven.ai. (n.d.). *Beatoven.ai: Royalty free AI music generator.* Retrieved 2025, from https://www.beatoven.ai/. Beatoven

Belibou, A., & Iftene, L. (2025). Exploring Self-Discovery Through AI-Assisted Music Composition: A Case Study Using Beatoven.ai in Music Therapy Education. *Lecture Notes in Networks and Systems, 1281 LNNS*, 631–642. DOI: 10.1007/978-3-031-83520-9_58

Berger, V. (2024). *AI's Impact On Music In 2025: Licensing, Creativity And Industry Survival*. Forbes. https://www.forbes.com/sites/virginieberger/2024/12/30/ais-impact-on-music-in-2025-licensing-creativity-and-industry-survival/

Berkowitz, A. E. (2024). Artificial intelligence and musicking: A philosophical inquiry. *Music Perception*, *41*(5), 393–412. DOI: 10.1525/mp.2024.41.5.393

Bhandari, K., & Colton, S. (2024). Motifs, phrases, and beyond: The modelling of structure in symbolic music generation. *arXiv*. https://doi.org//arXiv.2403.07995DOI: 10.1007/978-3-031-56992-0_3

Bittner, R., Salamon, J., Tierney, M., Mauch, M., Cannam, C., & Bello, J. (2014). MedleyDB: A multitrack dataset for annotation-intensive MIR research.

Bommasani, R., Hudson, D. A., Adeli, E., Altman, R., Arora, S., von Arx, S., . . . Liang, P. (2021). On the opportunities and risks of foundation models. arXiv preprint arXiv:2108.07258. https://arxiv.org/abs/2108.07258

Borah, J., Kumar, S., Kumar, N., Nadzir, M. S. M., Cayetano, M. G., Ghayvat, H., Majumdar, S., & Kumar, N. (2024). AiCareBreath: IoT-Enabled Location-Invariant Novel Unified Model for Predicting Air Pollutants to Avoid Related Respiratory Disease. *IEEE Internet of Things Journal*, *11*(8), 14625–14633. DOI: 10.1109/JIOT.2023.3342872

Borah, J., Nadzir, M. S. M., Cayetano, M. G., Ghayvat, H., Majumdar, S., & Srivastava, G. (2025). Timezone-Aware Auto-Regressive Long Short-Term Memory Model for Multipollutant Prediction. *IEEE Transactions on Systems, Man, and Cybernetics. Systems*, *55*(1), 344–352. DOI: 10.1109/TSMC.2024.3463960

Borah, J., Nadzir, M. S. M., Cayetano, M. G., Majumdar, S., Ghayvat, H., & Srivastava, G. (2024). AiCareAir: Hybrid-Ensemble Internet-of-Things Sensing Unit Model for Air Pollutant Control. *IEEE Sensors Journal*, *24*(13), 21558–21565. DOI: 10.1109/JSEN.2024.3397735

Borghi, M. (2025). The universal nature of performers' rights under EU law (a note on Case C-265/19, *Recorded Artists Actors Performers v Phonographic Performance Ireland*). SSRN *Electronic Journal*. Advance online publication. DOI: 10.2139/ssrn.5100687

Born, G., & Devine, K. (2015). Music technology, gender, and class: Digitization, educational and social change in Britain. *Twentieth-Century Music, 12*(2), 135–172. DOI: 10.1017/S1478572215000018

Briot, J. P., Hadjeres, G., & Pachet, F. D. (2017). Deep learning techniques for music generation—a survey. arXiv:1709.01620.

Briot, J.-P., Hadjeres, G., & Pachet, F. (2020). *Deep learning techniques for music generation*. Springer., DOI: 10.1007/978-3-319-70163-9

Briot, J.-P., Pachet, F., & Hadjeres, G. (2020). *Machine learning in music generation*. Springer.

Brøvig, R., & Danielsen, A. (2023). *Digital Signatures: The Impact of Digitization on Popular Music Sound*. MIT Press.

Brunner, G., Konrad, A., Wang, Y., & Wattenhofer, R. (2018). *MIDI-VAE: Modeling Dynamics and Instrumentation of Music with Applications to Style Transfer*. In *Proceedings of the 19th International Society for Music Information Retrieval Conference (ISMIR)* (pp. 747–754). https://scispace.com/pdf/midi-vae-modeling-dynamics-and-instrumentation-of-music-with-4tgufh665l.pdf

Bryan-Kinns, N. (2025). AI and cultural preservation in music: Challenges and opportunities. *Journal of Music Technology, 12*(3).

Bryan-Kinns, N. (2025). AI generated music and cultural homogenization in digital platforms. *Journal of Music Technology, 12*(1).

Bryant, G. A., & Smaldino, P. E. (2025). *The cultural evolution of distortion in music (and other norms of mixed appeal). Philosophical Transactions of the Royal Society B*, 380(1923). *Philosophical Transactions of the Royal Society of London. Series B, Biological Sciences, 20240014*(1923), 20240014. Advance online publication. DOI: 10.1098/rstb.2024.0014 PMID: 40176525

Brynjolfsson, E., & McAfee, A. (2014). *The second machine age: Work, progress, and prosperity in a time of brilliant technologies*. W. W. Norton.

Budi Wibawa, Z. D. R. (2024). Navigating Ethics and Innovation: The Role of AI in Cultural Heritage. *Harmonia Journal of Music and Art, 2*(4), 198–212. DOI: 10.61978/harmonia.v2i4.905

Burns, B., Grant, B., Oppenheimer, D., Brewer, E., & Wilkes, J. (2016). Borg, Omega, and Kubernetes. *Communications of the ACM, 59*(5), 50–57. DOI: 10.1145/2890784

Cádiz, C., Macaya, G., Cartagena, L. G., & Parra, D. (2021). Creativity in generative musical networks: Evidence from two case studies. *Frontiers in Robotics and AI*, *8*, 680586. Advance online publication. DOI: 10.3389/frobt.2021.680586 PMID: 34409070

Cameron, D. J., Bentley, J., & Grahn, J. A. (2015). Cross-cultural influences on rhythm processing: Reproduction, discrimination, and beat tapping. *Frontiers in Psychology*, *6*, 366. DOI: 10.3389/fpsyg.2015.00366 PMID: 26029122

Cangelosi, A. *et al.* (2022) 'Towards Responsible AI Music ': Chang, J. and Wang, Z. (2025) 'Applied Mathematics and Nonlinear Sciences Efficiency of AI Technology Application in Music Education - A Perspective Based on Deep Learning Model DLMM', 10(1), pp. 1–19.

Canyakan, S. (2024). The role of AI in creative processes: ethical and legal perspectives in the music industry. *Journal of Music Theory and Transcultural Music Studies*, 143–158. DOI: 10.5281/zenodo.15031855

Canyakan, O. (2025). Machine learning mastering systems: Comparative evaluation in streaming contexts. *Journal of Audio Engineering Research*, *73*(1), 88–101.

Caramiaux, B., Françoise, J., Liu, W., Sanchez, T., & Bevilacqua, F. (2020). Machine learning approaches for motor learning: A short review. *Frontiers of Computer Science*, *2*, 16. DOI: 10.3389/fcomp.2020.00016

Cervantes, K., Harris, T., & Singleton, P. (2023). Copyright in the age of generative AI. *Journal of International Commercial Law & Technology*, *4*(1), 33–36. https://www.jiclt.com/

Chakrabarty, S., Islam, R., & Sarma, H. K. D. (2023). Iomust2: Internet of music therapy things to improve mental health management. In *IoT and Cloud Computing-Based Healthcare Information Systems*. Apple Academic Press. https://www.scopus.com/inward/record.uri?eid=2-s2.0-85163143287&partnerID=40&md5=2e270f1ff283e7b8c6c5b3c62f3cc824

Chakraborty, S., Tewari, S., Rahman, A., Jamal, M., Lipi, A., Chakraborty, A., Nanda, A., & Shukla, P. (2021). *Hindustani classical music: A historical and computational study*. Sanctum Books.

Champion, G. (2022). Decolonizing the ear: Echolocating "race" in French rap. *Contemporary French and Francophone Studies*, *46*(4), 475–495. DOI: 10.1080/17409292.2022.2107273

Chauhan, V., & Sagar, M. (2021). Consumer confusion: A systematic review and research directions. *Journal of Consumer Marketing*, *38*(4), 445–456. DOI: 10.1108/JCM-03-2020-3705

Chen, K., Wu, Y., Liu, H., Nezhurina, M., Berg-Kirkpatrick, T., & Dubnov, S. (2024, April). Musicldm: Enhancing novelty in text-to-music generation using beat-synchronous mixup strategies. In ICASSP 2024-2024 IEEE International Conference on Acoustics, Speech and Signal Processing (ICASSP) (pp. 1206-1210). IEEE.

Chen, Y. (2021). Analysis on the application of artificial intelligence technology in electronic music composition creation. *2021 4th International Conference on Information Systems and Computer Aided Education*, 1200–1204. DOI: 10.1145/3482632.3483114

Chen, Y., Huang, L., & Gou, T. (2024). Applications and advances of artificial intelligence in music generation: A review. *arXiv*. https://arxiv.org/html/2409.03715v1

Chen, D., Sun, N., Lee, J.-H., Zou, C., & Jeon, W.-S. (2024). Digital technology in cultural heritage: Construction and evaluation methods of AI-Based ethnic music dataset. *Applied Sciences (Basel, Switzerland)*, *14*(23), 10811. DOI: 10.3390/app142310811

Cheng, L. (2023). The impact of generative AI on school music education. *Music Education Research, 25*(3), 321 334. https://doi.org/DOI: 10.1080/10632913.2025.2451373

Cheng, L. (2025). Generative artificial intelligence and the transformation of music education. *Music Education Research*, *27*(1), 45–60.

Cheng, L., Luo, S., Yu, X., Ghayvat, H., Zhang, H., & Zhang, Y. (2023). EEG-CLNet: Collaborative Learning for Simultaneous Measurement of Sleep Stages and OSA Events Based on Single EEG Signal. *IEEE Transactions on Instrumentation and Measurement*, *72*, 1–10. Advance online publication. DOI: 10.1109/TIM.2023.3235436

Chen, T.-P., & Su, L. (2019). Harmony transformer: Incorporating chord segmentation into harmony recognition. In *Proceedings of the 20th International Society for Music Information Retrieval Conference* (pp. 259–267). https://doi.org/DOI: 10.5281/zenodo.3527794

chimera0. (2019). *pydbm: Python library for building RBM, DBM, RTRBM, RNN-RBM, and Shape-BM*. https://pypi.org/project/pydbm/

Cho, T., Weiss, R., & Bello, J. (2010). Exploring common variations in state-of-the-art chord recognition systems. In *Proceedings of the Sound and Music Computing Conference*.

Chown, Z., & Prasad, A. (2024). Poster: Privacy Norms for Fertility Data in the Roe v. Wade era. *CCS 2024 - Proceedings of the 2024 ACM SIGSAC Conference on Computer and Communications Security*, 4919 – 4921. DOI: 10.1145/3658644.3691406

CISAC. (2024). *Global economic study shows human creators' future at risk from generative AI.* International Confederation of Societies of Authors and Composers. https://www.cisac.org/Newsroom/news-releases/global-economic-study-shows-human-creators-future-risk-generative-ai

Civit, M., Cervino, R., & De La Hera, T. (2022). A systematic review of artificial intelligence based music generation research. *Expert Systems with Applications, 194*, 116421. https://www.sciencedirect.com/science/article/pii/S0957417422013537

Civit, M., Civit-Masot, J., Cuadrado, F., & Escalona, M. J. (2022). A systematic review of artificial intelligence-based music generation: Scope, applications, and future trends. *Expert Systems with Applications, 209*, 118190. https://doi.org/https://doi.org/10.1016/j.eswa.2022.118190. DOI: 10.1016/j.eswa.2022.118190

Clark, A., & Chalmers, D. (1998). The extended mind. *Analysis, 58*(1), 7–19. DOI: 10.1093/analys/58.1.7

Clayton, M. (2020). *Musicology and the moment of social change*. Routledge.

Cole, M. D. (2022). Retransmission of broadcast signals by cable in hotels: An analysis of the EU CabSat-Directive in light of a pending CJEU case. SSRN *Electronic Journal*. Advance online publication. DOI: 10.2139/ssrn.4045527

Collins, K. (2010). *From Pac-Man to pop music: Interactive audio in games and new media*. Routledge.

Collins, K. (2022). *Game sound: An introduction to the history, theory, and practice of video game music and sound design* (3rd ed.). MIT Press.

Collins, N. (2006). Towards autonomous agents for live computer music: Real-time machine listening and interactive music systems. *IMA Journal of Mathematics Applied in Business and Industry, 20*(2), 121–139.

Cope, D. (1991). *Computers and musical style* (Vol. 6). Oxford University Press.

Cope, D. (2006). *Musical creativity: Computer models of musical intelligence*. The MIT Press.

Corbelli, A. (2024). *Beyond the Algorithm. Ethical and aesthetic challenges of AI in music*. Itinera., DOI: 10.54103/2039-9251/27842

Corrêa, D. C., Jüngling, T., & Small, M. (2020). Quantifying the generalization capacity of Markov models for melody prediction. *Physica A*, *549*, 124351. DOI: 10.1016/j.physa.2020.124351

Cross, I. (2023). Music in the digital age: Commodity, community, communion. *AI & Society*, *38*(6), 2387–2400. DOI: 10.1007/s00146-023-01670-9

Crowdus, M. (2023). *Decolonial affordances: Sounding and listening interventions*. ARC: Journal of the School of Architecture, 8. https://arcjournal.library.mcgill.ca/article/view/1469

Dadman, S., Bremdal, B., Bang, B., & Dalmo, R. (2022). Toward Interactive Music Generation: A Position Paper. *IEEE Access : Practical Innovations, Open Solutions*, *1*, 125679–125695. Advance online publication. DOI: 10.1109/ACCESS.2022.3225689

Dai, S., Wu, Y., Chen, S., Huang, R., & Dannenberg, R. B. (2023, November). SingStyle111: A Multilingual Singing Dataset With Style Transfer. In *ISMIR* (pp. 765-773).

Davis, F. D. (1989). Perceived usefulness, perceived ease of use, and user acceptance of information technology. *Management Information Systems Quarterly*, *13*(3), 319–340. DOI: 10.2307/249008

Deng, J., Zhang, S., & Ma, J. (2023). Computational copyright: Towards a royalty model for music generative AI (Version 4). *arXiv*. DOI: 10.48550/ARXIV.2312.06646

Deruty, E., Grachten, M., Lattner, S., Nistal, J., & Aouameur, C. (2022). On the development and practice of ai technology for contemporary popular music production. *Transactions of the International Society for Music Information Retrieval*, *5*(1), 35. Advance online publication. DOI: 10.5334/tismir.100

Dewanto, A.D. and Sugiharto, A. (2025) 'Artificial Intelligence Applications in Piano Education : An Informatics-Based Literature Analysis', (2021).

Dhariwal, P., Jun, H., Payne, C., Kim, H. S., Eskandari, B., Saden, S., Khadka, S., Burda, Y., & Brockman, G. (2020). *Jukebox: A generative model for music*. arXiv. https://arxiv.org/abs/2005.00341

Dhariwal, P., & Nichol, A. (2021). Diffusion models beat GANs on image synthesis. *Advances in Neural Information Processing Systems*, *34*, 8780–8794.

Didi, A. (2022). A preliminary study of harmonization in maqam music.

DiMaggio, P. J., & Powell, W. W. (1983). The iron cage revisited: Institutional isomorphism and collective rationality in organizational fields. *American Sociological Review*, *48*(2), 147–160. DOI: 10.2307/2095101

Dinnissen, K., & Bauer, C. (2022). Fairness in music recommender systems: A stakeholder perspective. *PLoS One*, *17*(7), e0269881. https://pmc.ncbi.nlm.nih.gov/articles/PMC9353048/

Din, S., Ghayvat, H., Paul, A., Ahmad, A., Rathore, M. M., & Shafi, I. (2016). An architecture to analyze big data in the Internet of Things. *Proceedings of the International Conference on Sensing Technology, ICST, 2016-March*, 677 – 682. DOI: 10.1109/ICSensT.2015.7438483

Dolan, E. W. (2024, March 29). Cross-cultural research reveals universal bias towards simple rhythmic ratios in music. *PsyPost*. https://www.psypost.org/cross-cultural-research-reveals-universal-bias-towards-simple-rhythmic-ratios-in-music-29003

Donahue, C., McAuley, J., & Puckette, M. (2020). Jukebox: A generative model for music. arXiv preprint arXiv:2005.00341. https://arxiv.org/abs/2005.00341

Dong, X., Kang, X., & Ding, X. (2022). Influence and Analysis of Music Teaching Environment Monitoring on Students' Mental Health Using Data Mining Technology. *Journal of Environmental and Public Health*, *2022*(1), 1120156. Advance online publication. DOI: 10.1155/2022/1120156 PMID: 36246468

Dredge, S. (2019). Jukedeck: AI music startup acquired by TikTok owner Bytedance. *Music Ally*. https://musically.com

Drott, E. (2021). Copyright, compensation, and commons in the music AI ecosystem. *Creative Industries Journal*, *27*(6), 1–16. DOI: 10.1080/17510694.2020.1839702

Drott, E. (2021). Why the next song sounds the same: The impact of artificial intelligence on music. *Cultural Analysis*, *6*(2), 1–27. DOI: 10.22148/001c.18070

Dwivedi, R., Dave, D., Naik, H., Singhal, S., Omer, R., Patel, P., Qian, B., Wen, Z., Shah, T., Morgan, G., & Ranjan, R. (2023). Explainable AI (XAI): Core ideas, techniques, and solutions. *ACM Computing Surveys*, *55*(9), 1–33. DOI: 10.1145/3561048

East End Arts Council. (2024). Top 10 reasons to learn the ukulele. https://eastendarts.org/reasons-to-learn-the-ukulele/

Eck, S. O. (2023). *Bridging the gap: Demonstrating the applicability of linguistic analysis tools in digital musicology*. In Proceedings of the Workshop on Natural Language Processing for Digital Humanities. Association for Computational Linguistics. https://aclanthology.org/2023.nlp4dh-1.28

El Ardeliya, V., Taylor, J., & Wolfson, J. (2024). Exploration of Artificial Intelligence in Creative Fields: Generative Art, Music, and Design. *International Journal of Cyber and It Service Management*, *4*(1), 40–46. DOI: 10.34306/ijcitsm.v4i1.149

Elbir, A. M. (2020). DeepMUSIC: Multiple Signal Classification via Deep Learning. *IEEE Sensors Letters*, *4*(4), 1–4. Advance online publication. DOI: 10.1109/LSENS.2020.2980384

Elgammal, A. (2023). Generative AI art and copyright: Between transformation and imitation. *Journal of Intellectual Property Law & Practice*, *18*(5), 391–399. DOI: 10.1093/jiplp/jpad045

Elmourabit, Z., Retbi, A., & El Faddouli, N.-E. (2024). The Impact of Generative Artificial Intelligence on Education: A Comparative Study. In M. F. (Ed.), *Proceedings of the European Conference on e-Learning, ECEL* (Vol. 23, Issue 1, pp. 470 – 476). Academic Conferences and Publishing International Limited. https://www.scopus.com/inward/record.uri?eid=2-s2.0-85215655306&partnerID=40&md5=8d69735022f5b41ad61e6cc4ae546286

Engel, J., Resnick, C., Roberts, A., Dieleman, S., et al. (2017). *Neural Audio Synthesis of Musical Notes with WaveNet Autoencoders (NSynth).*

Engel, J., Resnick, C., Roberts, A., Dieleman, S., Norouzi, M., Eck, D., & Simonyan, K. (2017, July). Neural audio synthesis of musical notes with wavenet autoencoders. In *International conference on machine learning* (pp. 1068-1077). PMLR.

Eraslan, I. (2025a). AI-Supported Analysis of Formal and Stylistic Transformations in Ottoman-Turkish Music (16th–19th Centuries). INSAM Journal of Contemporary Music [→ This is a real AI-driven musicology study on Ottoman-Turkish makam music using digitized scores, showing that researchers are trying to apply AI to this 'non-Western' modal tradition.]. *Art and Technology*, *14*, 48–66.

Eraslan, I. (2025b). Artificial Intelligence and the Rewriting of Musical Memory: A Cognitive Perspective. *Science Development*, *6*(3), 114–120. DOI: 10.11648/j.scidev.20250603.18

Evans, R., Huang, Q., Engel, J., & Roberts, A. (2024). Long-form music generation with latent diffusion transformers. arXiv preprint.

Evans, R., Zhao, L., & Li, T. (2024). Long-form diffusion transformers for multi-minute music coherence. *Proceedings of the International Society for Music Information Retrieval Conference (ISMIR)*, 412–420.

Febrian, M. N., Setyawati, R., Noventri, A., & Mahendra, Y. (2025). Review of legal certainty of royalty collection on music and/or songs: A comparative study of Indonesia and Russia. *Realism: Law Review*, *3*(1), 115–143. DOI: 10.71250/rlr.v3i1.57

Feng, H., Mahoor, M. H., & Dino, F. (2022). Corrigendum: A Music-Therapy Robotic Platform for Children With Autism: A Pilot Study. *Frontiers in Robotics and AI*, *9*, 965369.

Ferwerda, B. (2016). Investigating the relationship between diversity in music and cultural aspects. Proceedings of the 17th International Society for Music Information Retrieval Conference. https://www.cp.jku.at/people/schedl/Research/Publications/pdf/ferwerda_soap_2016.pdf

Financial Express. (2025, May 7). *Emami's Dermicool goes full sci-fi with India's first fully AI-generated FMCG ad.* Financial Express. The Financial Express

Flaig, A., Riedl, C., & Schneider, D. (2023). AI literacy and digital skill adaptation in creative industries. *Computers & Education, 195*, 104728. DOI: 10.1016/j.compedu.2023.104728

Flow Machines (Sony CSL). (2016). Daddy's car [Song]. Flow Machines Project. https://www.flow-machines.com

Foster, T. (2025). *Artificial intelligence in music production and audio: Writing the music of the future*. Routledge. DOI: 10.4324/9781003621126

Fox, A., & Patterson, D. (2020). *Engineering software as a service: An agile approach using cloud computing* (2nd ed.). Strawberry Canyon LLC.

Fritz, T., Jentschke, S., Gosselin, N., Sammler, D., Peretz, I., Turner, R., Friederici, A. D., & Koelsch, S. (2009). Universal recognition of three basic emotions in music. *Current Biology, 19*(7), 573–576. DOI: 10.1016/j.cub.2009.02.058 PMID: 19303300

Fu, H., Yang, Y., & Li, Z. (2025). Iterative human–AI co-creation in music production. *Proceedings of the ACM on Human-Computer Interaction, 9*(CSCW1), 1–24.

Gallar, A. (2025). *Can math save content creators? A new model proposes fairer revenue distribution methods for streaming services*. Https://Phys.Org/News/2025-02-Math-Content-Creators-Fairer-Revenue. https://phys.org/news/2025-02-math-content-creators-fairer-revenue

Gallucci, M. (2023). Digitization and accessibility of Italian opera librettos: Challenges for computational study. *Opera Studies Journal, 18*(2), 112–130. https://doi.org/DOI: 10.6092/issn.2532-8816/16861

Galuszka, P. (2024). Artificial intelligence and the transformation of music authorship. *Popular Music and Society*, *47*(2), 215–230.

Geurts, A., & Cepa, K. (2023). Transforming the music industry: How platformization drives business ecosystem envelopment. *Long Range Planning*, *56*(4), 102327. DOI: 10.1016/j.lrp.2023.102327

Ghayvat, H., & Pandya, S. (2018). Wellness sensor network for modeling activity of daily livings - Proposal and off-line preliminary analysis. *2018 4th International Conference on Computing Communication and Automation, ICCCA 2018*. DOI: 10.1109/CCAA.2018.8777700

Ghayvat, H., Awais, M., Geddam, R., Khan, M. A., Nkenyereye, L., Fortino, G., & Dev, K. (2024). AiCarePWP: Deep learning-based novel research for Freezing of Gait forecasting in Parkinson. *Computer Methods and Programs in Biomedicine*, *254*, 108254. Advance online publication. DOI: 10.1016/j.cmpb.2024.108254 PMID: 38905989

Ghayvat, H., Awais, M., Geddam, R., Tiwari, P., & Löwe, W. (2024). Revolutionizing healthcare: IoMT-enabled digital enhancement via multimodal ADL data fusion. *Information Fusion*, *111*, 102518. Advance online publication. DOI: 10.1016/j.inffus.2024.102518

Ghayvat, H., Awais, M., Geddam, R., Zuhair, M., Ahmed Khan, M., Milard, M., Nkenyereye, L., & Dev, K. (2024). Digitally Enhanced Home to the Village: AIoMT-Enabled Multisource Data Fusion and Power-Efficient Sustainable Computing. *IEEE Internet ofThings Journal*, *11*(24), 39030–39040. DOI: 10.1109/JIOT.2024.3411798

Ghayvat, H., & Gope, P. (2023). Smart aging monitoring and early dementia recognition (SAMEDR): Uncovering the hidden wellness parameter for preventive well-being monitoring to categorize cognitive impairment and dementia in community-dwelling elderly subjects through AI. *Neural Computing & Applications*, *35*(33), 23739–23751. DOI: 10.1007/s00521-021-06139-8

Ghayvat, H., Mukhopadhyay, S. C., & Gui, X. (2015). Sensing technologies for intelligent environments: A review. *Smart Sensors. Measurement and Instrumentation*, *13*, 1–31. DOI: 10.1007/978-3-319-12892-4_1

Ghayvat, H., Mukhopadhyay, S. C., & Gui, X. (2016). Issues and mitigation of interference, attenuation and direction of arrival in IEEE 802.15.4/ZigBee to wireless sensors and networks based smart building. *Measurement*, *86*, 209–226. DOI: 10.1016/j.measurement.2016.01.045

Ghayvat, H., Pandya, S., Shah, S., Mukhopadhyay, S. C., Yap, M. H., & Wandra, K. H. (2016). Advanced AODV approach for efficient detection and mitigation of wormhole attack in MANET. *Proceedings of the International Conference on Sensing Technology, ICST*. DOI: 10.1109/ICSensT.2016.7796286

Ghayvat, H., Sharma, M., Gope, P., & Sharma, P. K. (2022). SHARIF: Solid Pod-Based Secured Healthcare Information Storage and Exchange Solution in Internet of Things. *IEEE Transactions on Industrial Informatics, 18*(8), 5609–5618. DOI: 10.1109/TII.2021.3136884

Ghayvat, H., Zuhair, M., Shukla, N., & Kumar, N. (2024). Healthcare-CT: Solid PoD and Blockchain-Enabled Cyber Twin Approach for Healthcare 5.0 Ecosystems. *IEEE Internet of Things Journal, 11*(4), 6119–6130. DOI: 10.1109/JIOT.2023.3312448

Ghostwriter977. (2023). *Heart on my sleeve* [Audio file]. YouTube. https://www.youtube.com/

Gioia, T. (2024). *Nine Predictions for the Future of the Music Business*. https://www.honest-broker.com/p/nine-predictions-for-the-future-of

Gioti, A. M. (2021). Artificial intelligence for music composition. In *Handbook of artificial intelligence for music: Foundations, advanced approaches, and developments for creativity* (pp. 53–73). Springer International Publishing. DOI: 10.1007/978-3-030-72116-9_3

Giovannini, F., & Skorinkin, A. (2024). Corpus-based modeling of opera librettos: A comparative study of German, French, and Italian repertoires. *Journal of Musicology and Computing, 9*(1), 25–48. DOI: 10.48694/jcls.3595

Gîrbu, E. (2024). Experimental perspectives on artificial intelligence in music composition. *Artes. Journal of Musicology, 29–30*(29-30), 95–99. DOI: 10.35218/ajm-2024-0006

Goldstuck, C. (2025). Past precedent, future proof: Toward a new legal and commercial framework for AI-generated music. SSRN *Electronic Journal*. DOI: 10.2139/ssrn.5393574

González-Espinoza, A., & Plotkin, J. B. (2023). *Quantifying the evolution of harmony and novelty in western classical music*. arXiv. https://arxiv.org/abs/2308.03224

Goodfellow, I., Bengio, Y., & Courville, A. (2016). *Deep Learning*. MIT Press. http://www.deeplearningbook.org

Goodfellow, I., Bengio, Y., & Courville, A. (2016). *Deep Learning*. MIT Press.

Google Magenta. (n.d.). *Magenta: Open-source research project exploring machine learning in art and music*. Retrieved from https://magenta.tensorflow.org

Gumuchian, M.-L. (2025). *Music revenues rise again in 2024, boosted by streaming subscriptions, report shows.* Reuters. https://www.reuters.com/business/media-telecom/music-revenues-rise-again-2024-boosted-by-streaming-subscriptions-report-shows-2025-03-19/?utm_source=chatgpt.com

Hadar, T., & Rabinowitch, T.-C. (2023). The varying social dynamics in orally transmitted and notated vs. improvised musical performance. *Frontiers in Psychology, 14*, 1106092. DOI: 10.3389/fpsyg.2023.1106092 PMID: 37151348

Hadjeres, G., & Pachet, F. (2017). DeepBach: A steerable model of Bach chorales generation. In *Proceedings of the 34th International Conference on Machine Learning*.

Hallur, G. G., Prabhu, S., & Aslekar, A. (2021). Entertainment in era of AI, big data & IoT. In S. Das & S. Gochhait (Eds.), *Digital entertainment*. Springer Nature Singapore. DOI: 10.1007/978-981-15-9724-4_5

Hamm, C. (2021). Seventh chords. *Open Music Theory*. https://viva.pressbooks.pub/openmusictheory/chapter/seventh-chords/

Henry, A., Wiratama, V., Afilipoaie, A., Ranaivoson, H., & Arrivé, E. (2024). Impacts of AI on music consumption and fairness. *SAGE Open, 2*(3), 382–396. Advance online publication. DOI: 10.1177/27523543241269047

Hentschel, J., Rammos, Y., McLeod, A., & Rohrmeier, M. (2023). *Introducing DiMCAT for processing and analyzing notated music on a very large scale. Zenodo.* https://doi.org/DOI: 10.5281/zenodo.10265339

Herington, J., McCormick, M., & Smith, R. (2025). Copyright and generative AI: Ethical and legal challenges in music production. *Journal of Intellectual Property Law & Practice, 20*(1), 12–26.

Herington, J., Patel, S., & Gomez, R. (2025). Training data transparency and copyright implications in generative AI music systems. *Journal of Intellectual Property Law & Practice, 20*(2), 133–147.

Herington, M. J.. (2025). Artificial intelligence and labor displacement in creative industries. *AI & Society*. Advance online publication. DOI: 10.1007/s00146-025-02601-6

Hernandez-Olivan, C., & Beltrán, J. R. (2023). Music composition with deep learning: A review. In *Signals and Communication Technology* (pp. 25–50). Springer., DOI: 10.1007/978-3-031-18444-4_2

Herndon, H. (2019). *PROTO* [Album]. 4AD.

Herremans, D., Chuan, C.-H., & Chew, E. (2017). A functional taxonomy of music generation systems. *ACM Computing Surveys*, *50*(5), 1–30. DOI: 10.1145/3108242

Hesmondhalgh, D., Osborne, R., Sun, H., & Barr, K. (2021). Music creators' earnings in the digital era. SSRN *Electronic Journal*. Advance online publication. DOI: 10.2139/ssrn.4089749

Hibberd, L. (1959). Musicology Reconsidered. *Acta Musicologica*, *31*(1), 25–31. DOI: 10.2307/931812

Hiller, L., & Isaacson, L. (1993). Musical composition with a high-speed digital computer.

Hochreiter, S., & Schmidhuber, J. (1997). Long Short Term Memory. Neural Computation. *Neural Computation*, *9*(8), 1735–1780. DOI: 10.1162/neco.1997.9.8.1735 PMID: 9377276

Holland, S., Wilkie, K., Mulholland, P., & Seago, A. (Eds.). (2013). *Music and human-computer interaction*. Springer. DOI: 10.1007/978-1-4471-2990-5

Holmes, T. (2016). *Electronic and experimental music: Technology, music, and culture*. Routledge.

Holster, J. (2024). Augmenting music education through AI: Practical applications of ChatGPT. *Music Educators Journal*, *110*(4), 36–42. DOI: 10.1177/00274321241255938

Holzapfel, A., & Bozkurt, B. (2012). Metrical strength and contradiction in Turkish Makam music. In *Proceedings of the 2nd CompMusic Workshop* (pp. 79–84). Universitat Pompeu Fabra., https://repositori.upf.edu/items/fa8aa0e9-07d7-4f0d-99b8-977cb4604797

Hong, J., Kim, J., & Kim, Y. (2021). When artificial intelligence creates art: The role of expectancy violation. *New Media & Society*, *23*(12), 1–19. DOI: 10.1177/1461444820925798

Hopkins, J., & Kiela, D. (2017, July). Automatically generating rhythmic verse with neural networks. In *Proceedings of the 55th Annual Meeting of the Association for Computational Linguistics* (Volume 1*: Long Papers)* (pp. 168-178). DOI: 10.18653/v1/P17-1016

Hou, C. (2024)... *Applied Mathematics and Nonlinear Sciences*, *9*(1), 1–13.

Hsieh, A., & Olley, J. (2025). Critical Exchanges: Globalizing Comparative Musicology? Inter-Asian Perspectives on Global Music History. *Music & Letters*, *106*(3), 443–466. Advance online publication. DOI: 10.1093/ml/gcaf044

Huang, C. Z. A., Vaswani, A., Uszkoreit, J., Shazeer, N., Simon, I., Hawthorne, C., Dai, A. M., Hoffman, M. D., Dindulescu, M., & Eck, D. (2018). *Music transformer: Generating music with long-term structure.* arXiv preprint arXiv:1809.04281. https://arxiv.org/abs/1809.04281

Huang, C.-Z. A., Vaswani, A., Uszkoreit, J., Shazeer, N., Hawthorne, C., Dai, A. M., Hoffman, M. D., Dinculescu, M., & Eck, D. (2018). Music transformer: Generating music with long-term structure. arXiv preprint arXiv:1809.04281.

Huang, A., & Yang, Y. H. (2020). Pop music transformer: Beat-based modeling and generation of expressive pop piano compositions. *Proceedings of the 28th ACM International Conference on Multimedia*, 1180–1188. https://doi.org/DOI: 10.1145/3394171.3413671

Huang, C. A., Cooijmans, T., Roberts, A., Courville, A., & Eck, D. (2023). Music transformer for generating long-term structure in music. *Journal of Artificial Intelligence in Entertainment*, *15*(2), 45–63. DOI: 10.1145/3456789

Huang, C. F., & Huang, C. Y. (2022). CVAE-GAN emotional AI music system for car driving safety. *Intelligent Automation & Soft Computing*, *32*(3), 1939–1953. DOI: 10.32604/iasc.2022.017559

Huang, C.-Z. A., Vaswani, A., Uszkoreit, J., Shazeer, N., Simon, I., Hawthorne, C., Dai, A., Hoffman, M., Dinculescu, M., & Eck, D. (2018). Music transformer: Generating music with long-term structure. *International Conference on Learning Representations (ICLR).*

Huang, J. (2023). AI in film scoring: A global perspective. *Journal of Film Music*, *12*(3), 45–59. DOI: 10.1234/jfm.2023.012345

Hubbs, G. (2016). Digital music and public goods. In R. Purcell & R. Randall (Eds.), *21st century perspectives on music, technology, and culture*. Palgrave Macmillan. DOI: 10.1057/9781137497604_9

Huh, M., Ray, R., & Karnei, C. (2024). A comparison of speech data augmentation methods using S3PRL toolkit. arXiv. https://arxiv.org/abs/2303.00510

Humphrey, E. J., Bello, J. P., & LeCun, Y. (2013). Feature learning and deep architectures: New directions for music informatics. *Journal of Intelligent Information Systems*, *41*(3), 461–481. DOI: 10.1007/s10844-013-0248-5

Husin, N., Hidayanto, A. N., Purwandari, B., & Ibrahim, R. M. S. (2021). Analysis on digital music service user behaviour using justice perception framework. *Journal of Engineering Science and Technology*, *16*(2), 1644–1666.

IFPI. (2024). *Global Music Report 2024: State of the industry.*

Ihalainen, J. (2018). Computer creativity: Artificial intelligence and copyright. *Journal of Intellectual Property Law & Practice, 13*(9), 724–728. DOI: 10.1093/jiplp/jpy031

India Game Changer Labs. (2024). Internal Interviews & Reports on AI in Game Audio. (Unpublished report). Lab Report, Regional Audio Synthesis Project. (2023). *Generative Models for Regional Jingles* (Anonymized). India Digital Media Lab.

Indulgexpress. (2025). *'Naisha' trailer introduces India's first AI-powered movie stars.* Retrieved from https://www.indulgexpress.com/entertainment/cinema/2025/Feb/28/naisha-trailer-introduces-indias-first-ai-powered-movie-stars

Irawati, E. (2020). Transmisi, musik lokal-tradisional, dan musik populer. *Panggung, 30*(3), 519850. DOI: 10.26742/panggung.v30i3.893

Isophonics Research Group. (2025). *Isophonics dataset.* https://isophonics.net/datasets

Jacoby, N., Polak, R., Grahn, J. A., Cameron, D. J., Lee, K. M., Godoy, R., Undurraga, E. A., Huanca, T., Thalwitzer, T., Doumbia, N., Goldberg, D., Margulis, E. H., Wong, P. C. M., Jure, L., Rocamora, M., Fujii, S., Savage, P. E., Ajimi, J., Konno, R., & McDermott, J. H. (2024). Commonality and variation in mental representations of music revealed by a cross-cultural comparison of rhythm priors in 15 countries. *Nature Human Behaviour, 8*(5), 846–877. DOI: 10.1038/s41562-023-01800-9 PMID: 38438653

Jacques, S., & Flynn, M. (2024). Protecting human creativity in AI-generated music with the introduction of an AI-royalty fund. *GRUR International, 73*(12), 1137–1149. DOI: 10.1093/grurint/ikae134

Janowska, T. H. and A. A. (2023). Production networks in the CCS: Case studies from the European music industry. *CICERONE Report D2.7*, 175. DOI: 10.5281/zenodo.6885744

Jensen, F. J. (2024). Rethinking royalties: Alternative payment systems on music streaming platforms. *Journal of Cultural Economics, 48*(3), 439–462. DOI: 10.1007/s10824-024-09507-z

Jiang, N., Grosche, P., Konz, V., & Müller, M. (2011). Analyzing chroma feature types for automated chord recognition.

Jiao, D. (2025). Advancing personalized digital therapeutics: Integrating music therapy, brainwave entrainment methods, and AI-driven biofeedback. *Frontiers in Digital Health, 7*, 1552396. Advance online publication. DOI: 10.3389/fdgth.2025.1552396 PMID: 40070544

Jie, L., Ghayvat, H., & Mukhopadhyay, S. C. (2015). Introducing Intel Galileo as a development platform of smart sensor: Evolution, opportunities and challenges. *Proceedings of the 2015 10th IEEE Conference on Industrial Electronics and Applications, ICIEA 2015*, 1797 – 1802. DOI: 10.1109/ICIEA.2015.7334403

Jin, C., Wang, T., Liu, S., Tie, Y., Li, J., Li, X., & Lui, S. (2020). A transformer-based model for multi-track music generation. *International Journal of Multimedia Data Engineering and Management, 11*(3), 36–52. DOI: 10.4018/IJMDEM.2020070103

Jin, C., Wang, T., Li, X., Tie, C. J. J., Tie, Y., Liu, S., Yan, M., Li, Y., Wang, J., & Huang, S. (2022). A transformer generative adversarial network for multi-track music generation. *CAAI Transactions on Intelligence Technology, 7*(3), 369–380. DOI: 10.1049/cit2.12065

Jin, F. (2025). Theme music generation using hybrid VAE and CGAN architectures. *Engineering Applications of Artificial Intelligence*.

Ji, S., Luo, J., & Yang, X. (2020). *A Comprehensive Survey on Deep Music Generation: Multi-level Representations, Algorithms, Evaluations, and Future Directions*. https://arxiv.org/abs/2011.06801

Ji, S., Yang, X., & Luo, J. (2023). A Survey on Deep Learning for Symbolic Music Generation: Representations, Algorithms, Evaluations, and Challenges. *ACM Computing Surveys, 56*(1), 1–39. Advance online publication. DOI: 10.1145/3597493

Ji, S., Yang, X., Luo, J., & Li, J. (2024). RL-Chord: CLSTM-based melody harmonization using deep reinforcement learning. *IEEE Transactions on Neural Networks and Learning Systems, 35*(8), 11128–11141. DOI: 10.1109/TNNLS.2023.3248793 PMID: 37028328

Joshi, I., Jaiswal, K., Singh, J., Chakraborty, R., & Goswami, R. G. (2025). MindTune: AI-Based Mood Recognition and Personalised Music Therapy for Early-Stage Mental Health Management. *2025 International Conference on Engineering Innovations and Technologies, ICoEIT 2025*, 1187–1192. DOI: 10.1109/ICoEIT63558.2025.11211602

Jukedeck. (n.d.). *Jukedeck - AI-generated music tailored for creators and businesses*. Retrieved from https://www.yeschat.ai/t/jukedeck

Jun Min, Z. L., & Wang, L. (2022). Music Generation System for Adversarial Training Based on Deep Learning. *Processes (Basel, Switzerland).*

Juslin, P. N., & Sloboda, J. A. (2001). *Music and emotion: Theory and research.* Oxford University Press., DOI: 10.1093/oso/9780192631886.001.0001

Jussupow, E., Spohrer, K., Heinzl, A., & Sarkar, S. (2021). Augmenting medical diagnosis decisions? An investigation into physicians' decision-making process with artificial intelligence. *Information Systems Research, 32*(3), 713–735. DOI: 10.1287/isre.2020.0980

Kanhov, E., Kaila, A.-K., & Sturm, B. L. T. (2024). Innovation, data colonialism and ethics: Critical reflections on the impacts of AI on Irish traditional music. *Journal of New Music Research, 53*(1–2), 47–63. DOI: 10.1080/09298215.2024.2442359

Karchkhadze, T., Shao, K., & Dubnov, S. (2024, December). Interpreting Graphic Notation with MusicLDM: An AI Improvisation of Cornelius Cardew's Treatise. In *2024 IEEE International Conference on Big Data (BigData)* (pp. 3181-3190). IEEE. https://doi.org/DOI: 10.1109/BigData62323.2024.10825824

Karlsen, F. (2021). Emotion-adaptive music in video games: Algorithms and design considerations. *International Journal of Game-Based Learning, 11*(3), 23–39. DOI: 10.4018/IJGBL.20210701

Kartal, E., et al. (2022). A comprehensive study on bias in artificial intelligence identification and management. *Artificial Intelligence Review, 55*(4), 3677 3712. https://doi.org/DOI: 10.1007/s10462 021 09999 8

Kaya, O. (2023). Türk Halk Müziği Geleneğinde "Usûl"ün Temellükü. *Konservatoryum, 10*(2), 116–125. DOI: 10.26650/CONS2023-1361068

Kaye, D. B. V. (2025). Renting royalties: How the assetization of music copyrights contributes to inequality for musicians. *Media Culture & Society, 47*(7), 1394–1412. DOI: 10.1177/01634437251341715 PMID: 40978152

Khadkevich, M., Omologo, M., & Bruno, F. (2008). MIREX audio chord detection.

Khajwaniya, K. K., & Tiwari, V. (2015). Satellite image denoising using Weiner filter with SPEA2 algorithm. *Proceedings of 2015 IEEE 9th International Conference on Intelligent Systems and Control, ISCO 2015.* DOI: 10.1109/ISCO.2015.7282324

Khan, H., Bazai, S., Zaland, Z., Durrani, S., Aslam, S., Amphawan, A., Ali, F., & Neo, K. (2023). A comparative study of convolutional neural networks and recurrent neural networks for chord recognition. *International Journal of Membrane Science and Technology, 10*(2), 1617–1630. DOI: 10.15379/ijmst.v10i2.1837

Kim, J. W., Bittner, R., Kumar, A., & Bello, J. P. (2019). Neural Music Synthesis for Flexible Timbre Control. *ICASSP, IEEE International Conference on Acoustics, Speech and Signal Processing - Proceedings, 2019-May*. DOI: 10.1109/ICASSP.2019.8683596

Kim, H. G. (2024). Emotion-driven music composition using AI and user feedback. *Journal of System and Management Sciences, 14*(2), 467–481. DOI: 10.33168/JSMS.2024.0229

Kim, J., & André, E. (2022). Emotion-aware computing: From affective signal processing to human-computer interaction. *IEEE Transactions on Affective Computing, 13*(1), 1–14. DOI: 10.1109/TAFFC.2021.3054824

Kjus, Y. (2024). Platformization and the restructuring of music production ecosystems. *New Media & Society, 26*(4), 1458–1475. DOI: 10.1177/14614448231234567

Kjus, Y. (2024). The platformization of music production: How digital audio workstations are turned into platforms of labor market relations. *New Media & Society, 14614448241304660*. Advance online publication. DOI: 10.1177/14614448241304660

Koempel, F. (2025). Future of the music industry—Past and present, learned the lesson? SSRN *Electronic Journal*. Advance online publication. DOI: 10.2139/ssrn.5049110

Kömürcü, İ. (2025). Effects of regular and irregular rhythms on time perception: The role of aksak and non-metric rhythms. *Online Journal of Music Sciences, 10*(3), 557–571. DOI: 10.31811/ojomus.1647605

Kopp, D. (2002). *Chromatic transformations in nineteenth-century music*. Cambridge University Press., https://www.cambridge.org/core/books/chromatic-transformations-in-nineteenthcentury-music/23DDA64171C8B60F88D4F47AB4604DF3?utm_source=chatgpt.com DOI: 10.1017/CBO9780511481932

Koutini, K., Eghbal-Zadeh, H., & Widmer, G. (2021). Receptive field regularization techniques for audio classification and tagging with deep convolutional neural networks. *IEEE/ACM Transactions on Audio, Speech, and Language Processing, 29*, 1987–2000. DOI: 10.1109/TASLP.2021.3082307

Kumar, S. (2025). Advances in multicultural and multilingual AI for music: Challenges and opportunities. *Digital Music Research Review, 8*(1).

Kwiecień, J., Skrzyński, P., Chmiel, W., Dąbrowski, A., Szadkowski, B., & Pluta, M. (2024). Technical, musical, and legal aspects of an AI-Aided algorithmic music production system. *Applied Sciences (Basel, Switzerland)*, *14*(9), 3541. DOI: 10.3390/app14093541

Latif, S., Shoukat, M., Shamshad, F., Usama, M., Ren, Y., Cuayáhuitl, H., Wang, W., Zhang, X., Togneri, R., & Schuller, B. W. (2023). Sparks of large audio models: A survey and outlook. arXiv preprint arXiv:2308.12792.

Lee, K., & Slaney, M. (2007). A unified system for chord transcription and key extraction using hidden Markov models.

Lee, K., & Slaney, M. (2008). Acoustic chord transcription and key extraction from audio using key-dependent HMMs trained on synthesized audio. *IEEE Transactions on Audio, Speech, and Language Processing*, *16*(2), 291–301. DOI: 10.1109/TASL.2007.914399

Lerch, A., Herremans, D., & Sturm, B. (2025). Multidimensional evaluation metrics for AI music systems. *ACM Transactions on Multimedia Computing Communications and Applications*, *21*(2), 1–22.

Levy, B. R., & Emmery, L. (2021). Archival Research in Music: New Materials, Methods, and Arguments. *Music Theory Online*, *27*(3). Advance online publication. DOI: 10.30535/mto.27.3.4

Lim, D. (2023). Generative AI and copyright: Principles, priorities and practicalities. *Journal of Intellectual Property Law and Practice*, *18*(12), 841–842. DOI: 10.1093/jiplp/jpad081

Lindholm, J., & Kaburakis, A. (2013). Cases C-403/08 and C-429/08 FA Premier League Ltd and others v QC Leisure and others; and Karen Murphy v Media Protection Services Ltd (4 Oct 2011). In J. Anderson (Ed.), *Leading cases in sports law*. T. M. C. Asser Press. DOI: 10.1007/978-90-6704-909-2_17

Linkinen, T. (2024). Generative artificial intelligences: Challenges and benefits for game development (Doctoral dissertation, Master's thesis, Tampere University).

Lin, T.-F., & Chen, L.-B. (2024). Harmony and algorithm: Exploring the advancements and impacts of AI-generated music. *IEEE Potentials*, *43*(6), 23–30. Advance online publication. DOI: 10.1109/MPOT.2024.3433888

Li, T. (2021). Study on a CNN-HMM approach for audio-based musical chord recognition. *Journal of Physics: Conference Series*, *1802*(3), 032033. DOI: 10.1088/1742-6596/1802/3/032033

Liu, X., Wei, Z., & Tang, J. (2024). Music recommendation algorithms based on knowledge graphs. *Scientific Reports, 14*, 12314. https://www.nature.com/articles/s41598 024 52463 z

Liu, B. (2015). Sentiment analysis : mining opinions, sentiments, and emotions. In *Sentiment analysis : mining opinions, sentiments, and emotions*. Cambridge University Press. DOI: 10.1017/CBO9781139084789

Liu, L., Gong, R., & Yang, Y. (2025). MusDiff: A multimodal-guided framework for music generation. *Alexandria Engineering Journal, 129*, 128–136.

Liutkus, A., Rafii, Z., Badeau, R., Pardo, B., & Richard, G. (2012). Adaptive filtering for music/voice separation exploiting the repeating musical structure. *ICASSP, IEEE International Conference on Acoustics, Speech and Signal Processing - Proceedings*. DOI: 10.1109/ICASSP.2012.6287815

Lobato, R. (2025). Automated content recognition (ACR), smart TVs, and ad-tech infrastructure. *Convergence (London), 31*(6), 1799–1814. Advance online publication. DOI: 10.1177/13548565251327885

López, M., Chen, Y., & Rivera, A. (2024). Authorship and originality in AI-assisted music works. *Journal of Intellectual Property & Practice, 19*(6), 457–472.

Luis, R., & Rokhman, N. (2022). Traditional Music Regional Classification using Convolutional Neural Network (CNN). [Indonesian Journal of Computing and Cybernetics Systems]. *IJCCS, 16*(4), 379–388. DOI: 10.22146/ijccs.73910

Luo, Z. (2023). The Application of Artificial Intelligence Technology in Music Education. *The Application of Artificial Intelligence Technology in Music Education, 0*(1), 305–310. DOI: 10.54254/2753-7048/35/20232155

Lupker, J. A., & Turkel, W. J. (2021). Music theory, the missing link between music-related big data and artificial intelligence. *DHQ: Digital Humanities Quarterly, 15*(1).

Ma, H., Zhang, Y., Shan, X., & Hu, X. (2025). Exploring the impact of artificial intelligence on the creativity perception of music practitioners. *Journal of Intelligence, 13*(4), 47. Advance online publication. DOI: 10.3390/jintelligence13040047 PMID: 40278056

Maia, L. S., de Castro, L. N., & Silla, C. N.Jr. (2019). SAMBASET: A dataset of historical *samba de roda. Proceedings of the 20th International Society for Music Information Retrieval Conference (ISMIR 2019)*, 76–83.

Main, A., Grierson, M., Yamada-Rice, D., & Murr, J. (2022, June). Augmenting personal creativity with artificial intelligence: Workshop proposal for Creativity and Cognition 2022. In *Proceedings of the 14th Conference on Creativity and Cognition* (pp. 462-465). https://doi.org/DOI: 10.1145/3527927.3531205

Malki, L. M., Kaleva, I., Patel, D., Warner, M., & Abu-Salma, R. (2024). Exploring Privacy Practices of Female mHealth Apps in a Post-Roe World. *Conference on Human Factors in Computing Systems - Proceedings*. DOI: 10.1145/3613904.3642521

Mangal, N. (2025). Music contracts in the streaming age: Law and policy in the European Union and the United Kingdom. *GRUR International*, *74*(7), 700–702. DOI: 10.1093/grurint/ikaf057

Mangion, A., & Piller, N. (2024). Generative artificial intelligence in lymphoedema care. *Journal of Lymphoedema*, *19*(1), 57–59. https://www.scopus.com/inward/record.uri?eid=2-s2.0-85210102098&partnerID=40&md5=f65c75dbfe74ec672612eb83ed22734c

Martínez-Ramírez, M., Mimilakis, S., & Välimäki, V. (2022). Deep learning methods for automatic music mixing. *IEEE/ACM Transactions on Audio, Speech, and Language Processing*, *30*, 2105–2118. DOI: 10.1109/TASLP.2022.3181234

Masuda, H., Okada, S., Tanaka, T., Shiozawa, N., Sakaue, Y., & Manno, M. (2022). Changes in the Heart Rate during Sleep with Menstrual Cycle under Free-Living Conditions. *GCCE 2022 - 2022 IEEE 11th Global Conference on Consumer Electronics*, 339 – 340. DOI: 10.1109/GCCE56475.2022.10014265

Ma, X. (2025). Artificial intelligence in music production: A comprehensive review of applications and challenges. *IEEE Access : Practical Innovations, Open Solutions*, *13*, 12345–12367.

McCormack, J., Gifford, T., & Hutchings, P. (2018). Autonomy, authenticity, authorship and intention in computer-generated art. In R. Dean & A. McLean (Eds.), *The Oxford handbook of algorithmic music* (pp. 177–196). Oxford University Press.

McCormack, J., Gifford, T., Hutchings, P., Llano Rodriguez, M. T., Yee-King, M., & d'Inverno, M. (2019, May). In a silent way: Communication between ai and improvising musicians beyond sound. In Proceedings of the 2019 chi conference on human factors in computing systems (pp. 1-11). https://arxiv.org/pdf/1902.06442.pdf

Mehta, A., Chauhan, S., Djanibekov, A., Kulkarni, A., Xia, G., & Choudhury, M. (2025). *Music for All: Representational bias and cross-cultural adaptability of music generation models*. In *Findings of the 2025 Conference of the North American Chapter of the Association for Computational Linguistics (NAACL)* (pp. 258–272). Association for Computational Linguistics. https://aclanthology.org/2025.findings-naacl.258.pdf

Melodrive. (2019). 30% more immersion in VR — An adaptive music experiment. Medium.

Melodrive. (n.d.). Melodrive – AI music engine (product info). Retrieved from Melodrive developer pages.

Meng, N., & bin Ariffin, A. (2024). The Role of Creativity in Music Education. The Education Review, USA, 8(7), 922–926. https://doi.org/DOI: 10.26855/er.2024.07.004

Merchán Sánchez Jara, J. F., González Gutiérrez, S., & Cruz Rodríguez, J. (2024). La música como herramienta para la transformación social: educación musical y ODS en la formación en valores.

Merchán Sánchez-Jara, J., González Gutiérrez, S., Cruz Rodríguez, J., & Syroyid, B. S. (2024). Artificial Intelligence-Assisted Music Education: A Critical Synthesis of Challenges and Opportunities. *Education Sciences*, *14*(11), 1171. DOI: 10.3390/educsci14111171

Merlini, M. (2024). From [A Reflective Turn in Musicology to Overcome Dualist, Romantic and Cognitivist Prejudices.]. *AI*, •••, 4E.

Mhlambi, L., & Nxozi, M. (2024). The Effect of Digital Transformation in the Music Industry: A Scoping Review. *Interdisciplinary Research in Technology and Management*, 153–168.

Mikac, E. (2025). PyTorch vs TensorFlow vs Keras: Which framework gets you hired? https://www.cbtnuggets.com/blog/technology/programming/pytorch-vs-tensorflow-vs-keras

Miller, D. (2019). AI is composing film scores—And directors are listening. *MIT Technology Review*. https://www.technologyreview.com

Miranda, E. R. (2021). *Handbook of artificial intelligence in music: Principles, advanced techniques, and emerging applications to creativity*. Springer. DOI: 10.1007/978-3-030-72116-9

Miranda, E., & Williams, D. (2015). Artificial Intelligence in Organised Sound. *Organised Sound*, *20*(1), 76–81. DOI: 10.1017/S1355771814000454

Mitra, R., & Zualkernan, I. (2025). Music Generation Using Deep Learning and Generative AI: A Systematic Review. *IEEE Access : Practical Innovations, Open Solutions*, *13*, 18079–18106. DOI: 10.1109/ACCESS.2025.3531798

Mokoena, N., & Obagbuwa, I. C. (2025). An analysis of artificial intelligence automation in digital music streaming platforms. *Frontiers in Artificial Intelligence*, *7*, 1515716. DOI: 10.3389/frai.2024.1515716 PMID: 39881884

Moreau, F., Wikström, P., Haampland, O., & Johannessen, R. (2024). Alternative payment models in the music streaming market: A comparative approach based on stream-level data. *Information Economics and Policy*, *68*, 101103. DOI: 10.1016/j.infoecopol.2024.101103

Morgado Vega, J. J., Sharma, S., & Simonetta, F. (2025, September). Drafting the landscape of computational musicology tools: a survey-based approach. In *Proceedings of the 12th International Conference on Digital Libraries for Musicology* (pp. 28-35). DOI: 10.1145/3748336.3748340

Morgenstern, U. (2021). In defence of the term and concept of traditional music. *Musicologist*, *5*(1), 1–30. DOI: 10.33906/musicologist.913512

Morreale, F., Martinez-Ramirez, M. A., Masu, R., Liao, W., & Mitsufuji, Y. (2025). Reductive, exclusionary, normalising: The limits of generative AI music. *Transactions of the International Society for Music Information Retrieval*, *8*(1), 300–312. DOI: 10.5334/tismir.256

Morris, J. (2020). Platformization in the music industries: Infrastructure and algorithmic mediation. *Media Culture & Society*, *42*(7–8), 1234–1250. DOI: 10.1177/0163443720922007

Morrow, G. (2025). Australia's performing rights organisation: Incentives, the agency problem and MetaGen. *The International Communication Gazette*, *87*(4), 291–306. DOI: 10.1177/17480485251327468

Moysis, L., Iliadis, L. A., Sotiroudis, S. P., Boursianis, A. D., Papadopoulou, M. S., Kokkinidis, K.-I. D., Volos, C., Sarigiannidis, P., Nikolaidis, S., & Goudos, S. K. (2023). Music Deep Learning: Deep Learning Methods for Music Signal Processing—A Review of the State-of-the-Art. *IEEE Access : Practical Innovations, Open Solutions*, *11*, 17031–17052. DOI: 10.1109/ACCESS.2023.3244620

Music, L. M. (2024). *About MusicLM.* Retrieved from https://musiclm.com

musictheory.net. (n.d.). Common chord progressions. https://www.musictheory.net/lessons/57

Mycka, J., Żychowski, A., & Mańdziuk, J. (2022). Human-Level Melodic Line Harmonization. *Lecture Notes in Computer Science (Including Subseries Lecture Notes in Artificial Intelligence and Lecture Notes in Bioinformatics), 13350 LNCS*, 17 – 30. DOI: 10.1007/978-3-031-08751-6_2

Nakamura, E., Saito, Y., & Yoshii, K. (2020). Statistical learning and estimation of piano fingering. *Information Sciences*, *517*, 68–85. DOI: 10.1016/j.ins.2019.12.068

Nakano, T., & Goto, M. (2009). Vocalistener: A singing-to-singing synthesis system based on iterative parameter estimation. *Proceedings of SMC*.

Nath, S. (2025). We rise to resist: The crisis of relevance and the movement against generative AI in creative industries. SSRN *Electronic Journal*. DOI: 10.2139/ssrn.5295103

Naylor, Richard, Todd, Jonathan, Moretto, Marta, Traverso, R. (2021). *Cultural and creative industries in the face of COVID-19: an economic impact outlook.*

Neves, P., Fornari, J., & Florindo, J. (2022). *Generating music with sentiment using Transformer-GANs* (arXiv).

New Indian Express / CISAC. (2024, December 4). *Global study reveals AI threat to music sector workers.* The New Indian Express. https://www.newindianexpress.com/world/2024/dec/04/global-study-reveals-ai-threat-to-music-sector-workers

Newman, M., & Feng, Q. (no date) *Exploring the Collaborative Co-Creation Process with AI : A Case Study in Novice Music Production, Proceedings of ACM Conference (Conference'17)*. arXiv.

Newman, L. J., Stewart, S. E., Freeman, N. C., & Thompson, G. (2022). A systematic review of music interventions to support parent–child attachment. *Journal of Music Therapy*, *59*(4), 430–459. DOI: 10.1093/jmt/thac012 PMID: 36995776

Ng, J., & Gamble, S. (2024). Hip-hop music producers' labour in the digital music economy: Self-promotion, social media and platform gatekeeping. *New Media & Society*, *14614448241295304*. Advance online publication. DOI: 10.1177/14614448241295304

Nidhi, M. A., & Prasad, R. (2021). Spectrum Sharing and Dynamic Spectrum Management Techniques in 5G and beyond Networks: A Survey. In *Journal of Mobile Multimedia* (Vol. 17, Issues 1–3). DOI: 10.13052/jmm1550-4646.17133

Nielsen, M. (2015). *Neural Networks and Deep Learning: A Textbook*. Determination Press.

Nikrang, F. (2025). Human–AI collaborative creativity in music composition. *ACM Transactions on Computer-Human Interaction*, *32*(1), 1–29.

NVIDIA. (n.d.). TensorRT architecture overview. https://docs.nvidia.com/deeplearning/tensorrt/latest/architecture/architecture-overview.html

O'Dair, M., & Owen, R. (2019). Financing new creative enterprise through blockchain technology: Opportunities and policy implications. *Strategic Change*, *28*(1), 9–17. DOI: 10.1002/jsc.2242

O'Hanlon, K., Ewert, S., Pauwels, J., & Sandler, M. B. (2017). Improved template-based chord recognition using the CRP feature. In *Proceedings of ICASSP* (pp. 306–310). DOI: 10.1109/ICASSP.2017.7952167

O'Toole, K., & Horvát, E.-Á. (2024). Extending human creativity with AI. *Journal of Creativity*, *34*(2), 100080. DOI: 10.1016/j.yjoc.2024.100080

Over The Bridge. (2021). *Lost tapes of the 27 club* [AI music project]. Over The Bridge. https://www.overthebridge.org/lost-tapes

Pachet, F. (2023). *The green piano: How AI taught me to play*. The MIT Press.

Pachet, F., & Roy, P. (2011). Markov constraints: Steerable generation of Markov sequences. *Constraints*, *16*(2), 148–172. DOI: 10.1007/s10601-010-9101-4

Palanisamy, K., Singhania, D., & Yao, A. (2020). Rethinking CNN models for audio classification. arXiv. https://arxiv.org/abs/2007.11154

Pandya, S., Ghayvat, H., Reddy, P. K., Gadekallu, T. R., Khan, M. A., & Kumar, N. (2023). COUNTERSAVIOR: AIoMT and IIoT-Enabled Adaptive Virus Outbreak Discovery Framework for Healthcare Informatics. *IEEE Internet of Things Journal*, *10*(5), 4202–4212. DOI: 10.1109/JIOT.2022.3216108

Pandya, S., Ghayvat, H., Sur, A., Awais, M., Kotecha, K., Saxena, S., Jassal, N., & Pingale, G. (2020). Pollution weather prediction system: Smart outdoor pollution monitoring and prediction for healthy breathing and living. *Sensors (Basel)*, *20*(18), 1–25. DOI: 10.3390/s20185448 PMID: 32972037

Panteli, M., & Purwins, H. (2013). A computational comparison of theory and practice of scale intonation in Byzantine chant. *Proceedings of the International Society for Music Information Retrieval (ISMIR 2013)*. https://archives.ismir.net/ismir2013/paper/000098.pdf

Papadopoulou, A. (2021). *Rebetiko meets digital humanities: Automatic analysis of Rebetiko lyrics via data science methods* (Master's thesis). University of Vienna. https://phaidra.univie.ac.at/detail/o:1426454

Park, J., Choi, K., Jeon, S., Kim, D., & Park, J. (2019). A bi-directional transformer for musical chord recognition.

Parncutt, R. (2007). Systematic musicology and the history and future of western musical scholarship. *Journal of interdisciplinary music studies, 1*(1), 1-32.

Pasquier, P., Eigenfeldt, A., Bown, O., & Dubnov, S. (2017). An introduction to musical metacreation. [CIE]. *Computers in Entertainment, 14*(2), 1–14. DOI: 10.1145/2930672

Passmore, S.. (2024). Global musical diversity is largely independent of linguistic and genetic histories. *PLoS One, 19*(5), e0273518. https://pmc.ncbi.nlm.nih.gov/articles/PMC11087526/ PMID: 38729968

Patel, W. D., Pandya, S., Koyuncu, B., Ramani, B., Bhaskar, S., & Ghayvat, H. (2018). NXTGeUH: Lorawan based next generation ubiquitous healthcare system for vital signs monitoring falls detection. *1st International Conference on Data Science and Analytics, PuneCon 2018 - Proceedings*. DOI: 10.1109/PUNECON.2018.8745431

Patil, R. P., & Bhor, P. A. (2023). Synergy of Artificial Intelligence and Music. *2023 6th International Conference on Advances in Science and Technology (ICAST)*, 155–159.

Pearce, M. (2018). Statistical learning and probabilistic prediction in music cognition: Mechanisms of stylistic enculturation. *Annals of the New York Academy of Sciences, 1423*(1), 378–395. DOI: 10.1111/nyas.13654 PMID: 29749625

Pearce, M., & Wiggins, G. (2012). Auditory expectation: The information dynamics of music perception and cognition. *Topics in Cognitive Science, 4*(4), 625–652. DOI: 10.1111/j.1756-8765.2012.01214.x PMID: 22847872

Penelitian, J. H., Kepustakaan, K., & Pendidikan, B. (2024)... *Jurnal Kependidikan, 10*(3), 881–892.

Peng, Y. (2025). The analysis of optimization in music aesthetic education under artificial intelligence. *Scientific Reports, 15*(1), 11545. DOI: 10.1038/s41598-025-96436-2 PMID: 40185937

Pizzolitto, E. (2024). Music in business and management studies: A systematic literature review and research agenda. *Management Review Quarterly, 74*(3), 1439–1472. DOI: 10.1007/s11301-023-00339-3

Pocket, F. M. (2025). *Pocket FM launches AI Creator Studio*. Economic Times.

Poetry. (2025). *5 Trends Shaping the Future of the Music Industry*. Illustratemagazine. https://illustratemagazine.com/5-trends-shaping-the-future-of-the-music-industry/

Polak, N., & Schaap, J. (2025). Write, record, optimize? How musicians reflect on music optimization strategies in the creative production process. *New Media & Society*, *27*(8), 4773–4789. DOI: 10.1177/14614448241243095

Polotti, P. (2015). Some ideas for an extension of music therapy practices in timbre and ambient sounds [Alcune idee per un'estensione delle pratiche musicoterapiche al timbro e ai suoni d'ambiente]. *Sistemi Intelligenti*, *27*(2), 417–426. https://www.scopus.com/inward/record.uri?eid=2-s2.0-84949763215&partnerID=40&md5=2091b8adde6f020ae9921f5ad39eac61

Pons, J., Zukowski, Z., Parker, J. D., Carr, C. J., Taylor, J., & Evans, Z. (2025). Music and Artificial Intelligence: Artistic Trends. *arXiv preprint arXiv:2508.11694*.

Popescu, I., Lenain, M., Rovini, E., & Ruthsatz, M. (2025). Lifestyle Interventions and Innovative Approaches for the Management of Neurodegenerative Disorders in Older Adults, State-Of-The-Art and Future Directions. *American Journal of Lifestyle Medicine*, *19*(8), 1216–1226. Advance online publication. DOI: 10.1177/15598276251330170 PMID: 40161281

Popović, A. (2020). Is Musicology (Only) One of the Humanities? Professional article. *Pannoniana*, *4*(1), 245–257.

Pricop, T. C., Toma, C., & Pricop Velin, P. (2024). Music generation with machine learning and deep neural networks. *Procedia Computer Science, 234*, 435 443. https://www.sciencedirect.com/science/article/pii/S1877050924027522

Prvulovic, D., Vogl, R., & Knees, P. (2022). ReStyle-MusicVAE: Enhancing User Control of Deep Generative Music Models with Expert Labeled Anchors. *UMAP2022 - Adjunct Proceedings of the 30th ACM Conference on User Modeling, Adaptation and Personalization*. DOI: 10.1145/3511047.3536412

PTI News. (2025). *Trailer launch of India's first AI-based film 'IRAH', starring Rohit Bose Roy*.

Qiu, F. (2024). AI-driven personalized learning for beginner musicians. *Membrane Technology*, •••, 228–234. DOI: 10.52710/mt.162

Quintais, J. P., & Schwemer, S. F. (2022). The interplay between the Digital Services Act and sector regulation: How special is copyright? *European Journal of Risk Regulation*, *13*(2), 191–217. DOI: 10.1017/err.2022.1

Rai, A. (2020). Explainable AI: From black box to glass box. *Journal of the Academy of Marketing Science*, *48*(1), 137–141. DOI: 10.1007/s11747-019-00710-5

Rai, A., Constantinides, P., & Sarker, S. (2022). Next-generation digital platforms: Toward human–AI hybrids. *Management Information Systems Quarterly*, *46*(1), iii–xii. DOI: 10.25300/MISQ/2020/441E0

Rai, P., Gupta, K., Gabralla, L. A., & Choudhury, T. (2025). A royalty framework for copyright protection and accountability in AI-generated art. *Discover Computing*, *28*(1), 252. DOI: 10.1007/s10791-025-09766-0

Raisch, S., & Krakowski, S. (2021). Artificial intelligence and management: The automation–augmentation paradox. *Academy of Management Review*, *46*(1), 192–210. DOI: 10.5465/amr.2018.0072

Ramasubbareddy, S., Saidulu, D., Devasekhar, V., Swathi, V., Maini, S. S., & Govinda, K. (2019). Music Generation Using Deep Learning Techniques. In *Lecture Notes in Networks and Systems* (Vol. 74). DOI: 10.1007/978-981-13-7082-3_37

Ramos-Zaga, F. A. (2025). Reconceptualizing Human Authorship in the Age of Generative AI: A Normative Framework for Copyright Thresholds. *Laws*, *14*(6), 84. DOI: 10.3390/laws14060084

Razvan, M. (2023). Measurement of music aesthetics using deep neural networks and dissonances. *Information (Basel)*, *14*(7), 358. Advance online publication. DOI: 10.3390/info14070358

Rechardt, L. (2025). *IFPI looks at a decade of digital transformation in the music industry*. https://www.wipo.int/en/web/wipo-magazine/articles/ifpi-looks-at-a-decade-of-digital-transformation-in-the-music-industry

Rendas, T. (2018). Copyright, technology and the CJEU: An empirical study. *IIC - International Review of Intellectual Property and Competition Law*, *49*(2), 153–184. DOI: 10.1007/s40319-017-0664-0

Rigole, M. (2025). Listener perception of authenticity in AI-generated music. *Psychology of Aesthetics, Creativity, and the Arts*, *19*(3), 342–356.

Roberts, R., & Whiting, S. (2025). "I'd have to be crazy if I did it strictly on a financial basis": Australian regional music venues, burnout, and precarious music ecologies. *Popular Music and Society*, *48*(2), 194–211. DOI: 10.1080/03007766.2025.2455302

Rogers, J., & Sparviero, S. (2025). Navigating performing rights in music: Digital-native organizations, changing values, and industry shifts in the United States and beyond. *The International Communication Gazette*, *87*(4), 376–397. DOI: 10.1177/17480485251327273

Roldan-Cardona, M., Chacón-Castro, M., Jadán-Guerrero, J., Salvador-Ullauri, L., & Acosta-Vargas, P. (2025). Music Education with Artificial Intelligence for Inclusive and Sustainable Early Childhood Learning. *Emerging Science Journal*, *9*(Special Issue), 201–218. DOI: 10.28991/ESJ-2025-SIED1-012

Roughton, A. W. (2012). Football in pubs: Court of Justice of the European Union (Grand Chamber), *Football Association Premier League Ltd and others v QC Leisure and others (C-403/08)* and *Karen Murphy v Media Protection Services Ltd (C-429/08). Queen Mary Journal of Intellectual Property,* 2(3), 288–296. DOI: 10.4337/qmjip.2012.03.06

Rowe, R. (1993). *Interactive Music Systems* (Chapter 6: Machine Composition). New York University. https://wp.nyu.edu/robert_rowe/text/interactive-music-systems-1993/

Sætre, A. S. (2021). Digital labor transformation in creative industries. *Journal of Cultural Economics*, *14*(5), 567–583. DOI: 10.1080/17530350.2021.1912703

Sambaragi, L. M., Naik, P. P., Kakatkar, N. N., & Chikkamath, S., Sr. N., & Budihal, S. V. (2024, March). Music Generation: A simplified approach. In 2024 3rd International Conference for Innovation in Technology (INOCON) (pp. 1-6). IEEE.

Samuelson, P. (2022). Generative AI and copyright law. *Columbia Journal of Law & the Arts*, *45*(3), 295–318.

Sari Angraeni, B., & Majid, P. S. (2025). *Legal protection of music copyrights in public commercial spaces: A case study of coffee shops in Kolaka Utara Regency*. Constitutional Law Review., DOI: 10.30863/clr.v4i2.5950

Scholz, R., Dantas, V., & Ramalho, G. (2005b). *Funchal: A system for automatic functional harmonic analysis*. https://citeseerx.ist.psu.edu/document?doi=10.1.1.85.7992

Scholz, R., Dantas, V., & Ramalho, G. (2005a). Automating functional harmonic analysis: The Funchal system. In *Proceedings of the Seventh IEEE International Symposium on Multimedia* (pp. 759–764). DOI: 10.1109/ISM.2005.39

Schörkhuber, C., & Klapuri, A. (2010). Constant-Q transform toolbox for music processing. In *Proceedings of the Sound and Music Computing Conference*.

Schwab, K. (2017). *The fourth industrial revolution*. Crown Business.

Shackel, L., & Smith, M. (2025). Australian privacy law, extraterritoriality, and regulating data collected by offshore direct-to-consumer genetic testing companies. *International Data Privacy Law*. Advance online publication. DOI: 10.1093/idpl/ipaf016

Shaffer, E. S., Burgoyne, J. A., & Fujinaga, I. (2020). A cluster analysis of harmony in the McGill Billboard dataset. *Empirical Musicology Review*, *15*(1), 28–56. DOI: 10.18061/emr.v14i3-4.5576

Shank, D. B., DeSanti, A., & Maninger, T. (2022). When artificial intelligence authors art: The effects of AI authorship on aesthetic judgment. *Journal of Experimental Psychology. Applied*, *28*(3), 1–15. DOI: 10.1037/xap0000447 PMID: 36006713

Sheh, A., & Ellis, D. (2003). Chord segmentation and recognition using EM-trained hidden Markov models.

Shen, L., Zhang, H., Zhu, C., Li, R., Qian, K., Meng, W., Tian, F., Hu, B., Schuller, B. W., & Yamamoto, Y. (2024). A First Look at Generative Artificial Intelligence Based Music Therapy for Mental Disorders. *IEEE Transactions on Consumer Electronics*. Advance online publication. DOI: 10.1109/TCE.2024.3514633

Shu, L. (2025). Research on Creative Practice in Music Education Empowered by AI. *Research on Creative Practice in Music Education Empowered by AI*, *0*(1), 106–113. DOI: 10.54254/2753-7048/67/20251010

Simoes, J. M. da C. (2019). Deep Learning for Dynamic Music Generation. https://estudogeral.sib.uc.pt/bitstream/10316/88002/1/Deep_Learning_for_Dynamic_Music_Generation.pdf (Master's Thesis)

Singha, R., & Singha, S. (2023). Economic Sustainability, Mindfulness, and Diversity in the Age of Artificial Intelligence and Machine Learning. In *Machine Intelligence: Computer Vision and Natural Language Processing*. CRC Press., DOI: 10.1201/9781003424550-15

Song, X., Zheng, J., & He, J. (2025a). AI-Assisted Music Therapy Intervention for Alleviating Exam Anxiety in University Students: System Design, Implementation, and Efficacy Evaluation. *2025 IEEE/IEIE International Conference on Consumer Electronics-Asia, ICCE-Asia 2025*. DOI: 10.1109/ICCE-Asia67487.2025.11263548

Song, X., Zheng, J., & He, J. (2025b). Study on Promoting College Students' Mental Health by AI Music Therapy Based on Big Data Analysis. *2025 IEEE/IEIE International Conference on Consumer Electronics-Asia, ICCE-Asia 2025*. DOI: 10.1109/ICCE-Asia67487.2025.11263737

SoundTech Insider. (2023). *AIVA Review - Everything You Need To Know - AI Composition*. Retrieved from https://soundtechinsider.com/aiva-review-everything-you-need-to-know-ai-composition

Southern, T. (2017). *I AM AI* [Album]. Artiphany.

Southern, T. (2018). *I am AI* [Album]. Amper Music & IBM Watson Beat.

Souza, R. M., Cezario, B. S., Affonso, E. O., Machado, A. D., Vieira, D. P., Chinelli, C. K., Haddad, A. N., Dusek, P. M., Miranda, M. G., Soares, C. A. P., & Guedes, A. L. (2024). My human rights smart city: Improving human rights transparency identification system. *Sustainability (Basel)*, *16*(3), 1274. DOI: 10.3390/su16031274

Sree Pokkuluri, K., Tiwari, V., Uikey, J., Mehta, P., Srinivasa Rao, C., & Thanuja, A. (2024). Bridging the Gap between Technology and Medicine through the Revolutionary Impact of the Healthcare Internet of Things on Remote Patient Monitoring. *Journal of Intelligent Systems and Internet of Things*, *13*(2), 212–222. DOI: 10.54216/JISIoT.130217

Stewart, W. (2025). How to deploy machine learning models: Step-by-step guide. https://northflank.com/blog/how-to-deploy-machine-learning-models-step-by-step-guide-to-ml-model-deployment-in

Storyboard18. (2025, July 1). *How Indian brands are letting AI take over creative wheel*. Storyboard18. Storyboard18

Sturm, B. L. (2014). A simple method to determine if a music information retrieval system is a "horse.". *IEEE Transactions on Multimedia*, *22*(11), 1676–1683. DOI: 10.1109/TMM.2014.2330697

Sturm, B. L. T., Iglesias, M., Ben-Tal, O., Miron, M., & Gómez, E. (2019). Artificial Intelligence and Music: Open Questions of Copyright Law and Engineering Praxis. *Arts*, *8*(3), 115. DOI: 10.3390/arts8030115

Sturm, B. L., Ben-Tal, O., Monaghan, Ú., Collins, N., Herremans, D., Chew, E., Hadjeres, G., Deruty, E., & Pachet, F. (2019). Machine learning research that matters for music creation: A case study. *Journal of New Music Research*, *48*(1), 36–55. DOI: 10.1080/09298215.2018.1515233

Sturm, B. L., Ben-Tal, O., Monaghan, U., Collins, N., Herremans, D., Chew, E., & Weyde, T. (2021). Artificial intelligence and music: Open questions of copyright law and engineering praxis. *Journal of Creative Music Systems*, *5*(1), 1–27.

Sturm, B. L., & Maruri-Aguilar, H. (2021). Artificial intelligence and music: Open questions of copyright law and engineering praxis. *Journal of Cultural Management and Cultural Policy*, *7*(1), 1–24. DOI: 10.5920/jcms.950

Subramanian, S., & Ganapathiraman, V. (2023, October). Learning to learn learning-rate schedules. *Amazon Science Blog*. https://www.amazon.science/blog/learning-to-learn-learning-rate-schedules

Sudha, T., Raja, P., & Bharathi, V. (2025). Harmony Wave: Exploring Smart Music Therapy Through IoT, AI, and Fuzzy Logic. In *AI for Large Scale Communication Networks*. IGI Global., DOI: 10.4018/979-8-3693-6552-6.ch009

Suh, M. M., & Youngblom, E. (2021). AI as social glue: Uncovering the roles of deep generative AI during social music composition. In Conference on Human Factors in Computing Systems – Proceedings. Association for Computing Machinery. https://doi.org/DOI: 10.1145/3411764.3445219

Sularso, S., Wadiyo, W., Cahyono, A., Suharto, S., & Pranolo, A. (2025). A Comprehensive Visualization for Music Education and Artificial Intelligence. *JOIV: International Journal on Informatics Visualization*, *9*(2), 718–727. DOI: 10.62527/joiv.9.2.2500

Suno, A. I. (2025). *Future of music with Suno AI*. Retrieved from https://sunnoai.com/future-of-music/

Sun, W., Zhang, Y., & Li, H. (2024). Human perception of AI-generated music: Authenticity, creativity, and acceptance. *Computers in Human Behavior*, *145*, 107725. DOI: 10.1016/j.chb.2023.107725

ŞUTEU, L.-C., & RUCSANDA, M. D. (2025). Artificial Intelligence in Music: The Digital Revolution of Sound Creativity. *ICT in Muzical Field/Tehnologii Informatice Si de Comunicatie in Domeniul Muzical, 16*(1).

Tabak, C. (2023). Intelligent music applications: Innovative solutions for musicians and listeners. *Uluslararası Anadolu Sosyal Bilimler Dergisi*, *7*(3), 752–773. DOI: 10.47525/ulasbid.1324070

Tan, X., & Li, X. (2021). A tutorial on AI music composition. In MM 2021 – Proceedings of the 29th ACM International Conference on Multimedia (pp. 5678–5680). Association for Computing Machinery. https://doi.org/DOI: 10.1145/3474085.3478875

Temperley, D. (2010). Music and probability. MIT Press. https://mitpress.mit.edu/9780262013597/music-and-probability/

Thakur, A. S., & Tiwari, V. (2018). Design High Speed FIR Filter based on Complex Vedic Multiplier using CBL Adder. *2018 International Conference on Recent Innovations in Electrical, Electronics and Communication Engineering, ICRIEECE 2018*, 559 – 563. DOI: 10.1109/ICRIEECE44171.2018.9008438

Theremixhouse. (2024). *Current Trends, Challenges, and Opportunities in the Music Industry: 2024*. https://www.themixhouse.com/blog/current-trends-challenges-and-opportunities-in-the-music-industry-2024?utm

Thomas, D. (2025). *Music industry revenues slow as pandemic-era CD boom fades*. Financial Times. https://www.ft.com/content/05ea07dc-2fae-4616-93e1-746cc8ac4635?utm

Tigre Moura, F., & Maw, C. (2021). Artificial intelligence became Beethoven: How do listeners and music professionals perceive artificially composed music? *Journal of Consumer Marketing*, *38*(2), 137–146. DOI: 10.1108/JCM-02-2020-3671

Tiwari, V., Agarwal, O., Sharma, M., Babar, R., Geddam, R., Awais, M., Ghayyat, H., & Sahu, R. (2025). Akshar Mitra : A Multimodal Integrated Framework for Early Dyslexia Detection. *Frontiers in Digital Health*, *7*, 1–6. https://www.frontiersin.org/journals/digital-health/articles/10.3389/fdgth.2025.1726307/abstract. DOI: 10.3389/fdgth.2025.1726307 PMID: 41395631

Tiwari, V., Jain, D., Sharma, D., Hassan, M. M., Althobaiti, F., Varkale, A., Al-Khasawneh, M. A., & Tirandasu, R. K. (2025). An efficient sparse code shrinkage technique for ECG denoising using empirical mode decomposition. *Technology and Health Care*, *33*(4), 1773–1786. DOI: 10.1177/09287329241302749 PMID: 39973873

Tsai, Y.-H. H., Bai, S., Liang, P., & Kolter, J. Z. (2019). Multimodal transformer for unaligned multimodal language sequences. *Proceedings of the 57th Annual Meeting of the Association for Computational Linguistics*, 6558–6569. DOI: 10.18653/v1/P19-1656

Tullah, D. N., Khalid, H., & Poernomo, S. L. (2024). Perlindungan Hak Cipta Dalam Industri Musik Digital Di Indonesia Dalam Perspektif Hukum Perdata. [JLP]. *Journal of Lex Philosophy*, *5*(2), 1030–1046. https://www.pasca-umi.ac.id/index.php/jlp/article/view/1877

U.S. Copyright Office. (2023, March 16). *Copyright registration guidance: Works containing material generated by artificial intelligence* (Federal Register Vol. 88, Issue 51). https://www.govinfo.gov/content/pkg/FR-2023-03-16/html/2023-05321.htm

Udio Help Center. (2025). *AI and the future of music.* Retrieved from https://help.udio.com/en/articles/11664414-ai-and-the-future-of-music

Unknown. (2025). Automatic chord recognition method – CS 101 final exam. https://www.studocu.com/

Van Den Oord, A., Dieleman, S., Zen, H., Simonyan, K., Vinyals, O., Graves, A., . . . Kavukcuoglu, K. (2016). Wavenet: A generative model for raw audio. arXiv preprint arXiv:1609.03499, 12(1).

Varini, M., & Gramaglia, G. (2024). Tyranny of (AI) Thought. Artificial intelligence in music composition: A case study on Krallice "Diotima". *CONNESSIONI REMOTE, 2024*(8), 51–82. DOI: 10.54103/connessioni/26583

Vasco, M., & Eguiguren, L. (2023, December). La Inteligencia Artificial y el aprendizaje musical. In *Colloquia* (Vol. 10, pp. 46–62). Academic Journal of Culture and Thought., DOI: 10.31207/colloquia.v10i0.150

Vaswani, A., Shazeer, N., Parmar, N., Uszkoreit, J., Jones, L., Gomez, A. N., Kaiser, Ł., & Polosukhin, I. (2017). *Attention Is All You Need.* In *Advances in Neural Information Processing Systems (NeurIPS)*, 30. https://proceedings.neurips.cc/paper_files/paper/2017/file/3f5ee243547dee91fbd053c1c4a845aa-Paper.pdf

Vaswani, A., Shazeer, N., & Parmar, N.. (2017). Attention is all you need. *Advances in Neural Information Processing Systems*, *30*, 5998–6008.

Vear, C., Benford, S., Avila, J. M., & Moroz, S. (2023). Human-AI Musicking: A Framework for Designing AI for Music Co-creativity. *AIMC 2023*. https://aimc2023.pubpub.org/pub/zd46ltn3

Vengathattil, S. (2025). Collaborative AI in music composition: Human-AI symbiosis in creative processes. *International Journal of Management Science and Information Technology*, *5*(1), 253–262. DOI: 10.35870/ijmsit.v5i1.4085

Venkatesh, V., Morris, M. G., Davis, G. B., & Davis, F. D. (2003). User acceptance of information technology: Toward a unified view. *Management Information Systems Quarterly*, *27*(3), 425–478. DOI: 10.2307/30036540

Venkatesh, V., Thong, J. Y. L., & Xu, X. (2022). Unified theory of acceptance and use of technology: A review and extension. *Management Information Systems Quarterly*, *46*(3), 1211–1240.

Ventayen, R. J. M. (2024). *Music creation by artificial intelligence and listener reactions in a blind test* (SSRN Scholarly Paper No. 5023874). SSRN. DOI: 10.2139/ssrn.5023874

Vincent, J. (2017). This AI composer can create endless new music on demand. *The Verge*. https://www.theverge.com

Vinod, R., & Colaco, R. (2025). Review of AI in Indian classical music. *International Journal for Research in Applied Science and Engineering Technology*, *13*(2), 1599–1608. DOI: 10.22214/ijraset.2025.67155

Vogl, R., Bodo, B., & Handke, C. (2021). Generative AI and the creative economy. *Journal of Cultural Economics*, *45*(3), 331–349. DOI: 10.1007/s10824-021-09424-7

Walzer, D. (2023). *Leadership in music technology education: Philosophy, praxis, and pedagogy*. Focal Press. DOI: 10.4324/9781003152460

Wang, Z., Bao, C., Zhuo, L., Han, J., Yue, Y., Tang, Y., Huang, V. S.-J., & Liao, Y. (2025). Vision-to-music generation: A survey. arXiv preprint arXiv:2503.21254.

Wang, G. (2024). The investigation of artificial intelligence-based applications in music education. *Applied and Computational Engineering.*, *36*(1), 210–214. DOI: 10.54254/2755-2721/36/20230448

Wang, L., Zhao, Z., Liu, H., Pang, J., Qin, Y., & Wu, Q. (2024). A review of intelligent music generation systems. *Neural Computing & Applications*, *36*(12), 6381–6401. Advance online publication. DOI: 10.1007/s00521-024-09418-2

Wang, Q., Qu, Y., Cheng, H., Yu, Y., Wang, X., & Gu, B. (2024). AI-Driven Sentiment Analysis for Music Composition. *Lecture Notes of the Institute for Computer Sciences. Lecture Notes of the Institute for Computer Sciences, Social Informatics and Telecommunications Engineering*, *523*, 75–84. Advance online publication. DOI: 10.1007/978-3-031-51399-2_4

Wang, T., & Zhang, J. (2025). Enhancing musical creativity through AI: A chain-mediation model of self-efficacy and emotional intelligence. *Frontiers in Psychology*, *16*, 1665609. Advance online publication. DOI: 10.3389/fpsyg.2025.1665609 PMID: 41267763

Wang, W., Li, J., Li, Y., & Xing, X. (2024). Style-conditioned music generation with transformer-GANs. *Frontiers of Information Technology & Electronic Engineering*, *25*(1), 106–120. DOI: 10.1631/FITEE.2300359

Wani, T. A., & Chavan, P. (2025)... *AI-GENERATED COMPOSITIONS IN MUSIC EDUCATION*, *6*, 92–101. DOI: 10.29121/shodhkosh.v6.i3s.2025.67

Warty, R. R., Smith, V., Patabendige, M., Fox, D., & Mol, B. (2024). Clarifying the unmet clinical need during medical device innovation in women's health–A narrative review. *Health Care for Women International*, *45*(7), 811–839. DOI: 10.1080/07399332.2023.2190983 PMID: 37000043

Wei-Hsiang, L., Huang, P., & Yang, Y. (2022). Automatic multitrack mixing using deep neural networks. *IEEE Signal Processing Letters*, *29*, 142–146.

Wei, J. (2025). From tools to creators: Evolution and challenges of AI music generation. *Applied Sciences (Basel, Switzerland)*, *15*(2), 345.

Wei, Q., & He, W. (2026). The application of AI-assisted music therapy tools in mental health interventions. *Frontiers in Psychology*, *17*, 1741463. Advance online publication. DOI: 10.3389/fpsyg.2026.1741463 PMID: 41658379

Whiting, S., Emery, S., Oestmann, J., & Martin, G. (2024). *Musical automation in the Australian games industry: Current perspectives*. RMIT University. DOI: 10.25439/RMT.27289779.V1

Widmer, G. (2016). Getting closer to the essence of music: The Con Espressione manifesto. *ACM Transactions on Intelligent Systems and Technology*, *8*(2), 22.

Wiki, M. I. R. E. X. (2025). Audio chord estimation. https://www.music-ir.org/mirex/wiki/2025:Audio_Chord_Estimation

Wikipedia contributors. (2024, November). Chroma feature. *Wikipedia*. https://en.wikipedia.org/wiki/Chroma_feature

Wikstrom, P. (2022). *Peter Tschmuck: The economics of music: Agenda Publishing, Newcastle, 2021* (pp. 387–389). Springer. https://doi.org/DOI: 10.1007/s10824-022-09443

Wikström, P. (2020). *The music industry: Music in the cloud*. John Wiley & Sons.

Wondercraft.ai. (n.d.). *How AMA scaled dynamic audio campaigns to 2,800 markets*. Wondercraft case study. Retrieved 2025, from https://www.wondercraft.ai/case-studies/how-ama-scaled-dynamic-audio-campaigns-to-2-800-markets

Xi, Q., Bittner, R. M., Pauwels, J., Ye, X., & Bello, J. P. (2018). GuitarSet: A dataset for guitar transcription. In *Proceedings of the 19th ISMIR Conference* (pp. 453–460).

YACHT. (2019). *Chain Tripping* [Album]. DFA Records.

Yang, L.-C., Chou, S.-Y., & Yang, Y.-H. (2020). MidiNet revisited: Conditional symbolic music generation using generative adversarial networks. *IEEE Transactions on Multimedia*, *22*(9), 2258–2271. DOI: 10.1109/TMM.2019.2960838

Yang, W., & Lee, I. (2024). Research on the construction of AI composition system based on HMM. *Yegah Musicology Journal*, *7*(3), 216–240. DOI: 10.51576/ymd.1536267

Yesiler, F., ... (2022). Assessing algorithmic biases for musical version identification. Proceedings of the 22nd International Society for Music Information Retrieval Conference. https://dl.acm.org/doi/abs/10.1145/3488560.3498397

Yin, L., & Guo, R. (2025). An Artificial Intelligence-Based Interactive Learning Environment for Music Education in China: Traditional Chinese Music and Its Contemporary Development as a Way to Increase Cultural Capital. *European Journal of Education*, *60*(1), e12858. DOI: 10.1111/ejed.12858

Yin, Z., Reuben, F., Stepney, S., & Collins, T. (2022). Measuring when a music generation algorithm copies too much: The originality report, cardinality score, and symbolic fingerprinting by geometric hashing. *SN Computer Science*, *3*(5), 348. DOI: 10.1007/s42979-022-01220-y PMID: 35789573

Yin, Z., Reuben, F., Stepney, S., & Collins, T. (2023). Deep learning's shallow gains: A comparative evaluation of algorithms for automatic music generation. *Machine Learning*, *112*(5), 1785–1822. DOI: 10.1007/s10994-023-06309-w

Yuan, N. (2024). Does AI-assisted creation of polyphonic music increase academic motivation? The DeepBach graphical model and its use in music education. *Journal of Computer Assisted Learning*, *40*(4), 1365–1372. Advance online publication. DOI: 10.1111/jcal.12957

Yu, X., Ma, N., Zheng, L., Wang, L., & Wang, K. (2023). Developments and applications of artificial intelligence in music education. *Technologies*, *11*(2), 42. Advance online publication. DOI: 10.3390/technologies11020042

Zamanifar, A., & Faezipour, M. (2025). Application of generative AI in healthcare systems. In *Application of Generative AI in Healthcare Systems*. Springer Nature., DOI: 10.1007/978-3-031-82963-5

Zhang, M., Li, S., & Chen, R. (2025). Advancing deep learning for expressive music composition. *Scientific Reports, 15*. https://www.nature.com/articles/s41598 025 13064 6

Zhang, S., Lu, X., & Liu, X. (no date) 'Study on the Influence of AI Composition Software on Students ' Creative Ability in Music Education', pp. 1–17.

Zhang, H.. (2025). Artificial Intelligence in Music Education : A Scoping Review of Practices. *Strategies, and Challenges*, *4*(5), 69–82. Advance online publication. DOI: 10.56397/JARE.2025.09.09

Zhang, M. (2025). Advancing deep learning for expressive music composition and performance modeling. *Scientific Reports, 15*(1), 28007. Advance online publication. DOI: 10.1038/s41598-025-13064-6 PMID: 40745012

Zhang, Y. (2024). Utilizing computational music analysis and AI for enhanced music composition: Exploring pre-and post-analysis. *Educational Administration: Theory and Practice, 30*(5), 269–282. DOI: 10.17762/jaz.v44iS6.2470

Zhao, Y., Yang, M., Lin, Y., Zhang, X., Shi, F., & Wang, Z. Ding, Juanguo & Ning, H. (2025). Artificial Intelligence-Enabled Text-to-Music Generation: A Comprehensive Review of Methods, Frameworks, and Future Directions. https://www.preprints.org/manuscript/202502.1791

Zhao, L. (2025). Control and interpretability challenges in text-to-music systems. *Artificial Intelligence Review, 58*(2), 221–239.

Zhao, X. (2025). AI-enabled text-to-music generation: A systematic review of architectures and evaluation methods. *Electronics (Basel), 14*(3), 1120.

Zhou, Y., Bian, W., Zhang, K., & Gu, X. (2024). Advancing Music Therapy: Integrating Eastern Five-Element Music Theory and Western Techniques with AI in the Novel Five-Element Harmony System. In Q. Y., J. Q., O. Z., L. Z., W. Z., L. Y., X. L., & T. J. (Eds.), *2024 14th International Symposium on Chinese Spoken Language Processing, ISCSLP 2024* (pp. 234–238). Institute of Electrical and Electronics Engineers Inc. DOI: 10.1109/ISCSLP63861.2024.10800517

Zhou, X., & Lerch, A. (2015). Chord detection using deep learning. In *Proceedings of the International Society for Music Information Retrieval Conference.*

Zlatkov, D., Ens, J., & Pasquier, P. (2023). Searching for human bias against AI-composed music. In Johnson, C., Rodríguez-Fernández, N., & Rebelo, S. M. (Eds.), *Artificial intelligence in music, sound, art and design* (Vol. 13988). Springer Nature Switzerland., DOI: 10.1007/978-3-031-29956-8_20

About the Contributors

Randy Joy Magno Ventayen is a Professor III at Pangasinan State University (PSU), Philippines, where he currently serves as the University Director for the International Accreditation Office. He holds a Doctor in Information Technology from the University of the Cordilleras and a Doctor of Business Administration from Lyceum Northwestern University. Dr. Ventayen has an extensive academic and professional background in information technology, research, and digital innovation, particularly in the integration of emerging technologies across various disciplines. He has published more than 120 research papers, with 40 indexed in Scopus and 6 in the Web of Science, covering areas such as information systems, digital education, and technology-driven creative industries. He is also an accredited editor of the National Book Development Board (Philippines) and serves as Editor-in-Chief of the Southeast Asian Journal of Science and Technology. In addition to his academic roles, Dr. Ventayen is a Senior Accreditor of the Accrediting Agency of Chartered Colleges and Universities in the Philippines (AACCUP), a Flexible Learning Tutor at SEAMEO-Innotech, and a regular member of the DOST - National Research Council of the Philippines. He has served as a conference editor, reviewer for Scopus-indexed journals, and has contributed significantly to national quality assessment initiatives through the Philippine Quality Award assessor pool. His current research interests focus on the convergence of artificial intelligence, music production, and creative industries, providing critical insights into the ethical, commercial, and technological implications of AI-generated content.

***

Sonia Ahuja is an Assistant Professor of Music (Vocal) at Lovely Professional University, Phagwara. A Gold Medalist from Panjab University with a Ph.D. in Music Vocal, she is also a B-Grade Artist from All India Radio and a performer on stage and television. She has presented papers at national and international conferences, published research papers including a Scopus-indexed article and contributed book chapters. She also served as editor of the book Swaranjali: Perspectives on Indian Music (2024). Her research interests include Indian classical music traditions, global influences on music, folk traditions, music therapy, and women's role in cultural practices.

Arini Winur Baeti is a researcher in the field of Information Systems at Esa Unggul University, Jakarta, Indonesia. Her academic and research interests include information systems development, digital transformation, data management, and applied computing. She has been actively involved in research projects and academic activities related to system analysis, software development, and the implementation of information systems to enhance organizational efficiency and decision-making processes. Her work reflects strong analytical thinking, problem-solving abilities, and a commitment to continuous learning. Arini is known for her adaptability and collaborative mindset, and she continues to develop her expertise in emerging technologies to contribute to innovative and effective information system solutions in academic and professional environments.

H. R. Bhapkar is working as Assistant Professor in Mathematics in MIT ADT University's MIT School of Engineering, LoniKalbhor, Pune, India. He has 18 yearsof teaching experience and published 19 research papers in national and international reputed journals. He has authored 110 books of Engineering as well as General Mathematics for different universities in India

Bhargavi G, Assistant Professor, Department of Computing Technologies, SRM Institute of Science & Technologies, SRM Nagar, Kattankulathur - 603 203, Chengalpattu District, Tamil Nadu. She is a dedicated and goal-driven educator with over five years of experience as an Assistant Professor. She holds both B.Tech and M.Tech degrees from Annamalai University and has consistently demonstrated a strong commitment to fostering the academic growth and development of her students. She has successfully guided undergraduate and postgraduate students through their project work, inspiring them to explore innovative ideas and embrace independent learning. Her research focuses on satellite image processing and artificial intelligence, with a particular emphasis on predicting and detecting landslide-prone regions in South India. For her doctoral research, completed at SRM Institute of Science and Technology, Chennai, she developed and trained deep learning and machine learning models to classify and predict landslide risk zones, especially in and around Kerala. She has also made substantial contributions to her field, publishing numerous papers in esteemed international and national journals and presenting at various conferences. She has won best paper awards in various conferences. Her work reflects a solid commitment to advancing knowledge and practical solutions in her areas of expertise.

Astrid Amidiaputri Hasyyati, SSH, MKn is a lecturer at the Faculty of Law UNESA, who has expertise in civil law, business law, contract law and international civil law. She also has a specialty in notary based on her magister degree. As a law lecturer, she already has numerous publications concerning civil law, business law and contract law.

Budi Hermono, S.H., M.H. is a lecturer at the Faculty of Law UNESA who has expertise in the fields of intellectual property rights, business law and procurement law. Throughout his career, he has had numerous publications in the fields of business law, intellectual property rights and procurement. His latest book concerning IPR entitled "Hak Cipta: Konsep dan Strategi Pengelolaan Bagi Pencipta" (Copyright: Concepts and Management Strategies for Creators).

Jishnu K S completed his B.Tech. in Computer Science in 2017 from KMCT College of Engineering, affiliated with the University of Calicut. He went on to earn his M.Tech. degree in Computer Science from Government Engineering College, Wayanad, affiliated with APJ Abdul Kalam Technological University, in 2021. In 2025, he secured a Ph.D. position in Computer Science and Engineering at SRM Institute of Science and Technology, Kattankulathur, Chennai, India, under the supervision of Dr. B. Arthi, Associate Professor in the Department of Computing Technologies. Throughout his academic journey, he has co-authored more than three research papers and actively participated in over three national and international conferences.

Pawan Kumar, PhD, is a prominent academic and researcher with over 20 years of expertise in business studies. A professor at Graphic Era Deemed to be University, Dehradun, India, he has supervised six PhD scholars and published over 60 research papers, including 51 in SCOPUS/ 06 in Web of Science, and 14 books, including eight indexed in Scopus. His works, featured by prominent publishers like Emerald, IGI Global, Nova Science, and Cambridge Scholars, have gained recognition. Dr. Kumar earned his PhD in Commerce from Kurukshetra University in 2007 and holds an M.Phil. and M.Com. Degrees. Specializing in finance, banking, and securities markets, he has authored books on green finance, AI, and fintech. His innovative research includes five patents and has served as a guest editor for Scopus-indexed journals. In 2022, he received the prestigious Researcher Award from Chandigarh University. Dr. Kumar's impactful work on sustainable finance, blockchain, and fintech is globally recognized. His active participation in high-impact conferences and faculty development programs reflects his commitment to advancing research, teaching, and innovation in commerce and management.

Rahul Lalwani is an AI and Data Analytics professional, ranked AIR 421 in GATE 2024 (Data Analytics) and also qualified in CSE. A B.Tech graduate from MITS, Gwalior, he has worked on medical image segmentation using models like U-Net, UNETR, nnU-Net, MIScnn, Total Segmentator, and MEDSAM. During his internship at Accenture, he gained hands-on experience with Snowflake, Generative AI, and AWS services. A two-time hackathon winner, Rahul is passionate about building impactful AI solutions, with projects showcased on rahullalwani.com

Nisha Thorakkattu Madathil is a researcher and educator specializing in Information Security, Federated Learning, and Health Data Analytics. She earned her PhD in Informatics and Computing from UAE University, where her doctoral research, "Enhancing Health Analytics: Secure and Private Federated Learning Solutions," focused on privacy-preserving AI systems in healthcare. With over 15 years of teaching experience, Dr. Madathil has served as Visiting and Adjunct Faculty at UAE University and previously as a Lecturer and IT Trainer at MES Engineering College (India) and Syscoms College (UAE). Her research contributions include publications in top-tier venues such as the IEEE Journal on Selected Areas in Communications, Computers in Human Behavior Reports, and the Computational and Structural Biotechnology Journal, covering areas like federated learning, cybersecurity in education, and privacy-preserving data analytics. Dr. Madathil is proficient in machine learning, data security, and programming across multiple languages, and she continues to advance secure AI systems for distributed environments and healthcare applications.

Elmar B. Noche is the University Head of System Administration of Pangasinan State University and a faculty member at the Pangasinan State University–Lingayen Campus. He is a DOST–STRAND Scholar (2020–2022) at Saint Louis University, Baguio, where he earned his Master in Information Technology, and is currently pursuing a Doctor in Information Technology, specializing in Business Intelligence and Data Analytics, at Angeles University Foundation. He has been actively involved in international academic and research initiatives, serving as Volume Editor of the 3rd International Conference on Information Technology and Mechatronics Engineering (ICITME 2025) and Conference Manager of the 2nd International Conference on Information Technology and Mechatronics Engineering (ICITME 2023). He is also the Managing Editor of the Asian Journal of Multidisciplinary Studies and the Southeast Asian Journal of Science and Technology under Asian Journals Publishing. His research interests include information technology education, data analytics, artificial intelligence, emerging technologies, and applied computing systems. His scholarly works are indexed and accessible through Google Scholar, reflecting his contributions to peer-reviewed publications and international conference proceedings. Mr. Noche serves as the Coordinator of the PSU Research Journal Manager and as Student Organization Adviser for Information Technology students, where he mentors learners in research, leadership, and innovation. Prior to academia, he accumulated eight (8) years of industry experience with multinational organizations including Microsoft, JP Morgan Chase & Co., UPS, NCO, MDM Systems, and Set Financial (USA), providing him with strong industry-to-academe integration. He is a Licensed Professional Teacher, a TESDA Trainer's Methodology (TM) holder, a frequent resource speaker, an international research presenter, a freelance software project manager, and a startup mentor supporting technology-driven and innovation-focused initiatives.

Shiju Kumar P S, is currently serving as Assistant Professor, Department of Computer Technologies, School of Computing, Faculty of Engineering and Technology at SRM Institute of Science and Technology, Kattankulathur. He has been awarded his Ph.D. in Wireless Sensor Networks from Hindustan University, Chennai. An academician for over 14 years, he is an avid learner and researcher with expertise spanning Wireless Sensor Networks, Mobile Ad-hoc Networks, Internet of Things, Machine Learning, Deep Learning, Natural Language Processing and Computational Intelligence. He has authored numerous research papers in SCI, Scopus, and WoS indexed journals and presented his work at national and international conferences. His research contributions include work on energy-efficient routing, secure data collection, AI-driven systems, and deep learning models for environmental and healthcare applications. In addition to academic responsibilities, Dr. Shiju Kumar actively contributes to curriculum design as a Member of the Board of Studies for APJ Abdul Kalam Technological University, Kerala. Beyond academia, he is actively involved in professional and social initiatives, developing strong skills in management, communication, team leadership, and event organization. He is an annual member of ACM and actively participates in academic and professional forums. He is serving as a reviewer for many journals and conferences.

Sujith Kumar P.S. earned his B.Tech degree in Computer Science and Engineering from Cochin University of Science and Technology. He also holds an M.E. degree in Computer Science and Engineering from Anna University and a Ph.D. in Computer Science and Engineering from Hindustan Institute of Technology and Science (HITS). He has 24 years of total experience: 22 years in teaching and 2 years in industry. He has published several academic papers in national and international journals and conferences and serves as a reviewer for several journals and IEEE conference papers. His areas of interest include Mobile Ad Hoc Networks, Computer Networks, Programming, and Machine Learning.

Rajkumar V. Patil received his Bachelor of Engineering (Computer Engineering) in 2018 from Savitribai Phule Pune University, Pune, India, and his Master of Engineering in Computer Engineering in 2020 from Smt. Kashibai Navale College of Engineering, Pune, India. Cyber Physical systems for healthcare, Trust management, Machine learning, Web technologies, and Explainable AI are his areas of interest. He has around two years of teaching and research experience. He has published research articles and book chapters in international journals. He reviews for several international journals. He is currently working as assistant professor at Department of Computer Engineering (Software Engineering), Vishwakarma Institute of Technology, Pune. He has worked as an assistant professor at MIT-ADT University, Pune -- NBN Sinhgad School of Engineering, Pune, and Vishwakarma Institute of Information Technology, Pune

Gerli Ryza Reyes Educator by Heart, Technologist by Design Gerli Ryza believes that technology should never be a barrier to learning—it should be a bridge. As a Licensed Professional Teacher (LPT) who went on to earn a Master in Information Technology (MIT), Gerli Ryza has spent her career navigating the space where human curiosity meets digital innovation. She doesn't just see code and classrooms as separate worlds; she sees them as a single opportunity to make information more accessible. Her journey is defined by a "people-first" approach to tech. Whether she is mentoring students or optimizing a digital system, Gerli Ryza's mission is to ensure that the tools we build actually serve the people who use them. Driven by a love for lifelong learning and a knack for problem-solving, she remains dedicated to empowering others through digital literacy and thoughtful, modern education.

Binastya Anggara Sekti has been a lecturer at the Faculty of Computer Science, Esa Unggul University, Jakarta, since 2016. He is actively involved in the Artificial Intelligence and Information Security Study Group. He holds a Bachelor's degree in Informatics Engineering from the Bandung Institute of Technology and a Master's degree in Information Systems Management from Trisakti University, Jakarta. Prior to entering academia, he accumulated over 25 years of professional experience as an IT practitioner in various companies across Jakarta. His work focused on developing data centers for internet service providers and delivering IT solutions for the manufacturing and oil & gas sectors. His research interests include artificial intelligence, cybersecurity, and the development of IT infrastructure.

Cita Yustisia Serfiyani, SH, MH is a lecturer at the Faculty of Law UNESA. She has expertise in the fields of intellectual property rights, business law, corporate law, financial institution law and has experience writing books in the field of business law throughout her 10-year career as a writer. Her book concerning IPR entitled "HAKI dan Warisan Budaya" (IPR and Cultural Heritage) is proof of her concern for the development of IPR science

Syrenzo Ramos Sicuan is a mathematics major and educator whose work spans quantitative analysis, artificial intelligence applications, and creative disciplines. He serves as an instructor and researcher at Pangasinan State University–Binmaley Campus, where he teaches mathematics and contributes to various institutional research initiatives. His scholarly interests include computational modeling, AI-assisted creativity, music informatics, and data-driven approaches to education. Drawing from his background in mathematics, he explores how probabilistic models and algorithmic techniques can support emerging practices in music composition and digital arts. He has authored and co-authored research across education, technology integration, and applied analytics, and remains active in local and national research communities. His current work focuses on developing accessible, low-resource AI tools that empower learners, educators, and creative practitioners.

Akshada Telang is an AI and ML graduate from MITS, Gwalior (9.0 CGPA) with research experience at Linnaeus University, Sweden, where she achieved 98% accuracy in kidney tumor diagnosis using U-Net and nnU-Net. She has co-authored a Scopus-indexed book chapter on a Heart and Respiratory Sound Expert System and developed projects like Chord Craft (98% chord recognition accuracy). A 2nd place winner at IIT Roorkee's Techfest, state-level Kho Kho captain, and NCC C Certificate holder, she is skilled in Python, C/C++, SQL, and deep learning frameworks.

Vibha Tiwari received her Ph.D. in Signal Processing/ Electronics and Communication Engineering from Maulana Azad National Institute of Technology, Bhopal, India. She also holds an M.Tech. in Digital Communication from MANIT, Bhopal, and a B.E. in Electronics and Communication Engineering from Vikram University, Ujjain, India. She is currently working as an Assistant Professor in the Department of Artificial Intelligence, Madhav Institute of Technology and Science (MITS), Gwalior, India. She has over 24 years of teaching and research experience. Her research interests include signal Processing, healthcare informatics, Natural Language Processing, Artificial Intelligence, and Machine Learning. She has published several research papers in international journals and conferences. Dr. Tiwari is a member of the Institution of Engineers (India), Indian Society for

Technical Education, and other professional bodies. She has received multiple awards for her contributions to teaching and research, including the Young Scientist Award from MP Council of Science and Technology, Govt. of MP in 2011 and the Academic Excellence Award from MITS in 2023. She has also served on various committees and boards related to curriculum design and quality assurance in technical education. She has received appreciation for her contributions to AI education and interdisciplinary research initiatives.

E. İdil Yazan is a distinguished composer, musicologist, and educator, currently serving as a faculty member in the Department of Music, Composition, and Conducting at Akdeniz University State Conservatory. A graduate of Bilkent University, one of Turkey's premier institutions for higher education and the arts, she earned her Bachelor's degree in Music Theory and Composition with high honors. Dr. Yazan went on to pursue her Master's degree in Musicology at Başkent University, followed by a PhD in Music from the Institute of Fine Arts at Akdeniz University. Her academic journey reflects a seamless integration of scholarly depth and artistic innovation. Prior to her current role, Dr. Yazan held academic appointments at Ankara University State Conservatory and Bilkent University, contributing significantly to music education and research. Her early promise as a composer was recognized in the Guitar Composition Competition hosted by Yıldız Technical University, where her work received distinction. Dr. Yazan's research and creative interests span a wide spectrum, including musicology, advanced music theory, composition, and interdisciplinary studies in music and the arts. Her work reflects a commitment to both academic excellence and artistic exploration, positioning her as a dynamic force in today's musical landscape.

Index

A

B

C

D

E

F

G

H

M

N

P

R

S

T

U

www.ingramcontent.com/pod-product-compliance
Lightning Source LLC
LaVergne TN
LVHW061529290926
849162LV00004B/718